Fluid Mechanics

Fluid Mechanics

William Bober
Rochester Institute of Technology

Richard A. Kenyon
Rochester Institute of Technology

JOHN WILEY & SONS
New York Chichester Brisbane Toronto

Library of Congress Cataloging in Publication Data:

Bober, William, 1930–
Fluid mechanics.

Bibliography: p.
Includes index.
1. Fluid mechanics. I. Kenyon, Richard A., joint author. II. Title.
TA357.B6 532 79-12977
ISBN 0-471-04886-0

Preface

This textbook is primarily intended for third-year engineering students who are taking their first course in fluid mechanics in those schools where elements of vector calculus are taught prior to or concurrent with fluid mechanics. Although vector calculus is used extensively throughout the text, it is possible, by a careful selection of topics, to emphasize the application side of the subject. Sections that emphasize both the theoretical and mathematical aspects of the subject are designated by an asterisk.

As we began to write a new book on the subject of fluid mechanics, we first established these governing principles: (1) This textbook should provide the student with both a thorough understanding of the fundamental concepts and a working knowledge of the subject matter. (2) It should use the latest mathematical methods and techniques that would be consistent with a student's background in a typical engineering program in the United States. (3) It should provide reinforcement of the concepts learned in prerequisite courses—these being statics, dynamics, strength of materials, thermodynamics, and mathematics, up to and including differential equations. (4) It should use both the English system and the International System (SI) of Units throughout. (5) Computer problems should be dispersed throughout the book.

To satisfy our goals, we have presented the material from a continuum mechanics viewpoint, using vector notation. Thus, the fundamental laws are presented in integral form, and the associated field equations are obtained by the use of vector calculus (in particular, the divergence theorem). To emphasize the fundamental concepts, each concept is presented both verbally and mathematically, with each verbal statement labeled by a statement number. The practical application of theories is illustrated by sample problems, homework problems, and computer problems. The homework problems range in difficulty;

some include problem statements that present information not discussed in the main body of the text. Computer problems appear both in the body of the text (Chapters 3, 4, and 8) and in the problem section (Chapters 5, 6, and 9). The authors' solution programs, written in FORTRAN, are given in the body of the text for the computer problems presented in Chapters 3 and 4. The authors' solution programs for the remaining computer problems are available in the solutions manual.

The presentation of the subject matter is done, wherever possible, with the mathematical detail, rigor, definition, and sophistication comparable to that experienced in mathematics courses. It is this mathematical detail that distinguishes this book from others on fluid mechanics. We firmly believe that the responsibility of our profession is to train engineering students both in the methods of problem formulation and in the methods of solution of engineering-type problems. Training students to substitute numbers into formulas developed by others, although necessary, is not sufficient for a proper engineering education. It is this guiding belief that motivated us to use the mathematical detail found in many parts of the book. It is hoped that students using this textbook will find that their ability to develop mathematical models of physical phenomena is much improved and that they are better prepared for more advanced study in engineering.

Chapters 1 to 5 should be covered in the order presented. However, Chapters 6 to 9, although dependent on the first five, are independent of each other and may be covered in any order desired by the instructor. Instructors wishing to emphasize the aerodynamic aspects of fluid mechanics may cover Chapter 9, "An Introduction to Subsonic Aerodynamics," immediately after Chapter 5.

Characteristics of a fluid and fluid properties are discussed in Chapter 1. A summary of elements of vector calculus is given in Chapter 2. The summary of vector calculus given in this chapter is probably adequate only when an operational capability of this material is desired. If a physical interpretation of terms such as the directional derivative, gradient divergence, and the like, is desired, then the summary of vector calculus should be replaced with Appendix A.1, where the same material is presented in much greater detail.

Fluid statics is discussed in Chapter 3. The sections in this chapter on buoyancy and the atmosphere are more extensive than what is commonly found in other first-course texts. The section on buoyancy contains a computer problem. Our objective in the section on the atmosphere is to introduce students to some elementary concepts in air pollution and weather development. The section entitled "A Fluid Having a Uniform Acceleration" is also quite different from what is commonly found in other textbooks. Here the use of two coordinate systems (a concept learned in dynamics) properly develops the equation of the free surface.

Chapter 4, which we consider the core of the text, covers the fundamental laws of a nonviscous fluid. Conservation of mass, the first law of thermodynamics as applied to a fluid, Newton's second law of motion, and the moment-of-momentum equation are all presented in integral form. The first three are also presented from a field-equation viewpoint. Linearly accelerating control volumes and rockets are also discussed in this chapter. The computer problem in Chapter 4 involves the motion of an accelerating rocket. This problem is extended in Chapter 5. The field equations based on Newton's second law result in Euler's equations that are then integrated along a streamline for steady-state conditions. Other conditions that allow for the integration of Euler's equations are also given. The chapter contains examples related to existing equipment in the mechanical engineering laboratory at the Rochester Institute of Technology.

Dimensional analysis, as developed by Langhaar, and similitude are discussed in Chapter 5; pipe flow, from an empirical viewpoint, is discussed in Chapter 6. Pipe networks are also discussed in Chapter 6, including the Hardy-Cross method. At the end of the problem section of Chapter 6, the student is asked to develop a computer program to solve a network problem by the Hardy-Cross method. Some guidance for developing the program is given.

Open-channel flow is covered in Chapter 7. Wherever possible, the subject is treated from a control-volume viewpoint.

Chapter 8 covers one-dimensional, compressible, isentropic flow through nozzles and diffusers. In the last section of this chapter, the problem of the quasisteady discharge from a tank through a converging nozzle has been formulated for a computer solution. But, here again, students are asked to write their own computer program for obtaining a solution to the problem.

An introduction to subsonic aerodynamics is given in Chapter 9. The first part of the chapter deals with irrotational, incompressible flow. The velocity potential and stream function for a few simple flow configurations are given. The two dimensional flow around a circular cylinder is solved by use of separation of variables, a method commonly used to solve a broad range of engineering-type problems. This is followed by a discussion of how the solution to the circular cylinder problems may be used to provide a solution to the flow around a Joukowski airfoil, including that of a flat plate. Because it is not common for a third-year student to have a background in complex variables, the velocity field around a flat plate is given, but not derived. The chapter ends with a discussion of skin-friction drag on a flat plate by the Von Kármán–Pohlhausen method. In the problem section, students are asked to have the computer plot airfoil shapes by the Joukowski transformation.

We are grateful to Professor Harold R. Jacobs of the University of Utah for his critical review of the manuscript. His many constructive suggestions convinced us to include a number of topics which otherwise would have been omitted.

We wish to thank Mrs. Candace Giandana, Mrs. Margaret Urckfitz, Ms. Anna M. Costa, and Ms. Beth O'Neil for their typing efforts; and Ms. Joan Knizeski, for editing the manuscript. Finally, we wish to express our deep gratitude to our wives for tolerating the many hours we spent on the preparation of this manuscript, time which otherwise would have been devoted to our families.

W. Bober
R. A. Kenyon

Contents

CHAPTER 3

CHAPTER 4

Chapter One

Physical Characteristics of a Fluid

1.1 INTRODUCTION

A first course in fluid mechanics is usually directed toward familiarizing the student with the terminology and the basic concepts involved in the science. Some applications are always given, and so it is with this text. The concepts learned here may later be applied in the design of certain devices, machines, or systems that in some way involve fluid mechanics. Some examples of applications are: airplanes, ships, hydraulic machinery, water supply, drainage and gas pipleine systems, turbines, pumps, jet engines, among others. In the first two examples the fluid flows *externally* or *around* the airplane or ship. In the other systems or devices the fluid flows *internally*. Both internal and external flows are discussed in this text.

1.2 FLUIDS AND THE CONTINUUM MODEL

In nature, matter exists in either of three states; these are the solid, liquid, or gaseous states. Solids have a specific shape and a significant degree of rigidity. From strength of materials, we know that solids may be measurably deformed by sufficiently large forces. If a liquid is placed into a container, it will take the shape of the container except for the top surface. If a gas is allowed to flow into an evacuated container, it will completely fill the container. It takes a relatively large force to change the total volume of a liquid. Thus, for many problems, liquids may be considered to be *incompressible*. That is, the total volume of an

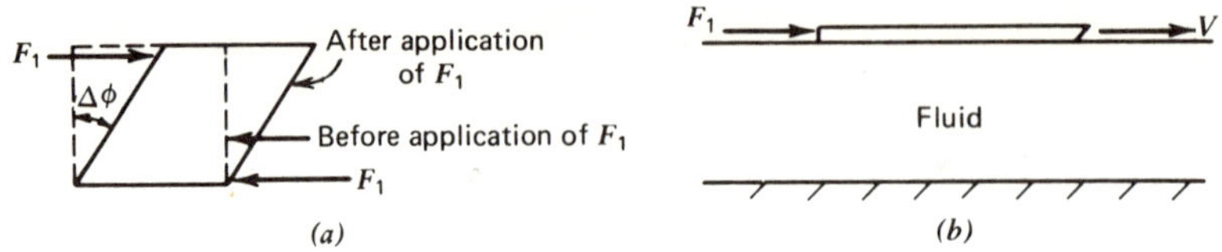

Fig. 1-1. Subjecting a solid and a fluid to a shear force. (*a*) Solid. (*b*) Fluid.

element of liquid remains unchanged as it flows through or past a device of interest. However, gases are readily compressed. Both gases and liquids are classified as fluids.

The response of a fluid to a shear force differs from the response of an elastic solid to a shear force. When an elastic solid is subjected to a shear force, it deforms as shown in Fig. 1-1*a*. The angular deformation, $\Delta\phi$, will be proportional to the magnitude of the shear force, F_1. Furthermore, as long as the shear force remains unchanged and the material has not been stressed beyond its yield point, the deformation will remain constant. However, this is not the case with a fluid. Because a flùid adheres to a solid boundary,[1] we are able to subject a fluid to a shear force by sliding a plate over a pool of the fluid of interest (see Fig. 1-1*b*). It has been experimentally observed that under such circumstances, the fluid will deform indefinitely. It is this characteristic that is used in the basic definition of a fluid.

> A fluid is a substance that continues to deform when subjected to a shear stress, no matter how small. Stm. 1.2-1[2]

However, it should be noted that there exist materials (for example, pastes) that behave in a dual fashion. That is, they exhibit a finite displacement when subjected to a small shear force but, when the shear force is increased above a certain level, these materials deform indefinitely. Such materials are classified as *Bingham fluids* or *Bingham plastics*. The treatment of such materials is beyond the scope of this text.

From the molecular viewpoint, the state of a substance results from the intermolecular forces within the material, from the spacing between molecules, and from the degree of freedom of motion of the molecules. The intermolecular forces in a solid are greater than the intermolecular forces in a liquid, and those of a liquid are greater than the intermolecular forces in a gas. The spacing

[1] This characteristic of a fluid has been established by experimental observation.
[2] Throughout the text, important verbal statements are designated by a number in the right margin.

between the molecules in a solid is less than the spacing between the molecules in a liquid, which is less than the spacing between the molecules in a gas.

For mechanical analysis, a fluid is treated as a *continuum*. That is, matter is assumed to exist continuously with mathematically continuous properties. Thus, we may take the value of a property at two adjacent points, subtract them, divide the result by the distance between the adjacent points, and take the limit as the distance between the points approaches zero, thus obtaining the derivative of that property with respect to distance. Of course, this is a mathematical abstraction since, as the distance between adjacent points gets as small as the intermolecular spacing, matter becomes discrete. That is, the region will consist of a molecule, a void, a molecule, a void, and so on. However, this does not disturb us, since the measuring instruments used in fluid mechanics cannot depict this fine structure. These instruments provide us with property measurements that result from the average behavior of many molecules, and they are therefore attuned to the continuum model. The interrelationship between molecular and continuum behavior is treated in the kinetic theory of gases and in statistical mechanics. However, the continuum model breaks down in the treatment of extremely rarefied gases (gases which are near vacuum pressures). Rarified gases are not discussed in this text.

The study of fluid mechanics is concerned with fluids both at rest and in motion. Fluids at rest are studied under the heading of fluid *statics*. When a fluid is static, it is in equilibrium; thus, the equations of equilibrium used in statics also apply for the case of a static fluid. This topic is covered in Chapter 3. However, the major portion of this textbook is concerned with fluid motion and the forces that cause this motion.

1.3 BASIC DIMENSIONS

Before it is possible to do any quantitative analysis in any field, it is necessary to select a set of basic dimensions associated with the properties involved in that field. In the United States, engineering textbooks have in the past invariably used the English system. Recently, by an international agreement, a decision was made to gradually adopt a worldwide system of units, designated as SI—*Système International d'Unités* or the International System of Units. Since we are now in a transitional stage, this book will employ both the English system and the SI system.

For both the English system and the SI system there are four fundamental quantities associated with fluid mechanics. The rest are based on these four. Standards for mass, length, time, and temperature have been established by the International Bureau of Weights and Measures in Sèvres, France, and by the

TABLE 1-1 System of Units

Quantity	English: Gravitational	English: Engineering	SI
Mass	slug-mass (slug)	pound-mass (lb_m)	kilogram (kg)
Length	foot (ft)	foot (ft)	meter (m)
Time	second (s)	second (s)	second (s)
Temperature	degrees Rankine (R)	degrees Rankine (R)	degrees Kelvin (K)
Force	pound (lb_f)	pound (lb_f)	newton (N)
Thermal energy	British thermal unit (Btu)	British thermal unit (Btu)	joule (J)

National Bureau of Standards in the United States. The dimensions for force are defined in terms of Newton's second law. A list of units for both systems is given in Table 1-1.

Note that in the English system, there are two different units for mass. Most books in the discipline of mechanics use *slugs*, whereas most books in the thermal sciences use lb_m. Both slug and lb_m are standard mass quantities. However,

$$1 \text{ slug} = 32.17 \text{ lb}_m$$

Force is defined in terms of Newton's second law. Newton's second law states that the unbalanced force, **F**, acting on a body, is proportional to both the mass of the body, m, and its acceleration, **a**; that is, $\mathbf{F} \propto m\mathbf{a}$, or

$$\mathbf{F} = \frac{m}{g_c}\mathbf{a} \qquad \text{Eq. 1.3-1}$$

where $1/g_c$ is the constant of proportionality.

Table 1-2 gives the definition of the force unit in each system and the value of g_c.

We can obtain the meaning of Table 1-2 by taking the first line as an example. The first line defines the unit of force in the English Engineering System. This unit of force can be defined by the following statement.

An unbalanced force of 1 lb_f will cause a 1 lb_m to accelerate 32.17 ft/s^2.
Stm. 1.3-1

TABLE 1-2 Definition of Force Unit and Proportionality Constant in Newton's Second Law

System	Force	Mass	Acceleration	g_c
English engineering	1 lb_f	1 lb_m	32.17 ft/s^2	$32.17 \frac{lb_m\text{-}ft}{lb_f\text{-}s^2}$
English gravitational	1 lb_f	1 slug	1 ft/s^2	$1 \frac{slug\text{-}ft}{lb_f\text{-}s^2}$
SI	1 N	1 kg	1 m/s^2	$1 \frac{kg\text{-}m}{N\text{-}s^2}$

Then g_c is evaluated from Eq. 1.3-1; that is,

$$1\ lb_f = \frac{1\ lb_m \times 32.17\ ft/s^2}{g_c}$$

or

$$g_c = 32.17 \frac{lb_m\text{-}ft}{lb_f\text{-}s^2}$$

It should be noted that when mass is expressed in either slugs or kilograms the constant of proportionality, $1/g_c$, is often omitted from Newton's second law, since its magnitude in both cases is 1. Then, 1 lb_f is equivalent to 1 slug-ft/s^2, and 1 Newton is equivalent to 1 kg-m/s^2.

If g is the acceleration of a free-falling body, then the relationship between weight and mass is

$$W = \frac{mg}{g_c} \qquad \text{Eq. 1.3-2}$$

Near the earth's surface, $g = 32.17\ ft/s^2$ or $9.806\ m/s^2$. But at large distances from the earth's surface, $g < 32.17\ ft/s^2$.

Inside the front cover many conversion constants between the SI and English systems are given.

1.4 CHARACTERISTICS OF A FLUID

1.4.1 DENSITY, SPECIFIC VOLUME, SPECIFIC WEIGHT, AND SPECIFIC GRAVITY

Since fluid properties may vary throughout a region, it is appropriate to define the fluid properties as field quantities. A *field quantity* is one that may be a function of position and time. To define the *density*, ρ, of a substance at a point, consider the amount of mass, Δm, contained in an infinitesimal region around that point. If the volume containing Δm is $\Delta V\!\!\!/$, then

$$\rho = \lim_{\Delta V\!\!\!/ \to 0} \frac{\Delta m}{\Delta V\!\!\!/} \qquad \text{Eq. 1.4-1}$$

The dimensions of ρ are lb_m/ft^3 in the English system and kg/m^3 in the SI system. The inverse of density is called the specific volume, $\tilde{v}$. This is a property used more often in thermodynamics than in fluid mechanics. The dimensions of specific volume are the inverse of those of density.

The *specific weight*, γ, of a substance is defined in a similar fashion as the density, except that the amount of mass contained in an infinitesimal region around a point is replaced by its weight; that is, if ΔW is the weight of the material contained in the region about the point, then

$$\gamma = \lim_{\Delta V\!\!\!/ \to 0} \frac{\Delta W}{\Delta V\!\!\!/} \qquad \text{Eq. 1.4-2}$$

A relationship between specific weight and density can be obtained by substituting Eq. 1.3-2 into Eq. 1.4-2 giving

$$\gamma = \lim_{\Delta V\!\!\!/ \to 0} \frac{g\,\Delta m}{g_c\,\Delta V\!\!\!/} = \frac{\rho g}{g_c} \qquad \text{Eq. 1.4-3}$$

The *specific gravity*, S, of a substance equals the ratio of its specific weight, γ, to the specific weight of water, γ_w, at standard conditions[3]; that is,

$$S = \frac{\gamma}{\gamma_w} \qquad \text{Eq. 1.4-4}$$

[3] Standard conditions are usually taken to be 14.7 lb_f/in^2 and 40°C, at which γ_w equals 62.43 lb_f/ft^3 or 9806 N/m^3.

Since $\gamma = \rho g / g_c$, S can also be expressed by

$$S = \frac{\rho}{\rho_w}$$

where ρ is the mass density of the substance, and ρ_w is the mass density of water at standard conditions.

1.4.2 STRESS, PRESSURE, AND VISCOSITY

The macroscopic concepts of internal forces and stress which are used in solid-body mechanics are carried over to fluid mechanics. The forces that act on a fluid element can be classified as either body forces or as surface forces. Body forces are forces that are distributed throughout the material and act at a distance. Gravitational and electromagnetic forces are classified as body forces. Surface forces are contact forces that act, as the name implies, on a surface.

To develop the concept of stress, we will examine the internal force that one part of the fluid exerts on the other across an infinitesimal surface. To develop the concept of an internal force, imagine a cutting plane within the fluid and, for the purpose of analysis, consider the material on one side of the plane to be removed (see Fig. 1-2). To provide an equivalent effect that the removed material had on the remaining material, we place a force distribution on the cutting surface and designate the total force resulting from the distribution as an internal force. In a solid this internal force would result from the bonding behavior between molecules. In a gas, it is postulated (in the kinetic theory of gases) that internal forces result from molecular transport across this surface. This results in a transfer of momentum. The net rate of momentum transfer is equivalent to a

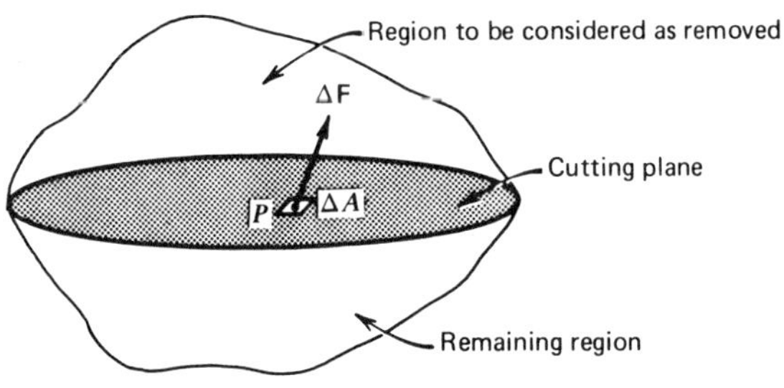

Fig. 1-2. Cutting plane through region for determination of internal force.

force. On the other hand, if a solid boundary is present, the molecules strike and rebound from the solid boundary. The resulting net rate of momentum change of the striking molecules is caused by the forces the boundary exerts on these molecules. The net force exerted on the wall is equal and opposite.

Since the internal force may vary from point to point, we subdivide the cutting surface into infinitesimal elements (as shown in Fig. 1-2) and consider the force, $\Delta\mathbf{F}$, that acts on area ΔA. Now the magnitude of $\Delta\mathbf{F}$ depends on the size of ΔA, and thus it cannot be considered a field quantity. A term called the *surface traction*, or stress vector, designated as $\boldsymbol{\tau}$, is a field quantity and is thus a more convenient term to use in continuum mechanics. The quantity $\boldsymbol{\tau}(P)$ does not depend on the size of ΔA and is defined as a limit, as $\Delta A \rightarrow 0$ around point P; that is,

$$\boldsymbol{\tau}(P) = \lim_{\Delta A \rightarrow 0} \frac{\Delta\mathbf{F}}{\Delta A} \qquad \text{Eq. 1.4-5}$$

It should also be pointed out that, if point P is held constant but the orientation of ΔA is changed, then the surface traction is also changed.

The surface traction, $\boldsymbol{\tau}$, is often decomposed into two components, one normal to the surface, designated as τ_n, and the other tangent to the surface, designated as τ_t, as shown in Fig. 1-3. Thus $\boldsymbol{\tau}(P)$ can be written as

$$\boldsymbol{\tau}(P) = \tau_n(P)\mathbf{e}_n + \tau_t(P)\mathbf{e}_t \qquad \text{Eq. 1.4-6}$$

where $\mathbf{e}_n$ and $\mathbf{e}_t$ are, respectively, the unit vectors normal and tangential to surface ΔA. The component τ_n is called the normal stress, and the component τ_t is called the shear stress. Normal stresses are classified as either tensile (pointing

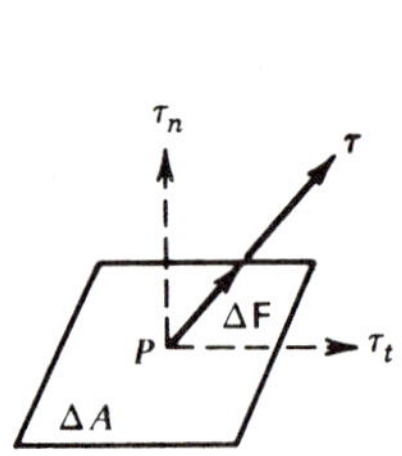

Fig. 1-3. Surface traction, normal stress, and shear stress.

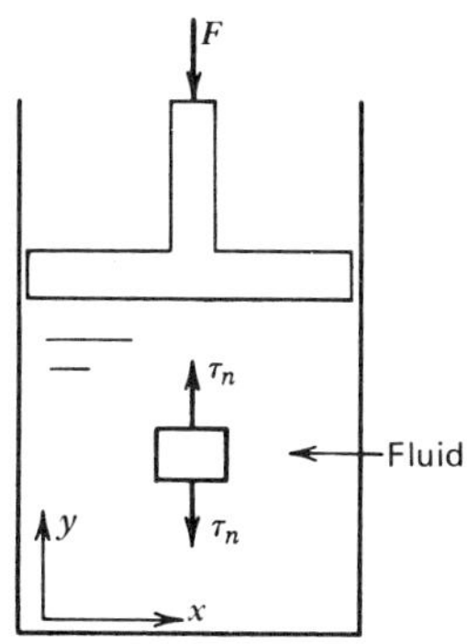

Fig. 1-4. A fluid subjected to a normal compressive stress.

outward from the retained region) or compressive (pointing inward toward the retained region). We retain the sign convention used in solid-body mechanics, taking tensile stresses as positive and compressive stresses as negative. Since a fluid cannot sustain any significant tensile stress, τ_n will normally be negative.

To further examine the behavior of normal stresses, let us turn to an experiment in which a fluid is contained without leakage between a cylinder and a moveable piston, as shown in Fig. 1-4.

Nearly everyone has sufficient experience to know that it would take an ever-increasing force, F, to continuously move the piston downward, even if the movement is done very, very slowly. A downward movement of the piston results in a normal strain of the fluid. We thus see that a fluid can sustain a normal strain, even if the strain occurs very, very slowly. However, as we will see later, this is not the case for a shear strain. Students of physics should recall that in a *static* fluid the force per unit area that acts normal to a surface is called *pressure*. Since the characteristics of pressure and normal stress are the same, we may take[4]

$$\tau_n = -p \qquad \text{Eq. 1.4-7}$$

Although the above expression is not exact in a general fluid-flow situation, it is sufficiently accurate for all cases treated in this text. It should be noted that

[4] It should be noted here, that under flow conditions the normal stresses are not the same in all directions, and some additional interpretation is required to relate the thermodynamic pressure to the normal stress components τ_{11}, τ_{22}, and τ_{33} (see Fig. 1-5).

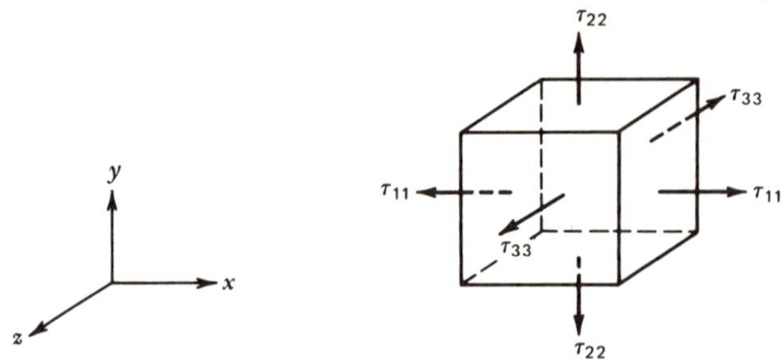

Fig. 1-5. Normal stress components, τ_{11}, τ_{22}, and τ_{33}.

Eq. 1.4-7 is exact in a static fluid. This is discussed in Section 3.2 (Chapter 3). Substituting Eq. 1.4-7 into Eq. 1.4-6 gives

$$\boldsymbol{\tau}(P) = -p(P)\mathbf{e}_n + \tau_t(P)\mathbf{e}_t \qquad \text{Eq. 1.4-8}$$

We now focus our discussion on τ_t and a fluid property called *viscosity*. Suppose that a flat plate is installed in the test section of a water tunnel. Assume that for $t \leq 0$, there is no flow. At this time, using a dye injector, we place a vertical streak line in the fluid perpendicular to the plate, as shown in Fig. 1-6. When the flow is started, the streak line gets distorted so that at t equal to Δt it appears (as shown in Fig. 1-6) as the curve labeled "Dye streak at $t = \Delta t$." We observe that the dye adheres to the solid boundary. This occurs for all values of V_∞. Experience indicates that this occurs even when the fluid is a gas at normal densities. We thus conclude that the fluid adheres to a solid boundary. We designate this nonslippage at a solid boundary as the *no-slip boundary condition.* In very rarefied gases, this no-slip condition does not hold; however, whenever the fluid behaves as a continuum, the no-slip boundary condition is valid.

We now turn to a second experiment. Let us consider a flow through a two-dimensional channel, having a constant cross-sectional area. For such a flow the fluid velocity, $\mathbf{v} = v_1(y)\mathbf{i}$. Suppose we insert a plug into the bottom wall, as

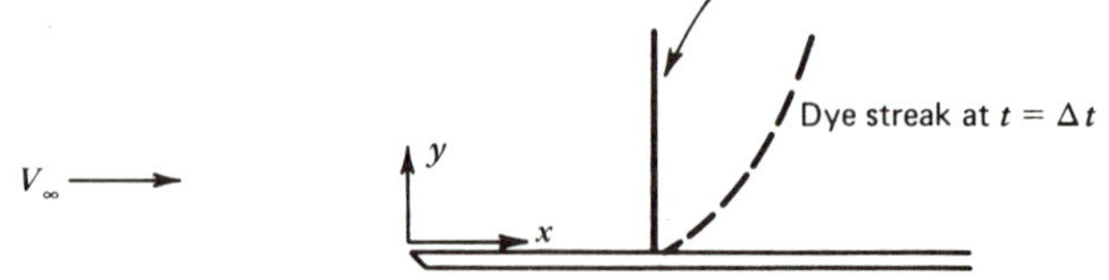

Fig. 1-6. Experiment indicating the no-slip boundary condition.

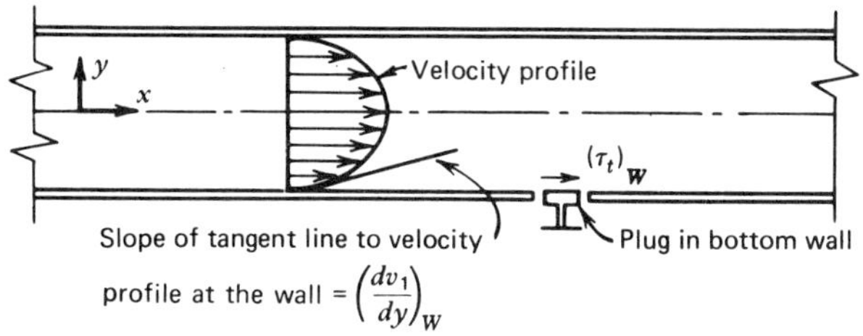

Fig. 1-7. Shear force measurement in a parallel flow.

shown in Fig. 1-7, and instrument the plug so that the shear force acting on it may be measured. Knowing the surface area of the plug, we can determine the shear stress on the plug $(\tau_t)_w$. Using an instrument capable of measuring the local fluid velocity,[5] we can obtain a velocity profile across the channel and determine the slope of this profile at the wall; that is $(dv_1/dy)_w$. If we measure $(\tau_t)_w$ and $(dv_1/dy)_w$ at different flow rates, we find that for a *wide class of fluids* (*fluids classified as Newtonian fluids*) (1) that $(\tau_t)_w \propto (dv_1/dy)_w$ and (2) that the velocity profile is parabolic.

For Newtonian fluids, we may now postulate that

$$\tau_t = \tilde{\mu}\left(\frac{dv_1}{dy}\right) \qquad \text{Eq. 1.4-9}$$

is also true everywhere in the flow. If we then work backwards, we can show that indeed the velocity profile is parabolic. We thus conclude that for *parallel flow*, Eq. 1.4-9 is valid everywhere in the flow.

The proportionality constant, $\tilde{\mu}$, is designated as the *viscosity* (or *absolute viscosity*) of the fluid. The dimensions of $\tilde{\mu}$ are $(\text{lb}_f\text{-s})/\text{ft}^2$ in the English system and $(\text{N-s})/\text{m}^2$ in the SI system. It can be shown for the condition of parallel flow, that (dv_1/dy) represents the rate of shear strain in the fluid (see Section 1.4.4). Thus, Eq. 1.4-9 can be interpreted to mean that the *shear stress is proportional to the rate of shear strain.*

In many thermal fluids texts, the proportionality constant, $\tilde{\mu}$, of Eq. 1.4-9 is replaced by μ/g_c. That is

$$\tau_t = \frac{\mu}{g_c}\left(\frac{dv_1}{dy}\right) \qquad \text{Eq. 1.4-10}$$

[5] An instrument capable of measuring the local fluid velocity near a wall is the hot-wire anemometer. For further details on this device the student is referred to any standard text on instrumentation.

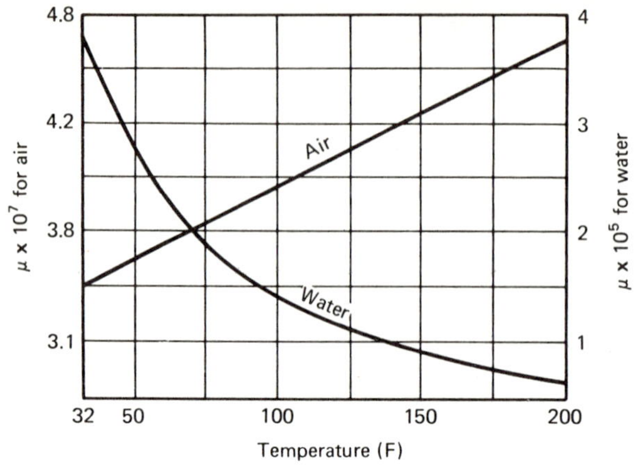

Fig. 1-8. Typical viscosity versus temperature dependence for both a gas and a liquid. (From *Principles of Fluid Mechanics*, 2nd ed., by S. Eskinazi, Copyright © 1968 by S. Eskinazi.)

Under these circumstances the dimensions for μ (also called viscosity) are $lb_m/(ft\text{-}s)$ in the English system and kg/(m-s) in the SI system. In this text, Eq. 1.4-10 is the form that is most often used. Since Eqs. 1.4-9 and 1.4-10 are equivalent, anything said about Eq. 1.4-10 is also applicable to Eq. 1.4-9.

A similar analysis can be made for a parallel flow through a cylindrical pipe. For this case,

$$\tau_t = \frac{\mu}{g_c}\left(\frac{dv_1}{dr}\right) \qquad \text{Eq. 1.4-11}$$

where r is the radial distance from the centerline of the pipe (r replaces y in Fig. 1-7). Experience has shown that the absolute viscosity of a fluid varies with changes in temperature but, in general, varies only slightly with changes in pressure. It is interesting to note that an increase in temperature causes the viscosity of a gas to increase, but causes the viscosity of a liquid to decrease. This behavior can be seen in Fig. 1-8. In this figure, the viscosity of both air and water is plotted as a function of temperature. We will now qualitatively explain the curves shown in Fig. 1-8.

In the kinetic theory of gases, it is theorized that the shear stress in a gas is mainly a result of molecular momentum transfer, from one layer of fluid to another. The force of attraction between molecules in a gas is small when compared with the force involved in the momentum transfer between layers. As the temperature of the gas is increased, the molecules are further agitated, and the rate of momentum transfer is increased. This serves to increase the magnitude of the shear stress for a given velocity gradient, which in turn serves to increase the value of the absolute viscosity.

In a liquid, on the other hand, the cohesive force between molecules is large when compared with the momentum transfer between adjacent fluid layers. This is just the opposite of the situation that exists in the case of a gaseous fluid. This liquid cohesive force has been found to decrease as the temperature of the liquid is raised and, hence, the shear stress and the absolute viscosity of the liquid tend to decrease under these conditions.

It turns out that μ and ρ appear together so often that a single term combining both quantities has been defined. This term is designated as *kinematic viscosity*, ν, and is defined by

$$\nu = \frac{\mu}{\rho} \qquad \text{Eq. 1.4-12}$$

The dimensions of ν are ft^2/s.

As was indicated earlier, not all fluids behave according to Eq. 1.4-10 (or its equivalent, Eq. 1.4-9). When a law is *not universal*, it is classified as a *constitutive relation*. For this reason, Hook's law in mechanics is also a constitutive relation, As stated earlier, fluids that behave according to Eq. 1.4-10 are classified as *Newtonian* fluids. All gases are Newtonian fluids. Most liquids, including solutions of inorganic salts and of sugar, are Newtonian. Complex mixtures such as paints, blood, polymer solutions, pastes, and slurries are examples of non-Newtonian fluids. Plots of some typical shear stress versus rate of shear strain curves for several types of non-Newtonian fluids are shown in Fig. 1-9. These curves can be represented by the *Ostwald-de Waele Equation*

$$\tau_t = K\left(\frac{dv_1}{dy}\right)^n$$

where K and n are constants; $n > 1$ for *dilatant fluids*, $n < 1$ for *pseudoplastic fluids*, and $n = 1$ for *Newtonian fluids*.

Although the viscosity of all Newtonian fluids and most non-Newtonian fluids is time independent, there exist fluids for which the viscosity varies with time. If the viscosity decreases with time, the fluid is said to be *thixotropic* and, if the viscosity increases with time, it is said to be *rheopectic*.

The general study of non-Newtonian fluids is classified as the field of *reology*. The flow of non-Newtonian fluids has been studied by W. L. Wilkinson and

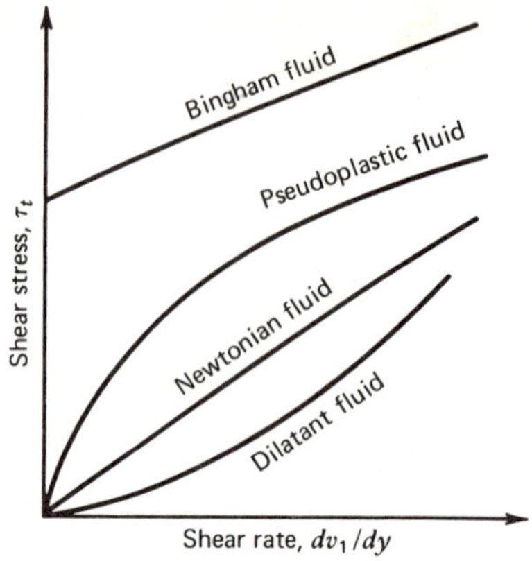

Fig. 1-9. Shear stress versus rate of strain for various types of fluids. (From *Essentials of Engineering Fluid Mechanics* by R. M. Olson, Copyright © 1973 by Harper and Row. Reprinted by permission of the publishers.)

A. B. Metzner, and their work is cited in the reference section in the back of this book.

Although our earlier discussion of the flow of Newtonian fluids centered around parabolic velocity profiles, there are situations where the velocity profile is something other than parabolic. The simplest case occurs when a weight or plate slides over a thin Newtonian fluid film. In this case, we may assume that the velocity profile is linear without introducing significant error (see Fig. 1-10). Then

$$v_1(y) = \frac{Vy}{\delta}, \qquad \frac{dv_1}{dy} = \frac{V}{\delta}$$

and

$$\tau_t = \frac{\mu V}{g_c \delta} = \frac{\tilde{\mu} V}{\delta} \qquad \text{Eq. 1.4-13}$$

To obtain some perspective on the magnitude of the shear force to be expected in some typical flow situations, let us consider the following example.

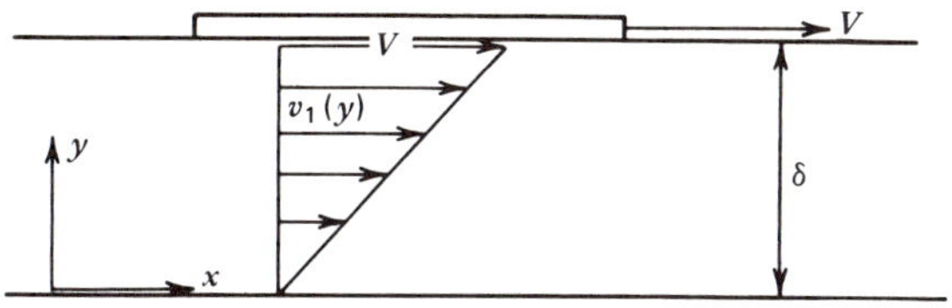

Fig. 1-10. Linear velocity profile.

Example 1-1

A thin plate moving at 50 ft/s slides over a thin fllm of water at 60 F. Find the shear force on the plate if the contact area is 1 ft^2 and the film thickness is 0.01 in.

SOLUTION

From Table A.6-1, $\nu = 1.217 \times 10^{-5}$ ft^2/s and $\rho = 62.37$ lb$_m$/ft^3. By Eq. 1.4-12

$$\mu = \rho\nu = 62.37 \times 1.217 \times 10^{-5} = 7.59 \times 10^{-4} \text{ lb}_m\text{/ft-s}$$

From Eq. 1.4-13

$$\tau_t = \frac{\mu V}{g_c \delta} = \frac{7.59 \times 10^{-4} \times 50 \times 12}{32.17 \times 0.01} = 1.416 \text{ lb}_f\text{/ft}^2$$

$$F_t = \tau_t A - 1.416 \text{ lb}_f$$

As in the above example, there are many flow situations where the shear forces are not very large. Under these circumstances, neglecting the viscous shear forces will not introduce significant error. On the other hand, situations do exist where the viscous forces are either dominant or at least significant. Several topics that fit into this category are discussed later in the text (Chapters 5 and 6).

It should be noted that the relationship between shear stress and velocity derivatives, as expressed in Eq. 1.4-10, is given for a parallel flow situation. For a general flow, the relationship between shear stress and the derivatives of the velocity components is more involved than that given by Eq. 1.4-10. This more general topic is developed in the book *Boundary Layer Theory*, by H. Schlichting (see reference section in the back of this book).

*1.4.3 SOME ADDITIONAL CONCEPTS RELATING TO STRESS

At any point in the continuum, the normal and shear stresses on an arbitrarily oriented surface can be related to the normal and shear stresses on those surfaces that are parallel to the coordinate planes (see Fig. 1-11*a*). Stresses on the surfaces parallel to the coordinate planes are usually labeled by two indices, such as τ_{11}, τ_{12}, τ_{22}, τ_{21}, and so on (see Fig. 1-11*b*). Stress problems are frequently solved in terms of τ_{11}, τ_{12}, τ_{22}, and so on. The meaning of the double indices and the sign convention is explained in the following discussion.

Let us look at Fig. 1-11*b* and particularly on the surface whose normal points in the $+y$ direction. On that surface there are two stress terms, τ_{21} and τ_{22}. The first index designates the direction of the normal to the surface. For example, index 2 implies that the normal to the surface is in either the $+y$ or $-y$ direction. The second index indicates the direction in which the stress points. That is, in the term τ_{21}, the index 1 implies the stress points in either the $+x$ or $-x$ direction. Now, if the normal to the surface points in the $+y$ direction, then the shear stress, τ_{21}, is positive if it points in the $+x$ direction, and the normal stress, τ_{22} is positive if it points in the $+y$ direction. On the other hand, if the normal to the surface points in $-y$ direction, then the shear stress, τ_{21}, is positive if it points in the $-x$ direction, and the normal stress, τ_{22}, is positive if it points in the $-y$ direction. Thus, all stresses shown in Fig. 1-11*b* are positive stresses. Positive stresses on the other surfaces are defined in a similar manner.

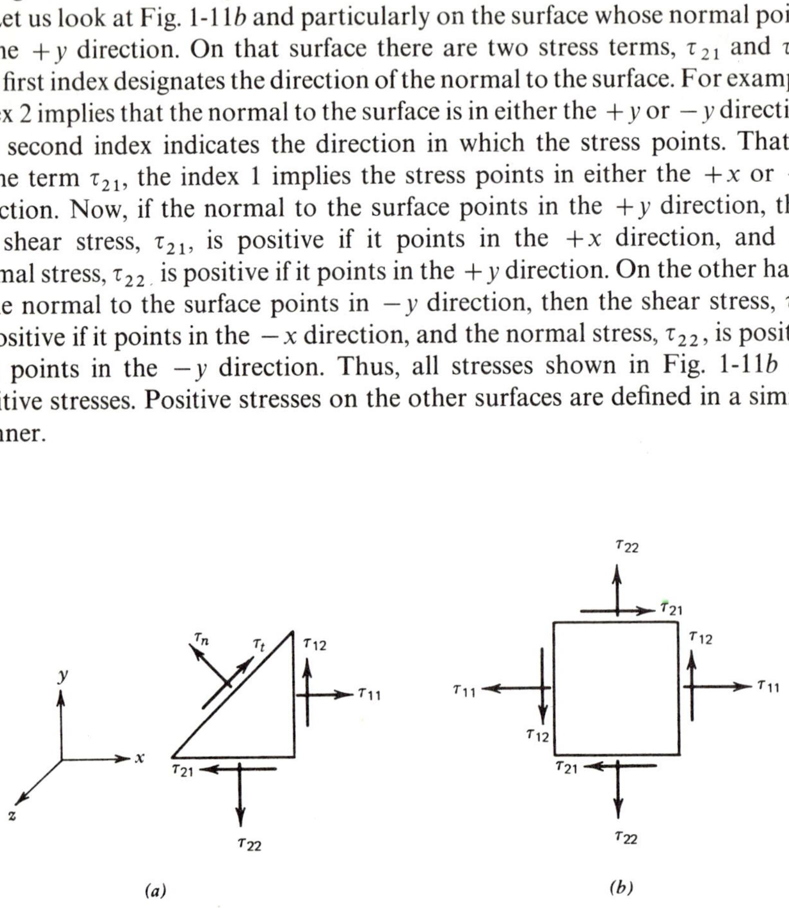

Fig. 1-11. Sign convention for stresses on planes parallel to the coordinate planes. (*a*) Element that gives τ_n and τ_t in terms of τ_{11}, τ_{12}, τ_{21}, τ_{22}. (*b*) Meaning of indices and sign convention.

Finally, as we indicated earlier, for all cases considered in this text, we may take[6]

$$\tau_{11} = \tau_{22} = \tau_{33} = \tau_n = -p \qquad \text{Eqs. 1.4-14}$$

*1.4.4 THE RATE OF SHEAR STRAIN IN A PARALLEL FLOW SITUATION

Consider a fluid element that at t appears as shown in Fig. 1-12. Take side $\overline{PQ}$ to be parallel to the flow direction and $\overline{PR}$ perpendicular to the flow direction; thus, $\measuredangle RPQ$ is a right angle. At $(t + \Delta t)$, points P, Q, and R move to points P', Q', and R', respectively, since $\mathbf{v} = v_1(y)\mathbf{i}$. The rate of shear strain is defined as $-(d\Gamma/dt)$; that is,

$$\text{Rate of shear strain} = -\frac{d\Gamma}{dt}$$

But

$$-\frac{d\Gamma}{dt} = -\lim_{\Delta t \to 0} \frac{\Gamma(t + \Delta t) - \Gamma(t)}{\Delta t}$$

$$= -\lim_{\Delta t \to 0} \frac{\left(\frac{\pi}{2} - \Delta\phi\right) - \frac{\pi}{2}}{\Delta t} = \lim_{\Delta t \to 0} \frac{\Delta\phi}{\Delta t}$$

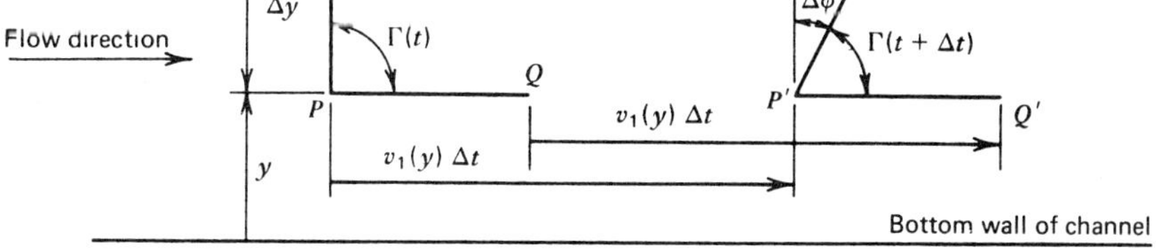

Fig. 1-12. Determination of rate of shear strain in parallel flow.

[6] For the case of a static fluid Eqs. 1.4–14 are exact. The proof of this fact is given in Chapter 3.

and

$$\tan(\Delta\phi) = \frac{\left(v_1(y) + \dfrac{dv_1}{dy}\Delta y\right)\Delta t - v_1(y)\,\Delta t}{\Delta y} = \left(\frac{dv_1}{dy}\right)\Delta t$$

Since Δt can be made as small as we like, $\Delta\phi$ can be taken to be very small. For small angles,

$$\tan \Delta\phi \approx \Delta\phi$$

or

$$\Delta\phi = \frac{dv_1}{dy}\,\Delta t$$

and

$$\text{Rate of shear strain} = \lim_{\Delta t \to 0} \frac{\dfrac{dv_1}{dy}\,\Delta t}{\Delta t} = \frac{dv_1}{dy} \qquad \text{Eq. 1.4-15}$$

The negative sign in the definition of the rate of strain is chosen so that a positive shear stress results in a positive rate of strain (see Fig. 1-13). The sign convention for stress was discussed in the previous section.

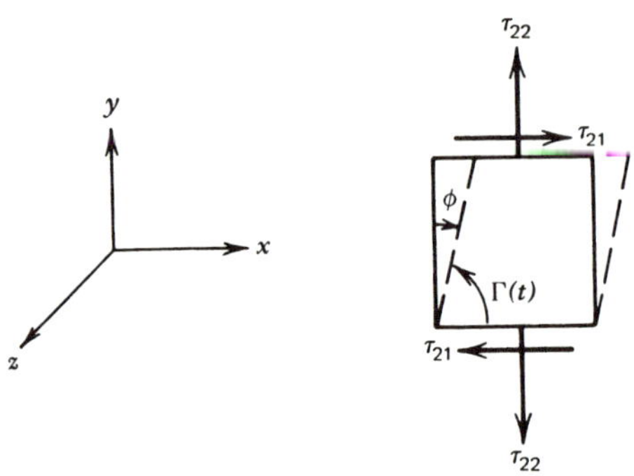

Rate of shear strain $= -\dfrac{d\Gamma}{dt} = \dfrac{d\phi}{dt}$

(A positive τ_{21} causes a positive rate of strain.)

Fig. 1-13. Sign convention for stresses on the planes parallel to the coordinate planes.

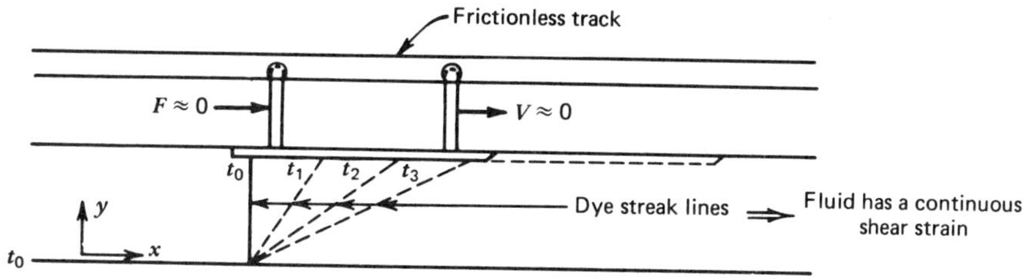

Fig. 1-14. Fluid subjected to a continuous shear strain.

By looking at Eqs. 1.4-10 and 1.4-15 and Fig. 1-14, we can see the reasoning behind the definition of a fluid as given in Stm. 1.2-1. We see that it would take nearly a zero force to move a thin plate very, very slowly across a fluid surface. Under these conditions, we can subject the fluid to a continuous shear strain with nearly zero force. In other words, the fluid has no resistance to very slow shear strain.

1.4.5 TEMPERATURE

Our intuitive concept of temperature is that it is a measure of the relative hotness or coldness of a substance. From a microscopic thermodynamics point of view, temperature represents a measure of the random motion of the molecules in a substance. The International Practical Temperature Scale is based on the effect that temperature has on the properties of materials. The Celsius scale established here is standardized according to the boiling and melting points of various pure substances[7] at specified pressures.

The properties of pressure, temperature, and specific volume determine the phase of a pure substance (gas, liquid, or solid). An analytical expression relating these three variables is denoted as an *equation of state*. Experiments indicate that gases at low densities behave according to the *perfect gas law*; that is,

$$p\tilde{v} = RT \qquad \text{Eq. 1.4-16}$$

or

$$p = \rho RT \qquad \text{Eq. 1.4-17}$$

[7] A pure substance is one that is homogeneous and has the same chemical composition in all phases.

where p is the pressure, ρ is the density, $\tilde{v}$ is the specific volume, T is the temperature, and R is the gas constant for that particular substance. T is measured on the absolute temperature scale. The dimensions of R are $(lb_f\text{-}ft)/(lb_m\text{-}R)$ in the English system and (N-m)/(kg-K) in the SI system. From the dimensions of R we see that absolute temperature is expressed in *degrees Rankine* (R) in the English system and *degrees Kelvin* (K) in the SI system. Associated with the Rankine and Kelvin scales are the Fahrenheit (F) and Celsius (C) scales, respectively. Temperature intervals on the Celsius and Kelvin scales are the same. That is, a ΔT of 1 C equals a ΔT of 1 K. Of course, temperature intervals on the Fahrenheit and Rankine scales are also the same. On both absolute temperature scales, the zero point represents the same physical state; and the zero point on each absolute scale can be shifted from its associated companion temperature scale; that is,

$$\text{Degrees Rankine (R)} = \text{degrees Fahrenheit (F)} + 459.7$$

$$\text{Degrees Kelvin (K)} = \text{degrees Celsius (C)} + 273.2$$

Eqs. 1.4-18

The relationship between the two absolute temperature scales is

$$1\text{ K} = \tfrac{9}{5}\text{ R}$$

In fluid mechanics the most commonly used expression of the perfect gas law is the form given by Eq. 1.4-17. Although air is not a pure substance, it obeys the perfect gas law over a wide range of (p, T, and ρ). The gas constant for air, R_a, is given by

$$R_a = \begin{cases} 53.34 \dfrac{lb_f\text{-ft}}{lb_m\text{-R}} \\[2ex] 286.9 \dfrac{\text{N-m}}{\text{kg-K}} \end{cases}$$

Many gases when subjected to even moderate pressures begin to deviate from the perfect gas behavior. The *compressibility factor*, Z, is often used to indicate the deviation from perfect gas behavior. The compressibility factor is defined by

$$Z = \frac{p\tilde{v}}{RT} \qquad \text{Eq. 1.4-19}$$

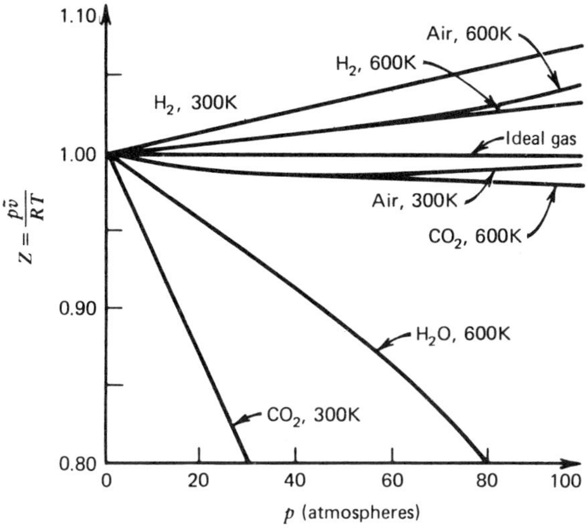

Fig. 1-15. The Z factor for several gases. (From *Engineering Thermodynamics* by W. C. Reynolds and H. C. Perkins, copyright © 1970 by McGraw-Hill. Used with permission of McGraw-Hill Book Company.)

Plots of Z versus p for a number of different gases at several temperatures are shown in Fig. 1-15. Perfect gas behavior is indicated by Z equal to 1.0. From Fig. 1-15, we see that air at 300 K exhibits nearly perfect gas behavior over a broad range of pressures. However, this is not the case for CO_2 at 300 K even at low pressures. Since, in this textbook, air is the gas with which we are mostly concerned, the perfect gas law covers our needs. Students interested in these types of real gas effects are referred to any standard text on thermodynamics.

1.4.6 VAPOR PRESSURE

All pure substances under certain pressure, temperature, and specific volume conditions will have their liquid and gas phases coexisting in equilibrium. In this equilibrium condition, the gas phase of the substance is called *saturated vapor*, and the pressure is designated as *vapor pressure*. The vapor pressure of a substance depends only on its temperature. Table A.6-7 gives the vapor pressure of several liquids at various temperatures. A laboratory apparatus illustrating the meaning of vapor pressure is shown in Fig. 1-16. Suppose we first evacuate Tank A and then allow some of the test liquid to flow into Tank A. When the

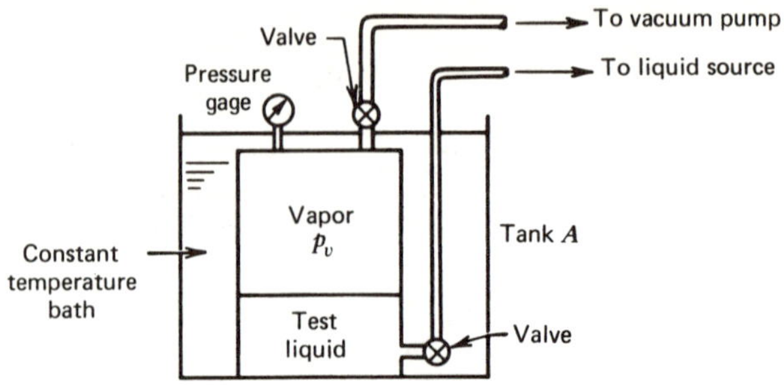

Fig. 1-16. Apparatus for measuring the vapor pressure of a liquid.

test liquid reaches thermal equilibrium with the constant temperature bath, the pressure gage reading will give p_v at the measured temperature. Several examples that involve vapor pressure effects are discussed later in this text.

1.4.7 COMPRESSIBILITY OF A FLUID

It is common knowledge that liquids are much less compressible than gases. A measure of the compressibility of a fluid is given by the *coefficient of compressibility* at constant temperature. Suppose a fluid is contained between a cylinder and a moveable piston, as shown in Fig. 1-17. As F is increased, the piston moves downward. Heat will have to be removed to keep the fluid temperature constant. By measuring the position of the piston and the force, F, the fluid volume, $\forall$, and

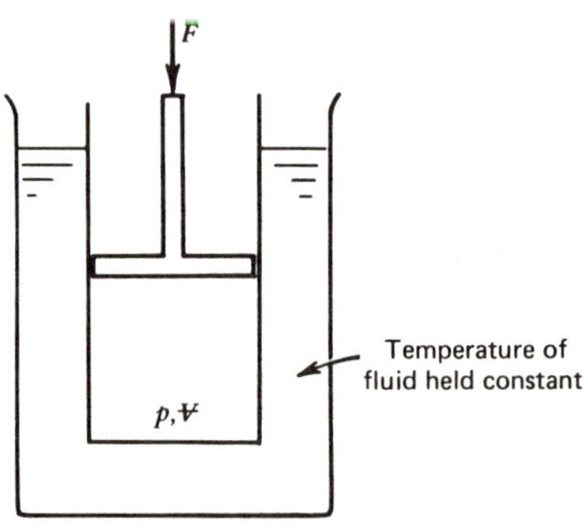

Fig. 1-17. Experiment for determining the coefficient of compressibility.

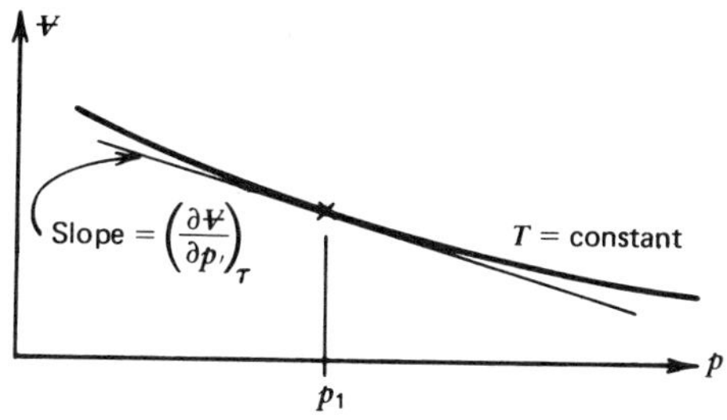

Fig. 1-18. Plot of volume versus pressure.

the pressure, p, can be determined and a plot made, as shown in Fig. 1-18. At any pressure, the slope of the curve, $(\partial \forall/\partial p)_T$, can then be determined. The coefficient of compressibility, α, is defined by

$$\alpha = -\frac{1}{\forall}\left(\frac{\partial \forall}{\partial p}\right)_T \qquad \text{Eq. 1.4-20}$$

Since the slope to the curve is negative, the minus sign is included in the definition to make α positive. The dimensions for α are the inverse of the dimensions of pressure. The larger the value of α, the greater is the compressibility of the fluid. Experiments indicate that for water at atmospheric pressure, $\alpha = 3.17 \times 10^{-6}$ in.^{2}/lb$_f$. The compressibility of an ideal gas is determined in the following example.

Example 1-2

Find an expression for the coefficient of compressibility at constant temperature for an ideal gas.

SOLUTION

For an ideal gas

$$p = \rho RT = \frac{mRT}{\forall} \quad \text{or} \quad \forall = \frac{mRT}{p}$$

Then

$$\left(\frac{\partial \forall}{\partial p}\right)_T = -\frac{mRT}{p^2}$$

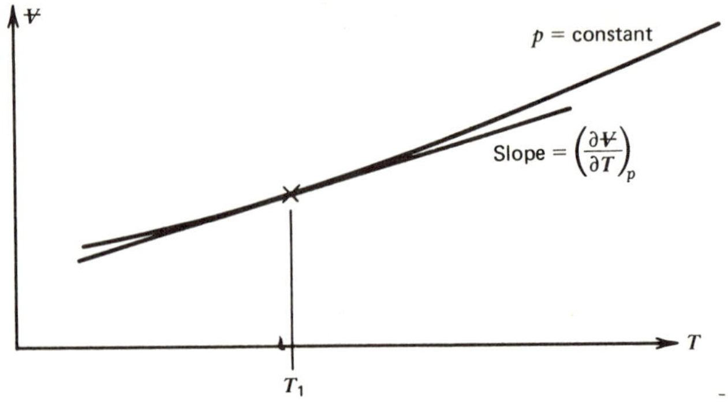

Fig. 1-19. Plot of volume versus temperature.

and by Eq. 1.4-20

$$\alpha = \frac{1}{\forall}\left(\frac{mRT}{p^2}\right) = \frac{1}{p}$$

We can see how much greater is the compressibility of air than water by observing the ratio of their coefficients of compressibility at standard atmospheric pressure (standard atmospheric pressure is 14.7 psia[8]). That is,

$$\frac{\alpha_{\text{air}}}{\alpha_{\text{water}}} = \frac{1}{14.7} \times \frac{1}{3.17 \times 10^{-6}} = 2.14 \times 10^4$$

The *bulk modulus of elasticity*, K, is defined as the reciprocal of the coefficient of compressibility; that is,

$$K = \frac{1}{\alpha} \qquad \text{Eq. 1.4-21}$$

The *coefficient of expansion* at constant pressure, β, is defined by the relation

$$\beta = \frac{1}{\forall}\left(\frac{\partial \forall}{\partial T}\right)_p \qquad \text{Eq. 1.4-22}$$

The dimensions for β are 1/R or 1/K. An experiment similar to the one shown in Fig. 1-17 may be used for determining β. However, in this case, F is held constant, and the temperature of the bath is varied. Under these circumstances a plot of $\forall$ versus T would be obtained as shown in Fig. 1-19. The slope to the obtained

[8] Here psia means $\text{lb}_f/\text{in.}^2$ absolute.

curve at any point would provide $(\partial V/\partial T)_p$ at that point. As an exercise, show that the coefficient of expansion at constant pressure is equal to $1/T$ for an ideal gas.

1.4.8 SURFACE TENSION AND CAPILLARITY

Virtually every student has witnessed the classic grade-school science experiment of floating a steel needle on the surface of a beaker of water. Since the density of the steel is acknowledged to be greater than that of the water, the needle should obviously sink to the bottom of the container. However, if carefully lowered to the water surface, the needle will remain at the surface, held in place by what appears to be an invisible film. The following discussion of adhesion, cohesion, and surface tension offers an explanation for this familiar phenomenon.

Cohesion is the term that describes the attractive forces that exist between molecules of the same type for one another, whereas *adhesion* is the term describing the attractive forces existing between molecules of different types. When a liquid forms a free surface in contact with a gas, the liquid molecules at the surface experience an attraction created by the molecules below the surface that is not balanced by the gas molecules above. This causes the surface molecules to, in effect, close together and act like a thin membrane in tension. This phenomenon is known as *surface tension*. Since work is required to alter the size or shape of the interface, the interface surface may be viewed as possessing potential energy. The potential energy per unit surface area is defined as the *surface-tension coefficient*, σ. The dimensions of σ are lb_f/ft or N/m.

Alternatively, the phenomenon may be viewed in terms of the thin-membrane concept. Suppose we make an imaginary cut across the interface surface, as shown in Fig. 1-20*c*. Since in the thin-membrane concept the membrane is in tension, a force distribution would exist along the line of the cut. If $d\mathbf{F}_\sigma$ is the surface-tension force acting on length, dl, then an alternative definition for the surface-tension coefficient, σ, is given by the following equation

$$d\mathbf{F}_\sigma = \sigma \, dl \, \mathbf{e}_t$$

where $\mathbf{e}_t$ is a unit vector tangent to the interface surface and perpendicular to the line of the cut. It can be shown that the two definitions for σ are equivalent. For a finite length cut in which $\mathbf{e}_t$ is a constant, the magnitude of the surface-tension force, F_σ, is given by

$$F_\sigma = \int_L \sigma \, dl$$

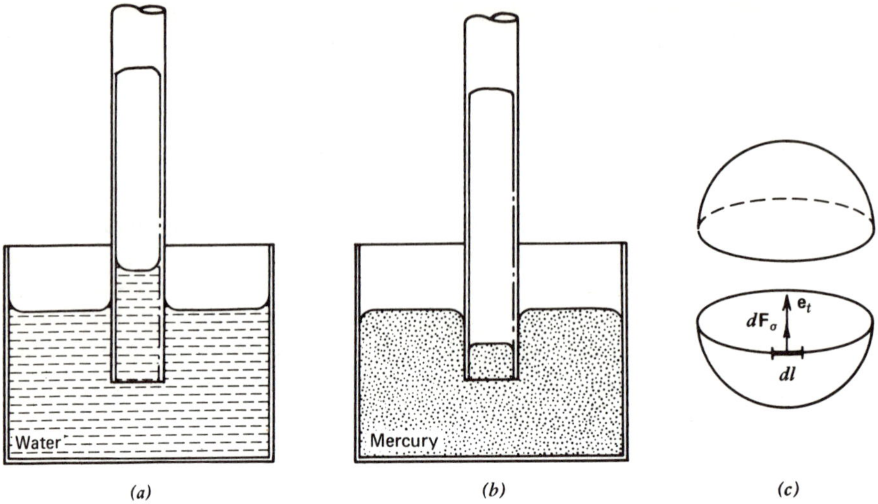

Fig. 1-20. The effect of capillarity on a liquid meniscus. (*a*) Adhesion > cohesion. (*b*) Cohesion > adhesion. (*c*) surface-tension force at interface of a cut section of a spherical bubble in a liquid.

It should be pointed out that the same surface-tension effect is observed at the interface of two liquids that are not miscible in one another (that do not mix). When the interface is a curved surface, a pressure jump exists across the interface because of surface tension. It can be shown (see Problem 3-30) that when the interface is a spherical surface (examples of which are the water droplet or an air bubble in a liquid) the pressure difference can be expressed by

$$p_2 - p_1 = \frac{2\sigma}{r} \qquad \text{Eq. 1.4-23}$$

where σ is the surface tension coefficient (see Table A.6-6), p_2 is the interior pressure, and p_1 is the exterior pressure. For a water droplet at 20 C in air with a radius of 0.005 ft (about $\frac{1}{8}$ in. in diameter), the pressure difference is approximately 1.0 psf. As a means of comparison, note that standard atmospheric pressure is 2116 psf. It is immediately apparent that surface-tension effects are significant only when the radius of curvature is extremely small, for example, in the study of aerosol sprays or colloidal particles. Surface tension accounts for the spherical liquid droplets that float about the zero-gravity confines of a spaceship cabin and plague astronauts.

The phenomenon of capillary attraction is a by-product of surface tension that depends on the relative magnitude of the cohesion of the liquid and the adhesion of the liquid for its container. When a small diameter tube is placed vertically in a container of liquid, the liquid surface is either raised or depressed from the level of the liquid outside the tube. If the adhesive tendency of the liquid for the walls of the tube is greater than the cohesion between the liquid molecules themselves, the liquid is said to "wet" the tube; and the liquid level in the tube is raised as in Fig. 1-20*a*. Water tends to wet the fibers of a paper towel or the material of a plastic sponge, thus accounting for the ability of the towel or sponge to soak up the water. Similarly, the liquid wax at the top of a candle wets the fibers of the wick, thus permitting the wax to rise up the wick and be burned.

On the other hand, if the cohesive force between liquid molecules is greater than the adhesive attraction for the material of the tube, the liquid does not wet the tube surface; and the liquid level in the tube is depressed as the tube is lowered into the liquid. Mercury is a fluid exhibiting this characteristic, as shown in Fig. 1-20*b*. Care must obviously be taken to allow for the effects of capillary attraction when small glass tubes are used in liquid measuring apparatus such as manometers, barometers, and draft gages. A clean glass tube with a diameter of 0.20 in. immersed in water will raise the level of water in the tube about one-tenth of an inch above the surrounding water surface outside the tube. In other words, when the glass tube is used as a manometer the effect of capillary attraction is to cause an apparent error in the pressure reading; in this case, the apparent error is equivalent to 0.10 in. of water. The same glass tube will depress a mercury column by approximately 0.05 in.

Problems

1-1 Compute the force in newtons required to provide a uniform linear acceleration of 50 ft/s^2 to a 3.7-slug mass.

1-2 What is the weight in lb_f of 10.0-kg mass at rest on the surface of the moon where the gravitational acceleration may be taken as one-sixth of that on earth?

1-3 A 1.5-lb_m block of steel is weighed on a spring scale at several different locations. The spring scale is calibrated to read 1 lb_f (weight) when weighing 1 lb_m at a point where $g = 32.174$ ft/s^2. What will the weight reading on the scale be for each of the following situations? Does the mass of the block change? (*a*) The 1.5-lb_m block is weighed where $g = 30.00$ ft/s^2. (*b*) The mass is weighed where $g = 35.00$ ft/s^2. (*c*) The 1.5-lb_m block is weighed on the moon (see Problem 1-2).

1-4 Repeat Problem 1-3, the only difference being that in this case, a beam balance is substituted for the spring scale. What are the resulting readings for parts (*a*), (*b*), and (*c*)?

1-5 (*a*) The potential energy of a mass is defined to be the work done in raising the mass in a gravitational field from elevation zero to elevation z, with nearly zero velocity. Determine an expression for the potential energy in (m-N) of a 100-lb_m block at an elevation of 1000 ft in a region where $g = 32.17$ ft/s^2.

(*b*) The kinetic energy of a mass is defined to be the work done in accelerating the mass from rest to velocity, v. Determine an expression for the kinetic energy of a mass. Using this expression, determine the kinetic energy in (ft-lb_f) of a 100-kg block that has a speed of 20 m/s in a region where $g = 9.806$ m/s^2.

1-6 A 10.0-kg mass occupies a volume of 0.16 ft^3 and is assumed to be of homogeneous composition. What is the density (in both English and SI units) of the material, (*a*) on the earth's surface, and (*b*) on the moon's surface? (On the moon's surface the gravitational acceleration is one-sixth of that on the earth's surface.) What is the specific weight in each of the above cases? (Again, in both sets of units.)

1-7 Find the specific gravity of the 10.0-kg mass of Problem 1-6 at a point where $g = 32.174$ ft/s^2. What is the specific gravity of the same material where $g = 25.76$ ft/s^2? Where $g = 37.6$ ft/s^2?

1-8 At a temperature of 70 F, the density of water is 62.3 lb_m/ft^3, and the absolute viscosity is 2.36 lb_m/(ft-hr). (*a*) Express the viscosity in units of (lb_f-s)/ft^2. (*b*) Express the viscosity in units of kg/(m-s). (*c*) Express the viscosity in poises [1 poise = 1 g/(cm-s)]. (*d*) What is the kinematic viscosity in ft^2/s? (*e*) Express the kinematic viscosity in centistokes (1 stoke = 1 cm^2/s).

1-9 If the absolute viscosity of air at 80 F is 3.85×10^{-7} (lb_f-s)/ft^2, find the kinematic viscosity when the pressure is 70 psia.

1-10 In the Fig. 1-10, the thin flat plate has an area of 0.05 m^2 and moves across the liquid surface with a constant velocity of 0.25 m/s. Take the liquid depth in the channel to be 0.05 m. Assuming that the liquid is water at 20 C, and that the velocity profile under the plate is linear, find the force, F (in SI units), required to keep the plate in motion.

1-11 Repeat Problem 1-9 for air at a temperature of 80 F and a pressure of two atmospheres.

1-12 A cube slides down an inclined plane on which there is a film of oil 0.5×10^{-4} m thick, see Fig. 1-21. The block weighs 200 N and has sides of 0.15 m. If the viscosity of the oil is 1.1×10^{-2} (N-s)/m^2, estimate the terminal velocity of the block,

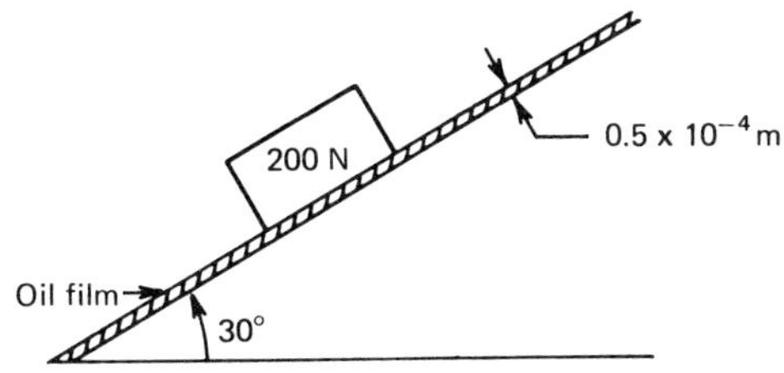

Fig. 1-21. Block sliding down inclined plane.

assuming that the block slides on top of the oil film and the velocity profile of the oil under the block is linear.

1-13 An element of a certain machine consists of a bar that moves up and down through guides, as shown in Fig. 1-22. The guides are lubricated by an oil film that completely

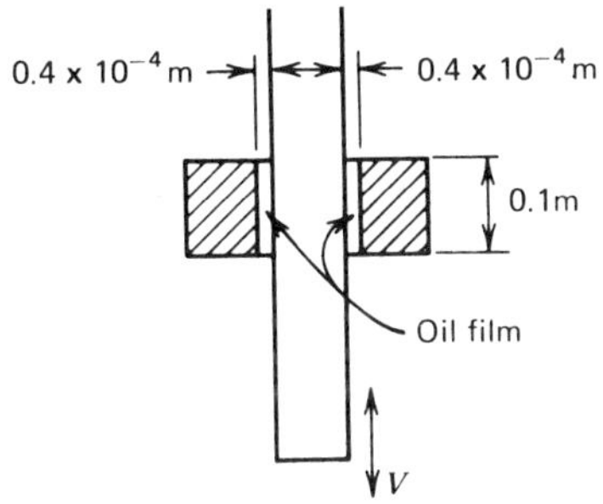

Fig. 1-22. Bar moving through lubricated guides.

fills the gap and has a kinematic viscosity of 5×10^{-6} m^2/s and a specific gravity equal to 0.86. What is the viscous drag force on the bar when the bar is moving upwards at 20 m/s? Assume that the velocity profile of oil is linear; that the cross-sectional area of the bar is 0.05 m by 0.05 m; and that the guides extend to the full width of the bar.

1-14 Water at 50 F flows through the two-dimensional channel shown in Fig. 1-7. A hot-wire anemometer traverse of the channel indicates that the velocity profile can be represented by

$$v_1(y) = 130(4 - y^2) \text{ in./s}$$

where y is measured in inches. If the surface area of the plug in the bottom wall is 0.8 ft^2 and the channel depth is 4 in., determine the shear force on the plug.

1-15 A vertical, circular cylinder with an inside cross-sectional area of 1 ft^2 is closed at the bottom and partially filled with water as shown in Fig. 1-23. A tight-fitting steel

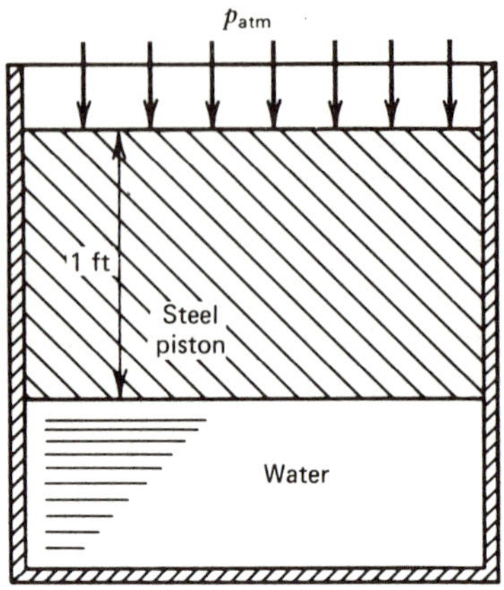

Fig. 1-23. Steel piston in cylinder.

piston, 1 ft high, rests on the water surface. If the steel has a density of 490 lb_m/ft^3, and if the atmospheric pressure is 14.7 psia, what is the pressure on the surface of the water when it comes in contact with the lower face of the piston? (*a*) In psia; (*b*) in psig; (*c*) in atmospheres; (*d*) in N/m^2 absolute.

1-16 Repeat Problem 1-15 for a cylinder with a cross-sectional area of 2.5 ft^2. All other data remain the same.

1-17 Repeat Problem 1-15 if the identical system is located in an environmental chamber on the moon, where the atomspheric pressure is 10.6 psia and g is 5.366 ft/s^2.

1-18 In chemistry and thermodynamics atomic weight, *molecular weight*, the mole, and molal mass are all frequently used terms. These quantities are defined as follows. The *atomic weight* is the ratio of the mass of an atom to $\frac{1}{16}$ the mass of an oxygen atom. The *molecular weight* is the sum of the atomic weights of the atoms which make up the molecule. Both of these terms represent dimensionless quantities. A *mole* is defined as that quantity of a substance that has a mass numerically equal to its molecular weight. By this definition, the amount of mass contained in a mole depends on the system of dimensions being used. Therefore, the mole is often preceded by a modifying term that indicates the mass system being used. For example, the mass of 1 g-mole of water is 18 g, but the mass of 1 kg-mole of water is 18 kg. The *molal mass*, $\mathcal{M}$, of a substance is the mass in a single mole of the substance. In the SI system, the dimensions of $\mathcal{M}$ are kg/kg-mole. Consider a particular substance. (*a*) If $\mathcal{N}$ equals the number of moles of the substance, and m equals its mass, express m in terms of $\mathscr{N}$ and $\mathscr{M}$. (b) Let N equal the number of molecules of the substance and $\hat{m}$ be the mass of one molecule. Express m in terms of N and $\hat{m}$. (*c*) Let M be the molecular weight of a substance and let $\hat{m}_0$ equal the mass of one molecule of oxygen. Express $\hat{m}$ in terms of $\hat{m}_0$ and M. (*d*) Noting that $\mathscr{M}/M$ is numerically equal to one, show that $N/\mathscr{N} = 16/\hat{m}_0$. Thus, the number of molecules in a mole is the same for every substance. It should be noted that $16/\hat{m}_0 = N_0 = 6.025 \times 10^{23}$ molecules/g-mole. N_0 is called *Avogadro's number*.

1-19 Avogadro's law states that equal volumes of gases at the same temperature and pressure have the same number of molecules. Consider two different gases of equal volumes, each at the same temperature and pressure, and both obeying the ideal gas law. Let ρ_1 and ρ_2 be the mass density of each gas. (*a*) Express ρ_1/ρ_2 in terms of their molal masses, $\mathscr{M}_1$ and $\mathscr{M}_2$. Justify your answer. (*b*) Using the result of part (*a*), show that the product of the molal mass, $\mathscr{M}$, and the gas constant, R, are the same for both gases. Thus, $\mathscr{M}R$ is universal. The product $\mathscr{M}R$ is called the universal gas constant, which we denote by the symbol, R^*. (*c*) Show that the dimensions of the universal gas constant, R^*, are (N-m)/(kg-mole-R) in the SI system. (*d*) If R^* has a numerical value of 8314.5 (N-m)/(kg-mole-R) in the SI system, determine the equivalent molecular weight of air. (*e*) Consider a gas that is contained. Express the perfect gas law in terms of its pressure, p, total volume, $\mathcal{V}$, temperature, T, and the universal gas constant, R^*.

1-20 What is the value of the coefficient of compressibility at constant temperature for air at 14.7 psia and 100 F? What is the value of the modulus of elasticity?

1-21 One cubic foot of water at 520 R and 14.7 psia is subjected to a pressure increase of 10,000 psi. What is the decrease in volume measured in cubic feet?

1-22 A field crew working for a water supply company discovers an old water line in a field. They wish to estimate the length of the line. To accomplish this they insert a device consisting of a piston inside a cylinder, with provisions for eliminating air pockets in the device and for determining both the travel distance of the piston and the pressure in the line. Assuming that the line is filled with water, estimate the length of the line if the piston moves 0.20 m when the pressure is increased from 1.4×10^5 N/m^2 to 2.8×10^5 N/m^2 (see Fig. 1-24).

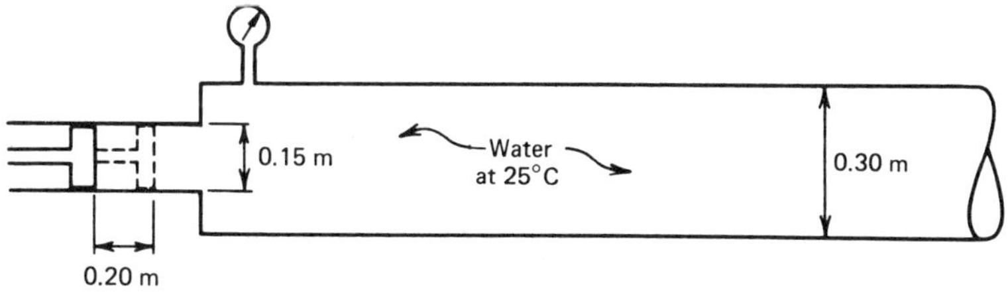

Fig. 1-24. Device for estimating pipe length.

1-23 (a) Find an expression for the coefficient of expansion, β, at constant pressure for an ideal gas, (b) Determine β for air at 14.7 psia and 70 F.

1-24 What is the coefficient of expansion at constant pressure for air at 3 atm and 40 C?

1-25 A glass bottle of drinking water is accidentally ruptured in the zero-gravity environment of a spaceship. The resulting spherical water droplets range in diameter from 0.005 to 0.001 ft. If the water is at 20 C and the cabin pressure is maintained at 8 psia, what is the range of internal pressures for the water droplets?

Chapter Two

Mathematical Summary

2.1 INTRODUCTION

In writing this chapter, we have presumed that the student has been previously exposed to the subject of vector calculus. The mathematical concepts used throughout this textbook are presented in a summary form here and *in detail* in Appendix A.1. At this point, it would be most helpful to review items (a) through (l), including their related problems (Problems 2-1 to 2-9); and items (m) through (s), including their related problems (Problems 2-10 to 2-27) before reading Chapter 9. If your mathematical background consists only of an operational capability in vector calculus, then we recommend that you now turn to Appendix A.1, where the physical interpretation of terms such as the directional derivative, gradient, divergence, and so on, is presented.

2.2 MATHEMATICAL SUMMARY

The following summary is listed in the topical order usually presented in a normal course in calculus.

a. Scalar and Vector Fields

If at each point (x, y, z) in a continuum space there corresponds a number $f(x, y, z)$, we say that f is a *scalar field*. Note that a scalar only exhibits magnitude. Fluid properties of mass density, temperature, pressure, internal energy, enthalpy, and entropy are all examples of scalar fields.

Similarly, a *vector field* exists if at each point (x, y, z) in a continuum space there corresponds a vector $\mathbf{F}(x, y, z)$. Note that a vector exhibits both magnitude and direction. Furthermore, vectors must also obey rules of vector addition and multiplication by scalars. Fluid velocity and acceleration are examples of vector fields.

b. The Mean Value Theorems

(1) $f(b) - f(a) = f'(x^*)(b - a)$ Eq. 2.2-1

(2) $\int_a^b f(x)\,dx = f(x^*)(b - a)$ Eq. 2.2-2

where $a < x^* < b$.

c. If $f = f(x, y)$, then the partial derivatives and the directional derivative are defined by

(1) $$\frac{\partial f}{\partial x}(x_0, y_0) = \lim_{x \to x_0} \frac{f(x, y_0) - f(x_0, y_0)}{x - x_0}$$ Eq. 2.2-3

(2) $$\frac{\partial f}{\partial y}(x_0, y_0) = \lim_{y \to y_0} \frac{f(x_0, y) - f(x_0, y_0)}{y - y_0}$$ Eq. 2.2-4

(3) $$\frac{\partial f}{\partial s}(x_0, y_0) = \lim_{s \to 0} \frac{f(x, y) - f(x_0, y_0)}{s}$$ Eq. 2.2-5

where s is the distance from point (x, y) to point (x_0, y_0). It is understood that the direction from (x, y) to (x_0, y_0) is fixed during the limiting process. As the direction changes the value of the directional derivative changes.

d. If $f = f(x, y, z)$, then the gradient of f, written as ∇f, is defined by

$$\nabla f = \frac{\partial f}{\partial x}\mathbf{i} + \frac{\partial f}{\partial y}\mathbf{j} + \frac{\partial f}{\partial z}\mathbf{k}$$ Eq. 2.2-6

It can be shown that

$$|\nabla f| = \left(\frac{\partial f}{\partial s}\right)_{\max}$$ Eq. 2.2-7

e. The directional derivative, $\partial f/\partial s$, in the direction of the unit vector $\mathbf{e}_s$ can also be obtained by taking the dot product of ∇f and $\mathbf{e}_s$; that is,

$$\frac{\partial f}{\partial s} = \nabla f \cdot \mathbf{e}_s \qquad \text{Eq. 2.2-8}$$

f. The definitions of the line and surface integrals are

$$\oint_c \mathbf{F} \cdot d\mathbf{r} = \lim_{\max \Delta \mathbf{r}_j \to 0} \sum_j \mathbf{F}_j \cdot \Delta \mathbf{r}_j \qquad \text{Eq. 2.2-9}$$

$$\oiint_S \mathbf{F} \cdot d\mathbf{S} = \lim_{\max \Delta \mathbf{S}_j \to 0} \sum_j \mathbf{F}_j \cdot \Delta \mathbf{S}_j \qquad \text{Eq. 2.2-10}$$

g. The divergence of a vector $\mathbf{A}$, written $\nabla \cdot \mathbf{A}$ or div $\mathbf{A}$, is defined at point P by

$$\text{div}\,\mathbf{A} = \nabla \cdot \mathbf{A} = \lim_{\Delta V \to 0} \frac{1}{\Delta V} \oiint_S \mathbf{A} \cdot d\mathbf{S} \qquad \text{Eq. 2.2-11}$$

where surface S encloses point P.

h. In Cartesian coordinates, the expression for $\nabla \cdot \mathbf{A}$ is

$$\nabla \cdot \mathbf{A} = \frac{\partial A_1}{\partial x} + \frac{\partial A_2}{\partial y} + \frac{\partial A_3}{\partial z} \qquad \text{Eq. 2.2-12}$$

i. The divergence theorem relates a volume integral to a surface integral

$$\iiint_V (\nabla \cdot \mathbf{A})\, dV = \oiint_S \mathbf{A} \cdot d\mathbf{S} \qquad \text{Eq. 2.2-13}$$

j. The Laplacian of a scalar, ϕ, written as $\nabla^2 \phi$, is defined by

$$\nabla^2 \phi = \nabla \cdot \nabla \phi \qquad \text{Eq. 2.2-14}$$

In Cartesian coordinates

$$\nabla^2 \phi = \frac{\partial^2 \phi}{\partial x^2} + \frac{\partial^2 \phi}{\partial y^2} + \frac{\partial^2 \phi}{\partial z^2} \qquad \text{Eq. 2.2-15}$$

k. Two useful vector identities are

$$\nabla \cdot (\phi \mathbf{A}) = \phi(\nabla \cdot \mathbf{A}) + \mathbf{A} \cdot \nabla\phi$$

$$\nabla \cdot (\mathbf{F} \times \mathbf{G}) = \mathbf{G} \cdot (\nabla \times \mathbf{F}) - \mathbf{F} \cdot (\nabla \times \mathbf{G}) \qquad \text{Eqs. 2.2-16}$$

l. Two other integral theorems, derivable from Eq. 2.2-13, are

$$\oiiint_{V} (\nabla\phi)\, dV = \oiint_{S} \phi\mathbf{n}\, dS$$

$$\oiiint_{V} (\nabla \times \mathbf{F})\, dV = \oiint_{S} (\mathbf{n} \times \mathbf{F})\, dS \qquad \text{Eqs. 2.2-17}$$

m. The curl of a vector **F**, written $\nabla \times \mathbf{F}$, is itself a vector and is defined by its components. If $\nabla \times \mathbf{F} = B_1\mathbf{i} + B_2\mathbf{j} + B_3\mathbf{k}$, then at point P,

$$B_1 = \lim_{\Delta S \to 0} \frac{1}{\Delta S} \oint_{C_1} \mathbf{F} \cdot d\mathbf{r} \qquad \text{Eq. 2.2-18}$$

where ΔS is an element of the plane surface enclosing point P and whose normal is the unit vector **i**. The bounding curve of ΔS is designated as C_1. Components B_2 and B_3 are defined in a similar manner (see Fig. A.1-11 for the surface used for evaluating B_3).

n. In Cartesian coordinates, the expression for $\nabla \times \mathbf{F}$ is

$$\nabla \times \mathbf{F} = \left(\frac{\partial F_3}{\partial y} - \frac{\partial F_2}{\partial z}\right)\mathbf{i} + \left(\frac{\partial F_1}{\partial z} - \frac{\partial F_3}{\partial x}\right)\mathbf{j} + \left(\frac{\partial F_2}{\partial x} - \frac{\partial F_1}{\partial y}\right)\mathbf{k} \qquad \text{Eq. 2.2-19}$$

o. Two vector identities involving the curl are

$$\nabla \times \nabla f = 0 \qquad \text{Eq. 2.2-20}$$

$$(\mathbf{v} \cdot \nabla)\mathbf{v} = \tfrac{1}{2}\nabla(v^2) - \mathbf{v} \times (\nabla \times \mathbf{v}) \qquad \text{Eq. 2.2-21}$$

p. Stokes Theorem relates a surface integral to a line integral.

$$\oiiint_{S} (\nabla \times \mathbf{F}) \cdot \mathbf{n}\, dS = \oint_{C} \mathbf{F} \cdot d\mathbf{r} \qquad \text{Eq. 2.2-22}$$

q. A necessary and sufficient condition for

$$\mathbf{F}(x, y, z) = \nabla f(x, y, z) \qquad \text{Eq. 2.2-23}$$

is that everywhere

$$\nabla \times \mathbf{F} = 0 \qquad \text{Eq. 2.2-24}$$

When Eq. 2.2-24 is valid everywhere, the field is said to be *irrotational.*

r. The relationship between the cylindrical components of a vector and its Cartesian components is given below. If

$$\mathbf{A} = A_1\mathbf{i} + A_2\mathbf{j} + A_3\mathbf{k} = A_r\mathbf{e}_r + A_\theta\mathbf{e}_\theta + A_Z\mathbf{e}_Z \qquad \text{Eq. 2.2-25}$$

then

$$A_r = A_1 \cos\theta + A_2 \sin\theta$$

$$A_\theta = A_1 \sin\theta + A_2 \cos\theta \qquad \text{Eqs. 2.2-26}$$

$$A_Z = A_3$$

s. The gradient, divergence, curl, and the Laplacian in *cylindrical* coordinates are given below.

(1) $$\nabla f = \frac{\partial f}{\partial r}\mathbf{e}_r + \frac{1}{r}\frac{\partial f}{\partial \theta}\mathbf{e}_\theta + \frac{\partial f}{\partial Z}\mathbf{e}_Z \qquad \text{Eq. 2.2-27}$$

(2) $$\nabla \cdot \mathbf{A} = \frac{1}{r}\frac{\partial}{\partial r}(rA_r) + \frac{1}{r}\frac{\partial A_\theta}{\partial \theta} + \frac{\partial A_Z}{\partial Z} \qquad \text{Eq. 2.2-28}$$

(3) $$\nabla \times \mathbf{A} = \left(\frac{1}{r}\frac{\partial A_Z}{\partial \theta} - \frac{\partial A_\theta}{\partial Z}\right)\mathbf{e}_r + \left(\frac{\partial A_r}{\partial Z} - \frac{\partial A_Z}{\partial r}\right)\mathbf{e}_\theta$$

$$+ \left(\frac{\partial A_\theta}{\partial r} + \frac{A_\theta}{r} - \frac{1}{r}\frac{\partial A_r}{\partial \theta}\right)\mathbf{e}_Z \qquad \text{Eq. 2.2-29}$$

(4) $$\nabla^2 f = \frac{1}{r}\frac{\partial}{\partial r}\left(r\frac{\partial f}{\partial r}\right) + \frac{1}{r^2}\frac{\partial^2 f}{\partial \theta^2} + \frac{\partial^2 f}{\partial Z^2} \qquad \text{Eq. 2.2-30}$$

Problems

2-1 If $f(x, y) = 3xy^2 - x^3e^y$, find $\partial f/\partial x$, $\partial f/\partial y$, and ∇f.

2-2 If $\phi(x, y, z) = y^2 \cos ax + e^x \sin z$, find: (*a*) $\nabla\phi(0, 2, \pi)$. (*b*) $\dfrac{\partial \phi}{\partial S}(0, 2, \pi)$ in the direction of the unit vector

$$\mathbf{e}_S = \frac{1}{\sqrt{3}}(\mathbf{i} + \mathbf{j} + \mathbf{k})$$

(*c*) The maximum directional derivative at $(0, 2, \pi)$.

2-3 If $\mathbf{F} = 3x^2\mathbf{i} + 14xy\mathbf{j}$, evaluate $\int_c \mathbf{F} \cdot d\mathbf{r}$ along the line from (0, 0) to (1, 2).

2-4 If $\mathbf{F} = x\mathbf{i} + z\mathbf{j} - 4x^2z\mathbf{k}$, evaluate $\oiint_S \mathbf{F} \cdot d\mathbf{S}$ where, (*a*) S is the cylindrical surface $x^2 + y^2 = 16$ between the planes $z = 0$ and $z = 2$. (*b*) S is the region enclosed by the circle, $x^2 + y^2 = 16$, $z = 0$. (*c*) S is the region enclosed by the circle, $x^2 + y^2 = 16$, $z = 2$.

2-5 If $\mathbf{F}$ is the same as that given in Problem 2-4, evaluate $\nabla \cdot \mathbf{F}$ at $(2, -3, 5)$.

2-6 If $\mathbf{F}$ is the same as that given in Problem 2-4, and $\mathcal{V}$ is the volume enclosed by the cylinder $x^2 + y^2 = 16$, and the planes $z = 0$ and $z = 2$, verify the divergence theorem for this special case of evaluating $\oiiint_{\mathcal{V}} (\nabla \cdot \mathbf{F})\, d\mathcal{V}$ and comparing your answer with the sum of the values obtained in Problems 2-4 (*a*), (*b*), and (*c*).

2-7 If $\phi(x, y, z)$ is the same as that given in Problem 2-2, find $\nabla^2\phi$.

2-8 Prove the identities given in Eqs. 2.2-16 by carrying out the specified operations on both sides of both equations and comparing the results.

2-9 (*a*) If $\phi = x^2z^2$ and $\mathcal{V}$ is the volume specified in Problem 2-6, verify the first of Eqs. 2.2-17 for this special case by carrying out the integrations on both sides of the equation. (*b*) Prove the second equation in the set of Eqs. 2.2-17. Use Eq. 2.2-13 and the second equation in the set of Eqs. 2.2-16. Take $\mathbf{G}$ to be a constant vector.

The following questions are concerned with orthogonal curvilinear coordinates and cylindrical coordinates in particular.

2-10 In any coordinate system, a point in space is obtained by the intersection of three surfaces. (*a*) What three surfaces are used in the cylindrical coordinate system? Sketch these surfaces. (*b*) Give equations for the surfaces mentioned in part (*a*).

2-11 A parametric curve associated with a variable q in a curvilinear coordinate system is obtained by holding two of the variables constant and varying the third. For a cylindrical coordinate system describe (or label in the sketch drawn for Problem 2-10) the parametric curve associated with (*a*) the variable r, (*b*) the variable θ, and (*c*) the variable Z.

2-12 A point in space can be represented either in Cartesian coordinates or in curvilinear coordinates. If (x, y, z) are the Cartesian coordinates of a point in space and (r, θ, Z) are the cylindrical coordinates of that same point, (*a*) express (x, y, z) in terms of (r, θ, Z), and (*b*) express (r, θ, Z) in terms of (x, y, z).

2-13 A set of unit vectors at any point P in space, associated with a curvilinear coordinate system, is the set of unit vectors tangent to the parametric curves passing through point P. If this set of unit vectors is orthogonal to each other, then the curvilinear coordinate system is said to be *orthogonal*. Let $(\mathbf{e}_r, \mathbf{e}_\theta, \mathbf{e}_Z)$ be the set of unit vectors associated with cylindrical coordinates. (*a*) Add this set of vectors to your sketch drawn for Problem 2-10. (*b*) Express $(\mathbf{e}_r, \mathbf{e}_\theta, \mathbf{e}_Z)$ in terms of $(\mathbf{i}, \mathbf{j}, \mathbf{k})$ where $(\mathbf{i}, \mathbf{j}, \mathbf{k})$ are the unit vectors associated with a Cartesian coordinate system. (*c*) Express $(\mathbf{i}, \mathbf{j}, \mathbf{k})$ in terms of $(\mathbf{e}_r, \mathbf{e}_\theta, \mathbf{e}_Z)$. (*d*) Is the cylindrical coordinate system an orthogonal set?

2-14 Any vector can always be expressed in either Cartesian coordinates or in a curvilinear system. If (v_1, v_2, v_3) are the (x, y, z) components of a vector in Cartesian coordinates, and (v_r, v_θ, v_Z) are the components in cylindrical coordinates, (*a*) express (v_r, v_θ, v_Z) in terms of (v_1, v_2, v_3), and (*b*) express (v_1, v_2, v_3) in terms of (v_r, v_θ, v_Z).

2-15 Eq. 2.2-6 gives the expression for ∇f in Cartesian coordinates. Using the chain rule and your answer to Problem 2-13*c*, express ∇f in cylindrical coordinates (see Eq. 2.2-27).

2-16 (*a*) Let (ds_1, ds_2, ds_3) be infinitesimal arc lengths along the parametric curves associated with an orthogonal curvilinear coordinate system (see Problem 2-11), with coordinates (q_1, q_2, q_3) and unit vectors $(\mathbf{e}_1, \mathbf{e}_2, \mathbf{e}_3)$. Define scale factors (h_1, h_2, h_3) such that $ds_1 = h_1\, dq_1$, $ds_2 = h_2\, dq_2$, and $ds_3 = h_3\, dq_3$ (see Fig. 2-1*a*). What are the scale factors for a cylindrical coordinate system? (*b*) The displacement vector, $d\mathbf{r}$, and the square of its length $|d\mathbf{r}|^2$ can be expressed as follows (see Fig. 2-1*b*):

$$d\mathbf{r} = h_1\, dq_1\, \mathbf{e}_1 + h_2\, dq_2\, \mathbf{e}_2 + h_3\, dq_3\, \mathbf{e}_3$$

and,

$$|d\mathbf{r}|^2 = h_1{}^2\, dq_1{}^2 + h_2{}^2\, dq_2{}^2 + h_3\, dq_3{}^2$$

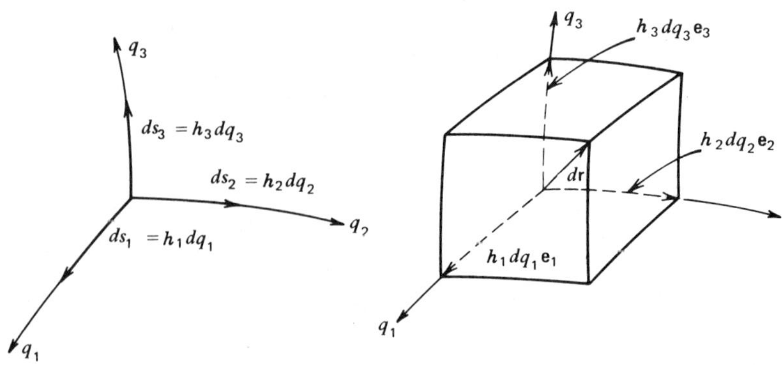

Fig. 2-1. Arc lengths and volume element in curvilinear coordinates. (*a*) Arc lengths. (*b*) Volume element.

Express $d\mathbf{r}$ and $|d\mathbf{r}|^2$ for a cylindrical coordinate system. (*c*) A volume element, $d\mathcal{V}$, is given by,

$$d\mathcal{V} = |h_1\, dq_1\, \mathbf{e}_1 \cdot (h_1\, dq_2\, \mathbf{e}_2 \times h_3\, dq_3\, \mathbf{e}_3)| = h_1 h_2 h_3\, dq_1\, dq_2\, dq_3$$

Express $d\mathcal{V}$ for a cylindrical coordinate system.

2-17 Using the definition of $\nabla \cdot \mathbf{A}$, as given by Eq. 2.2-11, and your answers to Problem 2-16 show that the expression for $\nabla \cdot \mathbf{A}$ in cylindrical coordinates is given by Eq. 2.2-28.

2-18 Using the definition of the components of $\nabla \times \mathbf{F}$ (the x-component of $\nabla \times \mathbf{F}$ in Cartesian coordinates is given by Eq. 2.2-18) show that in cylindrical coordinates the expression for the Z-compoucnt of $\nabla \times \mathbf{F}$ is given by the $\mathbf{e}_z$ component in Eq. 2.2-29.

2-19 Using Eqs. 2.2-14, 2.2-27, and 2.2-28 develop Eq. 2.2-30.

Chapter Three

Fluid Statics and Fluids Moving as a Rigid Body

3.1 INTRODUCTION

In this chapter we first examine the behavior of a fluid when it is in static equilibrium. The concepts of equilibrium for solid bodies carry over to fluids. As with a solid element, a fluid element is in equilibrium when the sum of all the forces acting on that element is zero. This occurs under two circumstances: (1) when all fluid elements in the medium are stationary and (2) when all fluid elements in the medium have a constant uniform velocity. The applications of fluid statics that are covered in this chapter are: the manometer used as a pressure-measuring device, hydrostatic forces acting on submerged bodies, buoyancy effects on submerged objects, a model atmosphere, and atmospheric stability. The treatment of a fluid accelerating as a rigid body requires only a slight modification from that of a static fluid. Thus, this topic is also discussed in this chapter.

3.2 HYDROSTATIC PRESSURE AT A POINT

In the study of solid-body mechanics, it was pointed out that the normal stresses acting on two infinitesimal surfaces of different orientation, but whose centers pass through the same point, are not generally the same. That is, if τ_{11} is the normal stress on a plane whose normal is in the x direction and τ_{22} is the normal stress on a plane whose normal is in the y direction, then τ_{11}, in general, does not equal τ_{22} (see Fig. 3-1). Furthermore, in a solid the surface traction, $\boldsymbol{\tau}$, is not usually normal to the surface. That is, the surface traction can be decomposed

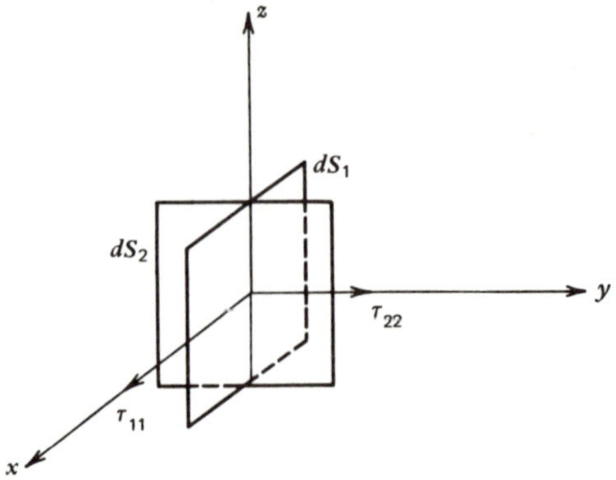

Fig. 3-1. Normal stresses on planes parallel to the coordinate planes.

into a normal stress and a shearing stress. It turns out that in a static fluid, nature is less complicated, since in a static fluid: (1) the surface traction is normal to the surface and (2) at a point the normal stress, which is a compressive stress, is the same in all directions. Point (2) is Pascal's law, which can readily be proved. Point (1) follows directly from our observation, discussed in Chapter 1, that the shear stress in a fluid is proportional to the derivatives of the fluid velocity components with respect to distance. Proof of Pascal's law is given below.

The student should recall that to obtain a relationship at a point, one first applies a physical law to an infinitesimal region. Next, the terms are algebraically manipulated and like terms are canceled. Finally, the $\lim_{\Delta\forall \to 0}$ (or $\lim_{\Delta S \to 0}$) is taken on both sides of the resulting equation. Consider the forces acting on an arbitrarily oriented prismatic element, as shown in Fig. 3-2. Take the normal stresses to be compressive, as shown; that is, let $\tau_{11} = -p_{11}$, $\tau_{22} = -p_{22}$, and so on. Since a fluid element in a static fluid is in equilibrium, the sum of the forces acting on the element must be zero,

$$\sum_{i=1}^{5} \mathbf{F}_i = p_{11}\,\Delta S_1\mathbf{i} + p_{22}\,\Delta S_2\mathbf{j} + p_{33}\,\Delta S_3\mathbf{k} - p_n\mathbf{n}\,\Delta S - \gamma\,\Delta\forall\mathbf{k} = 0 \qquad \text{Eq. 3.2-1}$$

Taking the dot product of Eq. 3.2-1 with **i**, **j**, and **k**, we obtain the three equations,

$$p_{11}\,\Delta S_1 - p_n\,\Delta\mathbf{S}\cdot\mathbf{i} = 0$$

$$p_{22}\,\Delta S_2 - p_n\,\Delta\mathbf{S}\cdot\mathbf{j} = 0 \qquad \text{Eqs. 3.2-2}$$

$$p_{33}\,\Delta S_3 - p_n\,\Delta\mathbf{S}\cdot\mathbf{k} - \gamma\,\Delta\forall = 0$$

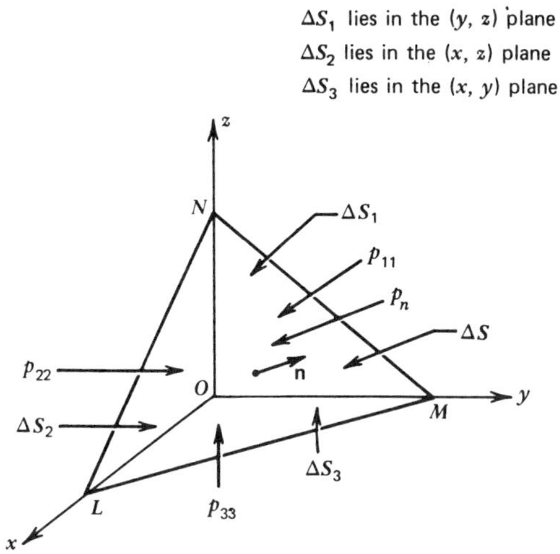

Fig. 3-2. Prismatic element in static equilibrium.

Now it can be shown (see Appendix A.2) that

$$\Delta \mathbf{S} \cdot \mathbf{i} = \text{projection of } \Delta S \text{ on the } (y, z) \text{ plane} = \Delta S_1$$

$$\Delta \mathbf{S} \cdot \mathbf{j} = \text{projection of } \Delta S \text{ on the } (x, z) \text{ plane} = \Delta S_2 \qquad \text{Eqs. 3.2-3}$$

$$\Delta \mathbf{S} \cdot \mathbf{k} = \text{projection of } \Delta S \text{ on the } (x, y) \text{ plane} = \Delta S_3$$

In fact, if ΔS is an arbitrarily oriented infinitesimal surface, then the projected area, ΔS_p, of ΔS on a plane whose normal is $\mathbf{e}_p$ is given by

$$\Delta S_p - \Delta \mathbf{S} \cdot \mathbf{e}_p \qquad \text{Eq. 3.2-4}$$

It can also be shown that the ΔV appearing in the last equation of the set of Eqs. 3.2-2 is given by

$$\Delta V = \frac{\Delta S_3}{3} \Delta z \qquad \text{Eq. 3.2-5}$$

The proof of Eq. 3.2-5 is left to the student as an exercise (see Problem 3-1). Substituting Eqs. 3.2-3 and Eq. 3.2-5 into Eqs. 3.2-2 gives

$$(p_{11} - p_n)\,\Delta S_1 = 0$$

$$(p_{22} - p_n)\,\Delta S_2 = 0$$

$$(p_{33} - p_n)\,\Delta S_3 - \frac{\gamma}{3}\,\Delta S_3\,\Delta z = 0$$

Dividing the first equation in the above set by ΔS_1, the second by ΔS_2 and the third by ΔS_3, then taking the $\lim_{\Delta X \to 0} \lim_{\Delta Y \to 0} \lim_{\Delta Z \to 0}$ of both sides of these equations, we obtain

$$p_{11} = p_{22} = p_{33} = p_n = p$$

Since ΔS is arbitrarily oriented, we have shown that at any point in a static fluid, the normal stresses are the same in all directions. We may thus drop the subscripting and designate the normal stresses by the symbol p. Then in a static fluid, the surface traction $\boldsymbol{\tau}$ is given by

$$\boldsymbol{\tau} = -p\mathbf{n} \qquad \text{Eq. 3.2-6}$$

The quantity p is the thermodynamic quantity we call *pressure.*

Now that we have established that pressure is the same in all directions, we can readily obtain the governing pressure-field equation in a static fluid. The most direct method is to consider the forces acting on the infinitesimal fluid element shown in Fig. 3-3. Select a fixed coordinate system denoted by (X, Y, Z). Since a fluid element in a static fluid is in equilibrium, we may apply the following vector equilibrium equation

$$\sum_{i=1}^{7} \mathbf{F}_i = 0 \qquad \text{Eq. 3.2-7}$$

This vector equation yields the three scalar equations, Eqs. 3.2-8. In this development we have allowed for the possibility that $p = p(X, Y, Z)$. Furthermore, to obtain an average force on the surface, the pressure on each surface is evaluated

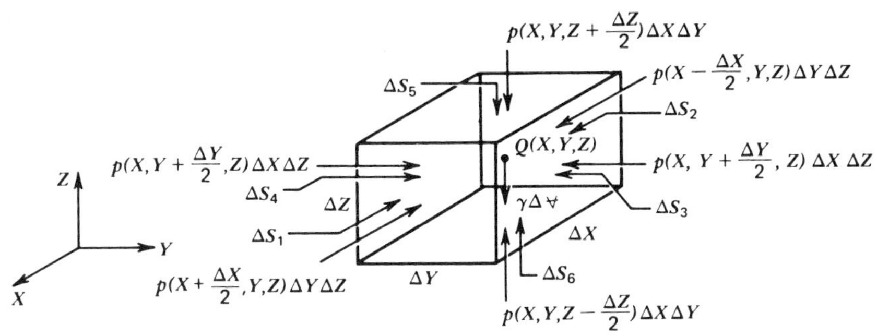

Fig. 3-3. Forces acting on the infinitesimal volume. $\Delta V = \Delta X\,\Delta Y\,\Delta Z$; $\Delta S_1 = \Delta S_2 = \Delta Y\,\Delta Z$; $\Delta S_3 = \Delta S_4 = \Delta X\,\Delta Z$; $\Delta S_5 = \Delta S_6 = \Delta X\,\Delta Y$; $\mathbf{n}_1 = -\mathbf{n}_2 = \mathbf{i}$; $\mathbf{n}_3 = -\mathbf{n}_4 = \mathbf{j}$; $\mathbf{n}_5 = -\mathbf{n}_6 = \mathbf{k}$.

at the center of that surface. Under these circumstances Eq. 3.2-7 gives the three scalar equations

$$-\left[p\left(X+\frac{\Delta X}{2}, Y, Z\right) - p\left(X-\frac{\Delta X}{2}, Y, Z\right)\right]\Delta Y\Delta Z = 0$$

$$-\left[p\left(X, Y+\frac{\Delta Y}{2}, Z\right) - p\left(X, Y-\frac{\Delta Y}{2}, Z\right)\right]\Delta X\,\Delta Z = 0$$

$$-\left[p\left(X, Y, Z+\frac{\Delta Z}{2}\right) - p\left(X, Y, Z-\frac{\Delta Z}{2}\right)\right]\Delta X\,\Delta Y - \gamma\,\Delta X\,\Delta Y\,\Delta Z = 0$$

Eqs. 3.2-8

By dividing through by $\Delta X\ \Delta Y\ \Delta Z$ in each equation and taking the $\lim_{\Delta X\to 0}\lim_{\Delta Y\to 0}\lim_{\Delta Z\to 0}$ of both sides in all three equations, we obtain, by the definition of the partial derivative,

$$\frac{\partial p}{\partial X} = 0$$

$$\frac{\partial p}{\partial Y} = 0 \qquad \text{Eqs. 3.2-9}$$

$$\frac{\partial p}{\partial Z} + \gamma = 0$$

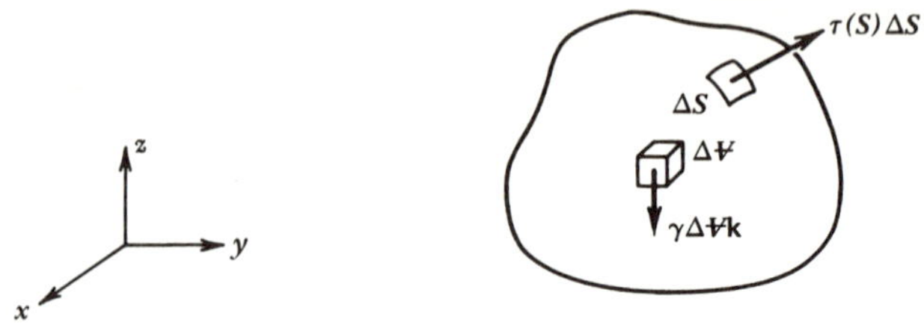

Fig. 3-4. Arbitrary system in the static continuum.

From the first two equations in this set, we can conclude that $p = p(Z)$. Thus, the partial derivative in the third equation can be replaced by the total derivative, giving, after some rearrangement of the terms,

$$\frac{dp}{dZ} = -\gamma \qquad \text{Eq. 3.2-10}$$

Equation 3.2-10 is the governing pressure-field equation in a static fluid. It can readily be seen that, if the fluid had a uniform acceleration, then only a small modification in the above development would be required to establish the governing pressure-field equation for this modified condition. This modified problem is discussed in Section 3.9.

Equations 3.2-9 can also be developed from a continuum mechanics viewpoint. Since concepts used in this second approach are extensively used in Chapter 4, it is appropriate to present this second viewpoint here. Take some *arbitrary* system occupying volume, $\forall$, in the static continuum. Consider all the forces acting on this system (see Fig. 3-4). Since the system is in static equilibrium, the sum of the forces acting on the system must be zero, that is,

$$-\oiiint_{\forall} \gamma \mathbf{k}\, d\forall + \oiint_{S} \boldsymbol{\tau}(S)\, dS = 0 \qquad \text{Eq. 3.2-11}$$

We showed that in a static fluid $\boldsymbol{\tau} = -p\mathbf{n}$ (see Eq. 3.2-6). Thus

$$\oiint_{S} \boldsymbol{\tau}\, dS = -\oiint_{S} p\mathbf{n}\, dS \qquad \text{Eq. 3.2-12}$$

But by Eq. 2.2-17,

$$\oiint_S p\mathbf{n}\, dS = \oiiint_{\forall} \nabla p\, d\forall \qquad \text{Eq. 3-2-13}$$

Substituting Eqs. 3.2-12 and 3.2-13 into Eq. 3.2-11 and combining the two integrals, we obtain

$$\oiiint_{\forall} (\nabla p + \gamma\mathbf{k})\, d\forall = 0 \qquad \text{Eq. 3.2-14}$$

Now this relationship was derived for any arbitrary region in the continuum. It is thus true for all possible limits of integration that are consistent with the limits of the continuum. The only way that this is possible is for the integrand to be identically equal to zero.[1] That is,

$$\nabla p + \gamma\mathbf{k} = 0 \qquad \text{Eq. 3.2-15}$$

or

$$\frac{\partial p}{\partial x}\mathbf{i} + \frac{\partial p}{\partial y}\mathbf{j} + \frac{\partial p}{\partial z}\mathbf{k} + \gamma\mathbf{k} = 0$$

Thus,

$$\frac{\partial p}{\partial x} = 0, \qquad \frac{\partial p}{\partial y} = 0, \qquad \text{and} \qquad \frac{\partial p}{\partial z} + \gamma = 0 \qquad \text{Eqs. 3.2-16}$$

By the same argument given earlier, we have

$$\frac{dp}{dz} = -\gamma \qquad \text{Eq. 3.2-17}$$

[1] This is similar to the statement: If

$$\int f(x)\, dx = 0$$

for all possible limits of integration, then $f(x) = 0$. It can readily be seen that if $f(x) \neq 0$ everywhere, then there would exist limits (a, b), where the initial statement was not true (see Fig. 3-5).

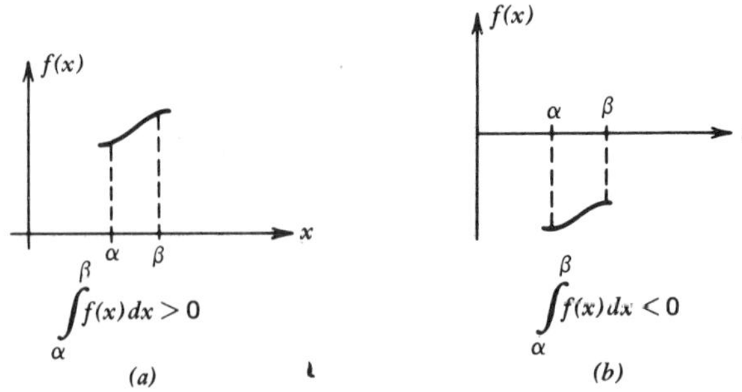

Fig. 3-5. Integral values for $f(x) \neq 0$, $\alpha \leq x \leq \beta$.

3.3 THE PRESSURE FIELD IN A STATIC LIQUID AND MANOMETRY

Let us first consider the integration of Eq. 3.2-17 in a liquid field. Since the specific weight, γ, of a liquid is only a mild function of temperature, we may take γ to be constant. Under these circumstances, Eq. 3.2-17 may immediately be integrated. If we take $p(z_1) = p_1$ and $p(z_2) = p_2$, then

$$\int_{p_1}^{p_2} dp = -\gamma \int_{z_1}^{z_2} dz$$

or

$$p_2 - p_1 = -\gamma(z_2 - z_1) \qquad \text{Eq. 3.3-1}$$

If we take the distance $(z_2 - z_1)$ to be h (see Fig. 3-6), then Eq. 3.3-1 with some rearranging becomes

$$p_2 = p_1 - \gamma h \qquad \text{Eq. 3.3-2}$$

From Eq. 3.3-2 we see that the pressure decreases as the elevation increases. This concept is used in the manometer, a device frequently employed for measuring pressure. The simplest form of a manometer is the *U*-shaped tube that is partially filled with an immiscible liquid relative to the fluid whose pressure is being measured. One end of the "U" tube may be opened, while the other end is connected to the reservoir containing the fluid whose pressure is desired (see Fig. 3-7). The specific weight γ_2 should be greater than specific weight γ_1 in order

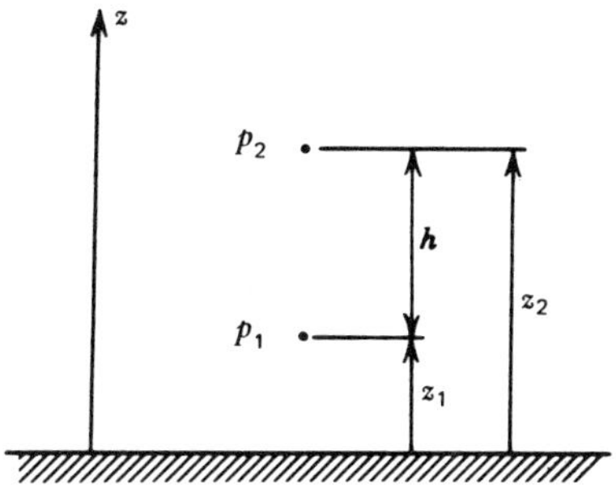

Fig. 3-6. Pressure at two points in a liquid.

that the fluid in the "U" tube remains at the bottom of the system. Mercury is often used as the immiscible liquid, but water may also be used if the fluid whose pressure is being measured is a gas, such as air.

To obtain an expression for p_0, apply Eq. 3.3-2 to the fluid whose specific weight is γ_1; that is,

$$p_0 = p_1 - \gamma_1 h_1$$

or

$$p_1 = p_0 + \gamma_1 h_1 \qquad \text{Eq. 3.3-3}$$

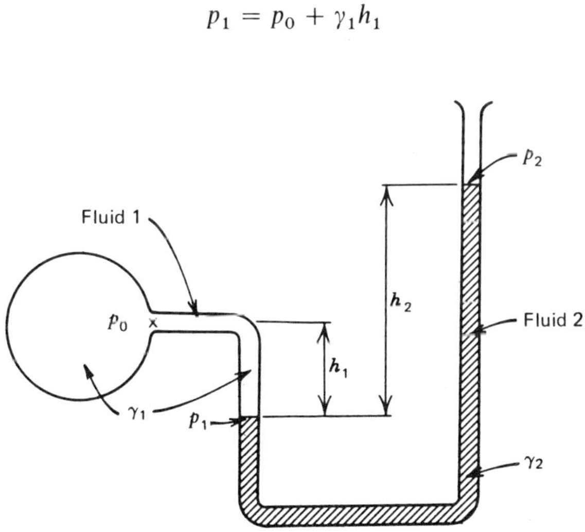

Fig. 3-7. A U-tube manometer.

Now apply Eq. 3.3-2 to the fluid whose specific weight is γ_2

$$p_2 = p_1 - \gamma_2 h_2$$

or

$$p_1 = p_2 + \gamma_2 h_2 \qquad \text{Eq. 3.3-4}$$

We have assumed that there is no pressure difference across the interface. This is only valid when the surface tension is negligible. Subtracting Eq. 3.3-4 from Eq. 3.3-3, the following equation is obtained:

$$p_0 + \gamma_1 h_1 - p_2 - \gamma_2 h_2 = 0$$

or

$$p_0 = p_2 + \gamma_2 h_2 - \gamma_1 h_1$$

If there are n different regions in the circuit, we can generalize the above expression into

$$p_0 = p_n \pm \sum_{i=1}^{n} \gamma_i h_i \qquad \text{Eq. 3.3-5}$$

That is, the pressure at one end of the circuit equals the pressure at the other end of the circuit $\pm \sum_{i=1}^{n} \gamma_i h_i$, where the (+) sign is used if one moves down in medium i while going through the circuit from one interface to another, and the (−) sign is used if one moves up in medium i. Since the right-hand tube of the manometer is open to the atmosphere, p_2 equals atmospheric pressure, p_{atm}. Thus,

$$p_0 = p_{atm} + \gamma_2 h_2 - \gamma_1 h_1 \qquad \text{Eq. 3.3-6}$$

In fact, most pressure measuring devices give the pressure above the atmospheric pressure. When the pressure is referenced to the local atmospheric pressure, it is designated as *gage pressure*. When the pressure is referenced to a complete vacuum, it is designated as *absolute pressure*. Thus,

Absolute pressure = atmospheric pressure
+ gage pressure reading

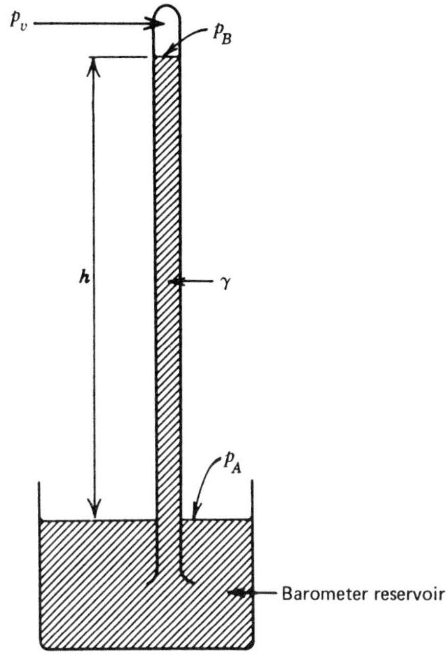

Fig. 3-8. A simple barometer.

Similarly, most vacuum gages give the pressure below atmospheric pressure. Thus,

$$\text{Absolute pressure} = \text{atmospheric pressure} - \text{vacuum gage pressure reading}$$

If Fluid 1 is a gas, then $\gamma_1 <<< \gamma_2$ and the third term in Eq. 3.3-6 will be negligible.

Another device that works on the same principle as the manometer is a simple-type barometer. A barometer is a device used to measure the local atmospheric pressure. The simple-type barometer consists of a straight tube inverted in a liquid reservoir, as shown in Fig. 3-8. The column must be sufficiently long so that a vacuum can be created in the region above the liquid column. However, evaporation will cause a small quantity of vapor from the barometer liquid to fill the region above the liquid column. Thus, the pressure above the column will equal the vapor pressure of the barometer liquid. The vapor pressure, p_v, of mercury at 68 F is 2.51×10^{-5} psia, whereas the vapor pressure for water at

70 F is 0.363 psia. To determine p_{atm} apply Eq. 3.3-2 between points A and B; that is,

$$p_B = p_A - \gamma h$$

Since $p_A = p_{atm}$ and $p_B = p_v$, we have

$$p_{atm} = p_v + \gamma h \qquad \text{Eq. 3.3-7}$$

If mercury is used as the barometer liquid, p_v is negligible.

3.4 FORCES ON SUBMERGED SURFACES AND THE CENTER OF PRESSURE

In the design of submerged structures, containers, and devices it is usually necessary to know the magnitude of the forces that act on their respective surfaces. The upstream face of a dam, a gate in a dam or lock, the hull of a stationary ship, the walls of a liquid-filled tank, all are examples of bodies that have submerged surfaces. In dealing with this type of problem, the quantities of interest are usually the *resultant force* and the *center of pressure*. The concept of the center of pressure is defined in the following statement:

> If a force equal to the resultant force was placed at the center of pressure, it would have the same force and moment effects as the distributed pressure force. Therefore, the resultant force located at the center of pressure represents an equivalent system to the distributed pressure-force system.
> Stm. 3.4-1

We will first consider submerged plane surfaces. Let us look at the forces acting on an area, A, embedded in one of the inclined walls of a closed tank containing a liquid, as shown in Fig. 3-9*a*. Assume the pressure above the liquid is p_0. Select the Z'-coordinate axis to point vertically down as shown. The governing equation for the pressure field in a static fluid when the z-axis is pointing vertically upward was given in Eq. 3.2-16; that is,

$$\frac{dp}{dz} = -\gamma \qquad \text{Eq. 3.2-17}$$

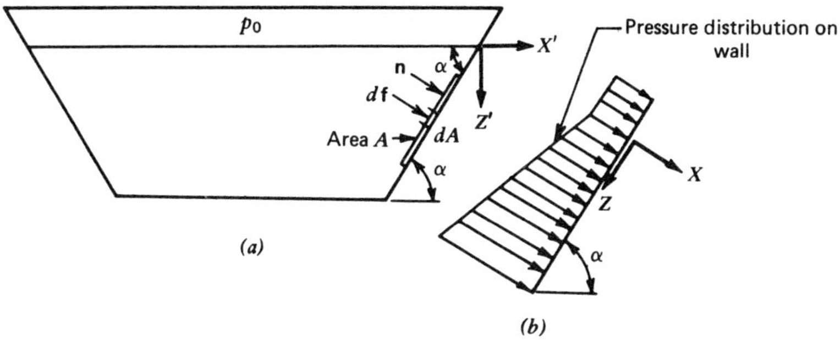

Fig. 3-9. Forces on a submerged plane wall. (*a*) Submerged plane surface. (*b*) Pressure distribution.

If $Z' = -z$, then by the chain rule

$$\frac{dp}{dZ'} = \frac{dp}{dz}\frac{dz}{dZ'} = -\frac{dp}{dz} = \gamma$$

or

$$\frac{dp}{dZ'} = \gamma \qquad \text{Eq. 3.4-1}$$

Since, for a liquid, γ can be considered to be constant, Eq. 3.4-1 can immediately be integrated giving

$$p = \gamma Z' + C$$

The constant C can be evaluated by applying the condition that $p(0) = p_0$. This gives that $C = p_0$ and that

$$p = p_0 + \gamma Z', \qquad \text{for } Z' \geq 0 \qquad \text{Eq. 3.4-2}$$

If the tank was open to the atmosphere, we would take $p_0 = 0$, since we are interested in the forces resulting from the pressure above atmospheric pressure. This is so because atmospheric pressure acts equally on both sides and thus its effects are canceled out. The pressure distribution on the surface for $Z' > 0$ will therefore be linear, as shown in Fig. 3.9*b*. The force, $d\mathbf{f}$, acting on the infinitesimal area, dA, is

$$d\mathbf{f} = p(Z')\,dA\,\mathbf{n} = (p_0 + \gamma Z')\,dA\,\mathbf{n} \qquad \text{Eq. 3.4-3}$$

The resultant force, $\mathbf{F}_R$ is obtained by integrating $d\mathbf{f}$ over the entire surface, A; that is,

$$\mathbf{F}_R = \oiint_A d\mathbf{f} = \mathbf{n} \oiint_A (p_0 + \gamma Z')\, dA \qquad \text{Eq. 3.4-4}$$

If we select an (X, Y, Z) coordinate system with (Y, Z) lying on the plane of the wall and X normal to the wall, as shown in Fig. 3.10, then

$$\mathbf{F}_R = \mathbf{i} \oiint_A (p_0 + \gamma Z')\, dA \qquad \text{Eq. 3.4-5}$$

where $dA = dY\, dZ$. But

$$Z' = Z \sin \alpha \qquad \text{Eq. 3.4-6}$$

Substituting Eq. 3.4-6 into Eq. 3.4-5 and carrying out the integration of the first term in the integrand gives

$$\mathbf{F}_R = \mathbf{i}\left(p_0 A + \gamma \sin \alpha \oiint_A Z\, dA\right) \qquad \text{Eq. 3.4-7}$$

where A is the area of interest. The student should recall from mechanics that

$$\oiint_A Z\, dA = Z_c A \qquad \text{Eq. 3.4-8}$$

where Z_c is the Z-coordinate of the centroid of the area. Substituting Eq. 3.4-8 into Eq. 3.4-7 gives

$$\mathbf{F}_R = A(p_0 + \gamma Z_c \sin \alpha)\mathbf{i} \qquad \text{Eq. 3.4-9}$$

Now, the vertically downward distance of the centroid from the free surface, Z_c, can be determined from Eq. 3.4-6; that is, $Z_c' = Z_c \sin \alpha$. Then by Eq. 3.4-2 the

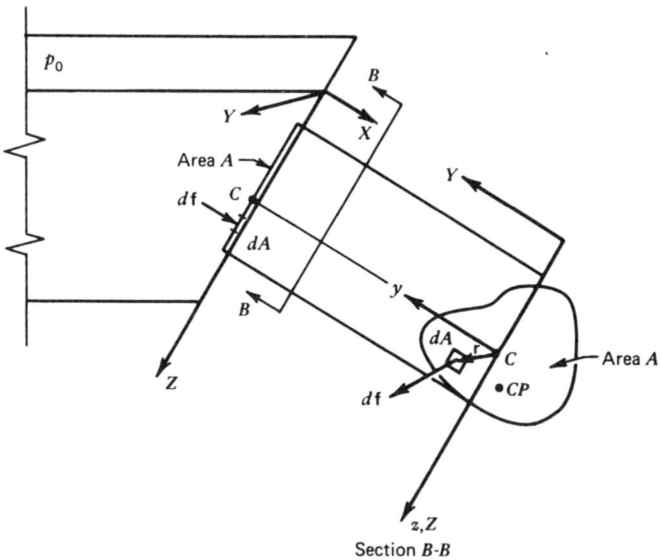

Fig. 3-10. Sketch for determining the center of pressure on a submerged area.

pressure at the centroid, p_c, is $p_c = p_0 + \gamma Z_c' = p_0 + \gamma Z_c \sin \alpha$. Therefore, we see that

$$\mathbf{F}_R = p_c A \mathbf{i} \qquad \text{Eq. 3.4-10}$$

We now seek to obtain an expression for the position of the center of pressure (see Statement 3.4-1). We will see that this will allow us to take advantage of the known geometric properties of areas of common shapes. Furthermore, these properties are usually given with respect to a coordinate system with its origin at the centroid.

Therefore, select an (x, y, z) coordinate system with its origin at the centroid of the area of interest, as shown in Fig. 3-10. The moment, $\mathbf{M}$, about the centroid caused by the distributed pressure force is given by

$$\mathbf{M} = \oiint_A \mathbf{r} \times d\mathbf{f} = \oiint_A (y\mathbf{j} + z\mathbf{k}) \times p\mathbf{i}\, dA$$

$$= -\mathbf{k} \oiint_A yp\, dA + \mathbf{j} \oiint_A zp\, dA \qquad \text{Eq. 3.4-11}$$

Substituting Eq. 3.4-2 into Eq. 3.4-11 gives

$$\mathbf{M} = -\mathbf{k}\left(\oiint_A y(p_0 + \gamma Z')\, dA + \mathbf{j} \oiint_A z(p_0 + \gamma Z')\, dA\right)$$

$$= -\mathbf{k}\left(p_0 \oiint_A y\, dA + \gamma \oiint_A yZ'\, dA\right)$$

$$+ \mathbf{j}\left(p_0 \oiint_A z\, dA + \gamma \oiint_A zZ'\, dA\right) \qquad \text{Eq. 3.4-12}$$

Now

$$\oiint_A y\, dA = Ay_c \qquad \text{and} \qquad \oiint_A z\, dA = Az_c \qquad \text{Eq. 3.4-13}$$

Since the (x, y, z) coordinate system has its origin at the centroid, y_c and z_c are zero; that is,

$$y_c = z_c = 0 \qquad \text{Eq. 3.4-14}$$

To obtain the desired expression for **M**, express Z' in terms of z,

$$Z' = Z \sin\alpha = (z + Z_c)\sin\alpha \qquad \text{Eq. 3.4-15}$$

Substitution of Eqs. 3.4-13, 3.4-14, and 3.4-15 into Eq. 3.4-12 gives

$$\mathbf{M} = -\mathbf{k}\gamma \oiint_A y(z + Z_c)\sin\alpha\, dA + \mathbf{j}\gamma \oiint_A z(z + Z_c)\sin\alpha\, dA$$

$$= \gamma \sin\alpha\left(-\mathbf{k} \oiint_A yz\, dA + \mathbf{j} \oiint_A z^2\, dA\right)$$

$$+ \gamma Z_c \sin\alpha\left(-\mathbf{k} \oiint_A y\, dA + \mathbf{j} \oiint_A z\, dA\right) \qquad \text{Eq. 3.4-16}$$

The student may recall from mechanics that $\oiint_A z^2\, dA$ is the moment of inertia of the area about the y-axis, designated as I_{yy}, and that $\oiint_A yz\, dA$ is the product of inertia, designated as I_{yz}. With these definitions and the substitution of Eqs. 3.4-13 and 3.4-14 into Eq. 3.4-16, Eq. 3.4-16 reduces to

$$\mathbf{M} = \gamma \sin \alpha(-I_{yz}\mathbf{k} + I_{yy}\mathbf{j}) \qquad \text{Eq. 3.4-17}$$

By the definition of the center of pressure (see Statement 3.4-1), the resultant force placed at the center of pressure, must produce the same moment effect as the distributed pressure force. Let the coordinates of and the position vector to the center of pressure be $(0, y^*, z^*)$ and $\mathbf{r}_{\text{c.p.}}$, respectively; then

$$\begin{aligned} \mathbf{M} &= \mathbf{r}_{\text{c.p.}} \times \mathbf{F}_R \\ &= (y^*\mathbf{j} + z^*\mathbf{k}) \times p_c A\mathbf{i} \\ &= p_c A(-y^*\mathbf{k} + z^*\mathbf{j}) \end{aligned} \qquad \text{Eq. 3.4-18}$$

Equating the $\mathbf{j}$- and $\mathbf{k}$-components in Eqs. 3.4-17 and 3.4-18, we obtain

$$y^* = \frac{\gamma I_{yz} \sin \alpha}{p_c A} \qquad \text{and} \qquad z^* = \frac{\gamma I_{yy} \sin \alpha}{p_c A} \qquad \text{Eqs. 3.4-19}$$

If the free surface is open to the atmosphere, then p_0 is zero and

$$p_c = \gamma Z_c \sin \alpha \qquad \text{Eq. 3.4-20}$$

Substitution of Eq. 3.4-20 into Eqs. 3.4-19 gives

$$y^* = \frac{I_{yz}}{AZ_c} \qquad \text{and} \qquad z^* = \frac{I_{yy}}{AZ_c} \qquad \text{Eqs. 3.4-21}$$

Now I_{yz} and I_{yy} for common geometric figures are tabulated in many textbooks on mechanics. This allows for quick computation of y^*, z^*, and $\mathbf{M}$. It should also be noted that for areas symmetrical about either the y- or z-axis, the product of inertia, I_{yz}, is zero. Two examples using these concepts are given below.

Example 3-1

A dam, which is used to control the water level in a small artificial lake, is constructed with a rectangular gate 10 ft by 15 ft, as shown in Fig. 3-11. The gate is

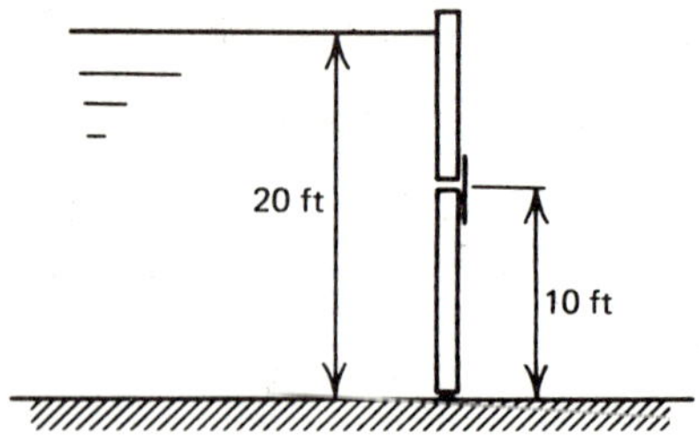

Fig. 3-11. Gate in a dam.

hinged at the bottom and locked into its closed position by a latch. Find the force that the latch exerts on the gate to keep it closed.

SOLUTION

A single force, equivalent in the equilibrium equations to the distributed pressure force, is $\mathbf{F}_R$, acting at the center of pressure, as shown in Fig. 3-12. By Eqs. 3.4-10 and 3.4-2

$$\mathbf{F}_R = \gamma Z_c A \mathbf{i}$$

or

$$|\mathbf{F}_R| = 62.4(15)(10 \times 15) = 140{,}400 \text{ lb}_f$$

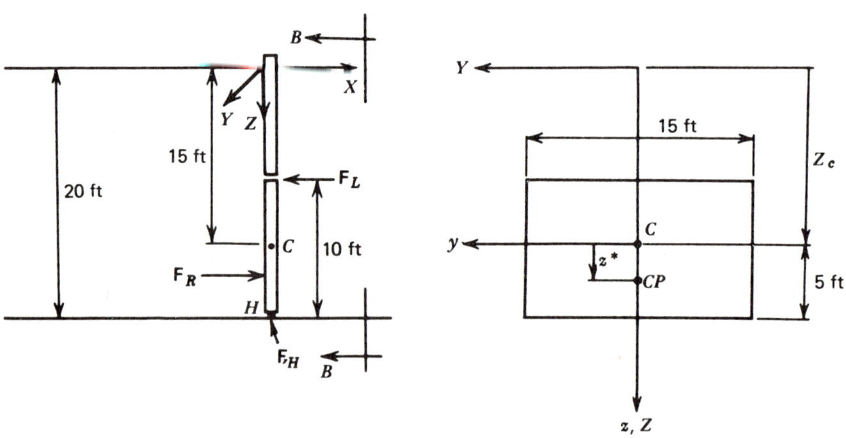

Fig. 3-12. Forces acting on a gate.

Section B–B

By Eqs. 3.4-21

$$z^* = \frac{I_{yy}}{AZ_c} \quad \text{and} \quad y^* = \frac{I_{yz}}{AZ_c}$$

It is left as a student exercise to show that for a rectangular section,

$$I_{yy} = \tfrac{1}{12} b\, d^3 \quad \text{and} \quad I_{yz} = 0 \qquad \text{Eqs. 3.4-22}$$

where b is the width of the rectangle and d is its depth. Therefore,

$$z^* = \frac{(15)(10)^3}{12(15)(10)(15)} = 0.555 \text{ ft}$$

and

$$y^* = 0$$

Taking moments about the hinge, H, and realizing that for equilibrium, the sum of moments about any point must be zero, we obtain

$$F_L(10) - F_R(5 - z^*) = 0$$

or

$$F_L = \frac{140{,}400(5 - 0.555)}{10} = 62{,}408 \text{ lb}_f$$

Example 3-2

Find the resultant force acting on the rectangular window in the configuration shown in Fig. 3-13 and the position of the center of pressure, z^*.

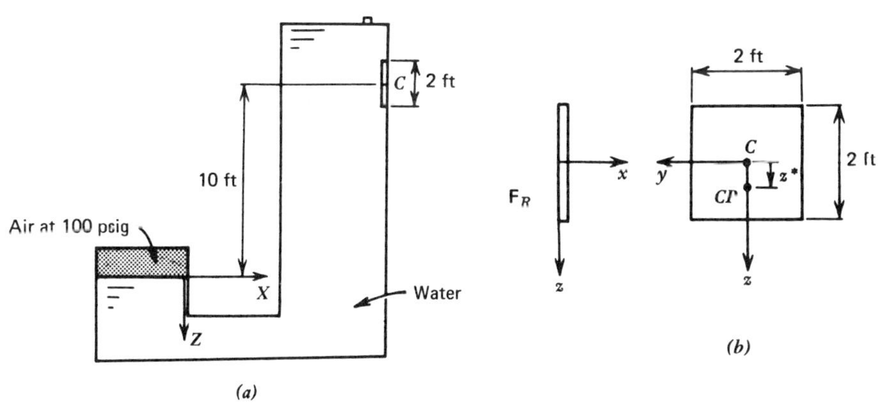

Fig. 3-13. Window in a pressurized system. (*a*) Overall system. (*b*) Enlargement of window.

SOLUTION

The magnitude of the resultant force is given by Eqs. 3.4-2 and 3.4-10, that is

$$p_c = p_0 + \gamma Z_c = 100(144) - 62.4(10) = 13{,}775\ \text{lb}_f/\text{ft}^2$$

and

$$|\mathbf{F}_R| = p_c A = 13{,}775(2)(2) = 55{,}100\ \text{lb}_f$$

The position of the center of pressure, z^*, is given by Eqs. 3.4-21 and Eq. 3.4-22. Also note that for a vertical wall, $\sin \alpha = 1$. Then

$$z^* = \frac{\gamma I_{yy}}{p_c A} = \frac{62.4(2)(2)^3}{12(55{,}100)} = 0.0015\ \text{ft}$$

and

$$y^* = 0$$

When dealing with problems involving forces on submerged curved surfaces, it is convenient to treat the horizontal and vertical components of the pressure force separately. Consider the forces acting on the submerged curved surface shown in Fig. 3-14. To obtain the Z-component of the resultant force, $\mathbf{F}_R$, first

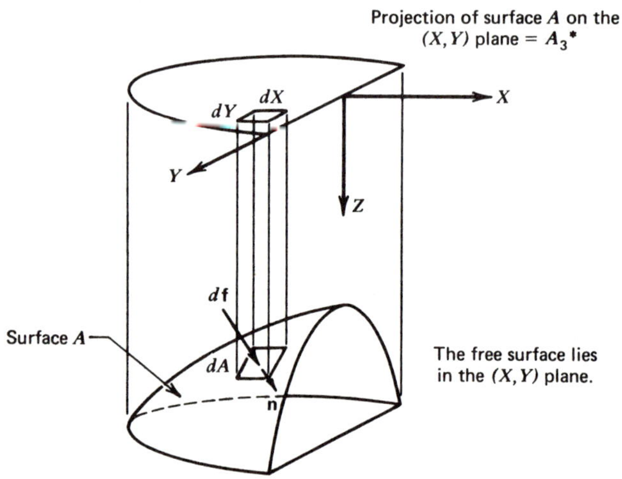

Fig. 3-14. Force on a submerged curved surface.

construct a grid on the surface by intersecting the surface with planes parallel to the (X, Z) and (Y, Z) planes. Then the force, $d\mathbf{f}$, acting on area dA, is given by

$$d\mathbf{f} = p\, dA\, \mathbf{n} = p\, d\mathbf{A} = df_1\, \mathbf{i} + df_2\, \mathbf{j} + df_3\, \mathbf{k} \qquad \text{Eq. 3.4-23}$$

The component df_3 can be obtained by taking the dot product of $d\mathbf{f}$ with the unit vector $\mathbf{k}$; that is,

$$df_3 = d\mathbf{f} \cdot \mathbf{k} = p\, d\mathbf{A} \cdot \mathbf{k} \qquad \text{Eq. 3.4-24}$$

Recall from Eqs. 3.2-3, that

$$\begin{aligned} d\mathbf{A} \cdot \mathbf{k} &= \text{the projection of } dA \text{ on the } (X,\ Y) \text{ plane} \\ &= dX\, dY \end{aligned}$$

Then,

$$df_3 = p \cdot [\text{projection of } dA \text{ on the } (X,\ Y) \text{ plane}]$$

Since $p = \gamma Z$,

$$df_3 = \gamma Z\, dX\, dY \qquad \text{Eq. 3.4-25}$$

But $(Z\, dX\, dY)$ is the volume in the column above the surface element, dA, and thus df_3 equals the weight of liquid in that column. The Z-component of the resultant force, $F_{R,3}$, is the sum of the forces on all the infinitesimal areas, which of course is an integral; that is,

$$\begin{aligned} F_{R,3} &= \oiint_{A_3^*} \gamma Z\, dX\, dY \\ &= \text{the weight of liquid directly above the surface } A \qquad \text{Eq. 3.4-26} \end{aligned}$$

where A_3^* is the projection of surface A on the $(X,\ Y)$ plane. In a similar fashion, the X- and Y-components of the resultant force $F_{R,1}$ and $F_{R,2}$, respectively, are

$$F_{R,1} = \oiint_{A_1^*} \gamma Z\, dY\, dZ$$

and

$$F_{R,2} = \oiint_{A_2^*} \gamma Z \, dX \, dZ \qquad \text{Eqs. 3.4-27}$$

where A_1^* is the projection of surface A on the (Y, Z) plane and A_2^* is the projection of surface A on the (X, Z) plane.[2] It should be observed that $F_{R,1}$ equals the force that would be exerted on a vertical plane surface coincident with the projected area A_1^*. Similarly, $F_{R,2}$ equals the force that would be exerted on a vertical plane surface coincident with A_2^*.

The force df_3 causes a moment about both the X- and Y-axis. Noting (1) the definition of the center of pressure as given in Statement 3.4-1, and (2) that forces are transmissible with respect to the moment equation, we may obtain the coordinates (X^*, Y^*) of the center of pressure of $F_{R,3}$. That is,

$$(X^*\mathbf{i} + Y^*\mathbf{j}) \times F_{R,3}\mathbf{k} = \oiint_{A_3^*} (X\mathbf{i} + Y\mathbf{j}) \times \mathbf{k} \, df_3 \qquad \text{Eq. 3.4-28}$$

Substituting for df_3 as given by Eq. 3.4-25 and equating the **i** and **j** components on both sides of the above equation, we obtain

$$Y^* F_{R,3} = \oiint_{A_3^*} \gamma YZ \, dX \, dY$$

and Eqs. 3.4-29

$$X^* F_{R,3} = \oiint_{A_3^*} \gamma XZ \, dX \, dY$$

Since $(\gamma Z \, dX \, dY)$ is the weight of water in the column above the surface element dA and $F_{R,3}$ equals the weight of liquid directly above the surface, we see that (X^*, Y^*) coincides with the (X, Y) coordinates of the center of gravity of the

[2] It should be noted that, if the curved surface is such that the projections of two elements on the surface overlap, then the integration has to be broken down into two parts.

liquid directly above surface A. In a similar fashion, we can obtain the coordinates of the center of pressure of $F_{R,1}$ and $F_{R,2}$. These are

$$Y^* F_{R,1} = \oiint_{A_1^*} \gamma YZ \, dY \, dZ$$

$$Z^* F_{R,1} = \oiint_{A_1^*} \gamma Z^2 \, dY \, dZ$$

Eqs. 3.4-30

$$X^* F_{R,2} = \oiint_{A_2^*} \gamma XZ \, dX \, dZ$$

$$Z^* F_{R,2} = \oiint_{A_2^*} \gamma Z^2 \, dX \, dZ$$

Eqs. 3.4-31

These concepts are illustrated in the following example.

Example 3-3

(a) Find the X- and Z-components of the resultant force acting on the submerged gate consisting of a $\frac{1}{4}$ circular cylinder, as shown in Fig. 3-15. (b) Find the coordinates for the center of pressure of the component of the resultant force, $F_{R,1}$. The cylinder is 4 ft long and has a radius of 3 ft.

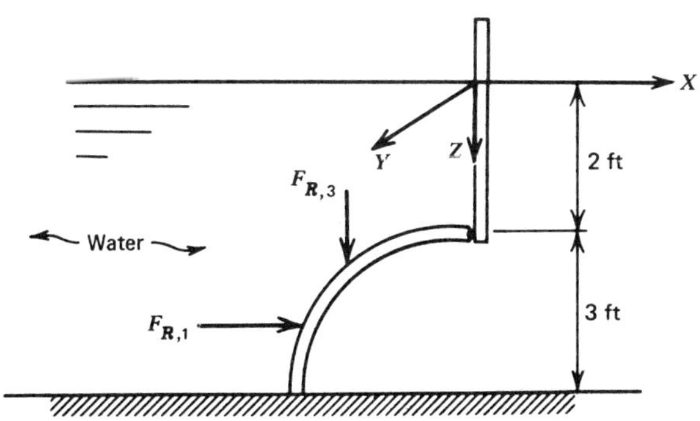

Fig. 3-15. Forces on a gate.

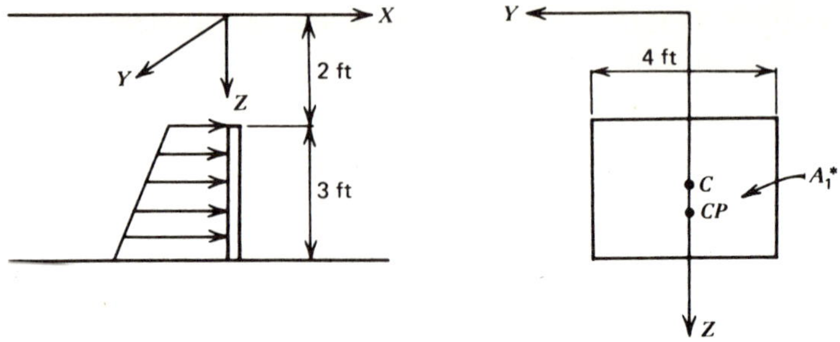

Fig. 3-16. Equivalent system for determining $F_{R,1}$.

SOLUTION

(*a*) $F_{R,1}$ is the same force that would exist on the vertical plane surface shown in Fig. 3-16. Its magnitude is given by the first equation in the set of Eqs. 3.4-27; that is,

$$F_{R,1} = \iint_{A_1^*} \gamma Z \, dY \, dZ = 4\gamma \int_2^5 Z \, dZ = \frac{4(62.4)}{2} [Z^2]_2^5 = 2621 \text{ lb}_f$$

The Z-component of the resultant force, $F_{R,3}$, equals the weight of liquid directly above the surface, as given by Eq. 3.4-26. Therefore,

$$F_{R,3} = \gamma \cdot (\text{volume of water directly above the gate})$$

$$= 62.4(4)\left[5(3) - \frac{\pi}{4}(3)^2\right] = 1980 \text{ lb}_f$$

(*b*) If (Y^*, Z^*) are the coordinates of the center of pressure of $F_{R,1}$, then

$$Z^* F_{R,1} = \iint_{A_1^*} \gamma Z^2 \, dY \, dX = 4\gamma \int_2^5 Z^2 \, dZ = \frac{4(62.4)}{3} [Z^3]_2^5 = 9734 \text{ lb}_f\text{-ft}$$

or

$$Z^* = \frac{9737}{2621} = 3.71 \text{ ft}$$

By symmetry, $Y^* = 0$.

3.5 ARCHIMEDES' PRINCIPLE AND BUOYANCY

Nearly all of us have experienced the fact that a body appears less heavy when it is submerged in water than when it is in air. This results from the buoyant force that a liquid exerts on a floating or submerged body. Buoyancy is the vertically upward component of the resultant surface force exerted on a floating or submerged body by the surrounding liquid. The surface forces result from the pressure that the liquid exerts on the surface of that portion of the body which is submerged. The magnitude of the buoyant force experienced by such a body is expressed in Archimedes' principle, which consists of two parts, and may be stated as follows: (1) the buoyant force exerted on a submerged or floating body is equal to the weight of the fluid displaced by the body, and (2) a floating body displaces a weight of fluid equal in magnitude to its own weight. The second part follows immediately from the first. Since a floating body is in equilibrium, the upward buoyancy force is balanced by the downward gravitational pull, which is the weight of the body (see Fig. 3-17). To prove the first part we first consider a totally submerged body, as shown in Fig. 3-18.

The total surface force, $\mathbf{F}_S$, acting on the body shown in Fig. 3-18 is given by

$$\mathbf{F}_S = \oiint_S \boldsymbol{\tau}(s)\, dS = -\oiint_S p\mathbf{n}\, dS \qquad \text{Eq. 3.5-1}$$

The governing equation for the pressure field in a static fluid when the Z-axis points vertically upward was given in Eq. 3.2-9. When the Z-axis points vertically downward a change of sign occurs, giving

$$\frac{dp}{dZ} = \gamma \qquad \text{Eq. 3.5-2}$$

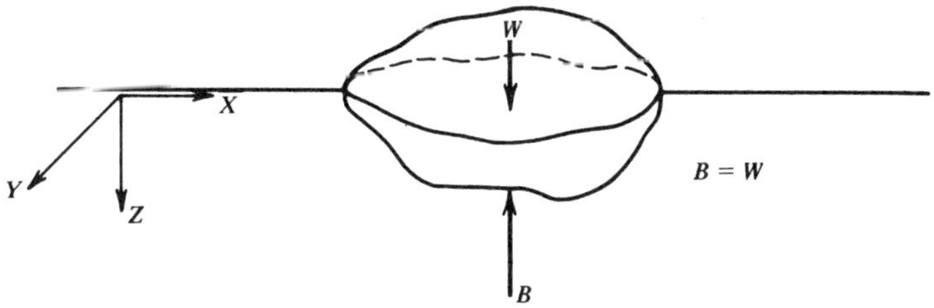

Fig. 3-17. Forces acting on a floating body.

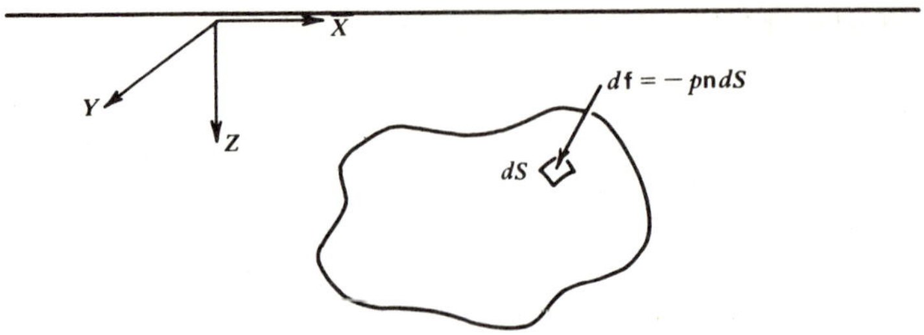

Fig. 3-18. Surface force on a totally submerged body.

Considering that: (1) γ is a constant and (2) the pressure at the free surface is zero, the integration of Eq. 3.5-2 with the application of the specified boundary condition gives

$$p = \gamma Z \qquad \text{Eq. 3.5-3}$$

Substituting Eq. 3.5-3 into Eq. 3.5-1 yields

$$\mathbf{F}_S = -\oiint_S \gamma Z\mathbf{n}\,dS = -\gamma \oiint_S Z\mathbf{n}\,dS \qquad \text{Eq. 3.5-4}$$

Converting the surface integral into a volume integral by Eq. 2.7-8 gives

$$\mathbf{F}_S = -\gamma \iiint_{V\!\!\!/} \nabla Z\,dV\!\!\!/ \qquad \text{Eq. 3.5-5}$$

Since $\nabla Z = \mathbf{k}$, the above integral reduces to

$$\mathbf{F}_S = -\gamma \iiint_{V\!\!\!/} \mathbf{k}\,dV\!\!\!/ = -\gamma V\!\!\!/\mathbf{k} \qquad \text{Eq. 3.5-6}$$

where $V\!\!\!/$ is the volume of the submerged body. We see that the resultant surface force is the same that would exist on a body having a uniform specific weight, γ, in a vertically upward gravitational field. The buoyancy force, B, is defined as the upward vertical component of the resultant surface force, $\mathbf{F}_S$. Thus,

$$B = (-\mathbf{k}) \cdot \mathbf{F}_S = \gamma V\!\!\!/ \qquad \text{Eq. 3.5-7}$$

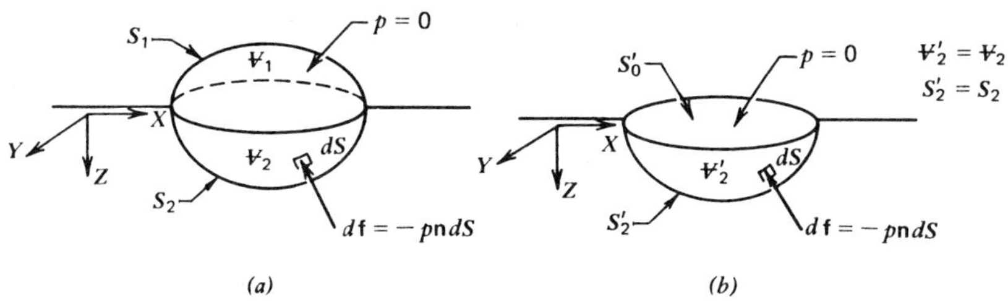

Fig. 3-19. Surface force on a floating body. (*a*) Actual floating body. (*b*) Equivalent floating body.

Since γ is the specific weight of the fluid displaced, we see that the bouyancy force equals the weight of the fluid displaced by the body. It should also be observed that the total surface force, $\mathbf{F}_S$, has no horizontal component.

To prove Archimedes' principle for floating bodies consider the surface force, $\mathbf{F}_S$, acting on the body shown in Fig. 3-19*a*. Take $\mathcal{V}_1$ to be the region of the body that lies above the free surface ($Z < 0$) and $\mathcal{V}_2$ to be the region of the body that lies on or below the free surface ($Z \geq 0$). A body consisting just of region $\mathcal{V}_2$, as shown in Fig. 3-19*b*, would experience the same surface force as the original body, since $p = 0$ for $Z \leq 0$. Thus, the second body is an equivalent body with respect to surface forces. This can be shown as follows. If $\mathbf{F}_S'$ is the total body force on the second body, then

$$\mathbf{F}_S' = -\oint\!\!\oint_{S_2'} p\mathbf{n}\,dS - \oint\!\!\oint_{S_0'} p\mathbf{n}\,dS \qquad \text{Eq. 3.5-8}$$

The surface force, $\mathbf{F}_S$, on the original body is

$$\mathbf{F}_S = -\oint\!\!\oint_{S_1} p\mathbf{n}\,dS - \oint\!\!\oint_{S_2} p\mathbf{n}\,dS \qquad \text{Eq. 3.5-9}$$

Since $p(S_0') = 0$, $p(S_1) = 0$, and $S_2 = S_2'$,

$$\mathbf{F}_S = \mathbf{F}_S' = -\oint\!\!\oint_{S_2} p\mathbf{n}\,dS \qquad \text{Eq. 3.5-10}$$

Substituting Eq. 3.5-3 into Eq. 3.5-10 and converting the surface integral into a volume integral by Eq. 2.2-17 yields

$$\mathbf{F}_S = -\gamma \oint\!\!\oint_{S_i} Z\mathbf{n}\, dS = -\gamma \oint\!\!\oint\!\!\oint_{V'_2} \nabla Z\, dV = -\gamma V'_2 \mathbf{k} \qquad \text{Eq. 3.5-11}$$

Since $V_2 = V'_2$,

$$\mathbf{F}_S = -\gamma V_2 \mathbf{k} \qquad \text{Eq. 3.5-12}$$

Again we see that the resultant surface force is the same that would exist on that portion of the body that lies on or below the free surface having a specific weight, γ, in a vertically upward gravitational field. Since the buoyancy, B, is defined to be the upward vertical component of the resultant body force, we have that

$$B = (-\mathbf{k}) \cdot \mathbf{F}_S = \gamma V_2 = \text{weight of liquid displaced} \qquad \text{Eq. 3.5-13}$$

The hydrometer, which is a device used for measuring the specific gravity of a liquid, is based on Archimedes' principle. A typical hydrometer consists of a glass or plastic bulb and stem, as shown in Fig. 3-20. The bulb region usually contains some heavy material. Specific gravity markings are etched into the stem. When the hydrometer is placed into distilled water, it floats at a level so that the free surface intersects the 1.0 marking on the stem, as shown in Fig. 3-20*a*.

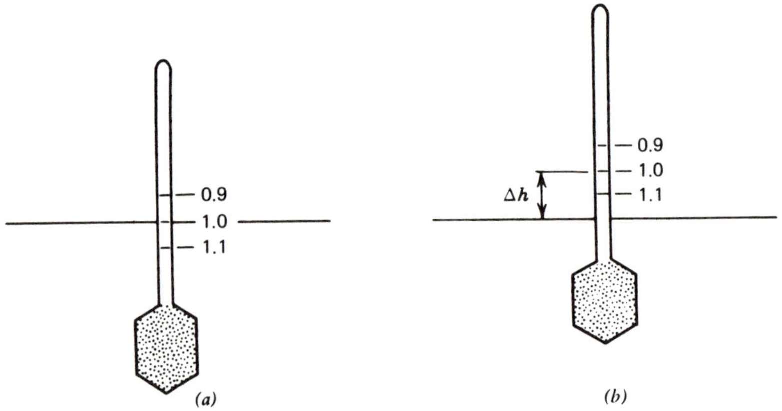

Fig. 3-20. A floating hydrometer. (*a*) Hydrometer in distilled water. (*b*) Hydrometer in an arbitrary liquid, $S > 1$.

When the hydrometer is placed into a liquid whose specific gravity, S, is greater than 1.0, the hydrometer will float with a larger portion of the stem protruding above the free surface than in the previous case (see Fig. 3-20*b*). Applying Archimedes' principle to both cases, we obtain

1. Hydrometer in distilled water:

$$B = \gamma_w \forall_0 = W \qquad \text{Eq. 3.5-14}$$

where $\forall_0$ is the volume of the hydrometer that is submerged (up to the 1.0 marking on the stem), W is the weight of the hydrometer, and γ_w is the specific weight of distilled water.

2. Hydrometer in a liquid with $S > 1$:

$$B = \gamma \forall = W \qquad \text{Eq. 3.5-15}$$

where $\forall$ is the volume of the hydrometer that is submerged and γ is the specific weight of the liquid.

From Eqs. 3.5-14 and 3.5-15, we have

$$\gamma \forall = \gamma_w \forall_0$$

or

$$\forall = \frac{\gamma_w}{\gamma} \forall_0 \qquad \text{Eq. 3.5-16}$$

From Eq. 1.4-4,

$$\frac{\gamma_w}{\gamma} = \frac{1}{S} \qquad \text{Eq. 3.5-17}$$

Furthermore, if A is the cross-sectional area of the stem, then

$$\forall = \forall_0 - A\,\Delta h \qquad \text{Eq. 3.5-18}$$

Substituting Eqs. 3.5-17 and 3.5-18 into Eq. 3.5-16 gives

$$\forall_0 - A\,\Delta h = \frac{\forall_0}{S}$$

Solving for S, we obtain

$$S = \frac{V_0}{V_0 - A\,\Delta h} = \frac{1}{1 - \dfrac{A\,\Delta h}{V_0}} \qquad \text{Eq. 3.5-19}$$

If $A\,\Delta h/V_0 << 1$, then the S scale is nearly linear with Δh. This follows from the Taylor Series expansion of the function $(1 - x)^{-1}$ when $x < 1$; that is,

$$\frac{1}{1 - x} = 1 + x + x^2 + \cdots \qquad \text{Eq. 3.5-20}$$

If $A\,\Delta h/V_0 << 1$, then the application of Eq. 3.5-20 to Eq. 3.5-19 gives

$$S = 1 + \frac{A\,\Delta h}{V_0} + \text{higher order terms}$$

3.6 THE CENTER OF BUOYANCY

In Section 3.4, it was demonstrated that the resultant surface force acting on a submerged or floating body is the same that would exist on a body having a uniform specific weight, γ, in a vertically upward gravitational field. The student should recall that in mechanics, the concept of center of gravity was introduced as a matter of convenience for problem solving. For the same reason, the center of buoyancy is also commonly used. The concept is expressed in the following statement: If a force equal to $\mathbf{F}_S$ was placed at the center of buoyancy, it would have the same moment effect as the distributed force. To obtain the position of

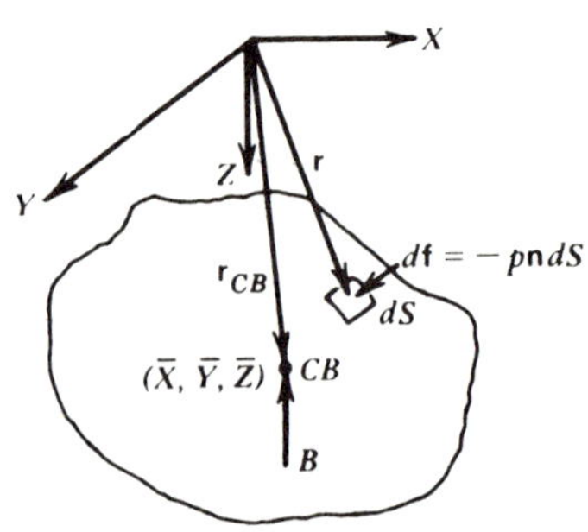

Fig. 3-21. Geometry for determining center of buoyancy.

the center of buoyancy, $(\bar{X}, \bar{Y}, \bar{Z})$, take the moments of the distributed force (see Fig. 3-21).

$$\mathbf{M}_B = \oiint_S \mathbf{r} \times d\mathbf{f} = -\oiint_S (X\mathbf{i} + Y\mathbf{j} + Z\mathbf{k}) \times p\mathbf{n}\, dS$$

$$= -\mathbf{i} \times \oiint_S Xp\mathbf{n}\, dS - \mathbf{j} \times \oiint_S Yp\mathbf{n}\, dS - \mathbf{k} \times \oiint_S Zp\mathbf{n}\, dS \qquad \text{Eq. 3.6-1}$$

Substituting Eq. 3.5-3 into the above equation yields

$$\mathbf{M}_B = -\mathbf{i}X\gamma \oiint_S XZ\mathbf{n}\, dS - \mathbf{j}X\gamma \oiint_S yZ\mathbf{n}\, dS - \mathbf{k}X\gamma \oiint Z^2\mathbf{n}\, dS$$

Converting the surface integrals to volume integrals by Eq. 2.2-17, we obtain

$$\mathbf{M}_B = -\mathbf{i} \times \gamma \oiiint_{\mathcal{V}} \nabla(XZ)\, d\mathcal{V} - \mathbf{j} \times \gamma \oiiint_{\mathcal{V}} \nabla(YZ)\, d\mathcal{V}$$

$$- \mathbf{k} \times \gamma \oiiint_{\mathcal{V}} \nabla(Z^2)\, d\mathcal{V} \qquad \text{Eq. 3.6-2}$$

Noting that

$$\nabla(XZ) = (X\,\nabla Z + Z\,\nabla X) = (X\mathbf{k} + Z\mathbf{i})$$

$$\nabla(YZ) = (Y\,\nabla Z + Z\,\nabla Y) = (Y\mathbf{k} + Z\mathbf{j})$$

$$\nabla(Z^2) = 2Z\,\nabla Z = 2Z\mathbf{k}$$

Equation 3.6-2 becomes

$$\mathbf{M}_B = -\mathbf{i} \times \gamma \oiiint_{\mathcal{V}} (X\mathbf{k} + Z\mathbf{i})\, d\mathcal{V} - \mathbf{j} \times \gamma \oiiint_{\mathcal{V}} (Y\mathbf{k} + Z\mathbf{j})\, d\mathcal{V}$$

$$- \mathbf{k} \times \gamma \oiiint_{\mathcal{V}} 2Z\mathbf{k}\, d\mathcal{V}$$

$$= \mathbf{j}\gamma \oiiint_{\mathcal{V}} X\, d\mathcal{V} - \mathbf{i}\gamma \oiiint_{\mathcal{V}} Y\, d\mathcal{V} \qquad \text{Eq.3.6-3}$$

By definition of the center of the buoyancy and Eq. 3.5-6,

$$\begin{aligned} \mathbf{M}_B &= \mathbf{r}_{CB} \times \mathbf{F}_S = (\bar{X}\mathbf{i} + \bar{Y}\mathbf{j} + \bar{Z}\mathbf{k}) \times (-\gamma \forall \mathbf{k}) \\ &= \gamma \bar{X} \forall \mathbf{j} - \gamma \bar{Y} \forall \mathbf{i} \end{aligned} \qquad \text{Eq. 3.6-4}$$

Equating the **i**- and **j**-components of Eqs. 3.6-3 and 3.6-4 gives, after some rearranging,

$$\bar{X} = \frac{1}{\forall} \iiint_{\forall} X \, d\forall$$

$$\bar{Y} = \frac{1}{\forall} \iiint_{\forall} Y \, d\forall \qquad \text{Eqs. 3.6-5}$$

We did not obtain an expression for $\bar{Z}$ because forces are transmissible as far as moments are concerned. The same situation exists in the definition of the center of gravity. Just as in the definition of the center of gravity, the Z-coordinate of the center of buoyancy is defined to maintain the symmetry in the final equations; that is,

$$\bar{Z} = \frac{1}{\forall} \iiint_{\forall} Z \, d\forall \qquad \text{Eq. 3.6-6}$$

The equations for the position of the center of buoyancy for floating bodies are the same as those above, except that the volume, $\forall$, is replaced by the region that lies on and below the free surface. *It should be observed that the position of the center of buoyancy is the same as the position of the centroid of the submerged region.* Furthermore, in order for a floating body to be in static equilibrium, the net moment on the body must be zero. This requires that the center of buoyancy and the center of gravity lie on the same vertical line as shown in Fig. 3-22*a*. Otherwise, an unbalanced moment would be created causing an angular acceleration.

3.7 STABILITY OF FLOATING BODIES

In the previous section we saw that for a floating body to be in equilibrium in a static fluid, (1) the buoyancy force, which acts vertically upward, must be equal to the weight of the body, and (2) the center of buoyancy must lie along the same vertical line as the center of gravity. Now, if a floating body is displaced slightly upward from its equilibrium position, the volume of water displaced will decrease causing the buoyancy force to decrease, which results in a downward acceleration. Similarly, when the body is displaced slightly downward from its equilibrium position, the volume of water displaced is increased, causing the buoyancy force to increase, which results in an upward acceleration. In each case the acceleration is in the opposite direction to the displacement and, thus, there is always a tendency for the body to return to its equilibrium position. This situation will result in the body oscillating about that position. When a body behaves in this manner, under the conditions mentioned above, then it is said to be in stable equilibrium with respect to a vertical displacement.

If a floating body, subjected to a slight rotation from its equilibrium position, behaves in a similar manner to that described in the previous paragraph, then it is said to be in stable equilibrium with respect to rotation. In this case the center of buoyancy and the center of gravity must shift to provide a couple that will tend to rotate the body back to its equilibrium position. This is shown in Fig. 3-22*b*. In that figure we assumed that the exterior shell was symmetrical and

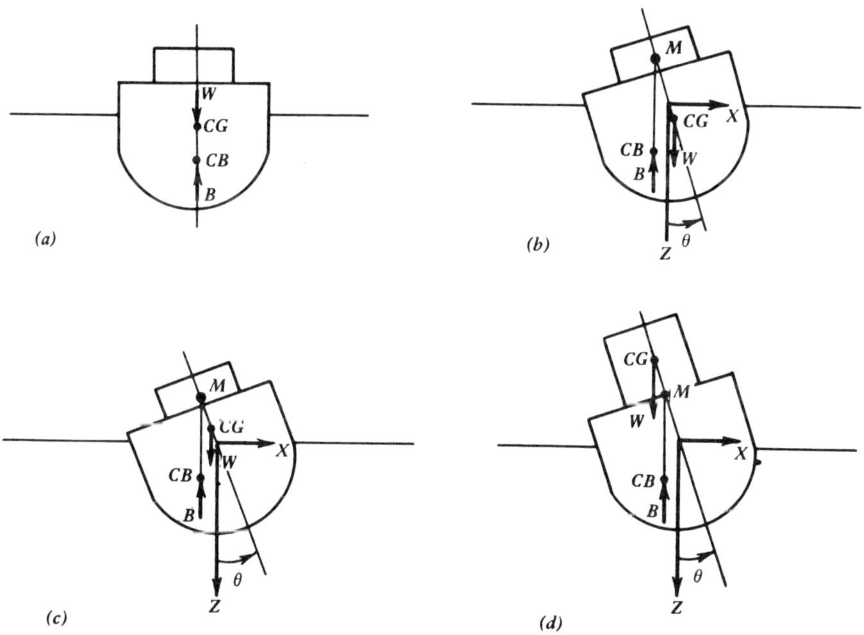

Fig. 3-22. Rotational stability of floating bodies. (*a*) Equilibrium position. (*b*) Net moment tends to rotate body back toward equilibrium position. (*c*) Net moment tends to rotate body back toward equilibrium position. (*d*) Net moment tends to further rotate body away from equilibrium position.

that the center of gravity of the loaded vessel was located on the center line of the body. We can see that the center of gravity and the center of buoyancy are located such that both tend to produce a moment that would accelerate the body back toward its equilibrium position, making the body stable. In Fig. 3-22*c* the center of gravity lies above the axis of rotation. Thus, in this case the gravitational field produces a moment that would tend to give the body an angular acceleration in the same direction as the displacement. On the other hand, the buoyancy force still produces a moment that would tend to rotate the body back toward its equilibrium position. Since the perpendicular distance from the axis of rotation to the line of action of the resultant buoyancy force is greater than the corresponding distance to the line of action of the resultant gravitational force, and since B equals W, the net moment will tend to rotate the body back toward its equilibrium position. The body will again be stable. In the above case, it is assumed that the region of the body moving below the free surface due to the rotation is equal to the region of the body moving above the free surface. This is a necessary condition if the rotation of the body is to be pure rotation.

The question of rotational stability is often discussed in terms of the *metacenter*. The metacenter, point M in Fig. 3-22*c*, is the point of intersection of the vertical line passing through the center of buoyancy and the center line of the body. By trying all possible combinations in the positions of the center of buoyancy and the center of gravity, it can be seen that if point M lies above the center of gravity for all small angles of θ, then the body is stable with respect to rotation. In Fig. 3-22*d* the metacenter lies below the center of gravity; therefore, here the body would be unstable. That is, the net moment on the body would tend to further increase its angular displacement and would thus tip the body. Two examples illustrating the concepts discussed in this section follow.

Example 3-4

A solid cube with sides of length b floats in still water, as shown in Fig. 3-23. The material of the cube has a specific gravity, S, where $S < 1$. The cube is displaced

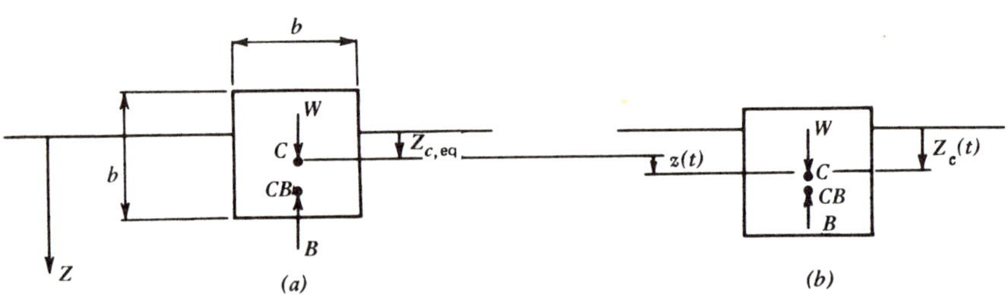

Fig. 3-23. A bobbing, floating cube. (*a*) Equilibrium position. (*b*) Nonequilibrium position.

downward from its equilibrium position by a distance α, where $\alpha < (b/2 - Z_{c,\text{eq}})$. At $t = 0$, it is released from rest. Determine the position of the centroid from its equilibrium position as a function of time.

SOLUTION

First we establish the equilibrium position of the centroid, $Z_{c,\text{eq}}$ (see Fig. 3.23*a*). At equilibrium

$$B = W$$

or

$$\gamma b^2\left(\frac{b}{2} + Z_{c,\text{eq}}\right) = S\gamma b^3 \qquad \text{Eq. 3.7-1}$$

where γ is the specific weight of water. Dividing through by γb^2, we obtain

$$\left(\frac{b}{2} + Z_{c,\text{eq}}\right) = Sb \qquad \text{Eq. 3.7-2}$$

Next we apply Newton's second law to the cube when it is displaced from its equilibrium position (see Fig. 3-23*b*). That is,

$$W - B = \frac{M}{g_c}\frac{d^2Z_c}{dt^2} \qquad \text{Eq. 3.7-3}$$

or

$$S\gamma b^3 - \gamma b^2\left(\frac{b}{2} + Z_c\right) = \frac{S\rho b^3}{g_c}\frac{d^2Z_c}{dt^2} \qquad \text{Eq. 3.7-4}$$

When the cube is displaced downward from its equilibrium position, the buoyancy force will be greater than the weight of the cube. Thus, the acceleration will be upward tending to restore the body back to its equilibrium position. This is the type of behavior that is required for stability. Substituting for ρ/g_c in the above equation ($\rho/g_c = \gamma/g$), then dividing through by γb gives

$$Sb - \left(\frac{b}{2} + Z_c\right) = \frac{Sb}{g}\frac{d^2Z_c}{dt^2}$$

or

$$\frac{Sb}{g}\frac{d^2Z_c}{dt^2} + Z_c = Sb - \frac{b}{2} \qquad \text{Eq. 3.7-5}$$

The above differential equation is nonhomogeneous, since the right-hand side of the equation is not zero. To make the equation homogeneous, let

$$z(t) = Z_c(t) - Z_{c,\,\mathrm{eq}} \qquad \text{Eq. 3.7-6}$$

Then

$$\frac{d^2z}{dt^2} = \frac{d^2Z_c}{dt^2} \qquad \text{Eq. 3.7-7}$$

Substituting Eqs. 3.7-6 and 3.7-7 into Eq. 3.7-5 gives

$$\frac{Sb}{g}\frac{d^2z}{dt^2} + (z + Z_{c,\,\mathrm{eq}}) = Sb - \frac{b}{2}$$

or

$$\frac{Sb}{g}\frac{d^2z}{dt^2} + z = Sb - \left(\frac{b}{2} + Z_{c,\,\mathrm{eq}}\right)$$

From Eq. 3.7-2, we see that the right-hand side of the above equation is zero; therefore,

$$\frac{d^2z}{dt^2} + \frac{g}{Sb}z = 0 \qquad \text{Eq. 3.7-8}$$

Letting $\omega^2 = g/Sb$, the above differential equation becomes

$$\frac{d^2z}{dt^2} + \omega^2 z = 0 \qquad \text{Eq. 3.7-9}$$

Recall that a solution to a differential equation is a function which satisfies that equation. Any function that satisfies this equation will have the property that its second derivative will have to be the function itself, multiplied by a constant. The function that has this property is the exponential function. Therefore, assume that

$$z = e^{mt} \qquad \text{Eq. 3.7-10}$$

Then

$$\frac{dz}{dt} = me^{mt} \quad \text{and} \quad \frac{d^2z}{dt^2} = m^2e^{mt} \qquad \text{Eq. 3.7-11}$$

Substituting Eqs. 3.7-10 and 3.7-11 into Eq. 3.7-9 gives

$$(m^2 + \omega^2)e^{mt} = 0$$

or

$$m = \pm\omega i$$

where $i = \sqrt{-1}$. We see that there are two solutions to Eq. 3.7-9. The sum of each multiplied by a constant gives the general solution; that is,

$$z = c_1^* e^{i\omega t} + c_2^* e^{-i\omega t} \qquad \text{Eq. 3.7-12}$$

But

$$e^{i\omega t} = \cos \omega t + i \sin \omega t \quad \text{and} \quad e^{-i\omega t} = \cos \omega t - i \sin \omega t \qquad \text{Eqs. 3.7-13}$$

Substituting Eqs. 3.7-13 into Eq. 3.7-12 gives

$$z = (c_1^* + c_2^*)\cos \omega t + i(c_1^* - c_2^*)\sin \omega t$$

$$= c_1 \cos \omega t + c_2 \sin \omega t \qquad \text{Eq. 3.7-14}$$

and

$$\frac{dz}{dt} = \omega(-c_1 \sin \omega t + c_2 \cos \omega t) \qquad \text{Eq. 3.7-15}$$

The constants c_1 and c_2 are evaluated by the initial conditions which are

$$z(0) = \alpha$$

$$\frac{dz}{dt}(0) = 0 \qquad \text{Eq. 3.7-16}$$

Application of the second equation in the above set gives

$$0 = \omega[-c_1(0) + c_2(1)]$$

or

$$c_2 = 0$$

Application of the first initial condition in the set of Eqs. 3.7-16 gives

$$\alpha = c_1(1)$$

Thus, the solution is

$$z = \alpha \cos \omega t = \alpha \cos \sqrt{\frac{g}{Sb}}\, t \qquad \text{Eq. 3.7-17}$$

We see that the body will oscillate harmonically about its equilibrium position.

We saw in Example 3-4 that a floating body displaced downward from its equilibrium position and then released underwent an oscillatory bobbing motion. Suppose that a floating body was rotated a small angle from its equilibrium position: Would the body also undergo an oscillatory motion? An example covering this phenomenon is given below.

Example 3-5

A barge loaded with coal has a total weight of 49,920 lb_f. It has a constant rectangular cross section from bow to stern (see Fig. 3-24). It is 10 ft wide, 20 ft long, and draws 4 ft of water as loaded. The center of gravity of the combined barge and load is at the centerline and 1 in. above the waterline. A sudden disturbance rolls the barge clockwise by the angle θ_0 radians, about an axis coinciding with the longitudinal centerline at the waterline level. At this moment

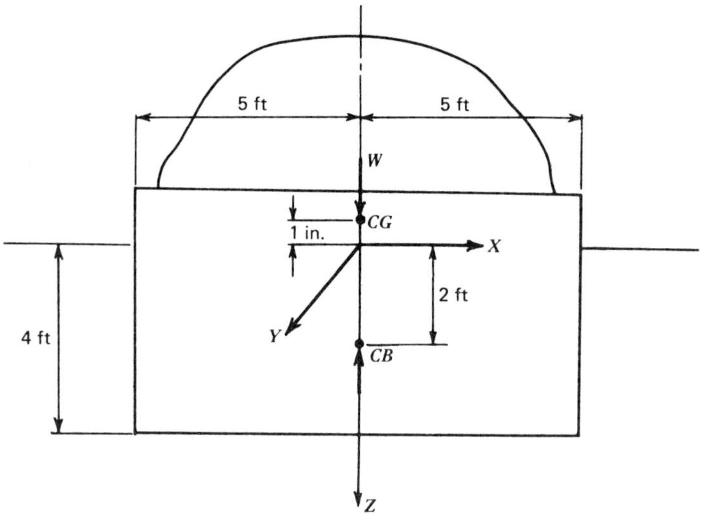

Fig. 3-24. The barge in its equilibrium position.

its rotational speed is zero. Neglecting viscous damping, determine the differential equation governing the motion. Assume that the moment of inertia, I_{yy}, of the barge and load is known.

SOLUTION

From Eq. 3.6-3, we see that the moment about the Y-axis, $M_{2,b}$, caused by the buoyancy force is

$$M_{2,b} - \gamma \iiint_{\mathcal{V}} X \, d\mathcal{V} \qquad \text{Eq. 3.7-18}$$

The easiest way to determine the X-coordinate of the centroid of the region of water displaced by the barge when it has rolled to some arbitrary position is to add to the original region, $\mathcal{V}_1$, the region $\mathcal{V}_2$, having the triangular cross section OAB, and to subtract region $\mathcal{V}_3$, having the triangular cross section OCD, as shown in Fig. 3-25. Then Eq. 3.7-18 is equivalent to Eq. 3.7-19

$$M_{2,b} = \gamma\left(\iiint_{\mathcal{V}} X \, d\mathcal{V} + \iiint_{\mathcal{V}_2} X \, d\mathcal{V} - \iiint_{\mathcal{V}_3} X \, d\mathcal{V}\right)$$

$$= \gamma(\mathcal{V}_1 \bar{X}_1 + \mathcal{V}_2 \bar{X}_2 - \mathcal{V}_3 \bar{X}_3) \qquad \text{Eq. 3.7-19}$$

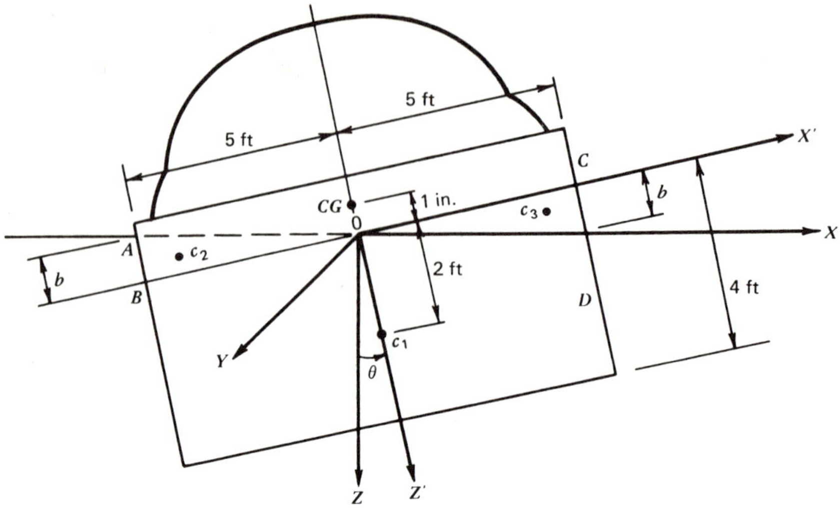

Fig. 3-25. The barge in an arbitrary rolled position.

Remember that the position of the centroid of a right triangular region is located a distance from each leg equal to $\frac{1}{3}$ the length of the leg, as shown in Fig. 3-26*a*. Thus, the position of the centroids in the (X', Y', Z') coordinate system is

$$(\bar{X}'_1, \bar{Z}'_1) = (0, 2), \qquad (\bar{X}'_2, \bar{Z}'_2) = \left(-\frac{10}{3}, -\frac{b}{3}\right)$$

and

$$(\bar{X}'_3, \bar{Z}'_3) = \left(\frac{10}{3}, \frac{b}{3}\right) \qquad \text{Eqs. 3.7-20}$$

To obtain a relationship of the position of a point in the (X, Y, Z) system and the (X', Y', Z') system, equate the position vector in each system

$$\mathbf{r} = X\mathbf{i} + Y\mathbf{j} + Z\mathbf{k} = X'\mathbf{i}' + Y'\mathbf{j}' + Z'\mathbf{k}' \qquad \text{Eq. 3.7-21}$$

Now express $(\mathbf{i}', \mathbf{j}', \mathbf{k}')$ in terms of $(\mathbf{i}, \mathbf{j}, \mathbf{k})$ (see Fig. 3-26*b*). Then,

$$\mathbf{i}' = \cos\theta\mathbf{i} - \sin\theta\mathbf{k}$$

$$\mathbf{j}' = \mathbf{j} \qquad \text{Eq. 3.7-22}$$

$$\mathbf{k}' = \sin\theta\mathbf{i} + \cos\theta\mathbf{k}$$

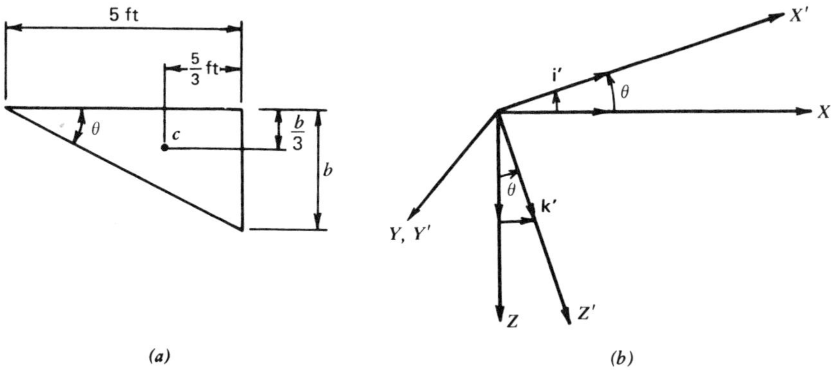

Fig. 3-26. Some related geometry for Example 3-4. (*a*) Position of centroid in a triangular region. (*b*) A rotation of the coordinate system.

Substituting Eqs. 3.7-22 into Eq. 3.7-21 gives

$$X\mathbf{i} + Y\mathbf{j} + Z\mathbf{k} = X'(\cos\theta\mathbf{i} - \sin\theta\mathbf{k}) + Y'\mathbf{j} + Z'(\sin\theta\mathbf{i} + \cos\theta\mathbf{k})$$

Equating the **i**, **j**, and **k**-components on both sides of the above equation gives

$$\begin{aligned} X &= X'\cos\theta + Z'\sin\theta \\ Y &= Y' \\ Z &= -X'\sin\theta + Z'\cos\theta \end{aligned} \qquad \text{Eq. 3.7-23}$$

Substituting Eqs. 3.7-20 into the first equation in the set of Eqs. 3.7-23 gives

$$\begin{aligned} \overline{X}_1 &= 2\sin\theta \\ \overline{X}_2 &= -\frac{10}{3}\cos\theta - \frac{b}{3}\sin\theta \\ \overline{X}_3 &= \frac{10}{3}\cos\theta + \frac{b}{3}\sin\theta \end{aligned} \qquad \text{Eqs. 3.7-24}$$

The volumes of the three regions are

$$\mathcal{V}_1 = 10(4)(20) = 800 \text{ ft}^3$$

$$\mathcal{V}_2 = \mathcal{V}_3 = \tfrac{1}{2}(5)(20)b = 50b \text{ ft}^3 \qquad \text{Eqs. 3.7-25}$$

Substituting Eqs. 3.7-24 and 3.7-25 into Eq. 3.7-19 gives

$$M_{2,b} = \gamma\left[1600 \sin\theta + (50b)\left(-\frac{10}{3}\cos\theta - \frac{b}{3}\sin\theta\right) - (50b)\left(\frac{10}{3}\cos\theta + \frac{b}{3}\sin\theta\right)\right]$$

$$= \gamma\left[1600 \sin\theta - \frac{1000}{3}b\cos\theta - \frac{100}{3}b^2\sin\theta\right] \text{lb}_f\text{-ft} \qquad \text{Eq. 3.7-26}$$

From Fig. 3-26*a*, we see

$$b = 5 \tan\theta \text{ ft}$$

Substituting the above relation into Eq. 3.7-26 we obtain

$$M_{2,b} = \gamma\left(1600 \sin\theta - \frac{5000}{3}\cos\theta\tan\theta - \frac{2500}{3}\tan^2\theta\sin\theta\right) \text{lb}_f\text{-ft}$$

Eq. 3.7-27

If the resulting motion is such that θ always remains small, then we may take $\cos\theta \approx 1$ and $\sin\theta \approx \tan\theta \approx \theta$; then after factoring out the $\frac{1}{3}$, Eq. 3.7-27 becomes

$$M_{2,b} = \frac{\gamma}{3}(4800\theta - 5000\theta - 2500\theta^3)$$

$$= -\frac{\gamma}{3}(200\theta + 2500\theta^3) = -4160\theta - 52{,}000\theta^3 \text{ lb}_f\text{-ft} \qquad \text{Eq. 3.7-28}$$

The student should recall from mechanics that for a body rotating about a stationary axis in an inertial reference frame, say for definitiveness the Y-axis, then,

$$M_2 = I_{YY} \frac{\ddot{\theta}}{g_c} \qquad \text{Eq. 3.7-29}$$

where M_2 is the moment about the Y-axis caused by the forces acting on the body, θ is the angle of rotation, $\ddot{\theta}$ is the second derivative of θ with respect to time, $I_{YY} = \iiint_{\mathcal{V}} \rho(X^2 + Z^2)\, d\mathcal{V}$, and ρ is the mass density of the body. The total moment M_2 results from both the buoyancy force and the gravitational force. The moment resulting from the gravitational field, $M_{2,G}$, is given by

$$M_{2,G} = -WX^*$$

where W is the total weight of barge and load, and X^* is the X-coordinate of the center of gravity, which can readily be obtained from the first equation in the set of Eqs. 3.7-23; that is,

$$X^* = -\frac{1}{12} \sin \theta \approx -\frac{\theta}{12} \qquad \text{Eq. 3.7-30}$$

Substituting the values of W and X^*, the equation for $M_{2,G}$ gives

$$M_{2,G} = \frac{49{,}920}{12} \theta = 4160\theta \text{ lb}_f\text{-ft} \qquad \text{Eq. 3.7-31}$$

Then

$$\begin{aligned} M_2 &= M_{2,G} + M_{2,B} = 4160\theta - 4160\theta - 52{,}000\theta^3 \\ &= -52{,}000\theta^3 \text{ lb}_f\text{-ft} \end{aligned} \qquad \text{Eq. 3.7-32}$$

Substituting the above expression for M_2 into Eq. 3.7-29 gives

$$-52{,}000\theta^3 = I_{YY} \frac{\ddot{\theta}}{g_c} \qquad \text{Eq. 3.7-33}$$

Equation 3.7-33 is the differential equation governing the motion. From Eq. 3.7-32 we see that for any small angle, θ, the moment on the body tends to restore the body back to its equilibrium position by causing an angular acceleration in the opposite direction to the displacement, θ. That is, if θ is positive, $\ddot{\theta}$ is negative; if θ is negative, $\ddot{\theta}$ is positive. This is the characteristic behavior of a body that is rotationally stable. It also should be noted that if the center of gravity for the barge and load were located more than 1 in. above the waterline, say for example $1\frac{1}{4}$ in. above the waterline, then

$$\begin{aligned} M_2 &= 1.25(4160)\theta - 4160\theta - 52{,}000\theta^3 \\ &= 1040\theta - 52{,}000\theta^3 \text{ lb}_f\text{-ft} \end{aligned} \qquad \text{Eq. 3.7-34}$$

Thus, a restoring moment would not occur until

$$52{,}000\theta^2 > 1040$$

or not until

$$\theta > 0.141 \text{ radians} = 8.1°$$

This would indeed be an unsatisfactory loading condition for the barge and should be avoided.

3.7.1 PERIOD OF OSCILLATION OF A FLOATING BODY AS A COMPUTER PROBLEM

Let us consider the barge described in Example 3-5. Suppose we were interested in determining the period of oscillation. The period is defined as the time it takes the barge to go through one cycle. It should be observed that the equation describing the motion of the barge, Eq. 3.7-33 is a second-order nonlinear differential equation. Despite the nonlinearity, the equation can be integrated once without too much difficulty. Let

$$\frac{d\theta}{dt} = u$$

$$\frac{d^2\theta}{dt^2} = \frac{du}{dt} = \frac{du}{d\theta}\frac{d\theta}{dt} = u\frac{du}{d\theta} \qquad \text{Eq. 3.7-35}$$

Substituting Eq. 3.7-35 into Eq. 3.7-33 gives

$$-52{,}000\theta^3 = \frac{I_{YY}}{g_c} u \frac{du}{d\theta}$$

Rearranging and integrating both sides of the above equation yields

$$\int u\, du = \frac{-52{,}000}{I_{YY}} g_c \int \theta^3\, d\theta$$

or

$$\frac{u^2}{2} = -\frac{52{,}000 g_c}{I_{YY}} \frac{\theta^4}{4} + C$$

When $t = 0$, $\theta = \theta_0$, and $d\theta/dt = u = 0$. Thus,

$$C = \frac{52{,}000 g_c}{4 I_{YY}} {\theta_0}^4$$

Then

$$\frac{d\theta}{dt} = \pm\sqrt{\frac{52{,}000 g_c}{2 I_{YY}}} \sqrt{{\theta_0}^4 - \theta^4} \qquad \text{Eq. 3.7-36}$$

The barge rotation from $\theta = \theta_0$ to $\theta = 0$ occurs in a time span equal to $\frac{1}{4}$ period. During this time interval, $d\theta/dt$ is negative. Thus, the period of oscillation, τ, can be determined from

$$\int_0^{\tau/4} dt = \frac{\tau}{4} = -\sqrt{\frac{2 I_{YY}}{52{,}000 g_c}} \int_{\theta_0}^{0} \frac{d\theta}{\sqrt{{\theta_0}^4 - \theta^4}}$$

or

$$\tau = \sqrt{\frac{32 I_{YY}}{52{,}000 g_c}} \int_0^{\theta_0} \frac{d\theta}{\sqrt{{\theta_0}^4 - \theta^4}} \qquad \text{Eq. 3.7-37}$$

The integral in the above equation cannot be classified as a simple integral and, therefore, numerical integration is appropriate. However, even by numerical methods, its evaluation requires some special treatment. This is so because the integral is improper. That is, the integrand is singular (it goes to ∞) at $\theta = \theta_0$. When the integrand is singular at one of its limits, the integral can be broken down into two parts as shown below. Let

$$I = \int_0^{\theta_0} \frac{d\theta}{\sqrt{\theta_0^4 - \theta^4}} = \int_0^{\theta^*} \frac{d\theta}{\sqrt{\theta_0^4 - \theta^4}} + \lim_{\varepsilon \to 0} \int_{\theta^*}^{\theta_0 - \varepsilon} \frac{d\theta}{\sqrt{\theta_0^4 - \theta^4}} = I_1 + I_2$$

Eq. 3.7-38

If $(\theta_0 - \theta^*)$ is made sufficiently small, then the second integral on the right-hand side of the above equation, I_2, can be evaluated by expanding $(\theta_0^4 - \theta^4)$ in a Taylor series about θ_0 and using only the first three terms. Under these circumstances, the integral can be analytically evaluated. The first integral on the right-hand side of Eq. 3.7-38, I_1, can be evaluated numerically by Simpson's rule. This procedure is carried out below. Let

$$g(\theta) = \theta_0^4 - \theta^4$$

then

$$g'(\theta) = -4\theta^3 \qquad \text{and} \qquad g''(\theta) = -12\theta^2$$

Expanding $g(\theta)$ in a Taylor series about θ_0 and neglecting terms of powers greater than two gives

$$g(\theta) = g(\theta_0) + g'(\theta_0)(\theta - \theta_0) + \frac{g''(\theta_0)}{2!}(\theta - \theta_0)^2$$

$$= 0 - 4\theta_0^3(\theta - \theta_0) - \frac{12\theta_0^2}{2!}(\theta - \theta_0)^2$$

$$= 2(-\theta_0^4 + 4\theta_0^3\theta - 3\theta_0^2\theta^2)$$

Then

$$I_2 = \lim_{\varepsilon \to 0} \int_{\theta^*}^{\theta_0 - \varepsilon} \frac{d\theta}{\sqrt{2}\sqrt{-\theta_0^4 + 4\theta_0^3\theta - 3\theta_0^2\theta^2}} = \frac{1}{\sqrt{2}} \lim_{\varepsilon \to 0} \int_{\theta^*}^{\theta_0 - \varepsilon} \frac{d\theta}{\sqrt{X}}$$

where $X = -\theta_0^4 + 4\theta_0^3\theta - 3\theta_0^2\theta^2 = a + b\theta + c\theta^2$. The integral,

$$\int (X)^{-1/2}\, d\theta$$

can be found in any standard table of integrals. That is

$$\int (X)^{-1/2}\, d\theta = \frac{-1}{\sqrt{-c}} \sin^{-1}\left(\frac{2c\theta + 6}{-q}\right)$$

where

$$q = 4ac - b^2 = 4(-\theta_0^4)(-3\theta_0^2) - (4\theta_0^3)^2 = -4\theta_0^6$$

Then

$$\begin{aligned} I_2 &= \frac{1}{\sqrt{2}} \lim_{\varepsilon \to 0} \frac{-1}{\sqrt{3\theta_0^2}} \left[\sin^{-1}\left(\frac{2(-3\theta_0^2)\theta + 4\theta_0^3}{2\theta_0^3}\right)\right]_{\theta^*}^{\theta_0 - \varepsilon} \\ &= -\frac{1}{\sqrt{6}\theta_0} \lim_{\varepsilon \to 0}\left(\sin^{-1} \frac{2\theta_0^3 - 3\theta_0^2(\theta_0 - \varepsilon)}{\theta_0^3} - \sin^{-1} \frac{2\theta_0^3 - 3\theta_0^2\theta^*}{\theta_0^3}\right) \\ &= -\frac{1}{\sqrt{6}\theta_0}\left[\sin^{-1}(-1) - \sin^{-1}\left(2 - \frac{3\theta^*}{\theta_0}\right)\right] \qquad \text{Eq. 3.7-39} \end{aligned}$$

We now turn our attention to the evaluation of I_1 by Simpson's rule. Let us consider a general integral, J, defined by

$$J = \int_{\theta_\alpha}^{\theta_\beta} f(\theta)\, d\theta$$

If the domain $(\theta_\beta - \theta_\alpha)$ is subdivided into N intervals, whose endpoints are $\theta_\alpha = \theta_1, \theta_2, \theta_3, \ldots, \theta_N, \theta_{N+1} = \theta_\beta$, and $f(\theta)$ is evaluated at these endpoints, giving $f_1, f_2, \ldots, f_{N+1}$, respectively, then by Simpson's rule

$$J = \frac{h}{3}(f_1 + 4f_2 + 2f_3 + 4f_4 + \cdots + 2f_{N-1} + 4f_N + f_{N+1}) \qquad \text{Eq. 3.7-40}$$

where $h = (\theta_\beta - \theta_\alpha)/N$, with N being an even integer. Proof of Simpson's rule can be found in any standard text on numerical methods. We have included a computer program for evaluating I_1 and τ. In this program we have taken $\theta_0 = 0.2$ radians, $I_{YY} = 5.2 \times 10^6$ lb_m-ft^2, and $\theta^* = 0.198$ radians. Substituting the above constants and Eq. 3.7-38 into Eq. 3.7-37, we obtain

$$\tau = \sqrt{\frac{3200}{32.17}}(I_1 + I_2) \qquad \text{Eq. 3.7-41}$$

where

$$I_1 = \int_0^{0.198} \frac{d\theta}{\sqrt{(0.2)^4 - \theta^4}}$$

and, by Eq. 3.7-39,

$$I_2 = \frac{1}{0.2\sqrt{6}}[\sin^{-1}(-0.97) - \sin^{-1}(-1)]$$

The authors' computer program for evaluating the period, τ, as described in this section, is given here.

```
C   COMPUTER PROGRAM FOR EVALUATING THE PERIOD OF
    AN OSCILLATING BARGE
       REAL INTGL,INTGL2
       THETAO=0.2
       THETAS=.198
       THETAI=0.0
       THETAF=THETAS
       N=50
       WRITE(108,6)
    6  FORMAT ('1 NUMERICAL INTEGRATION BY SIMPSONS RULE')
       DO5 K=1,4
       ZN=N
       H=(THETAF-THETAI)/ZN
       N1=N+1
       S=0.0
       ISN=-1
       DO 1 I=1,N1
       ZI=I-1
       THETA=ZI*H
       F=1.0/SQRT(THETAO**4-THETA**4)
       IF(I.EQ. 1 .OR. I.EQ. N1)GO TO 2
       ISN=-ISN
       IF(ISN .EQ.    1) ZJ=4.0
```

```
      IF(ISN .EQ.-1) ZJ=2.0
      GO TO 1
    2 ZJ=1.0
    1 S=S+ZJ*F
      INTGL=H*S/3.0
      WRITE(108,4)N,INTGL
    4 FORMAT(4HON= I3,5X,7HINTGL= E15.7)
    5 N=2*N
      WRITE(108,7)
    7 FORMAT('0DETERMINATION OF PERIOD OF OSCILLATION
     1OF BARGE')
      C1=SQRT(3200./32.17)
      INTGL2=1.0/SQRT(6.0)/THETAO*(ASIN(2.0-3.0*THETAS/
     1THETAO)-ASIN(-1.0))
      TAU=C1*(INTGL+INTGL2)
      WRITE(108,8)THETAO,THETAS,INTGL2,TAU
    8 FORMAT(9H0THETAO= E12.5,5X,8HTHETAS= E12.5,5X,
     18HINTGL2= E12.5,5X,5HTAU= E12.5)
      CALL EXIT
      END
```

Finally, the values obtained from the computer program are

$$I_1 = 6.0538, \qquad I_2 = 0.5013, \qquad \tau = 65.377 \text{ s}$$

3.8 STATIC COMPRESSIBLE FLUIDS AND THE ATMOSPHERE

The determination of the pressure field in a gaseous fluid over large vertical distances, as in the case of the atmosphere, by the integration of Eq. 3.2-16 is not quite as simple as in the previous cases. This is so because in the atmosphere $\gamma = \gamma(z)$. Thus, before the pressure field can be determined, the variation of the specific weight of the atmosphere with the altitude must be known.

3.8.1 THE MODEL ATMOSPHERE

Most of us are well aware that the pressure and density of the atmosphere decrease with altitude. Although the atmospheric pressure field slightly varies from day to day, it is convenient to have a model atmosphere that gives average conditions. The U.S. Standard Atmosphere developed in 1962 and extended in 1966 provides such a model. In this model, the variation of atmospheric temperature with altitude is given as shown in Fig. 3-27. Notice that this temperature

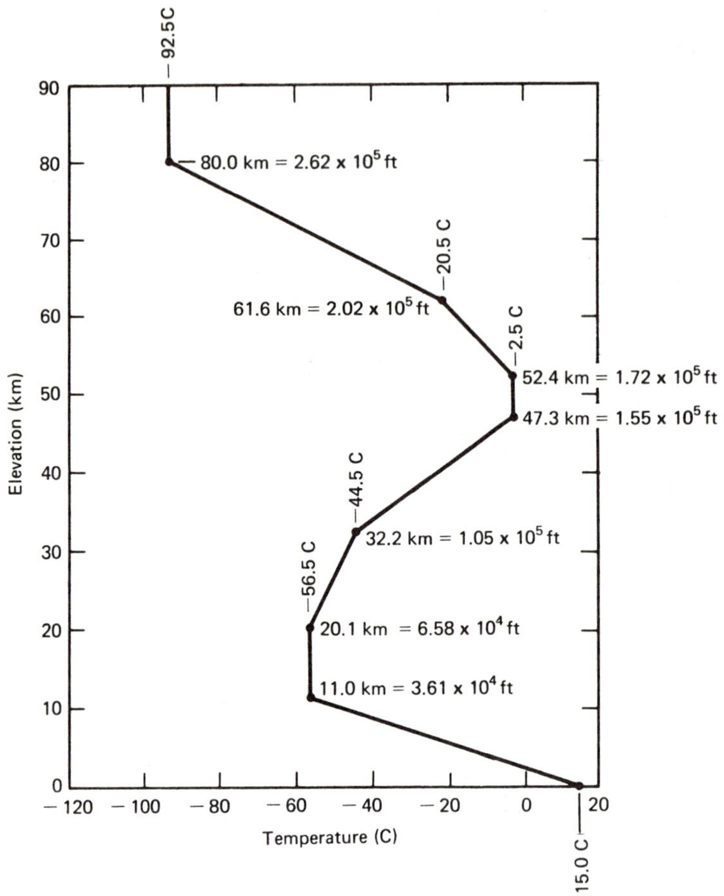

Fig. 3-27. Temperature variation with altitude in the U.S. Standard Atmosphere. (From *Introduction to Fluid Mechanics* by R. W. Fox and A. T. McDonald, Copyright © 1978 by John Wiley & Sons. Reprinted by permission of the publisher.)

profile can be broken down into several distinct regions. In each region, the temperature either varies linearly with altitude or is a constant. The lower region is designated as the *troposphere* and extends from sea level to approximately 36,200 ft. The first relatively constant temperature region is the *stratosphere*, which extends to approximately 66,000 ft. Above the stratosphere is the mesosphere, which extends to approximately 300,000 ft. It is in this region that the ozone layer absorbs the ultraviolet radiation from the sun.

The negative of the derivative of the temperature with respect to altitude, $-(dT/dz)$, is designated as the *lapse rate*. The symbol for the lapse rate in the actual atmosphere is λ, whereas that for the standard atmosphere is λ_s. The lapse rate for the standard atmosphere is tabulated in Table 3-1.

To determine the pressure field in the standard atmosphere, we utilize Eq. 3.2-16; that is,

$$\frac{dp}{dz} = -\gamma(z) \qquad \text{Eq. 3.2-16}$$

Since the atmosphere behaves as an ideal gas, we may take

$$p = \rho RT \qquad \text{Eq. 3.8-1}$$

Substituting for ρ from Eq. 1.4-3, we obtain

$$p = \frac{g_c}{g}\gamma RT = \gamma R'T \qquad \text{Eq. 3.8-2}$$

where R is the gas constant for air and

$$R' = \frac{Rg_c}{g}$$

Substituting for γ in Eq. 3.2-16, gives

$$\frac{dp}{dz} = -\frac{p}{R'T} \qquad \text{Eq. 3.8-3}$$

Since there are, basically, only two types of regions, one where the temperature varies linearly with altitude and the other where the temperature is a constant, there are only two variations for carrying out the integration of Eq. 3.8-3. First, we will consider the regions where λ_s is nonzero. We can separate the variables in Eq. 3.8-3 by the use of the chain rule

$$\frac{dp}{dz} = \frac{dp}{dT}\frac{dT}{dz} = -\lambda_S\frac{dp}{dT} \qquad \text{Eq. 3.8-4}$$

TABLE 3-1 Temperature, Lapse Rate, and the Gravitational Constant for the U.S. Standard Atmosphere

Altitude		Temperature		Lapse Rate		Gravitational Acceleration		Pressure	
ft	m	F	C	R/ft	K/m	ft/s^2	m/s^2	psia	N/m^2 abs.
0	0	59.0	15.0	3.566×10^{-3}	6.49×10^{-3}	32.17	9.807	14.696	1.013×10^{5}
36,150	11,020	−69.7	−56.5	0.0	0.0	32.06	9.773	3.281	2.262×10^{4}
65,820	20,060	−69.7	−56.5	-0.544×10^{-3}	-0.992×10^{-3}	31.97	9.745	0.7938	5.473×10^{3}
105,520	32,160	−48.1	−44.5	-1.52×10^{-3}	-2.80×10^{-3}	31.85	9.708	0.1258	8.675×10^{2}
155,350	47,350	27.5	−2.5	0.0	0.0	31.69	9.661	0.1648×10^{-1}	1.136×10^{2}
172,010	52,430	27.5	−2.5	1.08×10^{-3}	1.97×10^{-3}	31.60	9.646	0.8763×10^{-1}	6.042×10^{1}
202,070	61,590	−4.9	−20.5	2.15×10^{-3}	3.91×10^{-3}	31.55	9.617	0.2712×10^{-2}	1.870×10^{1}
262,450	80,000	−134.5	−92.5	0.0	0.0	31.36	9.561	0.1543×10^{-3}	1.064
299,520	91,290	−134.5	−92.5			31.25	9.526	0.1930×10^{-4}	0.1331

Substituting Eq. 3.8-4 into Eq. 3.8-3, we obtain

$$-\lambda_S \frac{dp}{dT} = -\frac{p}{R'T}$$

or

$$\frac{dp}{p} = \frac{1}{\lambda_S R'} \frac{dT}{T} \qquad \text{Eq. 3.8-5}$$

The gravitational acceleration, g, varies only slightly over the altitude range of the regions described in Table 3-1. Therefore for each region, we may take R' to be a constant. However, a more accurate solution can be obtained by the use of an averaged g value over the region of interest. Values for g are also tabulated in Table 3-1. If (p_i, T_i) are the pressure and *absolute* temperature at the lower-altitude limit of a region, and (p_i', T') are dummy variables of integration, then

$$\int_{p_i}^{p} \frac{dp'}{p'} = \frac{1}{\lambda_S R'} \int_{T_i}^{T} \frac{dT'}{T'} \qquad \text{Eq. 3.8-6}$$

Carrying out the integrations, we obtain

$$\ln \frac{p}{p_i} = \frac{1}{\gamma_S R'} \ln \frac{T}{T_i} = \ln\left(\frac{T}{T_i}\right)^{1/\lambda_S R'}$$

or

$$\frac{p}{p_i} = \left(\frac{T}{T_i}\right)^{1/\lambda_S R'} \qquad \text{Eq. 3.8-7}$$

The determination of temperature with respect to altitude is obtained by integrating the equation

$$\frac{dT}{dz} = -\lambda_S = \text{constant}$$

Again, if T' and z' are dummy variables of integration, then

$$\int_{T_i}^{T} dT' = -\lambda_S \int_{z_i}^{z} dz'$$

or

$$T = T_i - \lambda_S(z - z_i) \qquad \text{Eq. 3.8-8}$$

Substituting Eq. 3.8-8 into Eq. 3.8-7, we obtain

$$p = p_i\left[1 - \frac{\lambda_S(z - z_i)}{T_i}\right]^{1/\lambda_S R'} \qquad \text{Eq. 3.8-9}$$

In the altitude range $0 \leq z \leq 36{,}150$ ft, (p_i, T_i) become the pressure and temperature at sea level. In the standard atmosphere these are 14.696 psia and 59.0 F in the English system and 1.01325×10^5 N/m^2 abs. and 15.0 C in the SI system.

The integration of Eq. 3.8-3 for a constant temperature region is somewhat more direct. This is so because the variables are immediately separable. If T_i is the absolute temperature at the lower-altitude limit of the constant temperature region, then

$$\frac{dp}{dz} = -\frac{1}{R'T_i} \qquad \text{Eq. 3.8-10}$$

The variables are immediately separable, giving

$$\frac{dp}{p} = -\frac{1}{R'T_i}dz \qquad \text{Eq. 3.8-11}$$

If p_i and z_i are the pressure and elevation at the lower-altitude limit of the region, then

$$\int_{p_i}^{p}\frac{dp'}{p'} = \frac{1}{R'T_i}\int_{z_i}^{z}dz'$$

Carrying out the integrations, we obtain

$$\ln\frac{p}{p_i} = -\frac{1}{R'T_i}(z - z_i)$$

or

$$\frac{p}{p_i} = (e)^{-(z-z_i)/R'T_i} \qquad \text{Eq. 3.8-12}$$

Values for T_i and p_i can be obtained from Eqs. 3.8-8 and 3.8-9, for the region just below the constant temperature region. Values of p, T, and ρ, as functions of altitude for the U.S. Standard Atmosphere, are tabulated in Tables A.6-4*a* and A.6-4*b*.

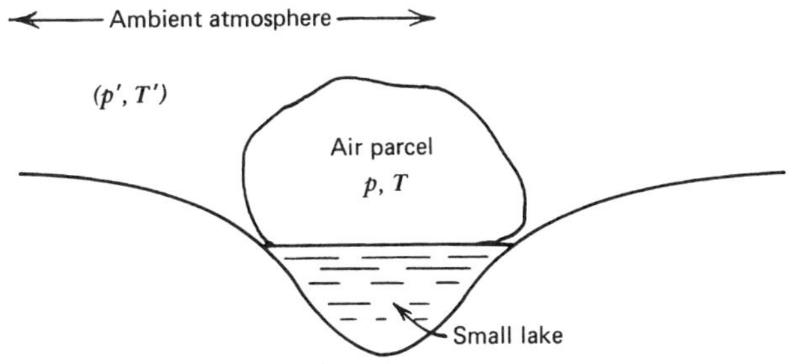

Fig. 3-28. An air parcel in an ambient atmosphere.

3.8.2 THE GASEOUS PARCEL

It should be remembered that the pressure and temperature fields just developed are for the standard atmosphere. The pressure and temperature fields in the real atmosphere vary from day to day. The actual lapse rate, λ, strongly influences the development of showers, thunderstorms, and the dispersion of pollutants. To develop this concept, the air parcel is introduced. An air parcel is some region of the atmosphere, usually near ground level, whose temperature develops differently from its surroundings. For example, a region directly over a small lake, where the lake is warmer than the surrounding ground, may be considered as an air parcel. Gases emitted from a smoke stack, in the simplest analysis, may also be considered as an air parcel. Such parcels will invariably develop a density which is less than that of the ambient atmosphere. Its lower density will result in a buoyant force that will lift the parcel to an altitude of equal density (see Fig. 3-28).

In the simplest type of analysis, the parcel is assumed to rise slowly and adiabatically without mixing with the surroundings. It is also assumed that the pressure in the parcel will equal the ambient pressure. This assumption is consistent with the experimental observation that under most circumstances pressure equalization occurs much quicker than temperature equalization. To obtain the governing temperature equation for the parcel we may consider the parcel as a thermodynamic system. The first law of thermodynamics for a system that only undergoes expansion work is expressed in terms of intensive variables by the following equation[3]

$$du = \delta q - p\, d\tilde{v} \qquad \text{Eq. 3.8-13}$$

[3] For a more extensive review of thermodynamics, see Appendix A.3.

For an ideal gas, du may be expressed by the equation

$$du = c_v \, dT \qquad \text{Eq. 3.8-14}$$

Since the atmosphere behaves as an ideal gas, Eq. 1.4-16 is applicable; that is,

$$p\tilde{v} = RT \qquad \text{Eq. 3.8-15}$$

Taking the differential of both sides of the above equation, we obtain

$$p \, d\tilde{v} + \tilde{v} \, dp = R \, dT \qquad \text{Eq. 3.8-16}$$

Substituting Eqs. 3.8-14 and 3.8-16 into Eq. 3.8-13, and rearranging, we obtain

$$\begin{aligned} \delta q &= c_v \, dT + R \, dT - \tilde{v} \, dp \\ &= (c_v + R) \, dT - \tilde{v} \, dp \end{aligned} \qquad \text{Eq. 3.8-17}$$

For an ideal gas,

$$c_p = c_v + R \qquad \text{Eq. 3.8-18}$$

In the above equation, dimensional homogeniety is required. Assuming that when the parcel rises it undergoes an adiabatic process, then $\delta q = 0$. Substituting this condition and Eqs. 3.8-18 and 1.4-16 into Eq. 2.8-17, we obtain

$$c_p \, dT = \frac{RT}{p} \, dp$$

Dividing both sides of the above equation by the differential dz, we obtain

$$\frac{dT}{dz} = \frac{R}{c_p} \frac{T}{p} \frac{dp}{dz} \qquad \text{Eq. 3.8-19}$$

If the ambient atmosphere has a temperature, pressure, and specific weight designated as (T', p', γ'), respectively, and the pressure of the parcel is everywhere the same as the ambient pressure, then

$$\frac{dT}{dz} = \frac{R}{c_p} \frac{T}{p'} \frac{dp'}{dz} = \frac{R}{c_p} \frac{T}{T'} \frac{T'}{p'} \frac{dp'}{dz} \qquad \text{Eq. 3.8-20}$$

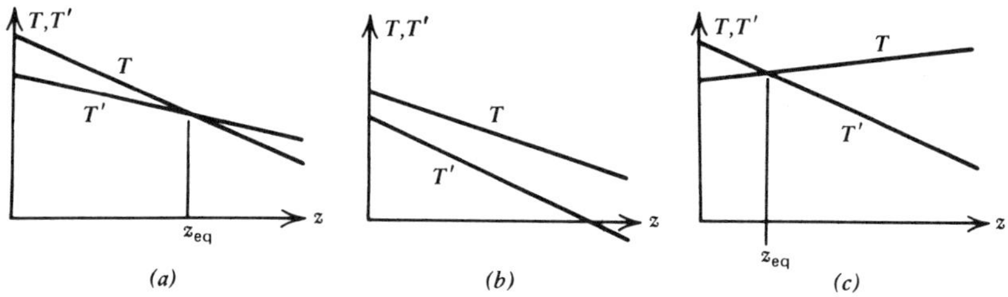

Fig. 3-29. Temperature profiles and atmospheric stability. (*a*) Stable atmosphere, $\lambda_d > \lambda' > 0$. (*b*) Unstable atmosphere, $0 < \lambda_d < \lambda'$. (*c*) Temperature inversion, $\lambda' < 0 < \lambda_d$.

But by Eqs. 3.2-17 and 3.8-2

$$\frac{dp'}{dz} = -\gamma' = -\frac{g}{g_c R}\frac{p'}{T'}$$

or

$$\frac{T'}{p'}\frac{dp'}{dz} = -\frac{g}{g_c R} \qquad \text{Eq. 3.8-21}$$

Substituting Eq. 3.8-21 into Eq. 3.8-20 gives

$$-\frac{dT}{dz} = \frac{g}{g_c c_p}\frac{T}{T'} = \lambda_d \qquad \text{Eq. 3.8-22}$$

where λ_d is called the dry adiabatic lapse rate. If λ' is the actual lapse rate of the atmosphere, then the ratio λ_d/λ' determines the convective stability of the air. To illustrate: consider a parcel of dry air that is in contact with a local hot spot. Because of heat transfer, it will attain a temperature higher than its surroundings. As a result, the parcel will be less dense than the surroundings. The buoyancy force on the parcel will be greater than the gravitational force, and thus the parcel will start to rise. As it rises, it cools at the adiabatic lapse rate, λ_d. If $\lambda_d/\lambda' > 1$, then the parcel will continue to rise until its density is the same as the density of the surroundings. We will designate this elevation as the equilibrium position, z_{eq} (see Fig. 3-29*a*). Because of the parcel's motion, it will overshoot the equilibrium position and $T < T'$ for $z > z_{eq}$. Since $p' = p$, the ideal gas law gives

$$\frac{\rho}{\rho'} = \frac{T'}{T} \qquad \text{Eq. 3.8-23}$$

and therefore $\rho > \rho'$. Under these circumstances the buoyancy force on the parcel will be less than its weight, and the parcel will move back toward its equilibrium position. Under these circumstances the atmosphere is said to be *stable*. In this type of atmosphere, pollutants leaving a stack will rise to a certain elevation and tend to stay there.

On the other hand, if $\lambda_d/\lambda' < 1$, then the parcel will rise and cool, but not as fast as the surrounding atmosphere. Therefore, the density of the parcel will remain less than the density of the surroundings, and thus the parcel will continue to rise without finding an equilibrium position (see Fig. 3-29*b*). Such an atmosphere is said to be *unstable*. Pollutants leaving a stack in such an atmosphere just continue to rise. This condition is very favorable for removal of pollutants. If $\lambda_d/\lambda' = 1$, then the atmosphere is said to be in *indifferent* equilibrium. Finally, if λ' is negative (which means that the ambient temperature increases with altitude), then again the parcel will rise to an equilibrium position, just as in the case of a stable atmosphere. However, in this case, the equilibrium altitude will be much lower than in the previous case (see Fig. 3-29*c*). Again, the rising parcel would overshoot the equilibrium position, causing the gravitational forces to exceed the buoyancy force, and causing the parcel to move back toward its equilibrium position. In this case, the restoring forces are relatively strong. When λ' is negative, the atmosphere is said to have a temperature inversion. Pollutants released into such an atmosphere tend to stay close to ground level.

Although, in above analysis, the presence of moisture in the parcel was neglected, there are some problems where its inclusion is essential, for example, in weather prediction (see Section 3.8.3). Students interested in this topic should see any standard text on meteorology.

Example 3-6

Consider the discharge from a 225-ft high smoke stack (base of stack is at sea level) that results from a combustion process where the air supply far exceeds the amount needed for the combustion process. Under these circumstances one may assume that the discharge from the stack is predominantly air. Take the temperature of the discharge at the exit of the stack to be at 140 F. If the ambient atmosphere is the same as the standard atmosphere, determine (*a*) λ_d at the exit of the stack; (*b*) T and T' at $z = 36{,}150$ ft; (*c*) z_{eq}; and (*d*) λ_d at z_{eq}. Neglect winds and mixing.

SOLUTION

(*a*) To determine λ_d from Eq. 3.8-22, we first need to calculate the ambient temperature at the stack exit using Eq. 3.8-8. Since ambient atmosphere is the same as the standard atmosphere, $\lambda' = \lambda_S = 3.566 \times 10^{-3}$ R/ft. Then,

$$T' = T_0 - \lambda_S z = 519 - (3.566 \times 10^{-3})(225) = 518.2\ R$$

Then by Eq. 3.8-22

$$\lambda_d = \frac{g}{g_c c_p}\frac{T}{T'} = \frac{600}{778(0.24)(519.2)}$$
$$= 6.20 \times 10^{-3}\ \text{R/ft}$$

Note that in the above equation, c_p was expressed in (ft-lb$_f$)/(lb$_m$-°R). Comparing the values of λ_d and λ', we see that

$$\lambda_d > \lambda'$$

(*b*) To obtain an expression for T, we must solve Eq. 3.8-22. Substituting for T' as given by Eq. 3.8-8 and separating variables, we obtain

$$\frac{dT}{T} = -\frac{1}{c_p}\frac{dz}{T_0 - \lambda' z}$$

To carry out the integration we will make the variable of integration z^* and integrate from $z^* = 225$ to z, obtaining

$$\int_{600}^{T}\frac{dT^*}{T^*} = \ln\frac{T}{600} = -\frac{1}{c_p}\int_{225}^{z}\frac{dz^*}{(T_0 - \lambda' z^*)}$$
$$= \frac{1}{c_p\lambda'}\ln\left(\frac{T_0 - \lambda' z}{518.2}\right)$$

or

$$T = 600\left(\frac{T_0 - \lambda' z}{518.2}\right)^{1/c_p\lambda'}$$

Let us evaluate the exponent in the above equation; that is

$$\frac{1}{c_p\lambda'} = \frac{1}{778(0.24)(3.566 \times 10^{-3})} = 1.502$$

Then at 36,150 ft

$$T = 600\left(\frac{519 - 3.566(36.15)}{518.2}\right)^{1.502}$$

$$= 391.6 \text{ R}$$

By Eq. 3.8-8,

$$T' = T_0 - \lambda_S z = 519 - 3.566(36.15) = 390.1 \text{ R}$$

Since at 36,150 ft, T' is still less than T,

$$z_{eq} > 36{,}150 \text{ ft}$$

(*c*) In order to solve for z_{eq} we need to obtain $T(z)$ for $z > 36{,}150$ ft. If $z_1 = 36{,}150$ ft, $T_1 = 391.6$ R, and $T'_1 = 390.1$ R, then by Eq. 3.8-22,

$$\int_{T_1}^{T} \frac{dT^*}{T^*} = -\frac{1}{c_p T'_1} \int_{z_1}^{z} dz^* = -\frac{1}{778(0.24)(390.1)} \int_{z_1}^{z} dz^*$$

$$\ln \frac{T}{T_1} = -1.373 \times 10^{-5}(z - z_1)$$

or

$$T = 391.6(e)^{-1.373 \times 10^{-5}(z - 36{,}150)}$$

and

$$T' = \text{constant} = 390.1 \text{ R}$$

At z_{eq}, $T = T'$. Thus,

$$390.1 = 391.6(e)^{-1.373 \times 10^{-5}(z_{eq} - 36.150)}$$

or

$$-1.373 \times 10^{-5}(z_{eq} - 36.150) = \ln\left(\frac{390.1}{391.6}\right) = -0.003838$$

and

$$z_{eq} = 36{,}150 + \frac{0.003838}{1.373} \times 10^5 = 36{,}429 \text{ ft}$$

(*d*) To obtain λ_d at z_{eq} note that $T = T'$ at $z = z_{eq}$. Then by Eq. 3.8-22, at z_{eq},

$$\lambda_d = \frac{g}{g_c c_p} = \frac{1}{778(0.24)} = 5.35 \times 10^{-3}\ R/\text{ft}$$

*3.8.3 SOME CONCEPTS IN WEATHER DEVELOPMENT

By extending the concept of the air parcel to an *air mass*, we can begin to understand some of the forces behind weather development. Air masses are *broad, deep bodies of air* whose temperature at ground level approaches the temperature of a contact surface. The temperature and humidity, within an air mass, are relatively uniform at a given elevation. In winter, the plains of the arctic region in North American and Eurasian areas are examples of contact surfaces that build cold air masses. The Sahara desert and the warm oceans in the summer are contact areas that build warm air masses. Eventually, pressure gradients cause air masses to move away from their building contact surfaces. If an air mass is warm and moist, and if by some mechanism it is sufficiently cooled, then rain will develop. Several mechanisms of cooling are discussed in this section.

(1) Convective Cooling

In the summer during the day, the land areas tend to be warmer than the oceans. If an air mass built over the ocean moves inland, surface heating of the moist air will result in a vertical instability, producing an upward convective current. A vertical movement of air is called a current, whereas a horizontal movement is called a wind. As the air mass rises, it expands with an accompanying drop in temperature. If the dew point is reached, showers develop.

(2) Cyclonic-Frontal Cooling

A second cooling mechanism results when a cool air mass meets a warm air mass. When two air masses of different temperatures meet, we have what is

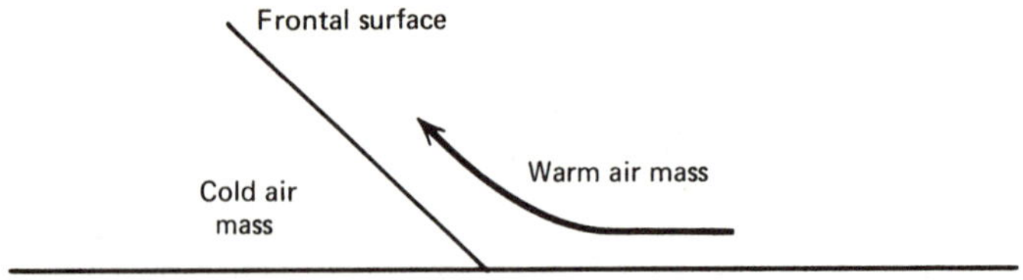

Fig. 3-30. Frontal surface.

called a *frontal surface*. The intersection of this surface at ground level is called a *surface front*. When a cool air mass moves into a warm sector it is called *cyclonic cooling*. The movement of cool air into the warm surroundings, forces the warm surroundings upward (see Fig. 3-30). These conditions tend to develop a pressure center, as shown in Fig. 3-31, which eventually moves horizontally. The curves labeled 1004, 1000, 996, and so on, are curves connecting points at ground level, where the pressure is the same. These curves are called isobars. Since meteorologists express pressure in milibars (mb),[4] the curve labeled 1004 indicates an isobar where the atmospheric pressure is 1004 mb. It should be observed that the isobar having the lowest pressure is at the center of the cyclonic front; the center is designated as a *low*. As the warm air moves up, a wind develops at ground level toward the center. In addition, as the warm air moves up it expands with an accompanying drop in temperature. If the dew point is reached, rain develops. This condition often results in a storm, which dies down as mixing occurs.

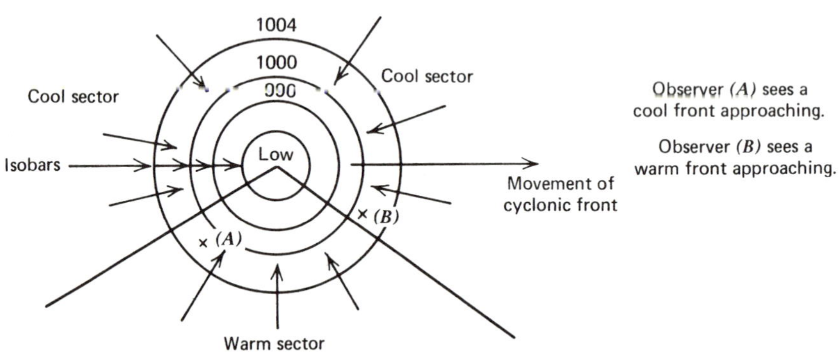

Fig. 3-31. Pressure pattern of a cyclonic front.

[4] 14.696 psia = 1013.2 mb.

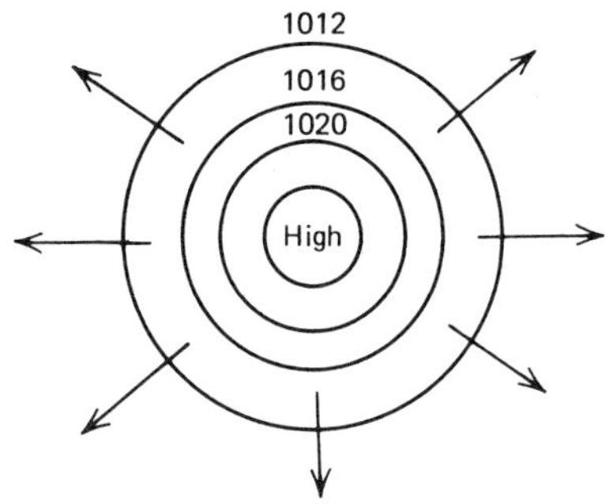

Fig. 3-32. Anticyclonic pressure pattern.

(3) Orographic Cooling

When air from the ocean flows over a mountain range, the air expands and cools. This is called *orographic cooling.* As the air mass moves up, it expands and cools; if the dew point is reached, precipitation results. If the mountain range is deep, regions on the leeward side tend to be very dry, since the air mass loses most of its moisture before it gets there.

Finally, when an air mass cooler than it surroundings falls toward ground it develops the anticyclonic pressure pattern, as shown in Fig. 3-32. This pressure pattern is called a *high.* As the air flows down, winds develop outward from the high. In addition, the downward moving air is compressed, causing a rise in its temperature which increases its capacity to hold moisture. This results in clear weather.

3.9 A FLUID HAVING A UNIFORM ACCELERATION

In Section 3.2, it was mentioned that the development of the governing equation for the pressure field when the fluid has a uniform linear acceleration requires only a slight modification from its development in a static fluid. For this reason, this problem is considered in this chapter. Suppose a tank, opened to the atmosphere, is partially filled with a liquid. Let the tank be secured to a vehicle. When the vehicle is motionless, the liquid level in the tank is z^*, as shown in Fig. 3-33*a*.

Suppose the vehicle on which the tank is secured is given a constant acceleration, a_1, in the X direction. Experiments indicate that after an initial adjustment period, the fluid moves as a rigid body with the free surface inclined, as shown in Fig. 3-33*b*. The forces acting on the infinitesimal volume, shown in the figure, can be expressed in the same manner as those applicable to the infinitesimal

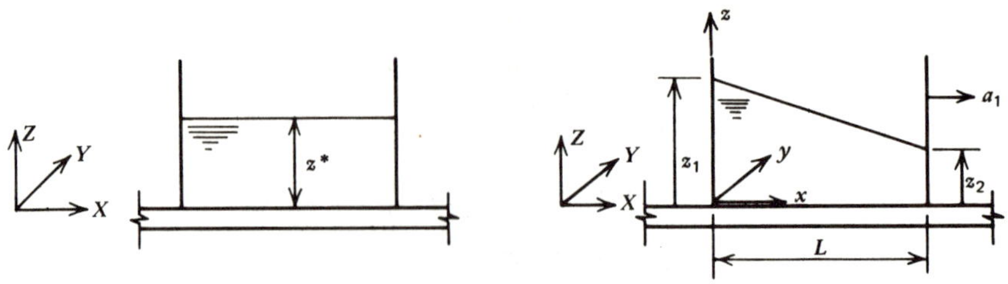

(a) (b)

Fig. 3-33. Liquid in a tank. (*a*) Vehicle motionless. (*b*) Vehicle accelerating.

volume shown in Fig. 3-3 in Section 3.2. We now apply Newton's second law to the infinitesimal volume; that is,

$$\sum_{i=1}^{7} \mathbf{F}_i = \frac{m}{g_c}\mathbf{a} \qquad \text{Eq. 3.9-1}$$

The left-hand sides of the three scalar equations obtained from this vector equation are the same as the left-hand sides of Eqs. 3.2-7. Since $\mathbf{a} = a_1\mathbf{i}$, Eq. 3.9-1 reduces to

$$-\left[p\left(X+\frac{\Delta X}{2}, Y, Z\right) - p\left(X-\frac{\Delta X}{2}, Y, Z\right)\right]\Delta Y\,\Delta Z = \frac{\rho}{g_c}a_1\,\Delta X\,\Delta Y\,\Delta Z$$

$$-\left[p\left(X, Y+\frac{\Delta Y}{2}, Z\right) - p\left(X, Y-\frac{\Delta Y}{2}, Z\right)\right]\Delta X\,\Delta Z = 0$$

$$-\left[p\left(X, Y, Z+\frac{\Delta Z}{2}\right) - p\left(X, Y, Z+\frac{\Delta Z}{2}\right)\right]\Delta X\,\Delta Y - \gamma\,\Delta X\,\Delta Y\,\Delta Z = 0$$

Eqs. 3.9-2

Dividing each equation by $\Delta X\,\Delta Y\,\Delta Z$ and taking the $\lim_{\Delta X\to 0}\lim_{\Delta Y\to 0}\lim_{\Delta Z\to 0}$, we obtain

$$-\frac{\partial p}{\partial X} = \frac{\rho a_1}{g_c}, \qquad -\frac{\partial p}{\partial Y} = 0, \qquad \text{and} \qquad -\left(\frac{\partial p}{\partial Z} + \gamma\right) = 0 \quad \text{Eqs. 3.9-3}$$

Since Newton's law must be written with respect to a nonaccelerating reference frame, we select the (X, Y, Z) coordinate system to be stationary with respect to the ground. Since we are interested in the pressure field within the tank, we want to express the pressure field relative to a coordinate system attached to the tank. Let (x, y, z) be parallel to (X, Y, Z) and attached to the tank with its origin located in the lower left-hand corner, as shown in Fig. 3-33*b*. In this way, the pressure at the point (x, y, z) will be steady (not vary with time), whereas the pressure at the point (X, Y, Z) will be unsteady (vary with time). For further discussion on the meaning of steady and unsteady conditions, see the first three paragraphs in Section 4.1.1. Assume that at $t = 0$, the two coordinate systems coincide. Then the position of a point in the fluid can be expressed in either system, the relationship being

$$X = a_1 \frac{t^2}{2} + x$$

$$Y = y \qquad \text{Eqs. 3.9-4}$$

$$Z = z$$

The partial derivatives of p with respect to (x, y, z) can be obtained by the use of the chain rule, giving

$$\frac{\partial p}{\partial x} = \frac{\partial p}{\partial X}, \quad \frac{\partial p}{\partial y} = \frac{\partial p}{\partial Y}, \quad \text{and} \quad \frac{\partial p}{\partial z} = \frac{\partial p}{\partial Z} \qquad \text{Eqs. 3.9-5}$$

Substituting Eqs. 3.9-5 into Eqs. 3.9-3 and multiplying each equation by (-1) gives

$$\frac{\partial p}{\partial x} = -\frac{\rho a_1}{g_c}$$

$$\frac{\partial p}{\partial y} = 0 \qquad \text{Eqs. 3.9-6}$$

$$\frac{\partial p}{\partial z} + \gamma = 0$$

A simple set of partial differential equations of this type can readily be integrated. The integration of the first equation can be accomplished by seeking a function that is the antiderivative of $\partial p/\partial x$. Such a function is obvious; that is,

$$p = -\frac{\rho a_1}{g_c} x + f(y, z) \qquad \text{Eq. 3.9-7}$$

where $f(y, z)$ is some arbitrary function. Equation 3.9-7 is a solution because, if we take the partial derivative of this equation with respect to x, we obtain the first equation in the set of Eqs. 3.9-6. The function $f(y, z)$ is determined by requiring Eq. 3.9-7 to satisfy the second and third equations in the set of Eqs. 3.9-6. Taking the partial derivative of Eq. 3.9-7 with respect to y, we obtain

$$\frac{\partial p}{\partial y} = \frac{\partial f}{\partial y} \quad \text{Eq. 3.9-8}$$

Substituting this equation into the second equation in the set of Eqs. 3.9-6, we obtain

$$\frac{\partial f}{\partial y} = 0 \quad \text{Eq. 3.9-9}$$

This means that f can at most be a function of z, giving

$$p = -\frac{\rho a_1}{g_c} x + f(z) \quad \text{Eq. 3.9-10}$$

Taking the derivative of this equation with respect to z gives

$$\frac{\partial p}{\partial z} = \frac{df}{dz} \quad \text{Eq. 3.9-11}$$

Substituting this equation into the third equation in the set of Eqs. 3.9-6, and bringing γ over to the right-hand side of the equation, we obtain

$$\frac{df}{dz} = -\gamma \quad \text{Eq. 3.9-12}$$

Integration of the equation gives

$$f = -\gamma z + c \quad \text{Eq. 3.9-13}$$

where c is a constant of integration. Substituting this equation into Eq. 3.9-10 gives

$$p(x, z) = -\frac{\rho a_1}{g_c} x - \gamma z + c \quad \text{Eq. 3.9-14}$$

To evaluate the constant c, assume that the z-coordinate of some point on the free surface is known. For example, suppose the z-coordinate of the free surface at $x = 0$ is z_1, as shown in Fig. 3-33b, then $p(0, z_1) = 0 = -\gamma z_1 + c$. Solving for c gives

$$c = \gamma z_1$$

Substituting this equation into Eq. 3.9-14 and noting that $\gamma = \rho g/g_c$, we obtain

$$p(x, z) = -\frac{\gamma a_1}{g} x + \gamma(z_1 - z) \qquad \text{Eq. 3.9-15}$$

To obtain the equation of the free surface, note that dp along the free surface is zero. This is so because p is zero everywhere on the free surface. Thus,

$$dp = 0 = \frac{\partial p}{\partial x} dx + \frac{\partial p}{\partial z} dz = -\frac{\gamma a_1}{g} dx - \gamma\, dz$$

or

$$\left(\frac{dz}{dx}\right)_{fs} = -\frac{a_1}{g} = \text{constant}$$

Therefore, the equation of the free surface is that of a straight line, or

$$z = -\frac{a_1}{g} x + c_1$$

Since $z(0) = z_1$, the constant c_1 can be evaluated; that is,

$$z_1 = c_1$$

or

$$z = -\frac{a_1}{g} x + z_1 \quad \text{(on the free surface)} \qquad \text{Eq. 3.9-16}$$

We can relate the elevation of the free surface at the front and rear of the tank in terms of the elevation of the free surface before acceleration by realizing that mass is conserved. That is, the cross-sectional area before and after acceleration must be equal. Thus,

$$z^*L = \tfrac{1}{2}(z_1 + z_2)L$$

or

$$z_2 = 2z^* - z_1 \qquad \text{Eq. 3.9-17}$$

But by Eq. 3.9-16

$$z(L) = z_2 = -\frac{a_1 L}{g} + z_1 \qquad \text{Eq. 3.9-18}$$

Adding Eqs. 3.9-17 and 3.9-18 gives

$$2z_2 = 2z^* - \frac{a_1 L}{g}$$

or

$$z_2 = z^* - \frac{a_1 L}{2g} \qquad \text{Eq. 3.9-19}$$

Subtracting Eq. 3.9-18 from Eq. 3.9-17 gives

$$0 = 2z^* - 2z_1 + \frac{a_1 L}{g}$$

or

$$z_1 = z^* + \frac{a_1 L}{2g} \qquad \text{Eq. 3.9-20}$$

For an open tank, as shown in Fig. 3-33, it is possible that a_1 will be sufficiently large and that some of the liquid will spill over the rear wall of the tank. When this occurs, mass is not conserved and Eqs. 3.9-17 through 3.9-20 are no longer valid. However, under these circumstances, the elevation z_1 becomes known, since it is equal to the height of the rear wall of the tank. Then z_2 can be determined from Eq. 3.9-16.

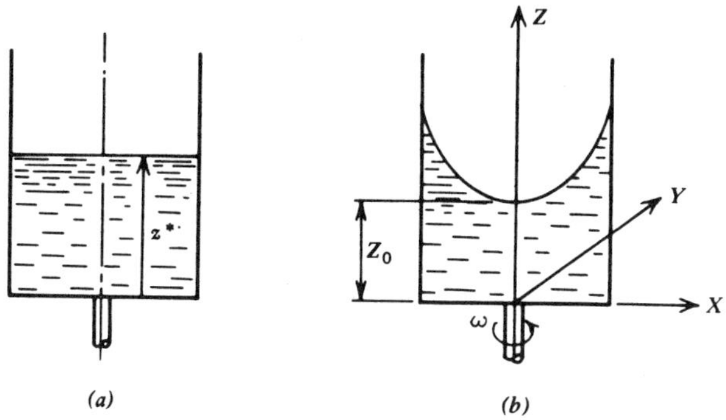

Fig. 3-34. Liquid in a rotating tank. (*a*) Tank motionless. (*b*) Tank rotating.

A liquid contained in a cylindrical tank that rotates at a constant angular speed behaves in a similar manner. Suppose that when the tank is motionless, the liquid level in the tank is z^*, as shown in Fig. 3-34*a*.

Suppose the tank is set rotating at a constant angular velocity. Again, experiments indicate that after an initial adjustment period the liquid will move as a rigid body with the free surface taking a parabolic shape as shown in Fig. 3-34*b*. In this problem, only a stationary coordinate system relative to the ground, with its origin on the axis of rotation, need be considered. This is so because $p \neq f(\theta)$. It is left as an exercise for the student to show that the pressure field is given by

$$p(r, Z) = -\gamma(Z - Z_0) + \frac{\gamma\omega^2}{2g} r^2 \qquad \text{Eq. 3.9-21}$$

where ω is the rotational speed and z_0 is the elevation of the free at $r = 0$ (see Fig. 3-34*b*). The equation of the free surface is

$$Z_{fs}(r) = Z_0 + \frac{\omega^2}{2g} r^2 \qquad \text{Eq. 3.9-22}$$

Problems

3-1 A cylindrical tank is filled to a depth of 30 ft with water of density 62.4 lb_m/ft^3. At sea-level conditions find the absolute and gage pressures in psi at the bottom of the tank.

3-2 If the tank of Problem 3-1 is moved to the top of Mount Everest, where the value of g is 30 ft/s^2, what is the gage pressure at the bottom of the tank?

3-3 Figure 3-35 shows a hydraulic lift used to raise an automobile. Piston A has a cross-sectional area of 2.6×10^{-3} m^2 and receives a downward force of 1110 N.

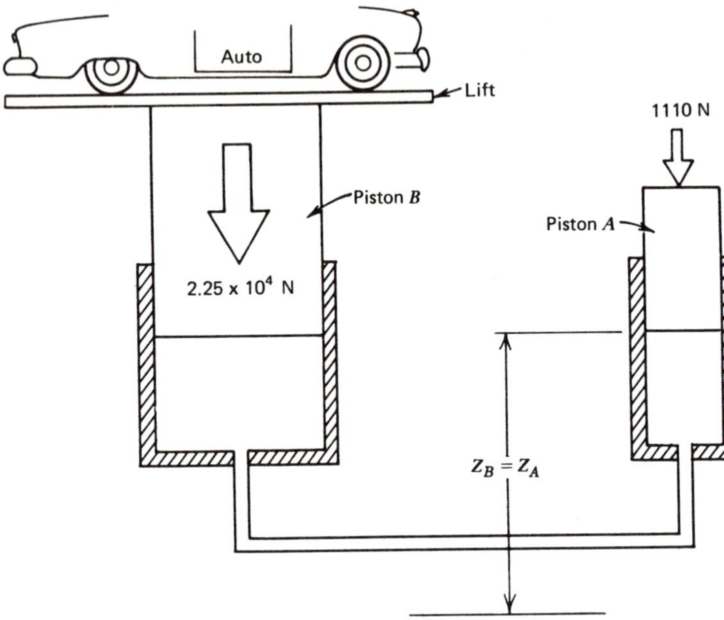

Fig. 3-35. Hydraulic lift.

What is the gage pressure on the face of piston A, neglecting the weight of the piston? What is the required area of piston B if the combined weight of the car and lift is 2.25×10^4 N.

3-4 What is the gage pressure on the face of piston B in Problem 3-3 if the fluid in the cylinders is oil with a specific gravity of 0.85?

3-5 A barometer similar to the one shown in Fig. 3-8 is constructed, using water as the barometer fluid. If the room temperature is at 70°F, what is the height of the column of water when the atmospheric pressure is (a) 14.0 psi and (b) 1.013×10^5 N/m^2?

3-6 Water is pumped from a well, using a vacuum pump. The pump is located at ground level and can maintain a pressure of 500 mm of mercury vacuum at the pump inlet. What is the maximum depth of the water surface below the ground level for which this pump will work?

Hint. Assume that at the maximum depth, the water velocity in the suction line is essentially zero.

3-7 If the water surface is twice as far below the ground level, as in Problem 3-6, can you suggest a system that will bring it to the surface?

3-8 What is the gage pressure at point *A* (see Fig. 3-36) if Fluid 1 in the pipe is water and Fluid 2 in the manometer is mercury with a specific gravity of 13.6? What is

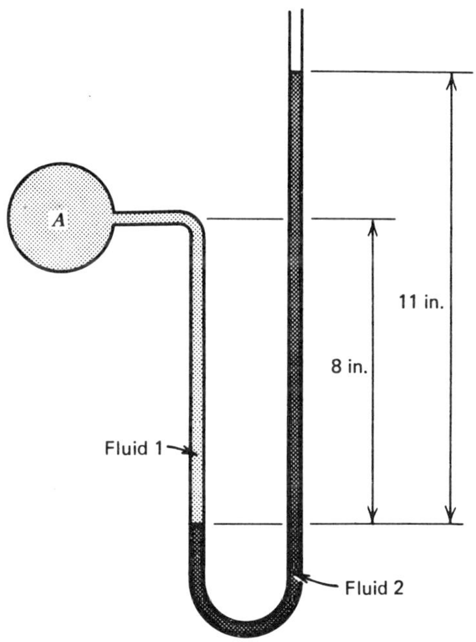

Fig. 3-36. Simple manometer.

the absolute pressure at *A* when the barometer reads 29.80 in. of mercury? Give your answers in psig and psia.

3-9 Repeat Problem 3-8 with Fluid 1 as air and Fluid 2 as water.

3-10 Repeat Problem 3-8 with Fluid 1 as air and Fluid 2 as oil of specific gravity 0.80.

3-11 Repeat Problem 3-8 with both Fluid 1 and Fluid 2 as water.

3-12 Flasks (1) and (2) are connected through a differential manometer, as shown in Fig. 3-37. What is the pressure difference between points A and B if Fluid 1 is oil

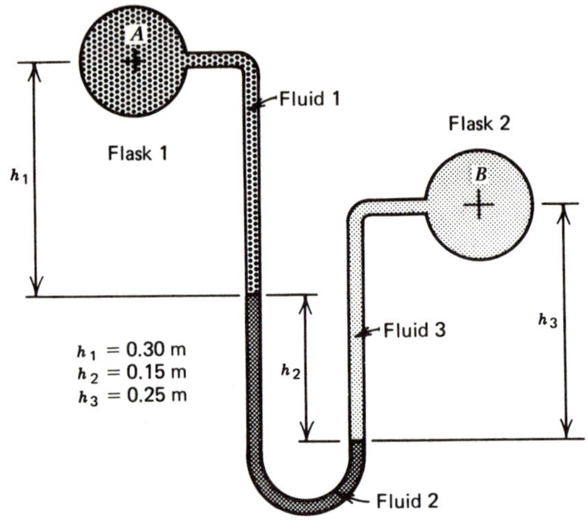

Fig. 3-37. Flasks connected by manometer.

with specific gravity 0.80, Fluid 2 is mercury, and Fluid 3 is water. Give your answer in psi.

3-13 Repeat Problem 3-12 with Fluid 1 and Fluid 3 both as water and Fluid 2 as mercury.

3-14 Repeat Problem 3-12 with Fluid 1 and Fluid 3 as air and Fluid 2 as oil with specific gravity 0.80.

3-15 Repeat Problem 3-14 with $h_1 = 609.6$ mm, $h_2 = 254.0$ mm, and $h_3 = 863.6$ mm.

3-16 What is the absolute pressure at point A in psia for the system shown in Fig. 3-38?

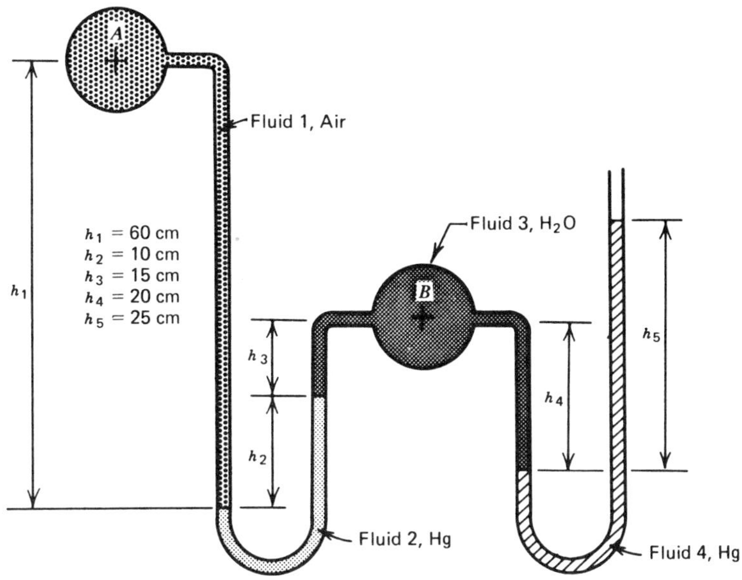

Fig. 3-38. Manometers connected to flasks.

3-17 Repeat Problem 3-16 where Fluid 1 and Fluid 3 are oil of specific gravity 0.80 and Fluid 2 and Fluid 4 are mercury.

3-18 A micromanometer is a device that provides a means for the measurement of small pressure differences. A sketch of the micromanometer is shown in Fig. 3-39. The principle behind the device is as follows: When ends (*c*) and (*d*) are open to the atmosphere, as shown in Fig. 3-39*a*, the manometer is filled with 2 immiscible liquids so that the levels of Fluids 1 and 2 are at the levels indicated

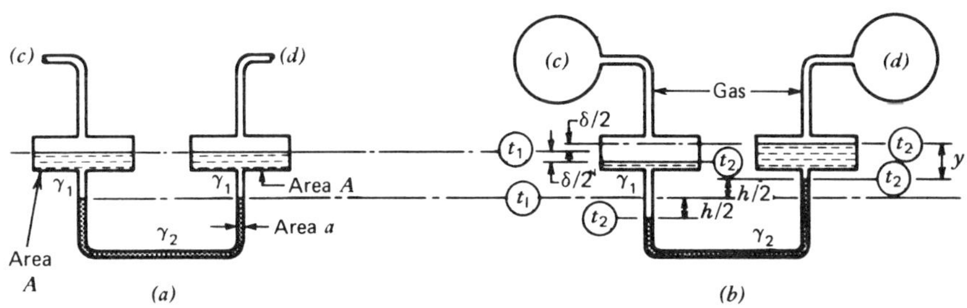

Fig. 3-39. The micromanometer. (*a*) Ends open to the atmosphere. (*b*) Ends connected to flasks.

by t_1. Now suppose ends (c) and (d) are connected to flasks, as shown in Fig. 3-39b, in which the pressures differ slightly. This will result in the fluid levels shifting to the levels indicated by t_2. If the cross-sectional areas of the reservoirs and tubes are A and a, respectively, (1) derive an expression for h in terms of $p_c, p_d, \gamma_1, \gamma_2, a, A$, justifying all steps, and (2) determine the specific weight, γ_2, for the following data: Diameter of reservoir = 3 in., diameter of U tube = $\frac{1}{2}$ in., $\gamma_1 = 55.00$ lb_f/ft^3, $p_c - p_d = 0.005$ psi, and $h = \frac{1}{5}$ in.

3-19 A tank 3 m high with a 1.5 m by 1.5 m base is filled with water to a depth of 2.45 m. Find the force exerted by the water on one side of the tank and the center of pressure.

3-20 Repeat Problem 3-19 with the tank filled to the top and under an initial pressure of 1.38×10^6 N/m^2 gage.

3-21 Find the normal force and the vertical and horizontal force components acting on the water side of the dam shown in Fig. 3-40. Also find the center of pressure.

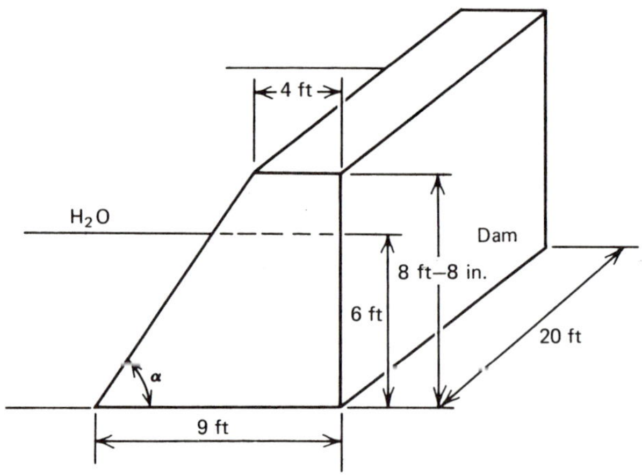

Fig. 3-40. A concrete dam.

3-22 A circular window, 1 ft in diameter, is located in a tank of water with its center 20 ft below the water surface. What is the force tending to push the window out? What is the position of the center of pressure?

3-23 Repeat Problem 3-22 but assume that the area of the circular window is 0.1 m^2 with its center 6 m below the water surface. Take the fluid to be oil with a specific gravity of 0.80.

3-24 A long rod is used to hold the gate in place, as shown in Fig. 3-41. What force must be applied to the rod to resist the water pressure? The gate is 5 ft wide.

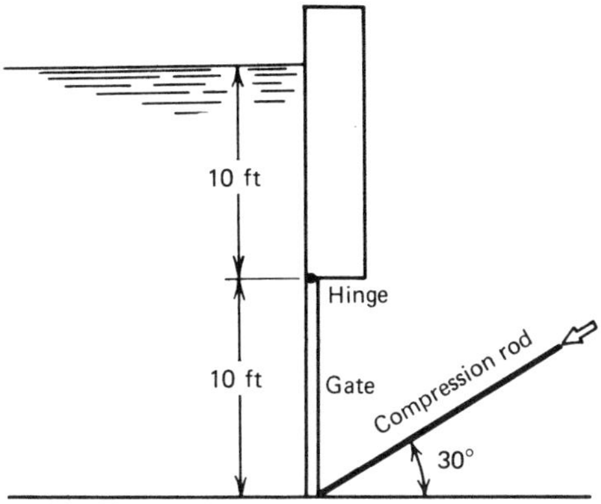

Fig. 3-41. Gate controlling water level.

3-25 What is the magnitude of the tensile force in the rod holding together the two sides of the aqueduct shown in Fig. 3-42 if the rods are 3 m on centers (one rod for each 3 m of aqueduct).

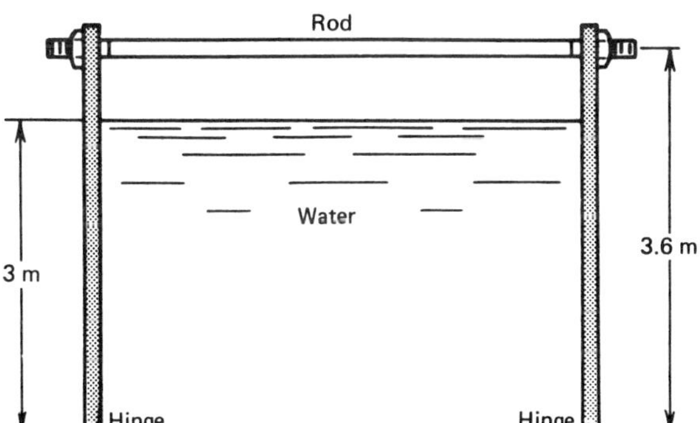

Fig. 3-42. Aqueduct filled with water.

3-26 Repeat Problem 3-25 but assume that the sides of the aqueduct are inclined at an angle of 45° with the horizontal. Take the vertical distances to be the same.

3-27 Find the force, **B**, necessary to keep the gate, shown in Fig. 3-43, closed.

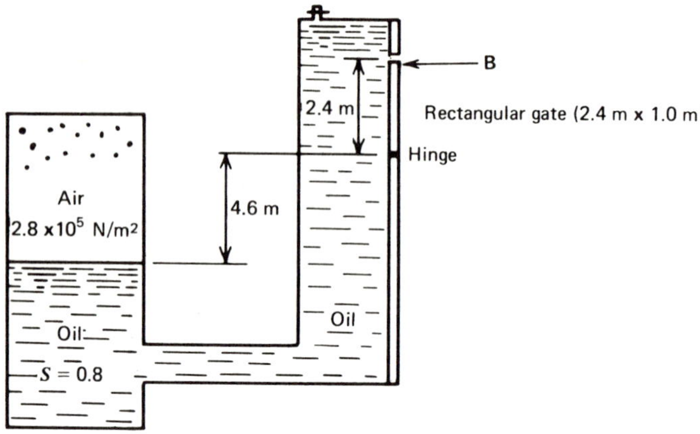

Fig. 3-43. Gate in a hydraulic device.

3-28 A circular gate, as shown in Fig. 3-44, is submerged in a liquid whose specific gravity is 0.8. The gate is hinged through its diameter and has a lock located at

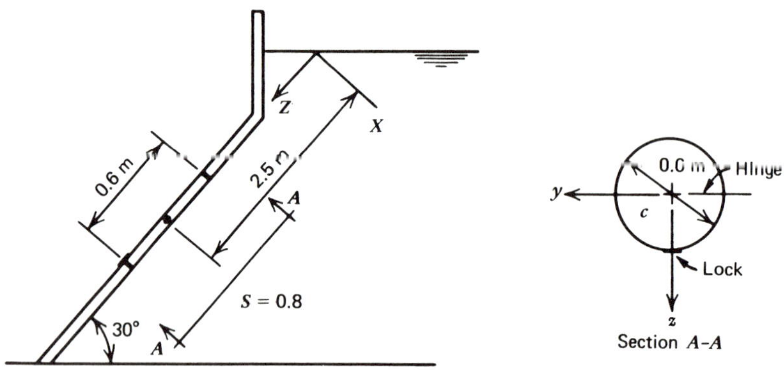

Fig. 3-44. Gate in a dam.

the very bottom of the gate, which keeps the gate closed. What force is exerted on the lock? Assume that the lock exerts a point load at the very bottom of the gate.

3-29 In the 1930s it was not uncommon for a sewage disposal system to be designed so that the storm sewer (line *A*) and the sanitary sewer (line *B*) both fed into a storage basin of the sewage disposal plant, as shown in Fig. 3-45*a*. The storage basin might have two outlets, one to the sewage treatment facilities (line *C*) and one directly to a lake (line *D*). Under normal conditions a flow through line *D* would be blocked by a gate located at the beginning of the line. Suppose the gate was constructed, as shown in Fig. 3-45*b*, so that it would automatically open when the water level in the basin reached a certain level, *h*. (*a*) For the dimensions shown in Fig. 3-45*b*, determine *h*. (*b*) What is your opinion regarding the design of such a system?

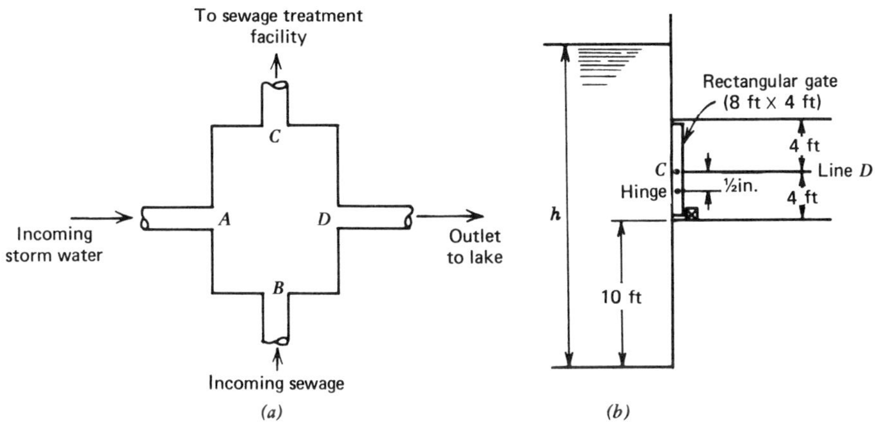

Fig. 3-45. Storage basin in sewage disposal plant. (*a*) Layout of storage basin. (*b*) Gate in line *D*.

3-30 Consider a small spherical water droplet falling very slowly through the atmosphere. By taking a section through the center of the droplet, as shown in Fig. 3-46, show that the difference in pressure between the interior of the droplet and

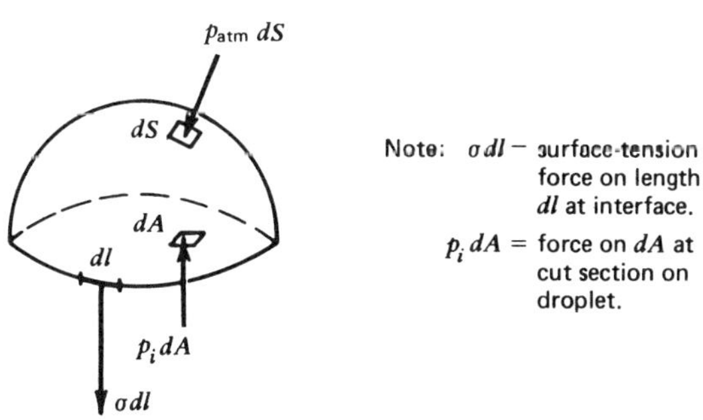

Fig. 3-46. Section of spherical water droplet.

the surrounding atmosphere, resulting from surface tension, is given by

$$p_i - p_{atm} = \frac{2\sigma}{r}$$

where σ is the surface-tension coefficient, p_i is the pressure inside the droplet, and p_{atm} is the atmospheric pressure. Neglect all dynamic effects.

3-31 A 6 m by 20 m barge, 1.5 m deep, has an average density of 320 kg/m^3 over its entire volume. To what depth will it sink in water?

3-32 If the barge of Problem 3-31 is loaded with 8×10^3 kg of coal evenly distributed, what will be the total depth to which it sinks?

3-33 The drum shown in the Fig. 3-47 weighs 220 N. The cross-sectional area is 0.372 m^2, and the wall thickness is assumed to be negligible. If it is filled with oil of

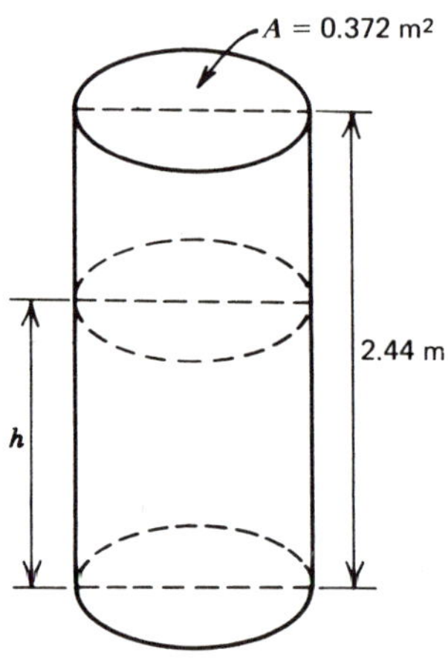

Fig. 3-47. Oil drum.

specific gravity 0.80, will it sink in water? If it will, to what depth, h, must it be filled to just float even with the surface? If not, assuming it to float upright, how much of the drum will be above water when it is filled with oil?

3-34 Repeat Problem 3-33, assuming the fluid in the drum to be water?

3-35 If the free surface of the surrounding water lies 0.60 m below the top of the cylinder, to what depth must the drum of Problem 3-33 be filled with water to float in an upright position?

3-36 A hydrometer is made up of two main regions (see Fig. 3-48). One region is a hollow sphere having a diameter, D, of 2 in. This region is to be filled with, N,

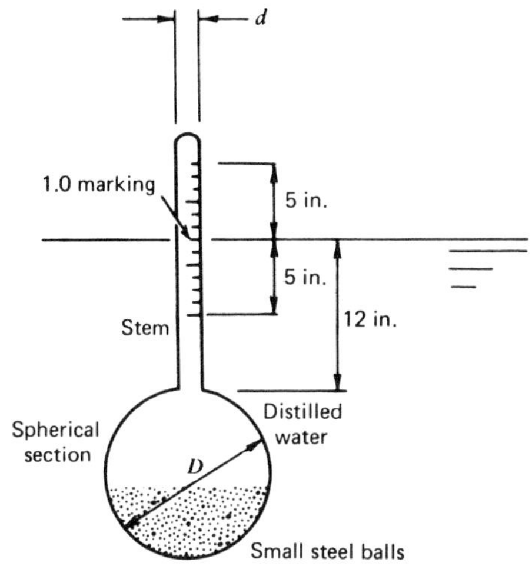

Fig. 3-48. A hydrometer.

number of small steel balls. Each ball is to have a diameter, b, of $\frac{1}{8}$ in. The second region, the stem, is a circular cylinder having a diameter, d, of $\frac{3}{8}$ in. Specific gravity markings, S, are to be etched on the stem. The 1.0 reading is 12 in. above the top of the sphere. (*a*) Find N. Neglect weight of glass. (*b*) The maximum and minimum S readings are to be 5 in. above and below the 1.0 reading. Find the maximum and minimum S readings. (*c*) State whether the maximum S reading will appear above or below the 1.0 reading. (*d*) Can the S scale be considered, for all practical purposes, as linear? Explain and demonstrate by evaluating an appropriate term. Note that the specific weight of steel is 490 lb_f/ft^3 and the volume of a sphere is $4\pi R^3/3$, where R is the radius of the sphere.

3-37 A solid circular cylinder, 20 cm in diameter and 10 cm tall has a specific gravity of 0.6. If the cylinder floats in still water and is then displaced downward 2 cm, determine (*a*) the governing equation of motion, (*b*) the displacement from the equilibrium position as a function of time, and (*c*) the frequency of oscillation.

3-38 If in Example 3-5, $\theta_0 = 0.2$ radians and $I_{YY} = 5.2 \times 10^6$ lb_m-ft^2, determine the maximum rotational speed of the rolling barge.

3-39 Suppose the floating barge in Example 3-5 was given a sudden pitching disturbance (a rotation about the X-axis) instead of the rolling disturbance. Neglecting viscous damping, determine the differential equation governing the motion.

3-40 A barge, having a uniform cross section, as shown in Fig. 3-49, is 20 ft long. (*a*) Find the weight of the barge as loaded. (*b*) When the barge is in its upright

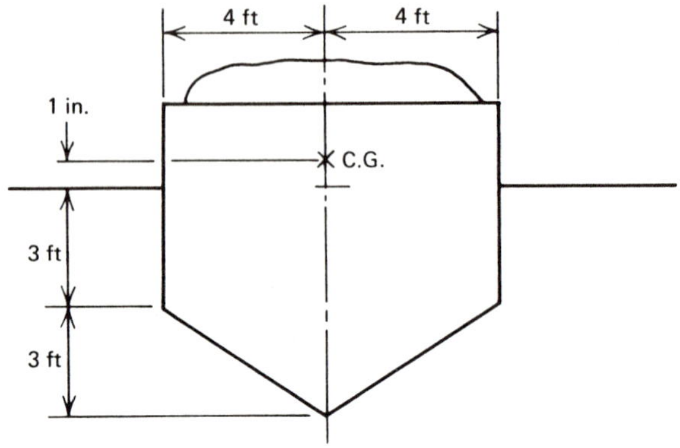

Fig. 3-49. Barge cross section.

position, its center of gravity is 1 in. above the waterline. If the barge is given a sudden and momentary rolling disturbance, determine the governing equation of motion.

3-41 This problem is a modification of the computer problem described in Section 3.7.1. Suppose in Example 3-5, the center of gravity of the barge and load were located $1\frac{1}{4}$ in. above the waterline, then the moment, M_2, acting on the barge would be given by Eq. 3.7-34. Assume the following initial conditions for the problem:

$$\theta(0) = 0 \qquad \text{and} \qquad \frac{d\theta}{dt}(0) = 0.01 \text{ rad/s}$$

and take $I_{YY} = 5.2 \times 10^6$ $\text{lb}_m\text{-ft}^2$. (*a*) Formulate an expression for the period of oscillation. Note that the oscillation is symmetrical about θ equal to zero. (*b*) Determine values for the maximum and minimum displacement angles, say θ_1 and θ_2, respectively. (*c*) For the improper integrals obtained in part (*a*), carry out the suggested method of treatment described in Section 3.7.1. (*d*) Develop a computer program for evaluating the period, τ.

3-42 Using Eqs. 3.8-8 and 3.8-9 (or 3.8-12) and the ideal gas law, find the pressure, temperature, and density for the U.S. Standard Atmosphere at an altitude of (*a*) 20,000 ft, (*b*) 15 km, and (*c*) 70 km.

3-43 The gravitational constant, g, varies with the altitude according to the following equation:

$$g = g_0\left(1 + \frac{z}{r_0}\right)^{-2} \approx g_0\left(1 - \frac{2z}{r_0}\right); \qquad \text{for } z << r_0$$

where r_0 = mean radius of the earth
= 2.090×10^7 ft = 6.3714×10^3 km
z = altitude measured from sea level
g_0 = gravitational constant at sea level = 32.174 ft/s²

(*a*) Show that if the above equation for g is used in developing the pressure-altitude equation for a nonzero lapse rate region, then Eq. 3.8-7 is replaced by

$$\ln \frac{p}{p_i} = \frac{g_0}{g_c R \lambda_S}\left[\left(1 - \frac{2(T_i + \lambda_S z_i)}{r_0 \lambda_S}\right) \ln\left(\frac{T}{T_i}\right) - \frac{2(z - z_i)}{r_0}\right]$$

where T is still given by Eq. 3.8-8. (*b*) Compare the pressure determined from the above equation at an altitude of 11,020 m to that given in Table 3-1.

3-44 A temperature inversion over a region of interest results in an actual atmospheric lapse rate of -1.5×10^{-3} R/ft over an altitude range from sea level to 10,000 ft. At an altitude of 10,000 ft the lapse rate abruptly changes to 3.6×10^{-3} R/ft, and stays at this value to an altitude of 50,000 ft. (*a*) Write a relationship for pressure versus altitude and temperature versus altitude for $0 \leq z \leq 10{,}000$ ft. (*b*) determine the actual pressure and temperature at $z = 25{,}000$ ft. Take $p_0 = 14.696$ psia and $T_0 = 519$ R.

3-45 A tank, 4 ft long, 2 ft wide, and 1 ft deep, is filled with water to a depth of 6 in. If the tank is given an acceleration of 8 ft/s² in the direction of its length, determine: (*a*) The angle the free surface makes with the horizontal. (*b*) The quantity of water spilled. (c) The forces the water exerts on the front and rear walls.

3-46 Repeat parts (*a*) and (*b*) of Problem 3-45; however, take the tank to be 2.5 ft deep, the acceleration to be 16.1 ft/s², and the water depth before acceleration to be 2.0 ft.

3-47 A tank open to the atmosphere at the top and only at one edge, as shown in Fig. 3-50, is given an acceleration of 5.661 m/s² in the X direction. Determine the pressures at points A, B, and C.

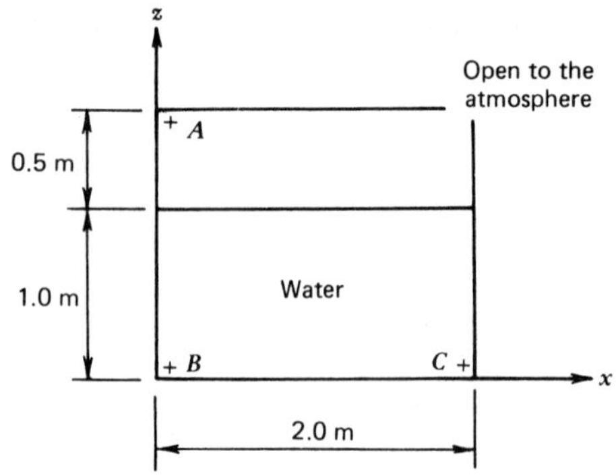

Fig. 3-50. Tank with opening on top surface.

3-48 A cylindrical tank 1 ft in diameter and 3 ft high is half full of water. At what angular speed must it be rotated about its axis to spill water.

Chapter Four

Flow Fields and the Fundamental Laws

4.1 INTRODUCTION

In this chapter we are mainly concerned with the study of the fundamental laws of fluid mechanics and their application to engineering problems. However, the presentation of the fundamental laws is preceded by the introduction of some flow-field terminology, including the concepts of a thermodynamic system, a control volume, and their mathematical significance.

4.1.1 SOME FLOW-FIELD TERMINOLOGY

An *ideal fluid* is a nonviscous incompressible fluid. Many flow situations may be approximated by considering the fluid to be an ideal fluid. A *flow field* is a region where the fluid properties (density, ρ; fluid velocity, $\mathbf{v}$; pressure, p; temperature, T; etc.) may be functions of position and time. There are two viewpoints for describing a flow field, *Lagrangian* and *Eulerian*. In the *Lagrangian viewpoint*, a single fluid particle or element is observed as it moves through the flow field. This approach is invariably used in the study of particle dynamics. In the *Eulerian viewpoint*, fluid particles are observed as they move into a particular point in space. It is the latter that is used more often in the study of fluid mechanics and the viewpoint used in this book.

If the fluid properties at every point in space do not vary with time, then the flow is said to be *steady*. The determination of whether a flow is steady depends on the frame of reference from which the fluid is being observed. For example,

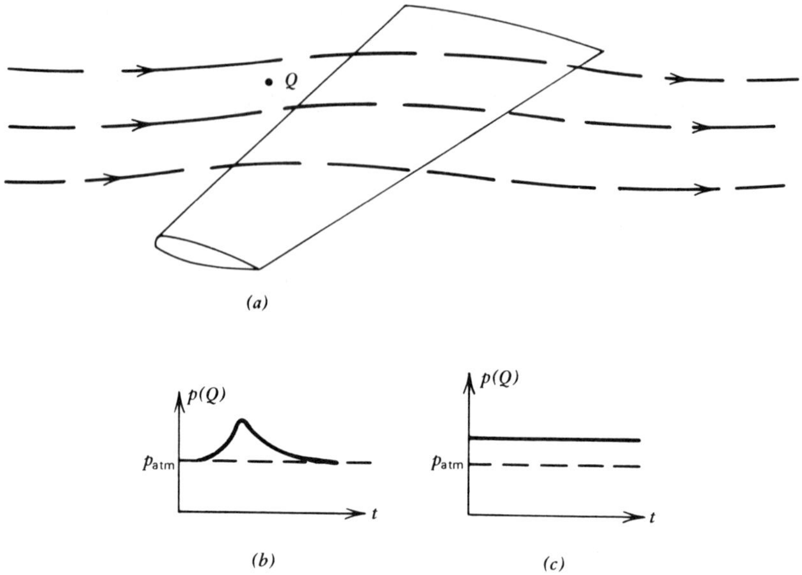

Fig. 4-1. (*a*) Flow around a wing. (*b*) *Q* fixed to ground. (*c*) *Q* fixed to wing.

suppose the fluid properties are observed at a fixed point, *Q*, relative to the ground as an airplane moves past it at constant speed. Clearly the properties at *Q* will vary with time. As seen from this coordinate system, the flow is unsteady (see Figs. 4-1*a* and 4-1*b*). On the other hand, if a coordinate system is attached to the airplane and the fluid properties are observed at a fixed point, relative to this coordinate system, they would not vary with time. The flow as seen from this coordinate system, is steady (see Figs. 4-1*a* and 4-1*c*).

If the fluid properties are the same everywhere in the flow field, the flow is said to be *uniform*. Nearly all real flows are three-dimensional. However, sufficient accuracy is often obtained by considering the flow to be either two-dimensional or one-dimensional. For example, consider an airflow around a wing. This type of flow is often treated as a two-dimensional problem. Although a two-dimensional description may be inaccurate near the wing tip, sufficient accuracy is obtained over the greater portion of the wing to justify its use.

A *streamline* is a curve in the flow field whose tangent points in the direction of the fluid-velocity vector at a given instant of time. If the flow is steady then a fluid particle will move along a streamline, since it moves in the direction of its velocity vector. To visualize the flow, smoke, dye, or minute hydrogen bubbles may be injected into a flow field. These particles are sufficiently small to be

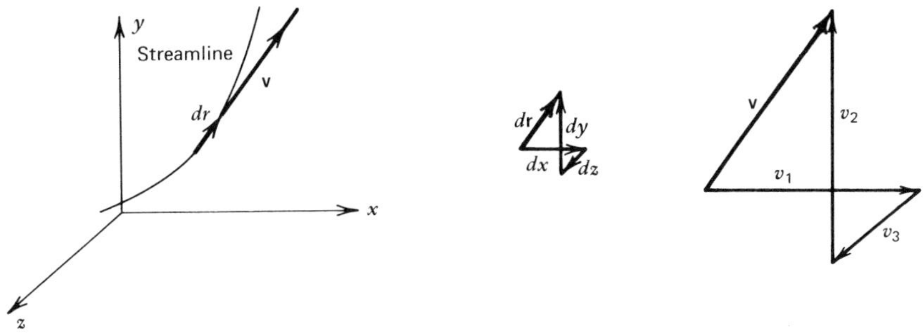

Fig. 4-2. Displacement along streamline and the velocity vector.

carried along with the surrounding stream. Their visibility provides a means of observing the flow pattern.

For a steady flow, the equation for a streamline in differential form can be obtained in terms of the velocity components. If $d\mathbf{r} = dx\,\mathbf{i} + dy\,\mathbf{j} + dz\,\mathbf{k}$ is a displacement vector along the streamline and $\mathbf{v} = v_1\mathbf{i} + v_2\mathbf{j} + v_3\mathbf{k}$ is the fluid velocity, then $d\mathbf{r}$ equals some scalar multiple of $\mathbf{v}$, since both vectors point in the same direction. Thus, $\mathbf{v} = \alpha\, d\mathbf{r}$, where $\alpha =$ constant and

$$\frac{dx}{v_1} = \frac{dy}{v_2} = \frac{dz}{v_3} = \frac{1}{\alpha} \qquad \text{Eq. 4.1-1}$$

See Fig. 4-2.

Example 4-1

Suppose a two-dimensional velocity field exists in the region $x \geq 0$ and $y \geq 0$. The velocity field is $v_1 = x$ and $v_2 = -y$. Find the equation of the streamline passing through the point (3,4).

SOLUTION

From Eq. 4.1-1, the differential equation for the streamline can be obtained; that is,

$$\frac{dy}{dx} = \frac{v_2}{v_1} = -\frac{y}{x}$$

The variables can be separated giving

$$\frac{dy}{y} = -\frac{dx}{x}$$

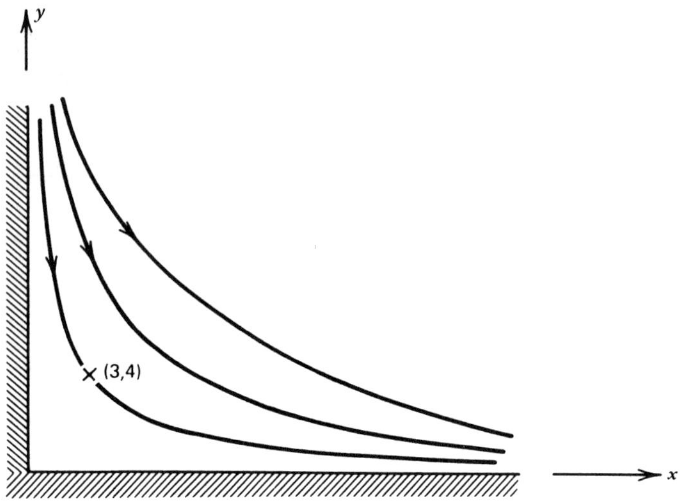

Fig. 4-3. Flow around a 90° bend.

Integration gives

$$\ln y = -\ln x + \ln C$$

or

$$yx = C$$

By taking different numerical values for C, a family of curves is obtained, as shown in Fig. 4-3. To obtain the value of C for the curve passing through the point (3,4) substitute (3,4) for (x,y) in the above equation. This gives $C = 12$. Thus, the equation of the streamline is $y = 12/x$. The family of streamlines gives the flow pattern. The direction of the flow can be determined by noting that $v_1 \geq 0$ and $v_2 \leq 0$.

It should also be realized that as the value of C gets smaller and smaller, the streamlines move closer and closer to the solid wall. Finally, it should be noted that if the solid wall was to be moved to a streamline, then the streamlines outside the wall would remain intact.

4.1.2 THE FLUX VECTOR

When a fluid element moves, it carries with itself mass, momentum, and energy. Thus, when there is a flow of fluid, there is a transport of these three important quantities. A convenient way of expressing this transport is through the use of an appropriate flux vector. This is described below.

Let surface ΔS be a surface element that is perpendicular to the flow direction of some physical quantity, such as mass, heat, momentum component, energy, and the like. Let $\mathbf{A}$ (x, y, z, t) be a vector whose magnitude equals the $\lim_{\Delta t \to 0}$ $\lim_{\Delta S \to 0}$ {of the amount of the physical quantity of interest, which in Δt flows through the surface area, ΔS, per unit time per unit surface area}. Also let $\mathbf{A}$ point in the direction of the flow of that physical quantity. *Then* $\mathbf{A}$ is the flux vector associated with that physical quantity of interest.

This concept can most readily be understood when $\mathbf{A}$ is the mass-flux vector. Let us determine the mass-flux vector. Let $\mathbf{v}$ = fluid velocity and ρ = mass density of the fluid. The fluid-velocity vector written in terms of its (x, y, z) components is $\mathbf{v} = v_1\mathbf{i} + v_2\mathbf{j} + v_3\mathbf{k}$. Construct lines in the direction of flow and take ΔS to be perpendicular to the flow direction (see Fig. 4-4).

The amount of mass crossing ΔS in $\Delta t = \rho\ \Delta V\!\!\!\!-$, where

$$\Delta V\!\!\!\!- = \Delta S|\mathbf{v}|\Delta t \qquad \text{Eq. 4.1-2}$$

By definition, $|\mathbf{A}| = \lim_{\Delta t \to 0} \lim_{\Delta S \to 0}$ {the amount of mass that crosses ΔS in Δt per unit area per unit time}; that is,

$$|\mathbf{A}| = \lim_{\Delta t \to 0} \lim_{\Delta S \to 0} \frac{\rho\ \Delta V\!\!\!\!-}{\Delta S \cdot \Delta t} = \lim_{\Delta t \to 0} \lim_{\Delta S \to 0} \frac{\rho\ \Delta S|\mathbf{v}|\Delta t}{\Delta S \cdot \Delta t} = \rho|\mathbf{v}| \qquad \text{Eq. 4.1-3}$$

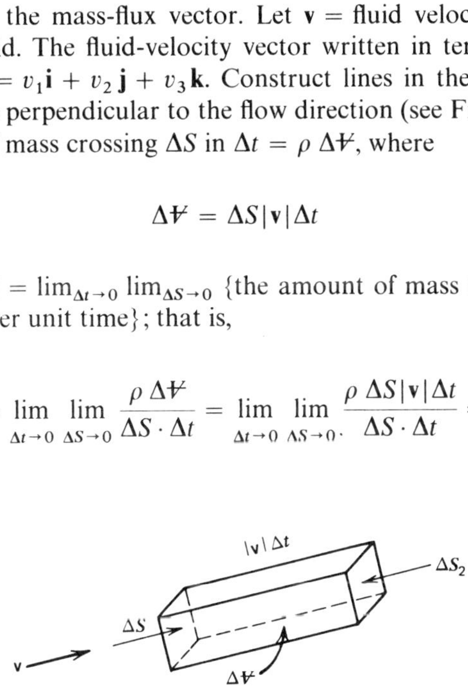

Fig. 4-4. Flow through surface that is perpendicular to the flow direction. At t, the leading edge of an element of mass is located at surface ΔS. At $(t + \Delta t)$, the leading edge of that same element of mass is located at ΔS_2.

Note that the rate that mass crosses ΔS is the same as the $\lim_{\Delta t \to 0}$ {of the amount of mass that in Δt crosses ΔS divided by Δt}. Also note that $\mathbf{A}$ is to point in the direction of flow. The direction of flow is in the $\mathbf{v}$ direction thus

$$\text{Mass-flux vector} = \rho \mathbf{v} \qquad \text{Eq. 4.1-4}$$

Now suppose we were interested in determining the energy-flux vector. A more detailed discussion of energy and the first law of thermodynamics are given later in this chapter. It is expected that students will be able to understand the following concepts based on their studies in thermodynamics. We will consider an element of fluid to have internal, kinetic, and potential energy. Let u = the internal energy per unit mass, z = the elevation in the opposite direction to the gravitational field, and g = the acceleration of a free-falling body. Now let $\mathbf{A}$ represent the energy-flux vector. Referring to Fig. 4-4, it can be seen that the rate that energy crosses ΔS per unit surface area is

$$|\mathbf{A}| = \lim_{\Delta t \to 0} \lim_{\Delta S \to 0} \frac{\rho\left(u + \dfrac{|\mathbf{v}|^2}{2g_c} + \dfrac{g}{g_c} z\right) \Delta \mathcal{V}}{\Delta t \cdot \Delta S}$$

$$= \lim_{\Delta t \to 0} \lim_{\Delta S \to 0} \frac{\rho\left(u + \dfrac{|\mathbf{v}|^2}{2g_c} + \dfrac{gz}{gc}\right) \cancel{\Delta S} |\mathbf{v}| \cancel{\Delta t}}{\cancel{\Delta t} \cdot \cancel{\Delta S}} \qquad \text{Eq. 4.1-5}$$

The direction of energy flow is in the direction of $\mathbf{v}$. Thus,

$$\text{Energy-flux vector} = \rho\left(u + \frac{|\mathbf{v}|^2}{2g_c} + \frac{gz}{g_c}\right)\mathbf{v} \qquad \text{Eq. 4.1-6}$$

In a similar manner, it can be shown that

$$\text{Flux vector associated with the } \begin{bmatrix} x \\ y \\ z \end{bmatrix} \text{ component of momentum} = \begin{bmatrix} \rho v_1 \mathbf{v} \\ \rho v_2 \mathbf{v} \\ \rho v_3 \mathbf{v} \end{bmatrix}$$

Eqs. 4.1-7

Now let us consider the reverse problem. Suppose the mass-flux vector, $\mathbf{A}$, is known to be $\rho\mathbf{v}$. Find the rate at which mass crosses an element of surface, ΔS, when ΔS is arbitrarily oriented in space (see Fig. 4-5). Observe that

$$\begin{matrix}\text{Rate at which mass} \\ \text{crosses the surface } \Delta S\end{matrix} = \lim_{\Delta t \to 0} \frac{\rho\, \Delta \mathcal{V}}{\Delta t} \qquad \text{Eq. 4.1-8}$$

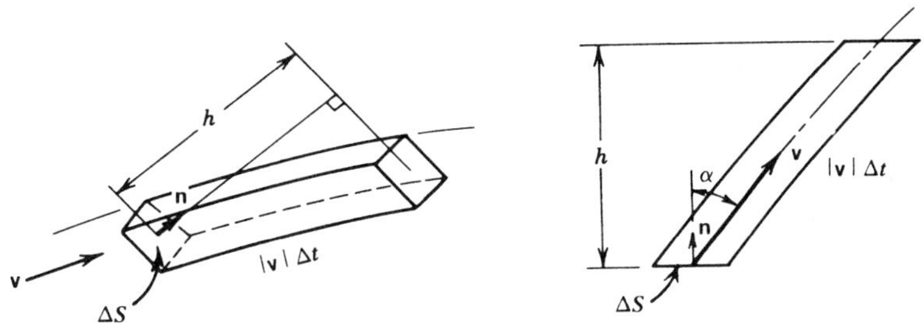

Fig. 4-5. Flow through an arbitrarily oriented surface. The volume is rotated to give a clearer picture for determining $\Delta\forall$.

But $\Delta\forall = \Delta S \cdot h = \Delta S|\mathbf{v}|\Delta t \cos\alpha$. Since $\mathbf{v}\cdot\mathbf{n} = |\mathbf{v}|\,|\mathbf{n}|\cos\alpha = |\mathbf{v}|\cos\alpha$, we see that

$$\mathbf{v}\cdot\Delta\mathbf{S} = \mathbf{v}\cdot\mathbf{n}\,\Delta S = |\mathbf{v}|\Delta S\cos\alpha$$

and

$$\Delta\forall = (\mathbf{v}\cdot\Delta\mathbf{S})\,\Delta t$$

Substituting the above expression into Eq. 4.1-8, we obtain

$$\begin{matrix}\text{Rate at which mass}\\ \text{crosses the surface } \Delta S\end{matrix} = \lim_{\Delta t\to 0}\frac{\rho\mathbf{v}\cdot\Delta\mathbf{S}\,\Delta t}{\Delta t} = \rho\mathbf{v}\cdot\Delta\mathbf{S}$$

and

$$\begin{matrix}\text{Rate at which mass}\\ \text{crosses a finite surface } S\end{matrix} = \oiint_S \rho\mathbf{v}\cdot d\mathbf{S} \qquad \text{Eq. 4.1-9}$$

In general, the rate (amount/unit time) at which flux crosses a surface area, ΔS, is $\mathbf{A}\cdot d\mathbf{S}$, and

$$\begin{matrix}\text{Net rate at which flux}\\ \text{crosses a finite surface } S\end{matrix} = \oiint_S \mathbf{A}\cdot d\mathbf{S} \qquad \text{Eq. 4.1-10}$$

The flow of thermal energy by conduction, often referred to as conductive-heat transfer, can also be represented by a flux vector. We designate this quantity as the heat-flux vector, **q**. An attempt to develop the heat-flux vector in the same manner as the previous flux vectors would fail, because the transport of thermal energy by conduction results from molecular collisions in gases, a drift of electrons in metals, and changes in the lattice vibrations in insulating materials; whereas the other flux vectors result from macroscopic fluid motion. Despite the difficulty in visualizing this quantity, it can still be defined in the same manner as the other flux vectors. That is, the heat-flux vector, **q**, is defined as follows: Let ΔS be an element of surface area that is perpendicular to direction of heat flow. Then,

$|\mathbf{q}| = \lim_{\Delta S \to 0}$ {of the rate at which heat crosses the element of surface area, ΔS, per unit surface area} and **q** points in the direction of heat flow.

Then, by Eq. 4.1-10,

$$\begin{matrix}\text{Net rate at which heat flows} \\ \text{by conduction through surface } S\end{matrix} = \oiint_S \mathbf{q} \cdot d\mathbf{S} \qquad \text{Eq. 4.1-11}$$

If the surface S encloses a region of interest, then

$$\begin{matrix}\text{Net rate at which heat flows } \textit{out} \\ \text{of the region of interest through } S\end{matrix} = \oiint_S \mathbf{q} \cdot d\mathbf{S}$$

and

$$\begin{matrix}\text{Net rate at which heat flows} \\ \text{into the region of interest through } S\end{matrix} = -\oiint_S \mathbf{q} \cdot d\mathbf{S}$$

The relationship of **q** to other thermodynamic properties is presented in courses concerned with heat transfer.

Finally, the student should recall from thermodynamics, that the second law of thermodynamics postulates the existence of a thermodynamic extensive variable called entropy, designated as $\mathscr{S}$ and defined by,

$$d\mathscr{S} = \frac{1}{T}\,dQ_{\text{rev}}$$

where dQ_{rev} is the amount of heat transferred to a system in a reversible process. Suppose we consider the infinitesimal system shown in Fig. 4-6, occupied by

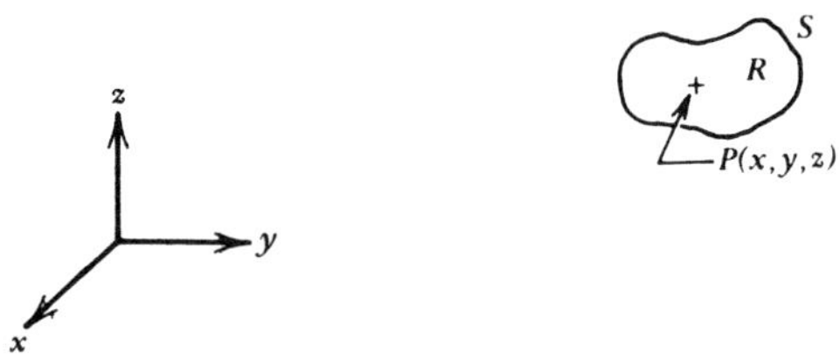

Fig. 4-6. An infinitesimal region enclosing point $P(x, y, z)$.

region, R, with bounding surface, S. If dQ_{rev} is the amount of heat transferred to the system in Δt in a reversible process, then

$$dQ_{\text{rev}} = -\Delta t \oiint_S \mathbf{q} \cdot d\mathbf{S}$$

The increase of entropy of the system, $d\mathscr{S}_{\text{rev}}$, is then

$$d\mathscr{S}_{\text{rev}} = -\frac{\Delta t}{T} \oiint_S \mathbf{q} \cdot d\mathbf{S}$$

For an infinitesimal region, T can be taken as a constant and brought inside the integral sign; that is,

$$d\mathscr{S}_{\text{rev}} = -\Delta t \oiint_S \frac{1}{T} \mathbf{q} \cdot d\mathbf{S}$$

Since $\left(\frac{1}{T} \mathbf{q} \cdot d\mathbf{S}\right)$ gives the rate at which entropy flows out of the region through the surface dS, we can take $\mathbf{q}/T$ to be the entropy-flux vector. For a general process, an additional term will have to be included to account for entropy generation. This is discussed more fully in Section 4.10.

4.2 FUNDAMENTAL LAWS, SYSTEMS, AND CONTROL VOLUMES AND THEIR MATHEMATICAL SIGNIFICANCE

A fundamental law is one which is always applicable, regardless of the material being considered. The fundamental laws are: (1) conservation of mass, (2) conservation of charge, (3) Newton's laws of motion, (4) the first law of thermodynamics, (5) Maxwell's electromagnetic equations, (6) the second law of thermodynamics. In the study of fluid flow, laws (2) and (5) are not relevant unless the fluid medium contains charged particles, as in a plasma. The study of a plasma medium is beyond the scope of this book.

The fundamental laws are most frequently stated for a *thermodynamic system.* A thermodynamic system is defined to be some definite quantity of matter that is separated from the rest of the continuum by a real or imaginary boundary. The selection of a thermodynamic system may be made solely for the purpose of convenience in problem formulation. The position and shape of a system may change with time.

The fundamental laws may, alternatively, be stated for a control volume. *A control volume* is some fixed region in space, as seen from a particular coordinate system. Matter within the control volume may change with time. Thus, the difference between a system and a control volume is that the latter specifies a region, whereas the former specifies an amount of matter. In dealing with some flow situations it may be more convenient to utilize the control volume approach, since it negates the necessity for identifying some definite quantity of matter.

In dealing with a particular fluid-flow problem, either approach may be used. Regardless of which approach is used, the end results must be the same. Each method has some advantage or disadvantage over the other. For example, a mathematical expression of the fundamental laws invariably involves a time derivative of some volume integral. This operation turns out to be more complicated when the systems approach is used. This is so because the limits of integration of a moving system vary with time. On the other hand, when the control volume approach is used, additional terms must be added to the conventional expression of the thermodynamic laws to account for the fact that the amount of matter within the control volume may vary with time. The meaning of this last paragraph will become more evident as the reader gets deeper into the chapter. Since all the fundamental laws involved in fluid mechanics contain a term that is the time derivative of an integral of a thermodynamic system, we examine this concept next.

Let $\phi(x, y, z, t)$ represent one of the following three physical intensive[1] scalar quantities, (1) ρ, (2) ρv_i, and (3) $\rho(u + v^2/2g_c + gz/g_c)$, where v_i is any of the three components of the velocity vector. Select an arbitrary system within the continuum. Let $\Phi(t)$ be the integral of $\phi(x, y, z, t)$ over $\mathcal{V}_{sys}$, the region occupied by the system at t. That is,

$$\Phi(t) = \iiint_{\mathcal{V}_{sys}} \phi(x, y, z, t)\, d\mathcal{V} \qquad \text{Eq. 4.2-1}$$

[1] Here intensive means a physical quantity per unit volume, such as mass per unit volume or internal energy per unit volume, and so on. For example, if ρ is the mass density in (lb_m/ft^3) and u is the internal energy in (Btu/lb_m), then (ρu) has units of (Btu/ft^3).

There the meaning of Φ, which corresponds to the three physical intensive quantities mentioned above, is: (1) the total mass in the system, (2) the ith component of the total momentum in the system, and (3) the total energy in the system. For definiteness let $\phi(x, y, z, t) = \rho(x, y, z, t)$ and $\Phi(t) =$ the total mass in the system, $M(t)$. Since the fundamental laws, as we will soon discover, involve some statement with respect to $d\Phi/dt$, we need to know how to evaluate this term. That is, we want to obtain an expression for

$$\frac{d}{dt} \iiint_{\mathcal{V}_{sys}} \rho(x, y, z, t)\, d\mathcal{V}$$

It should be observed that not only is the integrand a function of time, but so are the limits of integration. This is so because the system is moving in space. To determine an expression for this term, we turn to the definition of the derivative,

$$\frac{dM}{dt} = \lim_{\Delta t \to 0} \frac{M(t + \Delta t) - M(t)}{\Delta t} \qquad \text{Eq. 4.2-2}$$

At t the system occupies region $\mathcal{V}_1$ and $\mathcal{V}_2$, as shown in Fig. 4-7. At $(t + \Delta t)$ the system occupies regions $\mathcal{V}_2$ and $\mathcal{V}_3$. Note that region $\mathcal{V}_2$ is part of this system both at t and at $(t + \Delta t)$.

Let S_{sys} be the surface of the system at time t. Also let S^+ be a fixed surface in space that at time t coincides with that part of the system's surface, say S^+_{sys}, where $\mathbf{v}(S_{sys}) \cdot \mathbf{n}(S_{sys}) > 0$. Let S^- be a fixed surface in space that at time t

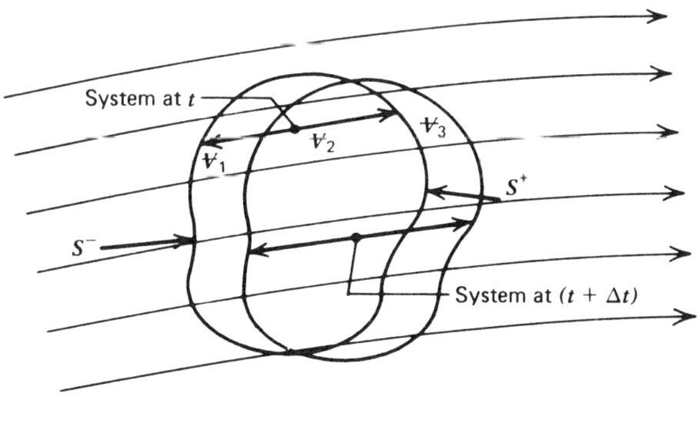

Fig. 4-7. Arbitrary system within the continuum.

coincides with that part of the system's surface, say $S_{\overline{sys}}$, where $\mathbf{v}(S_{sys}) \cdot \mathbf{n}(S_{sys}) < 0$. Now,

$$M(t + \Delta t) = \iiint_{\mathcal{V}_2} \rho(x, y, z, t + \Delta t)\, d\mathcal{V} + \iiint_{\mathcal{V}_3} \rho(x, y, z, t + \Delta t)\, d\mathcal{V},$$

$$M(t) = \iiint_{\mathcal{V}_1} \rho(x, y, z, t)\, d\mathcal{V} + \iiint_{\mathcal{V}_2} \rho(x, y, z, t)\, d\mathcal{V},$$

and

$$\frac{dM}{dt} = \lim_{\Delta t \to 0} \frac{1}{\Delta t} \left\{ \iiint_{\mathcal{V}_2} \rho(x, y, z, t + \Delta t)\, d\mathcal{V} + \iiint_{\mathcal{V}_3} \rho(x, y, z, t + \Delta t)\, d\mathcal{V} \right.$$
$$\left. - \iiint_{\mathcal{V}_1} \rho(x, y, z, t)\, d\mathcal{V} - \iiint_{\mathcal{V}_2} \rho(x, y, z, t)\, d\mathcal{V} \right\} \qquad \text{Eq. 4.2-3}$$

The first and fourth terms on the right-hand side of the equation are integrations over the same limits and can therefore be combined under one integral sign. Thus,

$$\frac{dM}{dt} = \lim_{\Delta t \to 0} \frac{1}{\Delta t} \iiint_{\mathcal{V}_2} [\rho(x, y, z, t + \Delta t) - \rho(x, y, z, t)]\, d\mathcal{V}$$
$$+ \lim_{\Delta t \to 0} \frac{1}{\Delta t} \iiint_{\mathcal{V}_3} \rho(x, y, z, t + \Delta t)\, d\mathcal{V} - \lim_{\Delta t \to 0} \frac{1}{\Delta t} \iiint_{\mathcal{V}_1} \rho(x, y, z, t)\, d\mathcal{V} \qquad \text{Eq. 4.2-4}$$

By the mean value theorem,

$$[\rho(x, y, z, t + \Delta t) - \rho(x, y, z, t)] = \frac{\partial \rho}{\partial t}(x, y, z, t^*)\, \Delta t \qquad \text{Eq. 4.2-5}$$

where $t < t^* < (t + \Delta t)$. In the $\lim_{\Delta t \to 0} t^* = t$ and $\mathcal{V}_2 = \mathcal{V}_{sys}$, therefore, the first term on the right-hand side of Eq. 4.2-4 after substitution of Eq. 4.2-5 can be replaced as shown below in Eq. 4.2-6; that is,

$$\lim_{\Delta t \to 0} \frac{1}{\Delta t} \iiint_{\mathcal{V}_2} [\rho(x, y, z, t + \Delta t) - \rho(x, y, z, t)]\, d\mathcal{V} = \iiint_{\mathcal{V}_{sys}} \frac{\partial \rho}{\partial t}(x, y, z, t)\, d\mathcal{V} \qquad \text{Eq. 4.2-6}$$

To obtain an expression for the second term on the right-hand side of Eq. 4.2-3, note that

$$\iiint\limits_{\mathcal{V}_3} \rho(x, y, z, t + \Delta t)\, d\mathcal{V}$$

approximately equals the amount of mass that has passed through surface S^+ in time Δt, which equals the rate at which mass has passed through surface S^+ times Δt. Then by Eq. 4.1-9 (and noting that $\lim_{\Delta t \to 0} S^+ = S^+_{sys}$),

$$\lim_{\Delta t \to 0} \frac{1}{\Delta t} \iiint\limits_{\mathcal{V}_3} \rho(x, y, z, t + \Delta t)\, d\mathcal{V} = \lim_{\Delta t \to 0} \frac{\Delta t}{\Delta t} \iint\limits_{S^+} \rho \mathbf{v} \cdot d\mathbf{S}$$

$$= \iint\limits_{S^+_{sys}} \rho \mathbf{v} \cdot d\mathbf{S}$$

In a similar fashion

$$\iiint\limits_{\mathcal{V}_1} \rho(x, y, z, t)\, d\mathcal{V}$$

approximately equals the amount of mass that has passed through surface S^- in time Δt, which equals the rate at which mass has passed through surface S^- times Δt. Then by Eq. 4.1-9,

$$\lim_{\Delta t \to 0} \frac{1}{\Delta t} \iiint\limits_{\mathcal{V}_1} \rho(x, y, z, t)\, d\mathcal{V} = -\lim_{\Delta t \to 0} \frac{\Delta t}{\Delta t} \iint\limits_{S^-} \rho \mathbf{v} \cdot d\mathbf{S}$$

$$= -\iint\limits_{S^-_{sys}} \rho \mathbf{v} \cdot d\mathbf{S}$$

We must insert a minus sign on the right-hand side of the above equation since $\mathbf{v}(S^-) \cdot \mathbf{n}(S^-) < 0$ and the integral on the left-hand side of the equation is positive. Thus,

$$\frac{d}{dt} \iiint\limits_{\mathcal{V}_{sys}} \rho(x, y, z, t)\, d\mathcal{V} = \iiint\limits_{\mathcal{V}_{sys}} \frac{\partial \rho}{\partial t}(x, y, z, t)\, d\mathcal{V} + \iint\limits_{S_{sys}} \rho(S)\mathbf{v}(S) \cdot d\mathbf{S} \quad \text{Eq. 4.2-7}$$

On the other hand, if the regions occupied by the thermodynamic system at t, regions $\mathcal{V}_1$ and $\mathcal{V}_2$ in Fig. 4-4, were considered as a control volume instead of a system, then

$$\frac{d}{dt} \iiint_{\text{C.V.}} \rho(x, y, z, t)\, d\mathcal{V} = \iiint_{\text{C.V.}} \frac{\partial \rho}{\partial t}(x, y, z, t)\, d\mathcal{V} \qquad \text{Eq. 4.2-8}$$

This is true because the control volume is fixed in space; thus, the limits of integration are not functions of time. Now, if a physical law contains a term of the form of the left-hand side of Eq. 4.2-7 (the time rate of change of a scalar integrated over the volume occupied by the system), and if the law is to be stated for a control volume, then by comparing Eq. 4.2-7 with Eq. 4.2-8 we see that the statement of the physical law for a control volume must contain an additional term to account for the second term on the right-hand side of Eq. 4.2-7. It should also be observed that the integral of this second term contains, for this case, the mass-flux vector discussed in Section 4.1.2, and thus represents *the net rate at which mass flows out of the region through the bounding surface*. Thus, we can formulate the following statement.

> The rate of increase of mass within the system translates into the rate of increase of mass within the control volume plus the net rate at which mass flows out of the control volume. Stm. 4.2-1

The generalization of these concepts follows. If

$$\Phi(t) = \iiint_{\mathcal{V}_{\text{sys}}} \phi(x, y, z, t)\, d\mathcal{V}$$

where Φ and ϕ are as described earlier (before and after Eq. 4.2-1), then

$$\frac{d\Phi}{dt} = \iiint_{\mathcal{V}_{\text{sys}}} \frac{\partial \phi}{\partial t}(x, y, z, t)\, d\mathcal{V} + \iint_{S_{\text{sys}}} \phi(S)\mathbf{v}(S) \cdot d\mathbf{S} \qquad \text{Eq. 4.2-9}$$

Thus the physical law from above can be restated.

> The rate of increase of Φ within the system translates to the rate of increase of Φ within the control volume plus the net rate at which Φ flows out of the control volume. Stm. 4.2-2

Specific examples of this will be seen later in the chapter.

Finally, the time derivative of integrals involving vectors can be treated by decomposing the vector into its components, bringing the unit vectors outside the volume integral signs, and then applying Eq. 4.2-9 to each term. When this procedure is carried out, it can be shown that

$$\frac{d}{dt}\iiint_{\mathcal{V}_{sys}} \mathbf{B}\, d\mathcal{V} = \iiint_{\mathcal{V}_{sys}} \frac{\partial \mathbf{B}}{dt}\, d\mathcal{V} + \oiint_{S_{sys}} \mathbf{B}(\mathbf{v}\cdot d\mathbf{S}) \qquad \text{Eq. 4.2-10}$$

4.3 CONSERVATION OF MASS IN INTEGRAL FORM

This section deals with the concept that mass can be neither created nor destroyed.[2] The law is stated from two points of view, from the systems viewpoint and from the control volume viewpoint. The former is presented first.

Systems Viewpoint

Select an arbitrary system within the continuum. Let the system have a volume, $\mathcal{V}$, bounded by surface, S (see Fig. 4-8). The law of conservation of mass, with respect to the system, is expressed in this statement.

The rate of increase of mass within the system is equal to zero. Stm. 4.3-1

The total mass, M, within the system is

$$M = \iiint_{\mathcal{V}} \rho\, d\mathcal{V} \qquad \text{Eq. 4.3-1}$$

Thus, the mathematical representation of the fundamental law of mass conservation is

$$\frac{dM}{dt} = \frac{d}{dt}\iiint_{\mathcal{V}} \rho\, d\mathcal{V} = 0 \qquad \text{Eq. 4.3-2}$$

[2] Of course, this is only true in a classical sense. In nuclear reactions, mass can be converted into energy. In this book we do not deal with conditions where such reactions can occur.

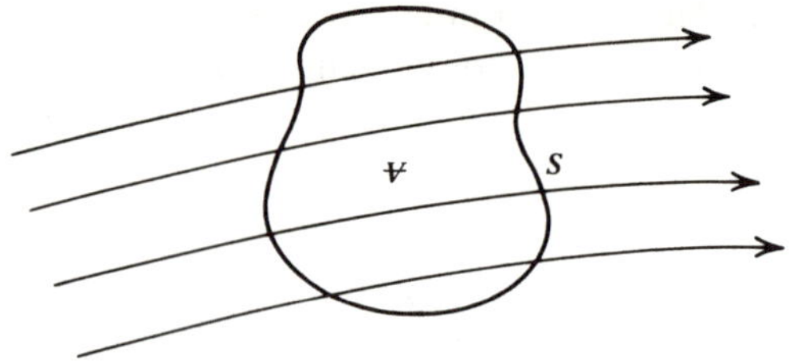

Fig. 4-8. Arbitrary system in the continuum.

By equation 4.2-9

$$\frac{d}{dt}\iiint_{\mathcal{V}} \rho \, d\mathcal{V} = \iiint_{\mathcal{V}} \frac{\partial \rho}{\partial t} \, d\mathcal{V} + \oint\!\!\oint_{S} \rho \mathbf{v} \cdot d\mathbf{S}$$

Thus,

$$\iiint_{\mathcal{V}} \frac{\partial \rho}{\partial t} \, d\mathcal{V} + \oint\!\!\oint_{S} \rho \mathbf{v} \cdot d\mathbf{S} = 0 \qquad \text{Eq. 4.3-3}$$

Control-Volume Viewpoint

Select a control volume to coincide at time, t, with the arbitrary system shown in Fig. 4-8. Then by applying Statement 4.2-1 to Statement 4.3-1, we obtain the statement expressing the law of conservation of mass with respect to the control volume.

> The rate of increase of mass within the control volume plus the net rate at which mass flows out of the control volume is equal to zero. Stm. 4.3-2

Recall that the mass-flux vector is $\rho\mathbf{v}$ and that the rate at which mass flows through the control surface, C.S., is $\oint\!\!\oint_{\text{C.S.}} \rho \mathbf{v} \cdot d\mathbf{S}$. Thus, the mathematical representation of the fundamental law of mass conservation is

$$\frac{d}{dt}\iiint_{\text{C.V.}} \rho \, d\mathcal{V} + \oint\!\!\oint_{\text{C.S.}} \rho \mathbf{v} \cdot d\mathbf{S} = 0 \qquad \text{Eq. 4.3-4}$$

The time derivative can be brought inside the integral, since the limits of integration for a control volume are not functions of time.[3] Thus,

$$\overbrace{\iiint\limits_{\text{c.v.}} \frac{\partial \rho}{\partial t}\, d\mathcal{V}}^{\substack{\text{Rate of}\\ \text{increase of}\\ \text{mass}\\ \text{within}\\ \text{c.v.}}} + \overbrace{\oiint\limits_{\text{c.s.}} \rho \mathbf{v} \cdot d\mathbf{S}}^{\substack{\text{Net rate}\\ \text{mass}\\ \text{flows out}\\ \text{of c.v.}}} = 0 \qquad \text{Eq. 4.3-5}$$

If Eqs. 4.3-3 and 4.3-5 are reexamined, it can be observed that both equations are the same, except that in the first case the integration is over a region considered to be fixed in space, whereas in the second case the integration is over a region that may move in space, but that always contains the same material. We see that once a fundamental law has been mathematically developed to the stage of either Eq. 4.3-3 or Eq. 4.3-5, no distinction need be made regarding whether the region over which the integration is to take place is that of a control volume or that of a system. Thus, it no longer becomes necessary to subscript the region of integration. Therefore, Eq. 4.3-5 then reads the same as Eq. 4.3-3; that is,

$$\iiint\limits_{\mathcal{V}} \frac{\partial \rho}{\partial t}\, d\mathcal{V} + \oiint\limits_{S} \rho \mathbf{v} \cdot d\mathbf{S} = 0 \qquad \text{Eq. 4.3-3}$$

One final note regarding Eq. 4.3-3. In many problems, the region of interest may contain more than one fluid during the time of observation. This is especially true when the control-volume viewpoint is used. Under these circumstances, Eq. 4.3-3 may be applied to either fluid, provided no chemical reaction takes place that affects the fluid of interest. This concept is used in Example 4-2.

[3] The one-dimensional version of this situation is as follows: If $g(t) = \int_a^b f(x, t)\, dx$, then

$$\frac{dg}{dt} = \int_a^b \frac{\partial f}{\partial t}(x, t)\, dx$$

Proof of this statement can be found in any standard calculus text.

4.4 ONE-DIMENSIONAL CHANNEL FLOW

Consider a fluid flow through a variable cross-sectional area channel. Assume that the fluid properties are uniform at each cross section, but vary in the axial direction. Take the region inside the dashed boundary to be either a control volume or a system at time t (see Fig. 4-9). Also assume that the flow is steady. Application of Eq. 4.3-5 gives

$$\iiint_V \frac{\partial \rho}{\partial t}\, dV + \oint\!\!\oint_{S_1} \rho \mathbf{v} \cdot d\mathbf{S} + \oint\!\!\oint_{S_2} \rho \mathbf{v} \cdot d\mathbf{S} + \oint\!\!\oint_{S_3} \rho \mathbf{v} \cdot d\mathbf{S} = 0 \qquad \text{Eq. 4.4-1}$$

Steady flow implies that $\partial \rho / \partial t = 0$. Also, since there is no mass flow through surface S_3, $\mathbf{v}(S_3) \cdot \mathbf{n}(S_3) = 0$. Thus, the first and fourth integrals are zero, giving

$$-\oint\!\!\oint_{S_1} \rho_1 V_1 \mathbf{n}_1 \cdot \mathbf{n}_1\, dS + \oint\!\!\oint_{S_2} \rho_2 V_2\, \mathbf{n}_2 \cdot \mathbf{n}_2\, dS = 0 \qquad \text{Eq. 4.4-2}$$

Now, ρ_1 and V_1 are constant over S_1. Similarly, ρ_2 and V_2 are constant over S_2. These terms may be brought outside the integral sign giving

$$-\rho_1 V_1 \oint\!\!\oint_{S_1} dS + \rho_2 V_2 \oint\!\!\oint_{S_2} dS = -\rho_1 V_1 A_1 + \rho_2 V_2 A_2 = 0 \qquad \text{Eq. 4.4-3}$$

or

$$\rho A V = \text{constant} = \dot{m} \text{ (mass flow rate)} \qquad \text{Eq. 4.4-4}$$

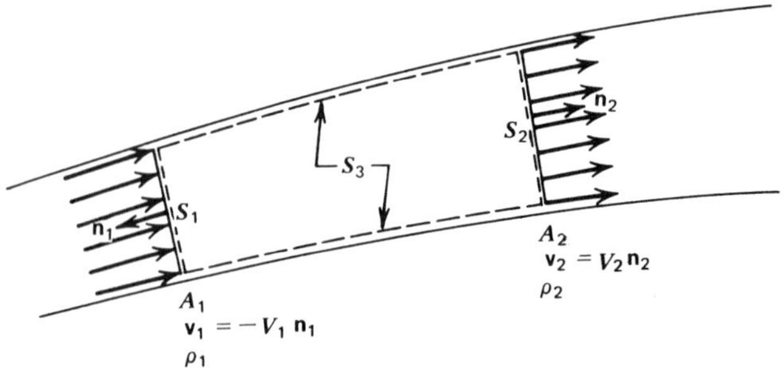

Fig. 4-9. Channel flow.

Also,

$$\gamma A V = \text{constant} = \dot{w} \text{ (weight flow rate)} \qquad \text{Eq. 4.4-5}$$

If the fluid is incompressible, then ρ = constant, giving

$$A V = \text{constant} = Q \text{ (volume flow rate)} \qquad \text{Eq. 4.4-6}$$

Remember that we started this development by assuming that the fluid properties are uniform at each cross section. Thus, it is appropriate to ask whether the above example represents a real flow situation. If a device that measures the local fluid velocity (such as a pitot tube or a hot-wire anemometer) was inserted into a flow within a pipe and a velocity profile was obtained, we would find that the velocity distribution is not uniform. The velocity distribution would appear more like that shown in Fig. 4-10. Thus, the assumption made in developing Eq. 4.4-4, that $\mathbf{v} = V(x)\mathbf{i}$, does not agree with reality. However, the concept is still a useful one, particularly when we are dealing with an incompressible liquid. $V(x)$, which is called the average velocity, or bulk velocity, is defined by the equation

$$\gamma A V(x) = \dot{w} = \oint\!\!\oint_S \gamma v(x, r)\, dS \qquad \text{Eq. 4.4-7}$$

where $\dot{w}$ is the actual weight flow rate passing through surface S at x and $v(x, r)$ is the actual velocity distribution at x.

In the laboratory, $\dot{w}$ can readily be determined when the flow is steady by allowing the pipe or channel to discharge into a collecting tank and determining the time it takes to collect a known weight of the liquid. The specific weight, γ,

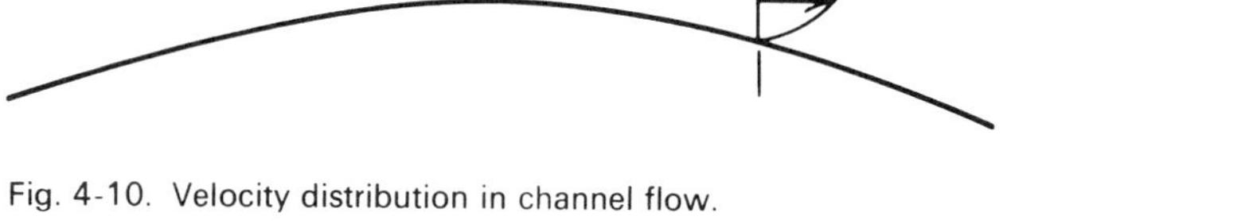

Fig. 4-10. Velocity distribution in channel flow.

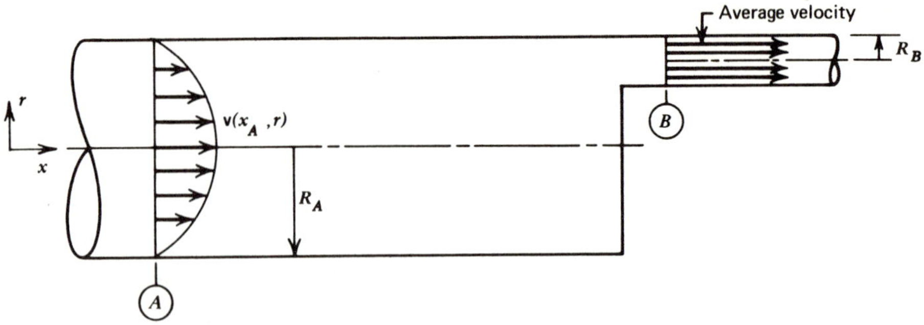

Fig. 4-11. One-dimensional pipe flow.

can also be determined by collecting a known volume of the liquid and obtaining its weight. Since $V(x)$ can readily be determined, it becomes useful in many ways as a reference velocity. We will see this as we get further into the chapter. Some of these concepts are illustrated in the following example.

Example 4-2

A circular pipe of radius R_B is connected to a larger circular pipe of radius R_A (see Fig. 4-11). A pitot tube is used to obtain the velocity profile in the larger pipe at section A. The measurements indicate that at section A

$$\mathbf{v} = 2\beta\left[1 - \left(\frac{r}{R_A}\right)^2\right]\mathbf{i}$$

(a) Find the average velocity, V_A, at section A. (b) Find the maximum velocity at section A. (c) Determine the volume flow rate, Q. (d) Determine the average velocity, V_B, at section B. (e) Sketch the region swept out, in Δt, by the fluid located at section A at $t = 0$, and determine its volume.

SOLUTION

(a) The average velocity, V_A, is given by

$$\pi R_A{}^2 V_A = \oiint_{S_1} \mathbf{v}\cdot d\mathbf{S} = \int_0^{R_A}\int_0^{2\pi} 2\beta\left[1 - \left(\frac{r}{R_A}\right)^2\right]\mathbf{i}\cdot\mathbf{i}r\,d\theta\,dr$$

$$= 2\beta(2\pi)\int_0^{R_A}\left(r - \frac{r^3}{R_A{}^2}\right)dr = 4\pi\beta\left[\frac{R_A{}^2}{2} - \frac{R_A{}^4}{4R_A{}^2}\right]$$

Eq. 4.4-8

Thus,

$$V_A = \beta$$

(*b*) By observation we see that the maximum velocity at section A occurs at $r = 0$. (For a nonobvious case, we would search for a relative extremum by setting $dv/dr = 0$). Thus,

$$v_{\max} = 2\beta$$

(*c*) The volume flow rate, Q, is

$$Q = A_A V_A = \pi R_A^{\,2} \beta$$

(*d*) The average velocity at section B is given by

$$Q = A_B V_B \qquad \text{or} \qquad V_B = \frac{\pi R_A^{\,2}\beta}{\pi R_B^{\,2}} = \beta\left(\frac{R_A}{R_B}\right)^2$$

(*e*) A sketch of the region swept out in Δt by the fluid located at $t = 0$ at section A is shown in Fig. 4-12*a*. The volume, $\mathcal{V}$, of this region is

$$\mathcal{V} = \oiint_A v(x_A, r)\,\Delta t\,dA = \Delta t \int_0^{R_A} \int_0^{2\pi} 2\beta\left[1 - \left(\frac{r}{R_A}\right)^2\right] r\,d\theta\,dr \qquad \text{Eq. 4.4-9}$$

The integral on the right-hand side of Eq. 4.4-9 is the same as the integral on the right-hand side of Eq. 4.4-8 and therefore

$$\mathcal{V} = \pi\beta R_A^{\,2}\,\Delta t = A_A V_A\,\Delta t = Q\,\Delta t$$

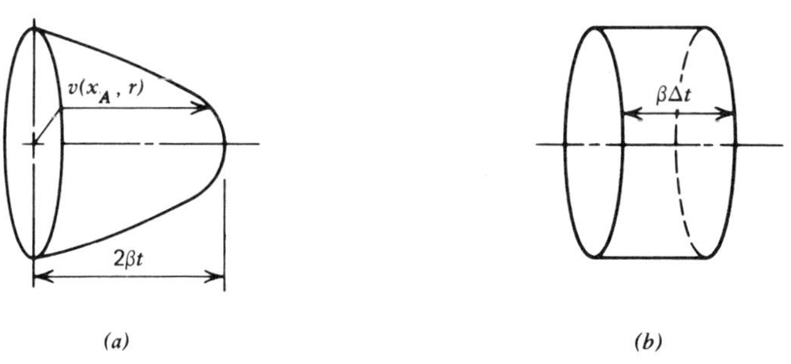

Fig. 4-12. Region swept out in Δt by fluid located at section A.

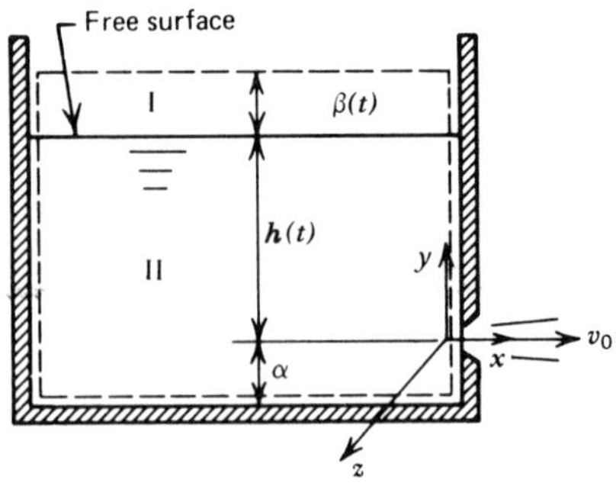

Fig. 4-13. Water discharging from a tank through an orifice.

It should be obvious that this volume is the same as that which would be swept out by the fluid located at $t = 0$ at section A, having a uniform velocity, V_A (see Fig. 4-12*b*).

Example 4-3

An open tank containing water has an orifice which is located near the bottom of the tank as shown in Fig. 4-13. At $t = 0$, the elevation of the free surface is h_0 about the datum line. For ideal flow, it can be shown[4] that the velocity at the orifice is uniform and equal to $\sqrt{2gh}$. Take A_t to be the base area of the tank and A_0 to be the area of the orifice. (*a*) Find the ideal volume flow rate, Q_I, through the orifice. (*b*) The actual flow rate, Q_A, through the orifice is somewhat less than Q_I; that is, $Q_A = c_D Q_I$. The discharge coefficient, c_D, is usually determined empirically. Find $h(t)$.

SOLUTION

(*a*) The integral $\oiint_{A_0} \mathbf{v} \cdot d\mathbf{S}$ gives the volume flow rate through the orifice; thus

$$Q_I = \oiint_{A_0} \mathbf{v} \cdot d\mathbf{S} = \oiint_{A_0} \sqrt{2gh}\,\mathbf{i} \cdot \mathbf{i}\, dS = \sqrt{2gh}\, A_0$$

[4] See Example 4-5.

(b) The actual volume flow rate through the orifice, Q_A, is given by

$$Q_A = c_D A_0 \sqrt{2gh}$$

If the region inside the dashed line is taken to be the control volume, then Eq. 4.3-3 can be applied to the liquid inside the control volume; that is,

$$-\frac{d}{dt} \oiint_{\text{C.V.}} \rho_L \, d\mathcal{V} = \oiint_{\text{C.S.}} \rho_L \mathbf{v} \cdot d\mathbf{S}$$

where ρ_L is the mass density of the liquid of interest. The only boundary surface that has liquid flowing through it is A_0. The above equation reduces to

$$-\frac{d}{dt} \{\rho_L A_t[h(t) + \alpha] + 0 \cdot A_t \beta(t)\} = c_D \rho_L A_0 \sqrt{2gh(t)}$$

In the above equation, the mass density in region I was set to zero, since the mass density of liquid in that region is zero. Also, the length α is a constant and has a zero time derivative. Thus, the above equation reduces to

$$-A_t \frac{dh}{dt} = A_0 c_D \sqrt{2gh}$$

Separating the variables and integrating

$$-\int \frac{dh}{\sqrt{h}} = \int \frac{A_0 c_D}{A_t} \sqrt{2g} \, dt + C$$

or

$$-2\sqrt{h} = \frac{A_0 c_D}{A_t} \sqrt{2g} t + C$$

Taking $h = h_0$ at $t = 0$, C can be evaluated

$$C = -2\sqrt{h_0}$$

Finally

$$\sqrt{h_0} - \sqrt{h} = \frac{c_D A_0}{2A_t} \sqrt{2g} t$$

4.5 THE CONTINUITY EQUATION

There are numerous occasions in the study of fluid mechanics where a one-dimensional description of the flow field would be insufficient to describe a particular phenomenon of interest. For example, a one-dimensional description of the flow field could not be used to describe the pressure field acting on the wings of an airplane in flight, nor could it be used to obtain the actual velocity distribution of a flow through a divergent channel. Under these circumstances, it becomes necessary to obtain the field equations, which are often referred to as the fundamental equations in differential form. The field equation based on the fundamental law of mass conservation is called the *continuity equation.* In this section the continuity equation is derived.

Although there are several approaches that may be used to derive this field equation, the most direct, though somewhat abstract, is to apply the divergence theorem to the second term of Eq. 4.3-3, giving

$$\iiint_{\mathcal{V}} \frac{\partial \rho}{\partial t}\, d\mathcal{V} + \iiint_{\mathcal{V}} (\nabla \cdot \rho \mathbf{v})\, d\mathcal{V} = \iiint_{\mathcal{V}} \left(\frac{\partial \rho}{\partial t} + \nabla \cdot \rho \mathbf{v}\right) d\mathcal{V} = 0 \qquad \text{Eq. 4.5-1}$$

Now this relationship is derived for any arbitrary region in the continuum. It is thus true for all possible limits of integration that are consistent with the limits of the continuum. The only way that this is possible is for the integral to be identically equal to zero. Recall that this concept was discussed in Section 3.2. Thus,

$$\frac{\partial \rho}{\partial t} + \nabla \cdot \rho \mathbf{v} = 0 \qquad \text{Eq. 4.5-2}$$

Equation 4.5-2 is called the continuity equation. Note that it is an equation that relates ρ and the components of $\mathbf{v}$ at every point in the continuum. The expression for the divergence in Cartesian coordinates, as given by Eq. 2.2-12, results in Eq. 4.5-2, reducing to

$$\frac{\partial \rho}{\partial t} + \frac{\partial}{\partial x}(\rho v_1) + \frac{\partial}{\partial y}(\rho v_2) + \frac{\partial}{\partial z}(\rho v_3) = 0 \qquad \text{Eq. 4.5-3}$$

Under many flow situations, the fluid may be considered to be incompressible. Under these circumstances, ρ can be considered to be a constant and Eq. 4.5-3 reduces to

$$\nabla \cdot \mathbf{v} = \frac{\partial v_1}{\partial x} + \frac{\partial v_2}{\partial y} + \frac{\partial v_3}{\partial z} = 0 \qquad \text{Eq. 4.5-4}$$

4.5.1 TWO ALTERNATE DERIVATIONS OF THE CONTINUITY EQUATION

Although the method just used in deriving the continuity equation has gained in popularity in recent times, it is not the most common. A field equation is most often derived by applying a fundamental law to an arbitrary infinitesimal volume in the continuum. After appropriate mathematical rearrangement or manipulation of terms, the field equation is obtained by taking the $\lim_{\Delta V \to 0}$, $\lim_{\Delta t \to 0}$, or both, for both sides of the equation. This method is essentially used in deriving the expression for the divergence in Cartesian coordinates from its definition (see Section A.1.4). The frequency of use of this technique by many authors requires that the student be familiar with it. The derivation of the continuity equation is repeated using two variations of this technique. The first variation relies on the definition of the divergence, the second does not.

Select an arbitrary control volume, ΔV, in the continuum. Let $P(x, y, z)$ be a point within ΔV (see Fig. 4-14). Now apply the law of conservation of mass, as defined in statement 4.3-2, to the control volume; that is, *the net rate of increase of mass within the control volume plus the rate at which mass flows out of the control volume equals zero.* Recall that the mass-flux vector is $\rho \mathbf{v}$. Then,

$$\text{Rate at which mass flows out of control volume} = \oiint_{\text{C.S.}} \rho \mathbf{v} \cdot d\mathbf{S} \qquad \text{Eq. 4.5-5}$$

To obtain the rate of decrease of mass within the control volume, note that

$$\text{Amount of mass in } \Delta V \text{ at } (t + \Delta t) = \iiint_{\Delta V} \rho(x, y, z, t + \Delta t)\, dV$$

$$= \rho(P', t + \Delta t)\, \Delta V \qquad \text{Eq. 4.5-6}$$

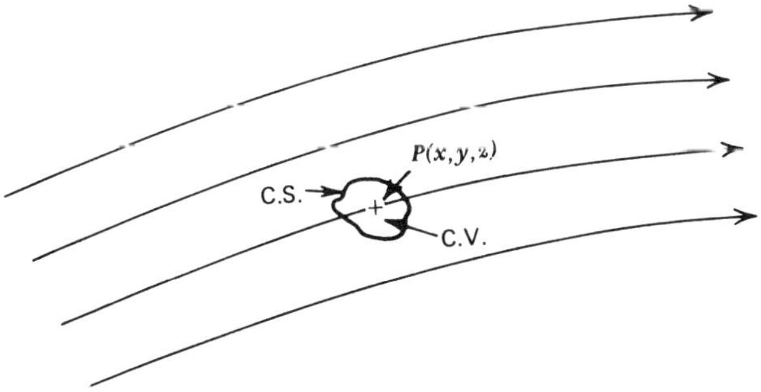

Fig. 4-14. Arbitrary infinitesimal control volume in the continuum.

and

$$\text{Amount of mass in } \Delta V \text{ at } t = \iiint_{\Delta V} \rho(x, y, z, t)\, dV = \rho(P'', t)\, \Delta V \qquad \text{Eq. 4.5-7}$$

where P' and P'' are somewhere within ΔV. The validity of this statement is based on the mean value theorem for the integral. See Eq. 2.2-2 for a statement of the one-dimensional version of this theorem. Then,

$$\text{Rate of increase of mass within } \Delta V = \lim_{\Delta t \to 0} \frac{\rho(P', t + \Delta t) - \rho(P'', t)}{\Delta t} \Delta V$$

Eq. 4.5-8

and

$$\lim_{\Delta t \to 0} \frac{\rho(P', t + \Delta t) - \rho(P'', t)}{\Delta t} + \frac{1}{\Delta V} \oiint_{\text{C.S.}} \rho \mathbf{v} \cdot d\mathbf{S} = 0 \qquad \text{Eq. 4.5-9}$$

Taking the $\lim_{\Delta V \to 0}$ of both sides of the above equation, the second term on the left-hand side, by definition of the divergence, becomes $\nabla \cdot \rho \mathbf{v}$, and the first term on the left-hand side becomes

$$\lim_{\Delta V \to 0} \lim_{\Delta t \to 0} \frac{\rho(P', t + \Delta t) - \rho(P'', t)}{\Delta t} = \lim_{\Delta t \to 0} \lim_{\Delta V \to 0} \frac{\rho(P', t + \Delta t) - \rho(P'', t)}{\Delta t}$$
$$= \lim_{\Delta t \to 0} \frac{\rho(P, t + \Delta t) - \rho(P, t)}{\Delta t} = \frac{\partial \rho}{\partial t}$$

Eq. 4.5-10

Thus

$$\frac{\partial \rho}{\partial t} + \nabla \cdot \rho \mathbf{v} = 0$$

which is the same as Eq. 4.5-2.

The second variation of the derivation of the continuity equation is given below. This method is the most common method presented in introductory textbooks on fluid mechanics. Apply the law of conservation of mass as stated in

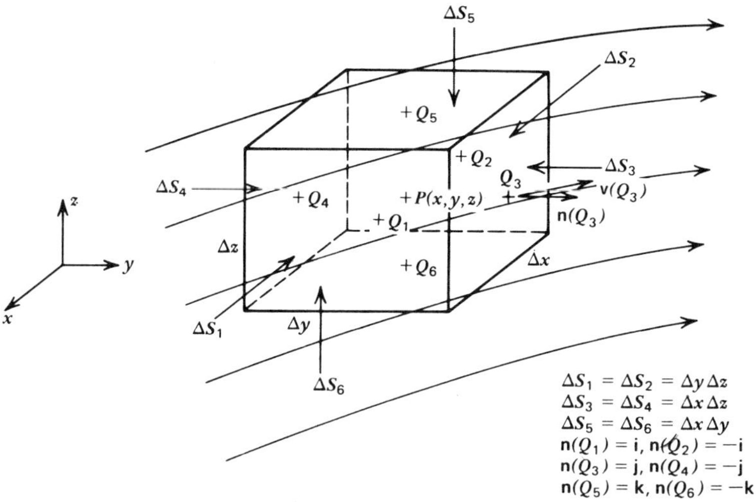

Fig. 4-15. Infinitesimal control volume in the continuum.

Section 4.3 to the infinitesimal control volume shown in Fig. 4-15. Let $P(x, y, z)$ be located at the center of $\Delta V\!\!\!\!/$.

Recall that $\rho\mathbf{v}\cdot\Delta\mathbf{S}$ equals the rate at which mass flows through a surface whose area is ΔS. To obtain the average rate of flow through the surface, ρ and $\mathbf{v}$ are to be evaluated at the center of the surface. Set Q_i to be the point located at the center of surface ΔS_i. Then, the rate at which mass flows out of control volume equals

$$\sum_{i=1}^{6} \rho(Q_i, t)\mathbf{v}(Q_i, t)\cdot\Delta\mathbf{S}_i$$

Now,

$$\begin{aligned}\rho(Q_1, t)\mathbf{v}(Q_1, t)\cdot\Delta\mathbf{S}_1 &= (\rho\mathbf{v})_{(x+\Delta x/2,\, y,\, z,\, t)}\cdot\mathbf{i}\,\Delta y\,\Delta z\\ &= (\rho v_1)_{(x+\Delta x/2,\, y,\, z,\, t)}\,\Delta y\,\Delta z\end{aligned}$$

and

$$\begin{aligned}\rho(Q_2, t)\mathbf{v}(Q_2, t)\cdot\Delta\mathbf{S}_2 &= (\rho\mathbf{v})_{(x-\Delta x/2,\, y,\, z,\, t)}\cdot(-\mathbf{i})\,\Delta y\,\Delta z\\ &= -(\rho v_1)_{(x-\Delta x/2,\, y,\, z,\, t)}\,\Delta y\,\Delta z\end{aligned}$$

Carrying out this type of operation on all six surfaces and taking the rate of mass increase in the control volume to be

$$\left(\frac{\partial \rho}{\partial t}\right)_{(x,\, y,\, z,\, t)} \Delta x \,\Delta y \,\Delta z$$

the law of conservation of mass becomes

$$\left(\frac{\partial \rho}{\partial t}\right)_{(x,\, y,\, z,\, t)} \Delta x \,\Delta y \,\Delta z + [(\rho v_1)_{(x+\Delta x/2,\, y,\, z,\, t)} - (\rho v_1)_{(x-\Delta x/2,\, y,\, z,\, t)}] \,\Delta y \,\Delta z$$
$$+ [(\rho v_2)_{(x,\, y+\Delta y/2,\, z,\, t)} - (\rho v_2)_{(x,\, y-\Delta y/2,\, z,\, t)}] \,\Delta x \,\Delta z$$
$$+ [(\rho v_3)_{(x,\, y,\, z+\Delta z/2,\, t)} - (\rho v_3)_{(x,\, y,\, z-\Delta z/2,\, t)}] \,\Delta x \,\Delta y = 0 \qquad \text{Eq. 4.5-11}$$

Dividing Eq. 4.5-11 by $(\Delta x \,\Delta y \,\Delta z)$ and taking $\lim_{\Delta x \to 0} \lim_{\Delta y \to 0} \lim_{\Delta z \to 0}$ of both sides of the equation, we obtain

$$\frac{\partial \rho}{\partial t} + \frac{\partial}{\partial x}(\rho v_1) + \frac{\partial}{\partial y}(\rho v_2) + \frac{\partial}{\partial z}(\rho v_3) = 0$$

which is the same as Eq. 4.5-3.

4.6 FLUID-FLOW SYSTEMS AND THE FIRST LAW OF THERMODYNAMICS

In this section, we study the first law of thermodynamics and its application to fluid-flow systems. The first law states that, in an overall sense, energy is conserved. The application of classical thermodynamics to fluid-flow problems involves a number of subtle theoretical points that are discussed in the next section. Students more concerned with the applications may skip the following section without loss of continuity.

*4.6.1 SOME CLASSICAL THEORETICAL THERMODYNAMIC CONSIDERATIONS

Thermodynamics deals with the relationships among heat, work, and the energy of thermodynamic systems. When a thermodynamic system undergoes a change in state, it is said to undergo a *process*. *Classical* thermodynamics concerns itself only with the states of systems that are in *thermodynamic equilibrium* and not with the dynamics of the changing state. A system is in thermodynamic equilibrium when there exists no spontaneous tendency toward a change in state. Such a condition would require each intensive property of the system to be uniform in value throughout the system. These last two statements may be taken as an expression of the *Zeroth law of thermodynamics.*

The basic problem of classical thermodynamics becomes, as a result of the Zeroth law, the determination and the description of the new equilibrium state that results from the interaction between the system and its surroundings following a modification of the constraints at the system's boundary.

The first law of thermodynamics, often expressed in the form of a general energy equation, is merely an energy accounting process that relates energy interactions with the surroundings, and then to the net change in energy within a given system. Energy interactions between a system and its surroundings take place either in the form of heat or work. Heat is energy, in transit across the boundary of a system, without the accompanying transfer of mass, and a result of a temperature difference between the system and the surroundings. Since classical thermodynamics is concerned only with energy differences, it is not necessary to establish any absolute zero on the energy scale, nor is it necessary in a given situation to concern oneself with those forms of energy within the system that do not change in value for a given process.

The forms of energy reposing in a given system's matter are many, but those of concern in any given thermodynamics problem are usually relatively few and readily evaluated. Among those energy forms of interest in the elementary study of fluid mechanics are kinetic energy, gravitational potential energy, and internal energy. The student should clearly understand from the outset that the energy forms are human constructs developed historically to preserve the conservation-of-energy concept embodied in the first law of thermodynamics.

The *internal energy* of a thermodynamic system may be considered as the sum total of the energies existing in and among the myriad of discrete particles making up the system. (For example, the translational, rotational, and vibrational kinetic energies of the particles, the various intraparticle binding energies, etc.). In classical thermodynamics we do not separate the component energies for individual consideration; instead, we view the macroscopic consequences of an internal energy change through properties that essentially average, in a statistical sense, the microscopic changes. Foremost among these macroscopic properties, resulting from the averaging process, is the construct known as thermodynamic temperature. We conclude, based on many years of experience, that macroscopic systems have a definite and precise internal energy subject to a conservation principle. Furthermore, through control of the boundaries of the system and by interaction with external systems, the internal energy of a given system may be maintained constant, increased or reduced by a specified amount, or measured relative to any fiducial state of the analyst's choice.

In classical thermodynamics, intensive properties such as pressure and temperature are defined, strictly speaking, only for equilibrium states. We have thus assumed "a priori" that the material undergoing a process from some initial state "i" to some final state "f" undergoes a process that is equivalent to a quasistatic process existing for an infinitely long time, from initial state "i" to final state "f," and passing through the identical series of intermediate equilibrium states as that experienced by the material under study. In fact, in classical thermodynamics, one does not consider the thermodynamic variables as being either a function of time or of space. Changes occur within a thermodynamic space. That is, thermodynamic variables are to be taken as functions of other thermodynamic variables. Problems of interest in both fluid mechanics and heat transfer rarely involve systems throughout which the properties are uniform. Therefore, in most cases of interest, we deal with a nonequilibrium situation. Experience tells us that for the situations dealt with in this textbook, we may make the following postulates: (1) that the thermodynamic properties defined for equilibrium states have real meaning even in nonequilibrium situations; (2) that equations of state, determined under equilibrium conditions, are valid for infinitesimal regions, even though the overall system is in a nonequilibrium condition (this is called local equilibrium); and (3) that the first law of thermodynamics is valid in an infinitesimal region at every instant of time, even though the overall system is in a nonequilibrium condition. Under these circumstances the intensive properties, such as pressure and temperature, may be viewed as a function of both physical space and time. The fundamental concept that thermodynamic properties change within thermodynamic space is unaltered. We must also point out that for a flowing system the first law is usually expressed as a rate equation. Since the first law involves heat transfer, a constitutive equation which describes how heat will flow within the continuum is required to complete the formulation of the problem. These concepts are discussed in greater detail in the next section.

4.6.2 APPLICATION OF THE FIRST LAW TO FLUID-FLOW SYSTEMS

We now state the first law for an arbitrary system or region in the continuum. Again, the law will be stated from both the systems and the control volume viewpoints. The former will be presented first, since it is the systems viewpoint that is presented in the conventional course in thermodynamics.

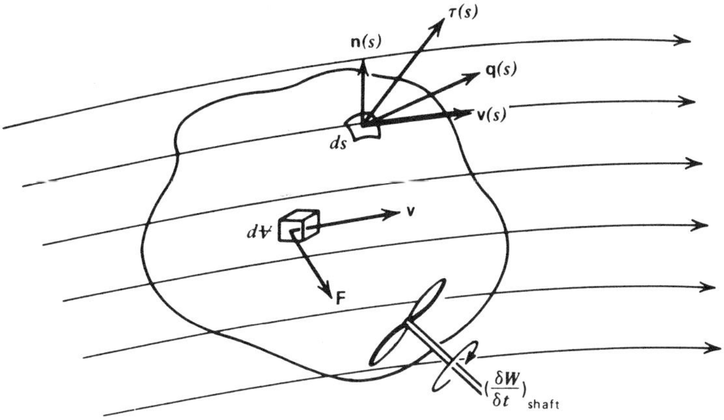

Fig. 4-16. Arbitrary system in the continuum.

Systems Viewpoint

Select an arbitrary system within the continuum, as shown in Fig. 4-16. We can express the first law of thermodynamics for the system in this way.

> The rate of increase in the total energy of a system equals the rate at which work is done on the system plus the net rate at which heat enters the system. Stm. 4.6-1

If E_T = Total energy in the system, in units of (length-force).

$\frac{\delta W}{\delta t}$ = Rate at which work is done on the system, in units of (length-force/unit time).

$\frac{\delta Q}{\delta t}$ = Net rate at which heat enters the system, in units of (thermal energy/unit time).

J = Joules constant, in units of (length-force/thermal energy).

then, a mathematical equation expressing the law is

$$\frac{dE_T}{dt} = \frac{\delta W}{\delta t} + J\frac{\delta Q}{\delta t} \qquad \text{Eq. 4.6-1}$$

Joules constant, J, is inserted so that each term will be dimensionally consistent when W and E_T are expressed in ft-lb$_f$ or m-N, and Q either in Btu or joules. (Since, in the SI system, W is in m-N and Q is in joules, J is unity in the SI system.) It should be noted that most authors on the subject of thermodynamics do not include Joules constant as part of the mathematical expression of the first law. When the first law is written without the J, it is assumed that E_T, W and Q are all expressed in same set of units. The student may also find that not all authors on the subject of thermodynamics use the same sign convention. As a result, the mathematical expression of the first law will most often take one of the two following forms:

$$\Delta E = JQ + W$$

or

$$\Delta E = JQ - W$$

In the first form, W represents the work done on the system whereas, in the second form, W represents the work done by the system. Since the work done by the system equals the negative of the work done on the system, both forms are correct. We prefer to use the former, since it is consistent with the view taken in Newton's second law. That is, Newton's second law is stated in terms of the forces that are exerted on the system by the surroundings. For flowing systems, the first law is invariably stated as a rate equation, as indicated in Eq. 4.6-1.

The rate-of-work and rate-of-heat flow terms are written as $\delta W/\delta t$ and $\delta Q/\delta t$, because work and heat are not perfect differentials. That is, they are not some *definite* function of the thermodynamic state variables. Instead, they are terms that are path dependent.

Since energy appears in several forms, E_T will consist of several terms. Furthermore, since the energy term in the first law appears only in the form of *a change or rate of change* in the energy state of the system, only those forms of energy that change as the system evolves need be included in the expression for the energy state. The sum of these energy states will be designated as E, and E will replace E_T in Eq. 4.6-1. For the fluid system under consideration, those forms of energy that may change within the system are kinetic energy, internal energy, and potential energy. For a given element of mass, kinetic energy is associated with its speed, internal energy is closely associated with its temperature, and potential energy is associated with its elevation. Since the potential energy of a given element of mass results from the gravitational field, one has a choice of either including a potential energy term in E, or adding a term representing the rate of work done by the gravitational field on the element of mass, but not both. The former is chosen in this presentation. Under these circumstances,

$$E = \iiint_{V} \left(Ju + \frac{v^2}{2g_c} + \frac{gZ}{g_c} \right) \rho \, dV \qquad \text{Eq. 4.6-2}$$

where u = the internal energy/unit mass.

To obtain useful expressions for the other terms in the first law, we introduce the following set of variables (see Fig. 4-16). Let

$\boldsymbol{\tau}(S)$ = Surface force acting on the boundary of the system/unit surface area (force/length^2).

$\mathbf{F}$ = Body force acting on a volume element of the system, other than that due to the gravitational field (force/length^3).

To obtain the rate-of-work term, recall that work is defined as the dot product of the force vector with a displacement vector; that is, the work done by the surroundings on a surface element, dS, of the system in Δt is $\boldsymbol{\tau}\, dS \cdot \Delta\mathbf{r}$, where $\Delta\mathbf{r}$ is the displacement of the surface element dS in Δt. To obtain rate of work, divide this term by Δt and take the $\lim_{\Delta t \to 0}$. Then the rate of work done on this surface element is $\boldsymbol{\tau}\, dS \cdot \mathbf{v}$, since $\mathbf{v}(S) = \lim_{\Delta t \to 0} \Delta\mathbf{r}(S)/\Delta t$. In a similar manner the rate of work done on a volume element, dV, by the body force is $\mathbf{F}\, dV \cdot \mathbf{v}$. Finally, if there is a moving device, such as either a turbine blade or a pump blade within the system, then a term accounting for the rate of work done by this device must be included. This type of work is designated as shaft work, W_{shaft}. If the device is a pump, W_{shaft} is positive. If it is a turbine, W_{shaft} is negative. Thus, the rate of work done on the system, $\delta W/\delta t$, is given by

$$\frac{\delta W}{\delta t} = \oiint_{S} (\boldsymbol{\tau} \cdot \mathbf{v})\, dS + \iiint_{V} (\mathbf{F} \cdot \mathbf{v})\, dV + \left(\frac{\delta W}{\delta t} \right)_{\text{shaft}} \qquad \text{Eq. 4.6-3}$$

Note that if we were dealing with a plasma (fluid containing charged particles), then the body-force term, $\mathbf{F}$, would also have to include terms representing the effects of both the electric and magnetic fields.

A term representing the rate at which heat leaves the system still needs to be determined. Students should recall from their studies of thermodynamics that heat represents a flow or transfer of internal energy from one region to another, or from one body to another. To represent this heat flow, the heat-flux vector,

q, in units of (thermal energy/unit time/unit area) is introduced. Recall that the heat-flux vector was discussed in Section 4.1.2. Also recall that

Net rate at which heat flows *out* of the system of interest through surface $dS = \mathbf{q} \cdot d\mathbf{S}$

whereas

Net rate at which heat flows *into* the system of interest through $dS = -\mathbf{q} \cdot d\mathbf{S}$

Then, $J\delta Q/\delta t$ is given by

$$J\frac{\delta Q}{\delta t} = -\oiint_S J\mathbf{q} \cdot d\mathbf{S} \qquad \text{Eq. 4.6-4}$$

The relationship of **q** to other thermodynamic properties is presented in courses concerned with heat transfer.

By substituting Eqs. 4.6-2, 4.6-3, and 4.6-4 into Eq. 4.6-1, we obtain

$$\frac{d}{dt}\iiint_{\Psi} \rho\left(Ju + \frac{v^2}{2g_c} + \frac{gZ}{g_c}\right) d\Psi = \oiint_S (\boldsymbol{\tau} \cdot \mathbf{v})\, dS + \iiint_{\Psi} (\mathbf{F} \cdot \mathbf{v})\, d\Psi - \oiint_S J\mathbf{q} \cdot d\mathbf{S} + \left(\frac{\delta W}{\delta t}\right)_{\text{shaft}} \qquad \text{Eq. 4.6-5}$$

Now

$$\frac{d}{dt}\iiint_{\Psi} \rho\left(Ju + \frac{v^2}{2g_c} + \frac{gZ}{g_c}\right) d\Psi = \iiint_{\Psi} \frac{\partial}{\partial t}\left[\rho\left(Ju + \frac{v^2}{2g_c} + \frac{gZ}{g_c}\right)\right] d\Psi + \oiint_S \rho\left(Ju + \frac{v^2}{2g_c} + \frac{gZ}{g_c}\right)\mathbf{v} \cdot d\mathbf{S} \qquad \text{Eq. 4.6-6}$$

Let us now examine the term $\oiint_S (\boldsymbol{\tau} \cdot \mathbf{v})\, dS$. For a system where viscous effects are negligible, $\boldsymbol{\tau} = -p\mathbf{n}$. Thus,

$$\oiint_S (\boldsymbol{\tau} \cdot \mathbf{v})\, dS = \oiint_S (-p\mathbf{n} \cdot \mathbf{v})\, dS = -\oiint_S p\mathbf{v} \cdot d\mathbf{S} \qquad \text{Eq. 4.6-7}$$

In this book, we only consider cases where the body forces result only from the gravitational field. Under these circumstances, **F** is a null vector, causing the second term on the right-hand side of Eq. 4.6-5 to equal zero. For our studies at this juncture, a convenient form for the energy equation is obtained by substituting Eqs. 4.6-6 and 4.6-7 into Eq. 4.6-5, giving

$$\iiint_{\mathcal{V}} \frac{\partial}{\partial t}\left[\rho\left(Ju + \frac{v^2}{2g_c} + \frac{gZ}{g_c}\right)\right] d\mathcal{V} + \oint\!\!\oint_S \rho\left(Ju + \frac{v^2}{2g_c} + \frac{gZ}{g_c}\right)\mathbf{v}\cdot d\mathbf{S}$$
$$= -\oint\!\!\oint_S p\mathbf{v}\cdot d\mathbf{S} - \oint\!\!\oint_S J\mathbf{q}\cdot d\mathbf{S} + \left(\frac{\delta W}{\delta t}\right)_{\text{shaft}} \qquad \text{Eq. 4.6-8}$$

Control-Volume Viewpoint

Select a control volume to coincide at time, t, with the arbitrary system shown in Fig. 4-16. By applying Statement 4.2-2 to Statement 4.6-1, a statement expressing the first law of thermodynamics with respect to this control volume can now be formulated.

> The rate of increase in the energy of the fluid inside the control volume plus the net rate at which energy flows out of the control volume by mass transport equals the rate at which work is done on the fluid inside the control volume plus the net rate at which heat flows into the control volume.
> Stm. 4.6-2

To obtain an expression for the rate at which energy flows out of the control volume by mass transport, recall that the energy-flux vector is

$$\rho\left(Ju + \frac{v^2}{2g_c} + \frac{g}{g_c}z\right)\mathbf{v}$$

Note that a potential energy term was included in the energy state and, therefore, a rate-of-work term resulting from the gravitational field is not to be included. The mathematical equation that represents this statement is

$$\frac{d}{dt}\iiint_{\text{C.V.}} \rho\left(Ju + \frac{v^2}{2g_c} + \frac{g}{g_c}z\right) d\mathcal{V} + \oint\!\!\oint_{\text{C.S.}} \rho\left(Ju + \frac{v^2}{2g_c} + \frac{g}{g_c}z\right)\mathbf{v}\cdot d\mathbf{S}$$
$$= \oint\!\!\oint_{\text{C.S.}} (\boldsymbol{\tau}\cdot\mathbf{v})\, dS + \left(\frac{\delta W}{\delta t}\right)_{\text{shaft}} - \oint\!\!\oint_{\text{C.S.}} J\mathbf{q}\cdot d\mathbf{S} \qquad \text{Eq. 4.6-9}$$

Again, we will consider a system where viscous effects are negligible, then $\boldsymbol{\tau} = -p\mathbf{n}$. Also, the time derivative of the volume integral appearing in the first term can be brought inside the integral sign, since the volume of a control volume is regarded as fixed in space. Under these circumstances, Eq. 4.6-9 becomes

$$\overbrace{\iiint\limits_{\text{C.V.}} \frac{\partial}{\partial t}\left[\rho\left(Ju + \frac{v^2}{2g_c} + \frac{gz}{g_c}\right)\right] d\mathcal{V}}^{\substack{\text{Rate of increase in the energy}\\ \text{of the fluid inside the control}\\ \text{volume}}} + \overbrace{\oiint\limits_{\text{C.S.}} \rho\left(Ju + \frac{v^2}{2g_c} + \frac{gz}{g_c}\right)\mathbf{v}\cdot d\mathbf{S}}^{\substack{\text{Net rate at which energy}\\ \text{flows out of control}\\ \text{volume by mass transport}}}$$

$$= \underbrace{-\oiint\limits_{\text{C.S.}} p\mathbf{v}\cdot d\mathbf{S} + \left(\frac{\delta W}{\delta t}\right)_{\text{shaft}}}_{\substack{\text{Rate of work done on}\\ \text{the fluid inside the}\\ \text{control volume}}} \underbrace{- \oiint\limits_{\text{C.S.}} J\mathbf{q}\cdot d\mathbf{S}}_{\substack{\text{Net rate}\\ \text{at which}\\ \text{heat flows}\\ \text{into the}\\ \text{control}\\ \text{volume}}} \qquad \text{Eq. 4.6-10}$$

We see that Eqs. 4.6-8 and 4-6-10 are identical. Furthermore, as argued earlier, at this stage no distinction need be made regarding whether the region of integration represents a control volume or a thermodynamic system. Such subscripting will now be dropped and Eq. 4.6-8 becomes the same as Eq. 4.6-10

$$\iiint\limits_{\mathcal{V}} \frac{\partial}{\partial t}\left[\rho\left(Ju + \frac{v^2}{2g_c} + \frac{gz}{g_c}\right)\right] d\mathcal{V} + \oiint\limits_{S} \rho\left(Ju + \frac{v^2}{2g_c} + \frac{gz}{g_c}\right)\mathbf{v}\cdot d\mathbf{S}$$

$$= -\oiint\limits_{S} p\mathbf{v}\cdot d\mathbf{S} - \oiint\limits_{S} J\mathbf{q}\cdot d\mathbf{S} + \left(\frac{\delta W}{\delta t}\right)_{\text{shaft}} \qquad \text{Eq. 4.6-11}$$

Although it might not be immediately obvious, Eq. 4.6-11 is essentially the same energy equation that the student has seen in thermodynamics, probably

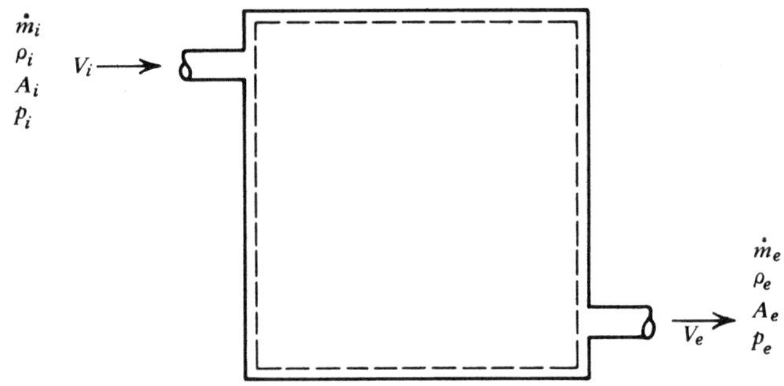

Fig. 4-17. Region with a pipe entrance and a pipe exit.

in a chapter entitled "Energy Analysis" or "Open Systems." To see that this is so, consider a region that has a pipe inlet and a pipe exit, as shown in Fig. 4-17, where the control volume is indicated by the dashed line. Let us now apply Eq. 4.6-11 to the control volume. The only places on the surface of the control volume where $\mathbf{v} \cdot d\mathbf{S} \neq 0$ are at the locations of the pipe entrance and pipe exit. Suppose we consider the fluid properties to be uniform on these surfaces; then Eq. 4.6-11, after dividing through by J, becomes

$$\iiint_{\forall} \frac{\partial}{\partial t}\left[\rho\left(u + \frac{v^2}{2g_c J} + \frac{gz}{g_c J}\right)\right] d\forall - \rho_i\left(u_i + \frac{V_i^{\,2}}{2g_c J} + \frac{gz_i}{g_c J}\right)V_i A_i$$

$$+ \rho_e\left(u_e + \frac{V_e^{\,2}}{2g_c J} + \frac{gz_e}{g_c J}\right)V_e A_e$$

$$= \frac{p_i V_i}{J} A_i - \frac{p_e V_e}{J} A_e - \oint\!\!\oint_S \mathbf{q} \cdot d\mathbf{S} + \frac{1}{J}\left(\frac{\delta W}{\delta t}\right)_{\text{shaft}} \qquad \text{Eq. 4.6-12}$$

The minus sign that appears with the second term on the left-hand side of the above equation and the plus sign that appears with the first term on the right-hand side result from the fact that the outward normal at the inlet area is in the opposite direction to the velocity vector on that surface. Now the mass rate of flow into the region, $\dot{m}_i$, equals $\rho_i A_i V_i$, and the mass rate of flow out of the region, $\dot{m}_e$, equals $\rho_e A_e V_e$. Bringing the time derivative in the first term on the left-hand side of the equation outside the integral sign, returning the rate of the heat-flow term to its original form given in Eq. 4.6-1, and with some rearranging, we obtain

$$\frac{d}{dt}\iiint_{\mathcal{V}} \rho\left(u + \frac{v^2}{2g_cJ} + \frac{gz}{g_cJ}\right) d\mathcal{V} = \frac{1}{J}\left(\frac{\delta W}{\delta t}\right)_{\text{shaft}} + \frac{\delta Q}{\delta t}$$

$$+ \dot{m}_i\left(u_i + \frac{V_i^2}{2g_cJ} + \frac{gz_i}{g_cJ} + \frac{p_i}{\rho_iJ}\right)$$

$$- \dot{m}_e\left(u_e + \frac{V_e^2}{2g_cJ} + \frac{gz_e}{g_cJ} + \frac{p_e}{\rho_eJ}\right) \quad \text{Eq. 4.6-13}$$

In thermodynamics, shaft work is usually designated as work done on the material inside the control volume and will therefore be redesignated as $(\delta W/\delta t)_{\text{C.V.}}$. If e is designated as the total energy/mass and E as the total energy in the control volume, then Eq. 4.6-13 can be written as

$$\left(\frac{dE}{dt}\right)_{\text{C.V.}} = \frac{1}{J}\left(\frac{\delta W}{\delta t}\right)_{\text{C.V.}} + \left(\frac{\delta Q}{\delta t}\right)_{\text{C.V.}} + \dot{m}_i\left(e_i + \frac{p_i}{\rho_iJ}\right) - \dot{m}_e\left(e_e + \frac{p_e}{\rho_eJ}\right)$$

Eq. 4.6-14

This equation, except perhaps for the sign convention, is in the form that is often presented in a course in thermodynamics.

4.6.3 THE FIRST LAW APPLIED TO ONE-DIMENSIONAL CHANNEL FLOWS

In this section, the energy equation, as given by Eq. 4.6-8, is applied to the problem treated in Section 4.4. Recall that in Section 4.4, a fluid flow through a variable cross-sectional area channel was considered. Furthermore, it was assumed that at each cross section the fluid properties were uniform, but varied in the axial direction. Again, take the region inside the dashed boundary, say region $\mathcal{R}$, to be either a control volume or a system at time t (see Fig. 4-18). Also, assume that the flow is steady; then the first term on the left-hand side of Eq. 4.6-8 would be zero. Also assume for the moment that no heat transfer takes place into $\mathcal{R}$ or within $\mathcal{R}$; then the second term on the right-hand side of Eq. 4.6-8 would be zero. Furthermore, since there is no mass flow through surface

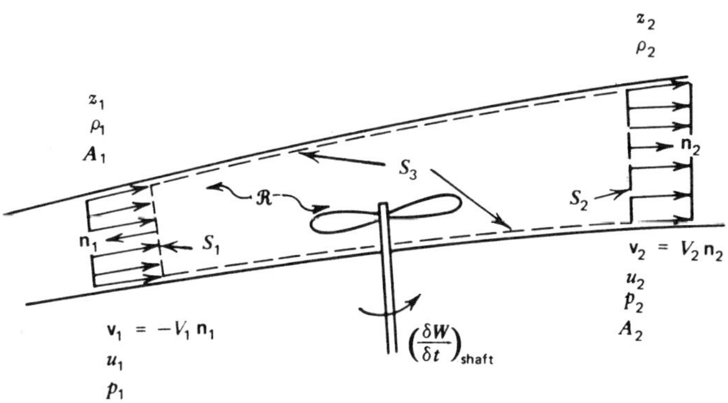

Fig. 4-18. Channel flow.

S_3, $\mathbf{v}(S_3) \cdot \mathbf{n}(S_3)$ is zero. Under these circumstances, application of Eq. 4.6-8 to region $\mathscr{R}$ gives

$$-\rho_1 A_1 V_1\left(Ju_1 + \frac{V_1^{\,2}}{2g_c} + \frac{gz_1}{g_c}\right) + \rho_2 A_2 V_2\left(Ju_2 + \frac{V_2^{\,2}}{2g_c} + \frac{gz_2}{g_c}\right)$$

$$= p_1 A_1 V_1 - p_2 A_2 V_2 + \left(\frac{\delta W}{\delta t}\right)_{\text{shaft}} \qquad \text{Eq. 4.6-15}$$

From Eq. 4.4-4, we see that

$$\rho_1 A_1 V_1 = \rho_2 A_2 V_2 = \rho A V = \dot{m}$$

Then dividing Eq. 4.6-15 by $\dot{m}$ and rearranging terms, we obtain

$$Ju_2 + \frac{p_2}{\rho_2} + \frac{V_2^{\,2}}{2g_c} + \frac{gz_2}{g_c} = Ju_1 + \frac{p_1}{\rho_1} + \frac{V_1^{\,2}}{2g_c} + \frac{gz_1}{g_c} + \frac{1}{\dot{m}}\left(\frac{\delta W}{\delta t}\right)_{\text{shaft}} \qquad \text{Eq.4.6-16}$$

When the continuum under consideration is that of a liquid and no phase changes occur within $\mathscr{R}$, then temperature variations within the system are usually negligible when the region under consideration is not in contact with a heat source or heat sink. Under these circumstances, $u_1 = u_2$ and may be canceled out of Eq. 4.6-16. In addition, ρ may be considered as a constant under most circumstances. At least three cases may still be distinguished.

(*a*) There exists no pump or turbine between sections (1) and (2). Then, Eq. 4.6-16 reduces to

$$\frac{p_2}{\rho} + \frac{V_2^2}{2g_c} + \frac{gz_2}{g_c} = \frac{p_1}{\rho} + \frac{V_1^2}{2g_c} + \frac{gz_1}{g_c} \qquad \text{Eq. 4.6-17}$$

If we multiply both sides of Eq. 4.6-17 by g_c/g and substitute γ for $\rho g/g_c$, we obtain a form that is sometimes more convenient, that is,

$$\frac{p_2}{\gamma} + \frac{V_2^2}{2g} + z_2 = \frac{p_1}{\gamma} + \frac{V_1^2}{2g} + z_1 = \text{total head} \qquad \text{Eq. 4.6-18}$$

The term $(p_2/\gamma + V_2^2/2g + z_2)$ is designated as the total head at section (2). Therefore, Eq. 4.6-18 indicates that the total head at any section in the channel between sections (1) and (2) is a constant.

(*b*) There exists a pump between sections (1) and (2). Under these circumstances

$$\left(\frac{\delta W}{\delta t}\right)_{\text{shaft}} = \left(\frac{\delta W}{\delta t}\right)_{\text{pump}} > 0$$

since work is done by the surroundings and is exerted on the system. Then Eq. 4.6-16 reduces to

$$\frac{p_2}{\rho} + \frac{V_2^2}{2g_c} + \frac{gz_2}{g_c} = \frac{p_1}{\rho} + \frac{V_1^2}{2g_c} + \frac{gz_1}{g_c} + \frac{1}{\dot{m}}\left(\frac{\delta W}{\delta t}\right)_{\text{pump}} \qquad \text{Eq. 4.6-19}$$

Multiplying both sides of Eq. 4.6-19 by g_c/g and substituting γ for $\rho g/g_c$, it becomes

$$\frac{p_2}{\gamma} + \frac{V_2^2}{2g} + z_2 = \frac{p_1}{\gamma} + \frac{V_1^2}{2g} + z_1 + \frac{1}{\dot{w}}\left(\frac{\delta W}{\delta t}\right)_{\text{pump}} \qquad \text{Eq. 4.6-20}$$

where $\dot{w} = \dot{m}g/g_c$ = weight flow rate.

We see that the total head at section (2) is greater than the total head at section (1) by the term $(1/\dot{w})(\delta W/\delta t)_{\text{pump}}$. We designate this term as h_p. It represents the total head developed by the pump; that is,

$$h_p = \frac{1}{\dot{w}}\left(\frac{\delta W}{\delta t}\right)_p = \left(\frac{p_2}{\gamma} + \frac{V_2^{\,2}}{2g} + z_2\right) - \left(\frac{p_1}{\gamma} + \frac{V_1^{\,2}}{2g} + z_1\right) \qquad \text{Eq. 4.6-21}$$

(*c*) There exists a turbine between section (1) and (2). Under these circumstances $(\delta W/\delta t)_{\text{shaft}} = -(\delta W/\delta t)_{\text{turbine}} < 0$, since work is done by the system.

$$\frac{p_2}{\rho_2} + \frac{V_2^{\,2}}{2g_c} + \frac{gz_2}{g_c} = \frac{p_1}{\rho_1} + \frac{V_1^{\,2}}{2g_c} + \frac{gz_1}{g_c} - \frac{1}{\dot{m}}\left(\frac{\delta W}{\delta t}\right)_{\text{turbine}} \qquad \text{Eq. 4.6-22}$$

On the other hand, if the continuum is a gas, then temperature variations within the region may be significant. Under these circumstances, it is convenient to introduce the enthalpy, h. Recall that

$$h = u + \frac{1}{J}\frac{p}{\rho} \qquad \text{or} \qquad Jh = Ju + \frac{p}{\rho}$$

Furthermore, gas-flow problems usually involve velocities of sufficient magnitude that the potential energy term (gz/g_c) may be neglected. Two cases can be distinguished.

(*d*) No shaft devices exist between section (1) and (2). Then Eq. 4.6-16 reduces to

$$Jh_2 + \frac{V_2^{\,2}}{2g_c} = Jh_1 + \frac{V_1^{\,2}}{2g_c} \qquad \text{Eq. 4.6-23}$$

(*e*) A fan exists between sections (1) and (2). Then

$$\left(\frac{\delta W}{\delta t}\right)_{\text{shaft}} = \left(\frac{\delta W}{\delta t}\right)_{\text{fan}}$$

since work is done on the system. And Eq. 4.6-16 reduces to

$$Jh_2 + \frac{V_2^{\,2}}{2g_c} = Jh_1 + \frac{V_1^{\,2}}{2g_c} + \frac{1}{\dot{m}}\left(\frac{\delta W}{\delta t}\right)_{\text{fan}} \qquad \text{Eq. 4.6-24}$$

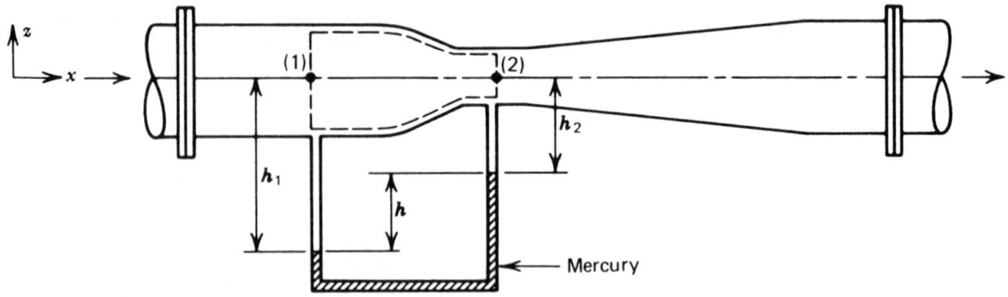

Fig. 4-19. The venturi meter.

The analysis of a venturi meter, when the substance flowing through the meter is a liquid, is an excellent example of case (*a*). A venturi meter is a device used to measure the volume flow rate through a pipe or channel. It is essentially a channel having a variable cross-sectional area. The section having the minimum cross-sectional area is called the throat. This area is assumed to be known. One pressure tap is provided at the throat and another upstream from the throat, as shown in Fig. 4-19.

Example 4-4

For the venturi meter shown in Fig. 4-19, determine the volume flow rate, Q, through the venturi in terms of the manometer reading, h.

SOLUTION

Applying Eqs. 4.6-18 and 4.4-6 and taking $z_1 = z_2$, we obtain

$$\frac{p_1}{\gamma} + \frac{V_1^2}{2g} = \frac{p_2}{\gamma} + \frac{V_2^2}{2g}$$

and

$$A_1 V_1 = A_2 V_2$$

Rearranging the terms, gives

$$V_2^2 - V_1^2 = \frac{2g}{\gamma}(p_1 - p_2)$$

Substituting for V_1, we obtain

$$V_2^2 - \left(\frac{A_2 V_2}{A_1}\right)^2 = V_2^2\left(1 - \frac{A_2^2}{A_1^2}\right) = \frac{2g}{\gamma}(p_1 - p_2)$$

Solving for V_2, we obtain

$$V_2 = \frac{1}{\sqrt{1 - (A_2/A_1)^2}} \sqrt{\frac{2g}{\gamma}(p_1 - p_2)} \qquad \text{Eq. 4.6-25}$$

Equation 4.6-25 gives V_2 in terms of the pressure difference $(p_1 - p_2)$. The ideal volume flow rate is given by

$$Q = A_2 V_2 = \frac{A_2}{\sqrt{1 - (A_2/A_1)^2}} \sqrt{\frac{2g}{\gamma}(p_1 - p_2)}$$

However, viscous effects and other minor losses cause the actual average velocity to be somewhat less than the ideal value, expressed by Eq. 4.6-25. An empirical coefficient, c_D, modifies this relationship. Thus,

$$Q = A_2 V_2 = \frac{c_D A_2}{\sqrt{1 - (A_2/A_1)^2}} \sqrt{\frac{2g}{\gamma}(p_1 - p_2)} \qquad \text{Eq. 4.6-26}$$

For a Herschel-type venturi meter,[5] a typical range for c_D is

$$0.93 < c_D < 0.99$$

Now $(p_1 - p_2)$ can be determined from the manometer readings; that is,

$$p_1 = p_2 + \gamma h_2 + \gamma_{hg} h - \gamma h_1$$

or

$$p_1 - p_2 = \gamma(h_2 - h_1) + \gamma_{hg} h$$

But

$$h_1 - h - h_2 = 0 \qquad \text{or} \qquad h_2 - h_1 = -h$$

[5] See *Fluid Meters, Their Theory and Application*, 5th ed., ASME, New York, 1959.

Thus

$$p_1 - p_2 = -\gamma h + \gamma_{hg} h = (\gamma_{hg} - \gamma)h \qquad \text{Eq. 4.6-27}$$

Substituting Eq. 4.6-27 into Eq. 4.6-26, we obtain

$$Q = \frac{c_D A_2}{\sqrt{1 - (A_2/A_1)^2}} \sqrt{2gh\left(\frac{\gamma_{hg}}{\gamma} - 1\right)} \qquad \text{Eq. 4.6-28}$$

At the end of Section 4.4 it was noted that, in reality, the velocity, **v**, at any one cross section was not uniform as was assumed in this analysis, but varied with r. However, if a similar experiment was carried out with respect to pressure, we would find that p would not significantly vary with r. It is a remarkable fact that the analysis, despite the error in the assumed velocity distribution, gives the correct volume-flow rate in terms of the pressure difference $(p_1 - p_2)$. Thus, the use of the average velocity at a cross section becomes a very useful concept.

A second example which fits into the category of case (a) is described in the following example.

Example 4-5

Consider the open tank described in Example 4-3. Show that for ideal flow, the discharge velocity at the orifice is $\sqrt{2gh}$.

SOLUTION

Consider the control volume shown in Fig. 4-20. Apply the energy equation, Eq. 4.6-18, to sections S_1 and S_2. Experiments indicate that the pressure at any

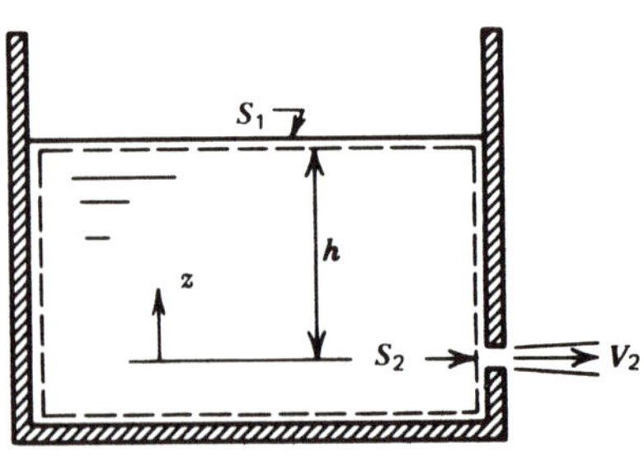

Fig. 4-20. Water discharging from a tank through an orifice.

section open to the atmosphere can be taken to be at atmospheric pressure. Therefore,

$$p_1 = p_2 = p_{atm}$$

Also, by Eq. 4.4-6, and by the fact that $A_1 \gg A_2$, we can take $V_1 \ll V_2$. Thus, V_1 can be neglected. Under these circumstances, Eq. 4.6-18 reduces to

$$\frac{V_2^2}{2g} = z_1 - z_2 = h$$

or

$$V_2 = \sqrt{2gh}$$

Strictly speaking, this problem is not a steady-flow problem, since h varies with time. However, if the time derivative term, the first term on the left-hand side of Eq. 4.6-11, is negligible compared to some of the other terms, then analyzing the problem as though it were a steady-flow problem does not introduce any significant error. Under these circumstances, the flow is said to be *quasisteady*. Experiments indicate that, for the problem under consideration, the assumption that the flow is quasisteady is a valid assumption.

A third example that fits into the category of case (a) is the siphon sketched in Fig. 4-21. To create a flow through the siphon tube, invert the tube, completely fill the tube with a liquid, cover the ends, and carefully replace the tube

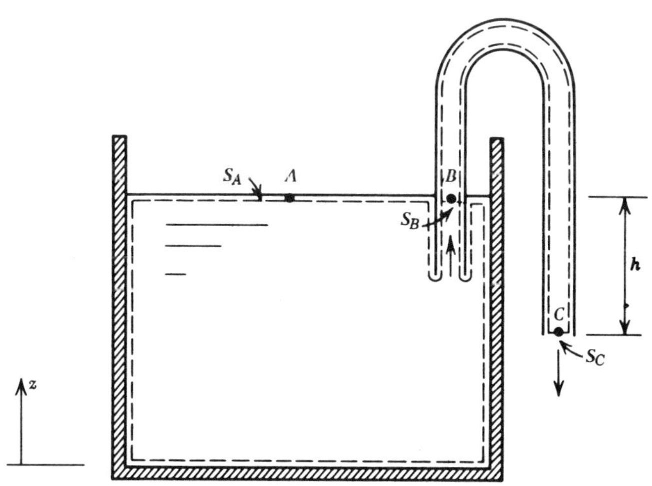

Fig. 4-21. The siphon tube.

to the position shown in the figure. Be careful to avoid the presence of air bubbles within the tube. To initiate a flow through the tube, remove the covers and tap the tube. The gravitational field tends to cause a tensile stress in the liquid. Since a liquid cannot sustain a tensile stress, fractures develop, creating a vacuum within the liquid in the right-hand column. Atmospheric pressure on the liquid surface in the tank then causes a flow through the tube. The analysis of the siphon is given in the following example.

Example 4-6

A siphon of uniform cross-sectional area is used to drain water from a tank (see Fig. 4-21). At the time of interest, let h be the elevation difference between the liquid level in the tank and the end of the tube. If the atmosphere is at standard atmospheric pressure and h is 6 ft, (*a*) Determine the average fluid velocity through the tube; (*b*) determine the pressure within the tube at point B; and (*c*) show that if the end of the tube is above the liquid level in the tank, then the liquid will not flow.

SOLUTION

(*a*) Select a control volume shown by the dashed lines in Fig. 4-21. First, write the energy equation, Eq. 4.6-18, between sections B and C. Since the tube is open to the atmosphere at section C, $p_C = p_{atm}$. Also, by Eq. 4.4-6 and the given information that the cross-sectional area of the tube is uniform, we have that $V_B = V_C$. Then Eq. 4.6-18 reduces to

$$\frac{p_C - p_B}{\gamma} = \frac{p_{atm} - p_B}{\gamma} = z_B - z_C = h \qquad \text{Eq. 4.6-29}$$

Now write the energy equation, Eq. 4.6-18, between sections A and B. Since the area at section A is much greater than the cross-sectional area of the tube, Eq. 4.4-6 gives that $V_A \ll V_B$ and may thus be neglected. Also, note that $z_A = z_B$ and that $p_A = p_{atm}$. Then Eq. 4.6-18 reduces to

$$\frac{p_A - p_B}{\gamma} = \frac{p_{atm} - p_B}{\gamma} = \frac{V_B^2}{2g} \qquad \text{Eq. 4.6-30}$$

Combining the last two equations, we obtain

$$V_B = \sqrt{2gh} \qquad \text{Eq. 4.6-31}$$

For $h = 6$ ft,

$$V_B = \sqrt{2(32.2)(6)} = 19.66 \text{ ft/s}$$

(*b*) From Eq. 4.6-29, we have

$$p_B = p_{\text{atm}} - \gamma h = 14.68 - \frac{62.4(6)}{144} = 12.08 \text{ psia}$$

(*c*) If the end of the tube is above the liquid level in the tank, then $(z_B - z_C)$ is negative. Then, by Eq. 4.6-31, no real solution exists for V_B.

An example of case (*b*) is given in the following example.

Example 4-7

A centrifugal pump has been set up in the laboratory so that its characteristics may be determined. A sketch of the laboratory arrangement is shown in Fig. 4-22. During a particular run, the volume flow rate through the pump was measured to be 20.3 gal/min. The pressure gage on the discharge side read 15 ft of water, while the vacuum gage on the suction side read 4.3 in. of Hg. The pipe diameter on the discharge side of the pump is $\frac{1}{2}$ in., while the pipe diameter on the suction side is $\frac{3}{4}$ in. Determine (*a*) the head developed by the pump, and (*b*) the power transmitted to the water by the pump.

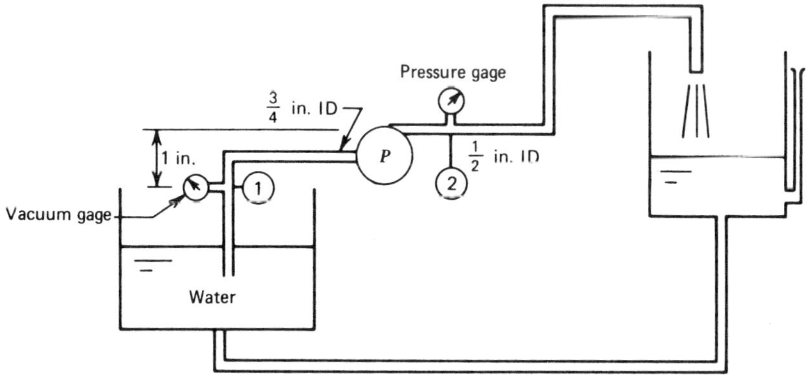

Fig. 4-22. Centrifugal pump set up in the laboratory.

SOLUTION

(*a*) Select a control volume containing the pump and the pipe lines between sections (1) and (2). Let us first determine pressures p_1 and p_2. Take section (1) to be at the vacuum gage and section (2) at the pressure gage; then

$$\frac{p_1}{\gamma} = \frac{p_{\text{atm}}}{\gamma} - \frac{4.3\gamma_{hg}}{12\gamma_w} = \frac{p_{\text{atm}}}{\gamma} - \frac{4.3(13.56)}{12}$$

$$= \frac{p_{\text{atm}}}{\gamma} - 4.86 \text{ ft of water}$$

and

$$\frac{p_2}{\gamma} = \frac{p_{\text{atm}}}{\gamma} + \frac{15\gamma}{\gamma} = \frac{p_{\text{atm}}}{\gamma} + 15 \text{ ft of water}$$

Now determine the velocities at sections (1) and (2).

$$Q = A_1 V_1 = A_2 V_2$$

Therefore,

$$V_1 = \frac{20.3 \text{ gal/min} \times (1/7.48)\text{ft}^3\text{/gal} \times (1/60)\text{min/s}}{\pi(\frac{3}{4})^2/(4 \times 144)} = 14.74 \text{ ft/s}$$

Similarly,

$$V_2 = \frac{20.3 \times 4 \times 144 \times 4}{7.48\pi \times 60} = 33.17 \text{ ft/s}$$

From Eq. 4.6-21, we have

$$h_p = \frac{p_2}{\gamma} - \frac{p_1}{\gamma} + \frac{V_2{}^2 - V_1{}^2}{2g} + (z_2 - z_1)$$

$$= (15 + 4.86) + \frac{(33.17)^2 - (14.74)^2}{64.4} + 1 = 34.57 \text{ ft}$$

(*b*) The power transmitted to the water by the pump = $(\delta W/\delta t)_p$. Applying Eq. 4.6-21

$$\left(\frac{\delta W}{\delta t}\right)_p = \dot{w}h_p = \gamma Q h_p = \frac{62.4 \times 20.3 \times 34.57}{7.48 \times 60}$$

$$= 97.57 \text{ ft-lb}_f/\text{s}$$

$$= 0.177 \text{ hp}$$

The determination of the wind speed in the test section of a subsonic wind tunnel from pressure and temperature measurements provides an example of case (*d*).

Example 4-8

A small subsonic wind tunnel is instrumented with a three-tube manometer and a thermocouple. One tube of the manometer is connected to the test section, one tube is connected to the plenum, and the third is open to the atmosphere. The thermocouple measuring junction is inserted in the plenum (see Fig. 4-23). (*a*) Develop a formula for determining the wind velocity, V, in the test section of the wind tunnel in terms of p, p_0, and T_0. (*b*) Determine the wind speed,

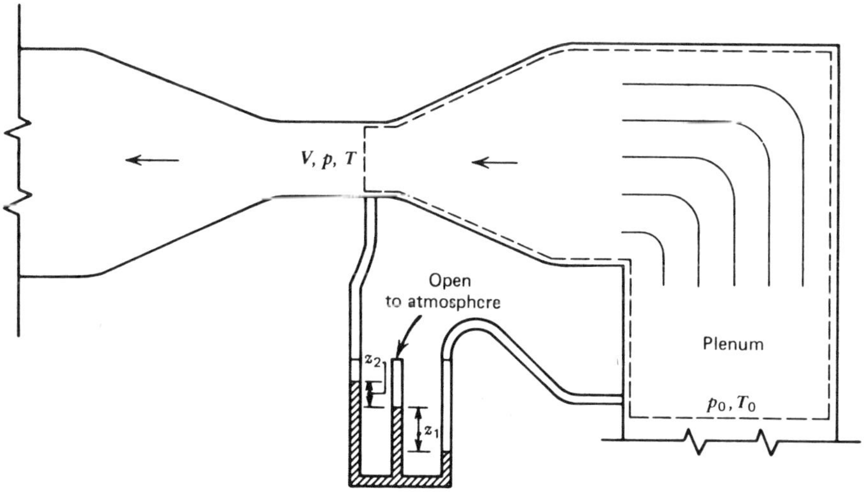

Fig. 4-23. Subsonic wind tunnel.

V, if the barometric pressure is 29.42 in. of Hg, the plenum temperature is 98 F, and the manometer readings are $z_1 = 10.1$ in. and $z_2 = 2.5$ in.

SOLUTION

(a) Consider the control volume enclosed by the dashed line in Fig. 4-23. The airflow through the wind tunnel can be considered to be isentropic flow. The velocity in the plenum is sufficiently small that it may be neglected. Therefore, the application of Eq. 4.6-23 to the control volume gives

$$\frac{V^2}{2g_c} + Jh = Jh_0$$

or

$$V = \sqrt{2g_c J(h_0 - h)} \qquad \text{Eq. 4.6-32}$$

For an ideal gas with constant specific heats, the following relations are valid (see Appendix A.3 for a review of some elements of thermodynamics).

$$u = c_v T \qquad \text{and} \qquad h = c_p T \qquad \text{Eqs. 4.6-33}$$

$$k = \frac{c_p}{c_v} \qquad \text{Eq. 4.6-34}$$

$$c_v = \frac{R}{J(k \quad 1)} \qquad \text{Eq. 4.6-35}$$

$$c_p = \frac{Rk}{J(k - 1)} \qquad \text{Eq. 4.6-36}$$

where the dimensions of c_v and c_p are, in the English system, Btu/($1b_m$-R) and, in the SI system, joules/(kg-K). In terms of these variables, the gas constant, R, has dimensions of (lb_f-ft)/(lb_m-R) in the English system and (N-m)/(kg-K) in the SI system. Substituting the second of Eqs. 4.6-33 and then Eq. 4.6-36 into Eq. 4.6-32, we obtain

$$V = \sqrt{2g_c J c_p (T_0 - T)} = \sqrt{\frac{2k}{k - 1} g_c R(T_0 - T)}$$

which, after factoring out T_0, becomes

$$V = \sqrt{\frac{2k}{k-1} g_c R T_0 \left(1 - \frac{T}{T_0}\right)} \qquad \text{Eq. 4.6-37}$$

For isentropic flow[6]

$$\frac{T}{T_0} = \left(\frac{p}{p_0}\right)^{(k-1)/k}$$

Substituting for T/T_0, we obtain

$$V = \sqrt{\frac{2k}{k-1} g_c R T_0 \left[1 - \left(\frac{p}{p_0}\right)^{(k-1)/k}\right]} \qquad \text{Eq. 4.6-38}$$

(*b*) From the barometer reading, determine p_{atm}.

$$p_{atm} = 29.42 \text{ in. of Hg} \times 0.4912 \frac{\text{psia}}{\text{in. of Hg}} = 14.45 \text{ psia}$$

Plenum and test section pressures are determined from the manometer readings; that is,

$$p_0 = 14.45 + \left(10.1 \times \frac{62.4}{1728}\right) = 14.815 \text{ psia}$$

$$p = 14.45 - \left(2.5 \times \frac{62.4}{1728}\right) = 14.360 \text{ psia}$$

The wind speed in the test section is determined from Eq. 4.6-38

$$V = \sqrt{\frac{2(1.4)(32.2)(53.35)(558)}{0.4}\left[1 - \left(\frac{14.360}{14.815}\right)^{0.2857}\right]}$$

$$= 244.0 \text{ ft/s}$$

$$= 166.4 \text{ mi/h}$$

[6] See Appendix A.3.

It should be noted that if V is to be determined to three significant figures, then

$$\left(\frac{p}{p_0}\right)^{(k-1)/k}$$

should be carried out to at least five significant figures. This is so because

$$\left(\frac{p}{p_0}\right)^{(k-1)/k}$$

is very near to one. Thus, several significant figures are lost when the two terms within the brackets of Eq. 4.6-38 are subtracted. The necessary accuracy can be obtained on a scientific hand calculator. If one chooses to avoid this problem, take

$$p = p_0 - \Delta p$$

then

$$\left(\frac{p}{p_0}\right)^{(k-1)/k} = \left(1 - \frac{\Delta p}{p_0}\right)^{(k-1)/k} \qquad \text{Eq. 4.6-39}$$

Applying the series expansion,

$$(1 - x)^n = 1 - nx + \frac{n(n-1)}{2} x^2 - + \cdots, \text{(valid for } x^2 < 1)$$

to Eq. 4.6-39, we have

$$\left(1 - \frac{\Delta p}{p_0}\right)^n = 1 - n\left(\frac{\Delta p}{p_0}\right) + \frac{n(n-1)}{2}\left(\frac{\Delta p}{p_0}\right)^2 - + \cdots$$

where $n = (k - 1)/k$. By neglecting terms of the order of $(\Delta p/p_0)^3$ and higher, Eq. 4.6-38 becomes

$$V = \sqrt{\left(\frac{2k}{k-1}\right) g_c R T_0 \left(\frac{k-1}{k}\right)\left(\frac{\Delta p}{p_0}\right)\left(1 + \frac{1}{2k}\frac{\Delta p}{p_0}\right)}$$

or

$$V = \sqrt{\frac{2 g_c R T_0}{p_0} \Delta p \left(1 + \frac{1}{2k}\frac{\Delta p}{p_0}\right)} \qquad \text{Eq. 4.6-40}$$

The wind speed, V, obtained from Eq. 4.6-40 is

$$V = \sqrt{\frac{2(32.2)(53.35)(558)(0.455)}{14.815}\left[1 + \frac{0.455}{2(1.4)(14.815)}\right]}$$

$$= 244.0 \text{ ft/s}$$

which is exactly the previously determined value to four places.

4.7 NEWTON'S SECOND LAW APPLIED TO A FLUID

In this section Newton's second law is presented from both the systems and control volume viewpoints. The law applied to a fluid system is presented first.

Systems Viewpoint

Select an arbitrary system within the continuum (see Fig. 4-24). Newton's law is expressed in the following statement.

> The sum of the force distributions acting on the system equals the time rate of change of the momentum in the system. Stm. 4.7-1

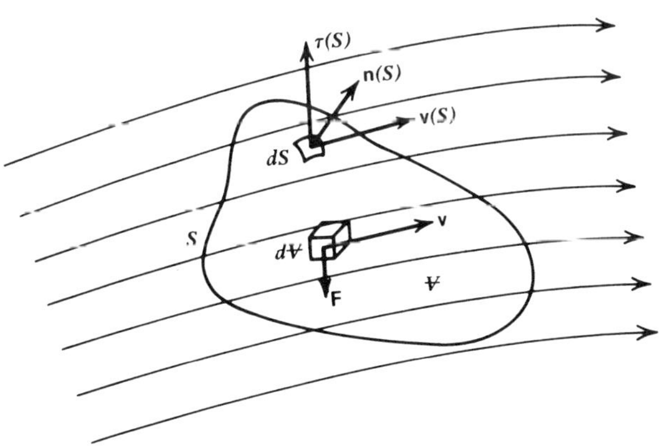

Fig. 4-24. Arbitrary system in the continuum.

The mathematical expression of the law is

$$\iiint_{\mathcal{V}} \mathbf{F}\, d\mathcal{V} + \oiint_{S} \boldsymbol{\tau}\, dS = \frac{d}{dt} \iiint_{\mathcal{V}} \frac{\rho}{g_c} \mathbf{v}\, d\mathcal{V} \qquad \text{Eq. 4.7-1}$$

Applying Eq. 4.2-10 to the right-hand side of Eq. 4.7-1, we obtain

$$\iiint_{V} \mathbf{F}\, d\mathcal{V} + \oiint_{S} \boldsymbol{\tau}\, dS = \iiint_{V} \frac{\partial}{\partial t}\left(\frac{\rho}{g_c}\mathbf{v}\right) d\mathcal{V} + \oiint_{S} \frac{\rho}{g_c}\mathbf{v}(\mathbf{v}\cdot d\mathbf{S}) \qquad \text{Eq. 4.7-2}$$

Here we set **B** in Eq. 4.2-10 equal to $(\rho/g_c)\mathbf{v}$. Again neglecting viscous effects, $\boldsymbol{\tau}$ reduces to $-p\mathbf{n}$ and Eq. 4.7-2 reduces to

$$\iiint_{V} \mathbf{F}\, d\mathcal{V} - \oiint_{S} p\mathbf{n}\, dS = \iiint_{\mathcal{V}} \frac{\partial}{\partial t}\left(\frac{\rho}{g_c}\mathbf{v}\right) d\mathcal{V} + \oiint_{S} \frac{\rho}{g_c}\mathbf{v}(\mathbf{v}\cdot d\mathbf{S}) \qquad \text{Eq. 4.7-3}$$

Equation 4.7-3 is often called the integral form of the linear momentum equation.

Control-Volume Viewpoint

Select a control volume to coincide at time t with the arbitrary system shown in Fig. 4-24. By applying Statement 4.2-2 to Statement 4.7-1, Newton's law with respect to this control volume can be expressed in this way.

> The sum of the force distributions acting on the fluid in the control volume equals the time rate of change of fluid momentum inside the control volume plus the rate at which momentum flows out of the control volume. Stm. 4.7-2

To obtain the rate at which momentum flows out of the control volume, recall that the flux vectors associated with the x, y, and z-components of the momentum vector are, respectively, $(\rho/g_c)v_1\mathbf{v}$, $(\rho/g_c)v_2\mathbf{v}$, and $(\rho/g_c)v_3\mathbf{v}$. Then the mathematical expression of Newton's law as applied to the control volume is

$$\underbrace{\iiint_V \mathbf{F}\,dV + \oint\!\!\oint_S \boldsymbol{\tau}\,dS}_{\substack{\text{Sum of the force}\\ \text{distributions acting}\\ \text{on the fluid in the}\\ \text{control volume}}} = \frac{d}{dt}\iiint_V \frac{\rho}{g_c}\mathbf{v}\,dV + \oint\!\!\oint_S \frac{\rho}{g_c}\mathbf{v}(\mathbf{v}\cdot d\mathbf{S})$$

$$= \underbrace{\iiint_V \frac{\partial}{\partial t}\left(\frac{\rho}{g_c}\mathbf{v}\right)dV}_{\substack{\text{Time rate of}\\ \text{change of fluid}\\ \text{momentum}\\ \text{inside control}\\ \text{volume}}} + \underbrace{\oint\!\!\oint_S \frac{\rho}{g_c}\mathbf{v}(\mathbf{v}\cdot d\mathbf{S})}_{\substack{\text{Rate at}\\ \text{which}\\ \text{momentum}\\ \text{flows out}\\ \text{control}\\ \text{volume}}} \qquad \text{Eq. 4.7-4}$$

We see that Eqs. 4.7-4 and 4.7-3 are identical and that at this stage of development no distinction need be made regarding whether the region of integration represents a control volume or a thermodynamic system.

4.7.1 SOME APPLICATIONS OF NEWTON'S SECOND LAW IN FLUID-FLOW PROBLEMS

In order to determine the fluid forces that act on the vanes of a turbine rotor or the walls of curved pipes and elbows, and the like, we use Newton's second law. The following examples illustrate the method.

Example 4-9

Determine the force exerted by the liquid on the curved pipe wall shown in Fig. 4-25. Take the pipe centerline to be horizontal and the flow through the section to be steady. Neglect viscous effects and consider the properties to be uniform at each section.

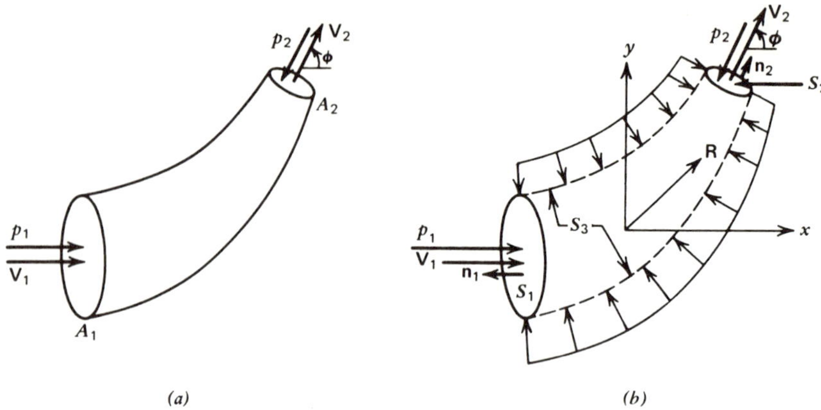

Fig. 4-25. Section of curved pipe wall and control volume. (*a*) Section of pipe wall. (*b*) Control volume.

SOLUTION

Select a control volume shown by the dashed line in Fig. 4-25*b*. Select a coordinate system so that the (x, y) plane is horizontal and z points in upward direction. Now apply Eq. 4.7-4 to the control volume. Since the flow is steady, the first integral on the right-hand side of the equation is zero. Since the only body force on the fluid is gravitational, $\mathbf{F} = -\gamma\mathbf{k}$. We may take the bounding surface of our control volume to consist of the three surfaces, S_1, S_2, and S_3. Then, taking $\boldsymbol{\tau} = -p\mathbf{n}$ on surface S_1 and S_2, we obtain

$$-\iiint_{V} \gamma\mathbf{k}\, dV - \oint\!\!\oint_{S_1} p\mathbf{n}\, dS - \oint\!\!\oint_{S_2} p\mathbf{n}\, dS + \oint\!\!\oint_{S_3} \boldsymbol{\tau}\, dS$$

$$= \oint\!\!\oint_{S_1} \frac{\rho}{g_c}\mathbf{v}(\mathbf{v}\cdot d\mathbf{S}) + \oint\!\!\oint_{S_2} \frac{\rho}{g_c}\mathbf{v}(\mathbf{v}\cdot d\mathbf{S}) + \oint\!\!\oint_{S_3} \frac{\rho}{g_c}\mathbf{v}(\mathbf{v}\cdot d\mathbf{S}) \qquad \text{Eq. 4.7-5}$$

Since there is no flow through the solid boundary, $(\mathbf{v}\cdot d\mathbf{S})$ on S_3 is zero. Thus, the last integral on the right-hand side of the above equation is zero. Taking γ to be a constant and the properties at each cross section to be uniform, we may move the properties outside the integral signs on the volume integral and those integrals on surfaces S_1 and S_2, giving

$$-\gamma \forall \mathbf{k} - p_1(-\mathbf{i})A_1 - p_2(\mathbf{n}_2)A_2 + \oiint_{S_3} \boldsymbol{\tau}\, dS$$

$$= \frac{\rho}{g_c}(V_1\mathbf{i})(-V_1A_1) + \frac{\rho}{g_c}(V_2\mathbf{n}_2)(V_2A_2) \qquad \text{Eq. 4.7-6}$$

The integration of $\boldsymbol{\tau}$ on surface S_3 gives the resultant force that the pipe wall exerts on the fluid. We designate this resultant force as $\mathbf{R} = R_1\mathbf{i} + R_2\mathbf{j} + R_3\mathbf{k}$. Taking $\mathbf{n}_2 = \cos\phi\mathbf{i} + \sin\phi\mathbf{j}$, we obtain the three scalar equations

$$p_1A_1 - p_2A_2\cos\phi + R_1 = -\frac{\rho}{g_c}A_1{V_1}^2 + \frac{\rho}{g_c}A_2{V_2}^2\cos\phi$$

$$-p_2A_2\sin\phi + R_2 = \frac{\rho}{g_c}A_2{V_2}^2\sin\phi$$

$$-\gamma\forall + R_3 = 0 \qquad \text{Eqs. 4.7-7}$$

or

$$R_1 = p_2A_2\cos\phi - p_1A_1 + \frac{\rho}{g_c}(A_2{V_2}^2\cos\phi - A_1{V_1}^2)$$

$$R_2 = p_2A_2\sin\phi + \frac{\rho}{g_c}A_2{V_2}^2\sin\phi$$

$$R_3 = \gamma\forall \qquad \text{Eqs. 4.7-8}$$

Example 4-10

A horizontal jet impinges on a vane attached to a wall, as shown in Fig. 4-26. The vane is so shaped that it contains the jet stream in a horizontal plane. If experiments indicate that the cross-sectional area of the jet stream is approximately constant, determine the x-component of the force the stream exerts on the vane. Assume that the flow is steady and neglect gravitational effects.

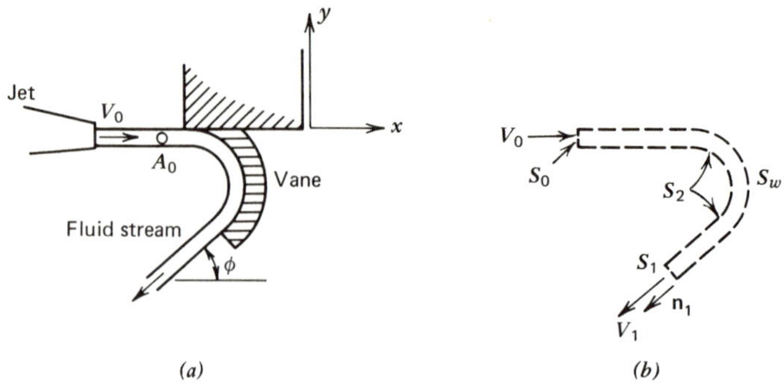

Fig. 4-26. Jet impinging on stationary vane. (*a*) Jet and vane. (*b*) Control volume.

SOLUTION

Apply Eq. 4.7-4 to the control volume shown in Fig. 4-26*b*; that is,

$$\underbrace{\oiint_S \boldsymbol{\tau}\, dS + \iiint_{\mathcal{V}} \mathbf{F}\, d\mathcal{V}}_{\text{Neglect}} = \underbrace{\iiint_{\mathcal{V}} \frac{\partial}{\partial t}\left(\frac{\rho}{g_c}\mathbf{v}\right) d\mathcal{V}}_{\text{Zero for steady flow}} + \oiint_S \frac{\rho}{g_c}\mathbf{v}(\mathbf{v}\cdot d\mathbf{S}) \qquad \text{Eq. 4.7-9}$$

Note that in this control volume, surfaces S_0, S_1, and S_2 are all open to the atmosphere and, therefore, we may take the pressures on these surfaces to be at atmospheric pressure. On the other hand, on S_w, the jet is in contact with the vane and, therefore, on S_w the pressure is not zero. Since atmospheric pressure acts on both sides of the vane, we may use gage pressure in our calculations. Then

$$\oiint_S \boldsymbol{\tau}\, dS = \oiint_{S_w} \boldsymbol{\tau}\, dS = R_1\mathbf{i} + R_2\mathbf{j} + R_3\mathbf{k} \qquad \text{Eq. 4.7-10}$$

where $\mathbf{R}$ is the resultant force that the vane exerts on the fluid. Since the cross-sectional area of the jet stream is a constant, $V_1 = V_0$, by mass conservation. Also notice that there is no flow through surfaces S_2 and S_w. Therefore,

$$\oiint_S \frac{\rho}{g_c}\mathbf{v}(\mathbf{v}\cdot d\mathbf{S}) = \frac{\rho}{g_c} V_0 \mathbf{i}(-V_0 A_0) + \frac{\rho}{g_c} V_1 \mathbf{n}_1 (V_1 A_1)$$

But

$$\mathbf{n}_1 = -\cos\phi\mathbf{i} - \sin\phi\mathbf{j}$$

giving

$$\oiint_S \frac{\rho}{g_c}\mathbf{v}(\mathbf{v}\cdot d\mathbf{S}) = \frac{\rho}{g_c}A_0V_0[-V_0\mathbf{i} - V_0(\cos\phi\mathbf{i} + \sin\phi\mathbf{j})] \quad \text{Eq. 4.7-11}$$

Substituting Eqs. 4.7-10 and 4.7-11 into Eq. 4.7-9 gives

$$R_1\mathbf{i} + R_2\mathbf{j} + R_3\mathbf{k} = \frac{\rho}{g_c}A_0V_0^2[(-1 - \cos\phi)\mathbf{i} - \sin\phi\mathbf{j}] \quad \text{Eq. 4.7-12}$$

or

$$\begin{aligned} R_1 &= -\frac{\rho}{g_c}A_0V_0^2(1+\cos\phi) = -\frac{\rho}{g_c}QV_0(1+\cos\phi) \\ R_2 &= -\frac{\rho}{g_c}A_0V_0^2\sin\phi = -\frac{\rho}{g_c}QV_0\sin\phi \\ R_3 &= 0 \end{aligned} \quad \text{Eqs. 4.7-13}$$

The x-component of the force that the jet exerts on the vane equals

$$-R_1 = \frac{\rho}{g_c}QV_0(1+\cos\phi) \quad \text{Eq. 4.7-14}$$

Control Volumes Moving at a Uniform Velocity

Suppose that in Example 4-10, the wall was moving in the x direction at a uniform velocity equal to U ft/s, while the nozzle remained stationary with respect to the ground. If in analyzing the problem, we selected a control volume as shown in Fig. 4-26*b*, but fixed to the ground, then there would be fluid flowing through surfaces S_2 and S_w, making the problem difficult to solve. However, if the selected control volume was considered attached to the vane, then again there would be no flow through surfaces S_2 and S_w. If we go back and examine the development of Statement 4.2-2, we find that the derivation holds for a control volume fixed with respect to a moving coordinate system, say (x, y, z), as long as the velocity in the $\mathbf{v} \cdot d\mathbf{S}$ and the time derivative are interpreted to be relative to (x, y, z); that is,

$$\frac{d}{dt_{xyz}} \iiint_{V_{sys}} \Phi(x, y, z, t)\, dV = \iiint_{C.V.} \frac{\partial \Phi}{\partial t_{xyz}}\, dV + \oiint_{C.S.} \Phi(\mathbf{v}_{xyz} \cdot d\mathbf{S})$$

$$= \frac{d}{dt_{xyz}} \iiint_{C.V.} \Phi\, dV + \oiint_{C.S.} \Phi(\mathbf{v}_{xyz} \cdot d\mathbf{S}) \quad \text{Eq. 4.7-15}$$

where the control volume is fixed in (x, y, z). A similar situation exists if Φ is replaced by a vector quantity $\mathbf{B}$. Under these circumstances Eq. 4.2-10 becomes

$$\frac{d}{dt_{xyz}} \iiint_{V_{sys}} \mathbf{B}\, dV = \iiint_{C.V.} \frac{\partial \mathbf{B}}{\partial t_{xyz}}\, dV + \oiint_{C.S.} \mathbf{B}(\mathbf{v}_{xyz} \cdot d\mathbf{S})$$

$$= \frac{d}{dt_{xyz}} \iiint_{C.V.} \mathbf{B}\, dV + \oiint_{C.S.} \mathbf{B}(\mathbf{v}_{xyz} \cdot d\mathbf{S}) \quad \text{Eq. 4.7.16}$$

Now Newton's second law as given by Eq. 4.7-1 must be written with respect to an *inertial reference frame*. However, we know from dynamics that, if (x, y, z) moves linearly at a constant speed with respect to the ground, then it can be considered to be an inertial reference. Therefore, Eq. 4.7-5 is valid for a control volume moving linearly at a constant speed.

Example 4-11

Suppose that in Example 4 10 the wall was moving in the X direction at a constant velocity equal to U ft/s. If the jet is stationary with respect to the ground, determine the force the jet stream exerts on the vane.

SOLUTION

Let (X, Y, Z) be a coordinate system that is fixed to the ground, and $(x\ y, z)$ be a coordinate system, parallel to (X, Y, Z), but fixed to the vane. Select a control volume that is fixed relative to (x, y, z) occupying the region shown in Fig. 4-27. Then Eq. 4.7-4 is still valid, except that $\mathbf{v}$, which appears on the right-hand side of the equation, is to be taken with respect to (x, y, z). From dynamics we know that

$$\mathbf{v}_{XYZ} = \mathbf{v}_{xyz} + U\mathbf{i}$$

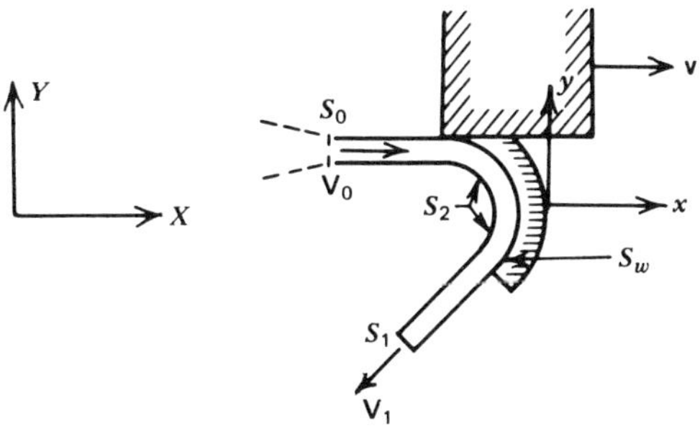

Fig. 4-27. Jet impinging on a moving vane. $\mathbf{V}_0$ = Velocity of jet relative to (X, Y, Z).

Thus,

$$\mathbf{v}_{xyz}(S_0) = V_0\mathbf{i} - U\mathbf{i} = (V_0 - U)\mathbf{i}$$

Again assuming that the cross-sectional area of the jet stream is a constant, we obtain, by mass conservation,

$$|\mathbf{v}_{xyz}(S_1)| = (V_0 - U)$$

It now becomes obvious that the components of the resultant force that the vane exerts on the fluid are the same as those given by Eqs. 4.7-13, except that V_0 is replaced by $(V_0 - U)$; that is,

$$R_1 = -\frac{\rho}{g_c} Q(V_0 - U)(1 + \cos\phi)$$

$$R_2 = -\frac{\rho}{g_c} Q(V_0 - U)\sin\phi \qquad \text{Eqs. 4.7-17}$$

$$R_3 = 0$$

where Q here is given by $A_0(V_0 - U)$. The force components that the jet stream exerts on the vane are equal and opposite to those given by Eqs. 4.7-17.

The determination of the thrust created by a jet engine provides another interesting application for Newton's second law. We may demonstrate the versatility of the basic momentum equation (Eq. 4.7-5) by applying it to a

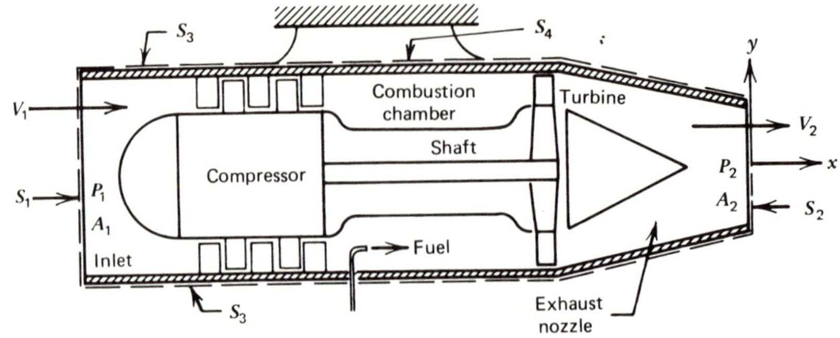

Fig. 4-28. Typical jet engine (plan view).

complex engineering system, namely the gas turbine-jet engine, shown in Fig. 4-28.

Despite its complex inner workings, the jet engine's function is very simply to accelerate the gas flow. To accomplish this, the gas is first compressed by a compressor. Then its temperature and pressure are raised in the combustion chamber. From there, the gas flows through a turbine, creating the power to drive the compressor. Finally, the gas flows through the exhaust nozzle at a much higher velocity than it entered. In order for the gas to accelerate, the engine walls must exert a net force on the gas in the same direction as its acceleration. An equal and opposite reaction force, known as *thrust*, is exerted on the engine.

The ram jet is merely a simplistic jet engine with few, if any, internal parts, but it will not operate until the vehicle to which it is attached has first been brought up to some minimum initial forward velocity. Similarly, the rocket is simply a jet engine with no inlet, since it carries its own supply of gas onboard. The engine-propeller combination is a jet engine with no casing. The airstream directly affected by the propeller's plane, known as the slip stream, is the relevant control volume otherwise defined by the jet engine's casing. From the standpoint of the momentum equation, their analyses are all similar.

In our analysis of the jet engine, we treat the system as a one-dimensional flow situation and further assume, for simplicity, that the engine's orientation is horizontal so that we may ignore gravitational forces. In addition, we simplify the analysis by considering the jet engine to be moving at a uniform speed under steady-state conditions only. Our reference coordinates, of course, are those attached to the engine itself. The control volume is the region inside the dashed line, shown in Fig. 4-28, and is considered to be fixed with respect to the engine.

At this point we consider a term-by-term review of Eq. 4.7-4, the basic momentum formulation of Newton's second law. For the selected control volume, Eq. 4.7-4 becomes

$$-\iiint_{\mathcal{V}} \gamma \mathbf{k}\, d\mathcal{V} - \oint\!\!\oint_{S_1} p\mathbf{n}\, dS - \oint\!\!\oint_{S_2} p\mathbf{n}\, dS + \oint\!\!\oint_{S_3} \boldsymbol{\tau}\, dS + \oint\!\!\oint_{S_4} \boldsymbol{\tau}\, dS$$

$$= \oint\!\!\oint_{S_1} \frac{\rho}{g_c} \mathbf{v}(\mathbf{v} \cdot d\mathbf{S}) + \oint\!\!\oint_{S_2} \frac{\rho}{g_c} \mathbf{v}(\mathbf{v} \cdot d\mathbf{S}) + \oint\!\!\oint_{S_3+S_4} \frac{\rho}{g_c} \mathbf{v}(\mathbf{v} \cdot d\mathbf{S}) \qquad \text{Eq. 4.7-18}$$

Looking first at the right-hand side of the above equation, we note that because of the horizontal orientation of the engine the velocities are all directed along the x-axis. The first two terms then become

$$\left(-\frac{\rho_1 V_1 A_1}{g_c} V_1 + \frac{\rho_2 V_2 A_2}{g_c} V_2\right)\mathbf{i}$$

where A_1 and A_2 are the areas, respectively, defined by

$$A_1 = \oint\!\!\oint_{S_1} dS \qquad \text{and} \qquad A_2 = \oint\!\!\oint_{S_2} dS$$

Note that $\rho_1 V_1 A_1 = \dot{m}_1$ and $\rho_2 V_2 A_2 = \dot{m}_2$, where $\dot{m}$ is the mass flow rate, steady with respect to time, at any point within the system. The mass flow rate through the exhaust nozzle, $\dot{m}_2$, exceeds the inlet mass flow, $\dot{m}_1$, by the fuel introduced in the combustion chambers. Thus,

$$\dot{m}_2 = (1 + m_f)\dot{m}_1$$

where m_f = engine fuel/air ratio.

The third term on the right-hand side of Eq. 4.7-18 is zero, since there is no flow through surfaces S_3 and S_4. Thus, the right-hand side reduces to

$$\frac{\dot{m}_1}{g_c}[(1 + m_f)V_2 - V_1]\mathbf{i}$$

Turning now to the left-hand side of Equation 4.7-18, we note that the first term

$$-\oint\!\!\oint_{\mathcal{V}} \gamma \mathbf{k}\, d\mathcal{V}$$

represents the weight of the fluid and other engine parts inside the engine casing. This vertical force component must be balanced by another term on the left-hand side of the equation, since there are no terms on the right-hand side directed vertically. (This balancing term is R_3, discussed a few lines down.) Because of the unidirectional nature of the flow through the jet engine, and because A_1 and A_2 are normal to the flow path, the second and third terms become simply

$$(p_1 A_1 - p_2 A_2)\mathbf{i}$$

It should be realized that the solid boundaries within the engine (the engine casing, the compressor blades, the walls of the combustion chambers, the turbine blades, and so on) all exert a force on the fluid inside the control volume. The equal and opposite force that the fluid exerts on these parts must then be balanced by the shear and normal stresses on the engine's support structure. We may then obtain the fluid's thrust on the engine by evaluating the internal force, **R**, existing in the support structure member. This internal force enters our analysis because our control surface cuts the member in question. That is,

$$\oiint_{S_4} \boldsymbol{\tau}\, dS = \mathbf{R} = R_1\mathbf{i} + R_2\mathbf{j} + R_3\mathbf{k}$$

The integration of $\oiint_{S_3} \boldsymbol{\tau}\, dS$ gives the resultant aerodynamic force acting on the engine casing. Its evaluation for an engine in flight is a complex problem and is usually neglected in this type of analysis. This is equivalent to assuming the $p(S_3)$ equals 0 psig.[7] Thus, we take

$$\oiint_{S_3} \boldsymbol{\tau}\, dS = 0$$

It should be noted that p_1 and p_2 are also to be measured in gage pressure. Then the x-component of Eq. 4.7-18 becomes

$$p_1 A_1 - p_2 A_2 + R_1 = \frac{\dot{m}_1}{g_c}\left[(1 + m_f)V_2 - V_1\right]$$

[7] A more thorough analysis of this aspect of the problem is given in Example 4-13.

In many cases our ultimate concern with a jet engine is the force, $\mathbf{F}_f$, exerted *by* the fluid *on* the engine (and, hence, on the aircraft). From Newton's basic laws of motion we recognize that $\mathbf{F}_f$ equals the force exerted by the fluid on the engine, which equals $-\mathbf{R}$. The component of $\mathbf{F}_f$ in the direction of flight is defined as the thrust, F_T, of the engine. For our coordinate system selection, the flight direction is in the $-\mathbf{i}$ direction. Therefore,

$$F_T = (-\mathbf{i})\cdot(-\mathbf{R}) = R_1$$

Thus, with some rearrangement, we finally see that the engine thrust, as derived from the basic momentum principle. Eq. 4.7-18, is

$$F_T = \frac{\dot{m}_1}{g_c}(1 + m_f)V_2 - \frac{\dot{m}_1}{g_c}V_1 + p_2 A_2 - p_1 A_1 \qquad \text{Eq. 4.7-19}$$

We identify the individual terms of 4.7-19 as

$$\frac{\dot{m}_1}{g_c}(1 + m_f)V_2 = \text{gross jet thrust}$$

$$\frac{\dot{m}_1}{g_c}V_1 = \text{ram drag (a resistance)}$$

$$(p_2 A_2 - p_1 A_1) = \text{relative pressure thrust}$$

It should be noted that the values of V_2 (or V_{jet}), p_1, m_f, and p_2 are complex functions of the internal aerodynamics and thermodynamics of the engine itself and are not predicted by the simple momentum analysis given here. We merely show that the thrust of the jet engine can be computed from the momentum analysis *if* we know the inlet and outlet conditions.

It can be easily shown, by a similar analysis to that above, that the thrust for a typical liquid-fueled rocket (see Fig. 4-29) is given as

$$F_T = \left(\frac{\dot{m}_f}{g_c}V_{\text{jet}} + p_2 A_2\right) \qquad \text{Eq. 4.7-20}$$

where $\dot{m}_f$ is the combined flow rate of fuel plus oxidizer.

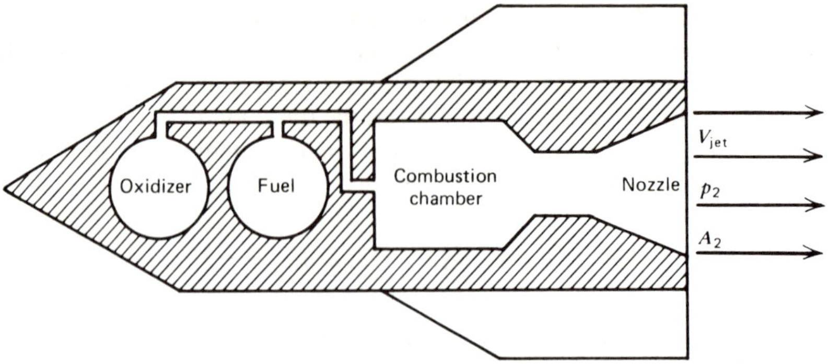

Fig. 4-29. Typical rocket.

Example 4-12

A typical rocket utilizes a liquid fuel–oxidizer combination, which it burns at the rate of 36,000 lb_m/h. The hot gases leave the nozzle with a velocity of 2000 ft/s relative to the rocket body. The static pressure in the rocket exhaust is the same as that in the surrounding atmosphere. The diameter of the exhaust nozzle (and thus the jet) is 8 in., and the rocket carries a payload of 200 lb_m and has (at the moment of concern) a total weight of 400 lb_f. The rocket velocity is 1000 ft/s. Find the engine thrust and the instantaneous horsepower developed.

SOLUTION

$$F_T = \left(\frac{\dot{m}_f}{g_c} V_{jet} + p_2 A_2\right)$$

Note that $p_2 = 0$ and thus $p_2 A_2 = 0$

$$F_T = \frac{10 \times 2000}{32.2} = 620 \text{ lb}_f$$

then the power developed, P, is

$$P = \frac{F_T V_{rocket}}{550} = \frac{620 \times 1000}{550} = 1127 \text{ hp}$$

Accelerating Control Volume

We have shown in the previous section that in analyzing certain types of problems, it is useful to select a control volume that moves in space. This is true even when the control volume is accelerating. Note: Eqs. 4.7-15 and 4.7-16 are valid even if the moving coordinate system (x, y, z) is accelerating with respect to our fixed coordinate system (X, Y, Z). However, Eq. 4.7-1, the equation expressing Newton's second law, must be written with respect to an *inertial* reference frame. To obtain an appropriate equation let us start with Eq. 4.7-1 written with respect to (X, Y, Z), where (X, Y, Z) is fixed in space and is therefore an inertial reference frame; that is,

$$\iiint_{V_{sys}} \mathbf{F}\, dV + \oint\!\!\oint_{S_{sys}} \boldsymbol{\tau}\, dS = \frac{d}{dt_{XYZ}} \iiint_{V_{sys}} \rho \mathbf{v}_{XYZ}\, dV \qquad \text{Eq. 4.7-1}$$

From dynamics we know that

$$\frac{d}{dt_{XYZ}}(\mathbf{A}) = \frac{d}{dt_{xyz}}(\mathbf{A}) + \boldsymbol{\omega} \times \mathbf{A} \qquad \text{Eq. 4.7-21}$$

$$\mathbf{v}_{XYZ} = \mathbf{v}_{xyz} + \mathbf{v}_{rf} + \boldsymbol{\omega} \times \boldsymbol{\zeta} \qquad \text{Eq. 4.7-22}$$

where $\boldsymbol{\omega}$ and $\mathbf{v}_{rf}$ are, respectively, the angular velocity and velocity of (x, y, z) as seen from (X, Y, Z); and $\boldsymbol{\zeta}$ is the position vector of a fluid particle as seen from (x, y, z). Applying Eq. 4.7-21 to Eq. 4.7-1, it becomes

$$\iiint_{V_{sys}} \mathbf{F}\, dV + \oint\!\!\oint_{S_{sys}} \boldsymbol{\tau}\, dS = \frac{d}{dt_{xyz}} \iiint_{V_{sys}} \rho \mathbf{v}_{XYZ}\, dV + \boldsymbol{\omega} \times \iiint_{V_{sys}} \rho \mathbf{v}_{XYZ}\, dV \qquad \text{Eq. 4.7-23}$$

If we now consider that at time t our control volume coincides with our thermodynamic system, we can utilize Eq. 4.7-16 giving

$$\iiint_{\text{C.V.}} \mathbf{F}\, dV + \oint\!\!\oint_{\text{C.S.}} \boldsymbol{\tau}\, dS = \frac{d}{dt_{xyz}} \iiint_{\text{C.V.}} \rho \mathbf{v}_{XYZ}\, dV + \oint\!\!\oint_{\text{C.S.}} \rho \mathbf{v}_{XYZ}(\mathbf{v}_{xyz} \cdot d\mathbf{S})$$

$$+ \boldsymbol{\omega} \times \iiint_{\text{C.V.}} \rho \mathbf{v}_{XYZ}\, dV \qquad \text{Eq. 4.7-24}$$

In addition, conservation of mass is valid, regardless from which coordinate system the phenomenon is observed. Therefore,

$$\frac{\partial}{\partial t} \underset{\text{C.V.}}{\oiiint} \rho \, d\forall + \underset{\text{C.S.}}{\oiint} \rho \mathbf{v}_{xyz} \cdot d\mathbf{S} = 0 \qquad \text{Eq. 4.7-25}$$

An example illustrating the use of Eq. 4.7-24 is given below.

Example 4-13

A rocket is launched vertically from rest. The initial total mass of the system (rocket plus fuel) is M_0 lb_m, and the exhaust mass flow rate is $\dot{m}_e$ $\text{lb}_\text{m}/\text{s}$. The exhaust gas gage pressure and velocity relative to the rocket are p_e and v_e, respectively. The exhaust area is A_e. Except for the exhaust area, neglect the aerodynamic pressure forces. Neglecting viscous effects, determine the velocity and elevation of the rocket as a function of time.

SOLUTION

Select a control volume attached to the rocket as shown by the dashed line in Fig. 4-30. Now apply Eq. 4.7-24 to the accelerating control volume, noting that $\boldsymbol{\omega} = 0$; then

$$\underset{\forall}{\oiiint} \mathbf{F} \, d\forall + \underset{S}{\oiint} \boldsymbol{\tau} \, dS = \frac{d}{dt_{xyz}} \underset{\forall}{\oiiint} \frac{\rho}{g_c} \mathbf{v}_{XYZ} \, dV + \underset{S}{\oiint} \frac{\rho}{g_c} \mathbf{v}_{XYZ}(\mathbf{v}_{xyz} \cdot d\mathbf{S}) \qquad \text{Eq. 4.7-26}$$

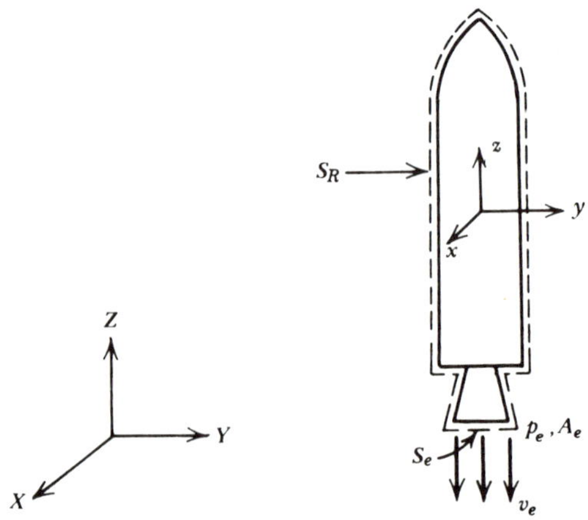

Fig. 4-30. Accelerating rocket and control volume.

The first term on the left-hand side of the above equation gives the gravitational effect on the rocket; that is,

$$\oiiint_{V} \mathbf{F}\, dV = \oiiint_{V} (-\gamma\mathbf{k})\, dV = W\mathbf{k} = -\frac{Mg}{g_c}\mathbf{k}$$

where M and W are, respectively, the total mass and weight of the rocket plus fuel. M and W are functions of time. The second term on the left-hand side of the equation gives the total surface forces acting on the rocket. The total surface area can be considered to be composed of two parts; first, the surface area at the engine exhaust, S_e, and second, the rest of the surface area, S_R. Thus,

$$\oiint_{S} \boldsymbol{\tau}\, dS = \oiint_{S_R} \boldsymbol{\tau}\, dS + \oiint_{S_e} \boldsymbol{\tau}\, dS \qquad \text{Eq. 4.7-27}$$

On S_e we may neglect viscous effects and take

$$\boldsymbol{\tau}(S_e) = p_e\mathbf{k}$$

Rocket engines are usually designed so that p_e will be greater than the base pressure, p_b. The base pressure is the pressure that would exist at surface S_e if the engine was turned off. Let the free-stream pressure (the pressure some distance away from the rocket) be p_∞; we may then write

$$p_e = p_b + (p_e - p_\infty) - (p_b - p_\infty)$$

and

$$\boldsymbol{\tau}(S_e) = [p_b + (p_e - p_\infty) - (p_b - p_\infty)]\mathbf{k}$$

Substituting the above equation into Eq. 4.7-27 gives

$$\oiint_{S} \boldsymbol{\tau}\, dS = \oiint_{S_R} \boldsymbol{\tau}\, dS + \oiint_{S_e} p_b\mathbf{k}\, dS + \oiint_{S_e} (p_e - p_\infty)\mathbf{k}\, dS - \oiint_{S_e} (p_b - p_\infty)\mathbf{k}\, dS$$

Eq. 4.7-28

The sum of the first two terms on the right-hand side of the above equation represents the total drag that the rocket would experience, if there was no exhaust from the rocket engine. We will designate this sum as $\text{Drag}_{\text{engine off}}$. Values for this term could be obtained through wind tunnel testing. Many rocket engines are frequently designed so that p_e is a constant. However, p_∞ varies with altitude and may be taken to be the pressure of the U.S. Standard Atmosphere, as discussed in Section 3.8. Values of $(p_b - p_\infty)$ could also be obtained through wind-tunnel testing and would usually be represented by a local pressure coefficient, $\tilde{c}_p$, which is defined by the following equation

$$\tilde{c}_p = \frac{p - p_\infty}{\dfrac{\rho V_\infty{}^2}{2g_c}}$$

At the higher altitudes $(p_b - p_\infty)$ would become negligible. Substituting these concepts into Eq. 4.7-28, it becomes

$$\oint\!\!\oint_S \boldsymbol{\tau}\, ds = \left(-\text{Drag}_{\text{engine off}} + (p_e - p_\infty)A_e - \frac{\rho V_\infty{}^2}{2g_c}\tilde{c}_{p,b}A_e\right)\mathbf{k}$$

where $\tilde{c}_{p,b}$ is the local pressure coefficient at the base of the rocket.

In evaluating the first integral on the right-hand side of Eq. 4.7-26, we must realize that the exact nature of the fluid-flow field inside the control volume is not known. Therefore, we approximate this integral by assuming that the velocity with respect to (x, y, z) of the gas inside the rocket is zero, but take the velocity at the exhaust boundary to be v_e. Thus, the velocity of all material inside the control with respect to (X, Y, Z) is $V_R\mathbf{k}$. Hence,

$$\frac{d}{dt_{xyz}}\oint\!\!\oint\!\!\oint_{\mathcal{V}} \frac{\rho}{g_c}\mathbf{v}_{XYZ}\, d\mathcal{V} = \frac{d}{dt_{xyz}}\left(\frac{V_R}{g_c}\mathbf{k}\oint\!\!\oint\!\!\oint_{\mathcal{V}} \rho\, d\mathcal{V}\right) = \frac{d}{dt_{xyz}}\left(\frac{M}{g_c}V_R\mathbf{k}\right)$$

But

$$\frac{d}{dt_{xyz}}\left(\frac{MV_R}{g_c}\mathbf{k}\right) = \frac{d}{dt_{XYZ}}\left(\frac{MV_R}{g_c}\mathbf{k}\right) = \frac{M}{g_c}\frac{dV_R}{dt}\mathbf{k} + \frac{V_R}{g_c}\frac{dM}{dt}\mathbf{k}$$

Since in the above case, the time derivative with respect to (X, Y, Z) is the same as that for (x, y, z), we have dropped the subscripting on the right-hand side of the above equation. In evaluating the second term on the right-hand side of Eq. 4.7-26, we note that fluid crosses the control surface only at S_e. Therefore

$$\mathbf{v}_{XYZ}(S_e) = V_R\mathbf{k} - v_e\mathbf{k}.$$

The minus sign occurs because the exhaust velocity relative to the rocket points in the $(-\mathbf{k})$ direction. Therefore,

$$\oiint_S \frac{\rho}{g_c} \mathbf{v}_{XYZ}(\mathbf{v}_{xyz} \cdot d\mathbf{S}) = \oiint_{S_e} \frac{\rho_e}{g_c} (V_R \mathbf{k} - v_e \mathbf{k})[-v_e \mathbf{k} \cdot (-\mathbf{k})\, dS]$$

$$= \left(\frac{\rho_e}{g_c} V_R v_e A_e - \frac{\rho_e}{g_c} v_e^2 A_e\right)\mathbf{k}$$

Collecting terms, we obtain

$$-\text{Drag}_{\text{engine off}} + (p_e - p_\infty)A_e - \frac{\rho V_\infty^2}{2g_c}\tilde{c}_{p,b} A_e - \frac{Mg}{g_c}$$

$$= \frac{M}{g_c}\frac{dV_R}{dt} + \frac{V_R}{g_c}\frac{dM}{dt} + \frac{\rho_e}{g_c} A_e v_e (V_R - v_e) \qquad \text{Eq. 4.7-29}$$

Note that $\rho_e A_e v_e$ equals the exhaust mass flow rate, $\dot{m}_e$, and that the mass of the rocket plus fuel is given by

$$M = M_0 - \dot{m}_e t, \qquad \text{with } \frac{dM}{dt} = -\dot{m}_e$$

Substituting these relations into Eq. 4.7-29 gives

$$-\text{Drag}_{\text{engine off}} + (p_e - p_\infty)A_e - \frac{\rho V_\infty^2}{2g_c}\tilde{c}_{p,b} A_e - \frac{Mg}{g_c}$$

$$= \frac{M}{g_c}\frac{dV_R}{dt} - \frac{V_R}{g_c}\dot{m}_e + \frac{V_R}{g_c}\dot{m}_e - \frac{\dot{m}_e}{g_c}v_e = \frac{M}{g_c}\frac{dV_R}{dt} - \frac{\dot{m}_e v_e}{g_c} \qquad \text{Eq. 4.7-30}$$

In general, the integration of Eq. 4.7-30 can only be accomplished by numerical methods. However, a closed-form solution of this equation can be obtained, if the first three terms on the left-hand side of the equation are neglected. This would provide us with a check on any numerical scheme that might be used to integrate Eq. 4.7-30. This is done below. Neglecting those terms mentioned above, Eq. 4.7-30 reduces to, after some rearranging, the following equation

$$\frac{dV_R}{dt} = \frac{\dot{m}_e v_e}{M} - g = \frac{\dot{m}_e v_e}{M_0 - \dot{m}_e t} - g = f_1(t) \qquad \text{Eq. 4.7-31}$$

This equation may be integrated giving

$$V_R = \dot{m}_e v_e \int \frac{1}{M_0 - \dot{m}_e t} dt - gt + C_1$$

$$= -v_e \ln (M_0 - \dot{m}_e t) - gt + C_1$$

At $t = 0$, $V_R = 0$. Therefore,

$$C_1 = v_e \ln M_0$$

and

$$V_R = v_e \ln \frac{M_0}{M_0 - \dot{m}_e t} - gt \qquad \text{Eq. 4.7-32}$$

We can integrate this expression and obtain the elevation of the rocket as a function of time. Noting that

$$\frac{dZ_R}{dt} = V_R \qquad \text{Eq. 4.7-33}$$

we have

$$Z_R = -\frac{gt^2}{2} + v_e\left[(\ln M_0)t - \int \ln (M_0 - \dot{m}_e t)\, dt\right] + c_2$$

Noting that $\int \ln u\, du = u \ln u - u$, the above equation becomes

$$Z_R = -\frac{gt^2}{2} + v_e\left\{t \ln M_0 + \frac{1}{\dot{m}_e}(M_0 - \dot{m}_e t)[\ln (M_0 - \dot{m}_e t) - 1]\right\} + c_2$$

$$= -\frac{gt^2}{2} + v_e\left\{t\left(\ln \frac{M_0}{M_0 - \dot{m}_e t} + 1\right) + \frac{M_0}{\dot{m}_e}[\ln (M_0 - \dot{m}_e t) - 1]\right\} + c_2$$

At $t = 0$, $Z_R = 0$, giving

$$c_2 = -v_e \frac{M_0}{\dot{m}_e} (\ln M_0 - 1)$$

Substituting for c_2, the equation for Z_R becomes

$$Z_R = -\frac{gt^2}{2} + v_e \left[t\left(\ln \frac{M_0}{M_0 - \dot{m}_e t} + 1 \right) - \frac{M_0}{\dot{m}_e} \ln \frac{M_0}{M_0 - \dot{m}_e t} \right] - \frac{gt^2}{2}$$

Eq. 4.7-34

A Computer Problem

In this section we apply the Runge–Kutta method for numerically solving the simplified problem given in Example 4-13, mainly, Eqs. 4.7-31 and 4.7-33. In Chapter 5, the student is asked to numerically solve the more general problem given in Example 4-13, mainly, Eqs. 4.7-30 and 4.7-33.

To apply the Runge–Kutta formula to our problem, subdivide the time axis into N intervals. Then, the values for the dependent variables at time step t_{i+1}, in terms of the values of the variables at time step t_i are

$$V_{R,i+1} = V_{R,i} + \frac{\Delta t}{6}(r_1 + 2r_2 + 2r_3 + r_4)$$

$$Z_{R,i+1} = Z_{R,i} + \frac{\Delta t}{6}(q_1 + 2q_2 + 2q_3 + q_4)$$

Eqs. 4.7-35

where

$$r_1 = f_1(t_i) \qquad r_2 = f_1\left(t_i + \frac{\Delta t}{2}\right)$$

$$r_3 = f_1\left(t_i + \frac{\Delta t}{2}\right) \qquad r_4 = f_1(t_i + \Delta t)$$

Eqs. 4.7-36

and where f_1 is defined by Eq. 4.7-31. The q values are defined as follows:

$$q_1 = V_{R,i} \qquad q_2 = V_{R,i} + \frac{\Delta t}{2} r_1$$
$$q_3 = V_{R,i} + \frac{\Delta t}{2} r_2 \qquad q_4 = V_{R,i} + \Delta t\, r_3 \qquad \text{Eqs. 4.7-37}$$

The initial conditions,

$$V_R(0) = 0 \qquad \text{and} \qquad Z_R(0) = 0$$

provide the means for starting the calculations. Let us assume that

$$\dot{m} = 276 \text{ lb}_\text{m}/\text{s}, \qquad v_e = 6540 \text{ ft/s}, \qquad \text{and} \qquad M_0 = 28{,}000 \text{ lb}_\text{m}$$

Also assume that g = constant = 32.17 ft/s^2. Take the power cutoff time, t_f, to be 66 s. The authors' computer program for solving the above problem by the Runge–Kutta formula is given below. The solution was obtained for N = 100. It is not anticipated that the student will have trouble in recognizing the terms in the program, since the Fortran variable symbols are very close to their corresponding written symbols used in this section. Values for velocity and altitude obtained from Eqs. 4.7-32 and 4.7-33 are also calculated in the program. These are represented by the symbols VRTH and ZRTH, respectively. The program follows.

```
C     COMPUTER PROGRAM FOR EVALUATING THE VELOCITY AND
C     ALTITUDE OF A ROCKET
      REAL MED,MO
      DIMENSION VR(2),ZR(2),R(4),Q(4),DTT(4)
      MED=276.
      MO=28000.
      VE=6540.
      DRAG=0.0
      TF=66.
      G=32.17
      GC=32.17
      N=100
      WRITE(108,50)
   50 FORMAT('1ROCKET PROBLEM')
      WRITE(108,54)MED,MO,VE,TF
   54 FORMAT(/'MASS FLOW RATE= ',F5.1,3X,'MASS OF ROCKET
     1LOADED= ',F7.1,//5X,'EXHAUST VELOCITY= ',F6.1.5X,'CUT
     2-OFF TIME= ',F6.2)
      DT=TF/N
```

```
      DTT(1)=0.0
      DTT(2)=DT/2.0
      DTT(3)=DT/2.0
      DTT(4)=DT
      VR(2)=0.0
      ZR(2)=0.0
      VRTH=0.0
      ZRTH=0.0
      WRITE(108,52)N
   52 FORMAT(22H0NUMBER OF INTERVALS= I5)
      WRITE(108,49)
   49 FORMAT(//11X,'NUMERICAL SOLUTION',7X,'CLOSED FORM
     1SOLUTION')
      WRITE(108,53)
   53 FORMAT(3X,'TIME',3X,'VELOCITY',5X,'ALTITUDE',5X,'VELOC
     1ITY',5X,'ALTITUDE')
      T=0.0
      N1=0
      WRITE(108,51)T,VR(2),ZR(2),VRTH,ZRTH
   51 FORMAT(F7.2,F11.2,E15.7,F11.2,E15.7)
      DO 2 I=1,N
      VR(1)=VR(2)
      ZR(1)=ZR(2)
      N1=N1+1
      DO 3 J=1,4
      TT=T+DTT(J)
      R(J)=(MED*VE-DRAG*GC)/(MO-MED*TT)-G
      IF(J.EQ.1)GO TO 4
      Q(J)=VR(1)+DTT(J)*R(J-1)
      GO TO 3
    4 Q(J)=VR(1)
    3 CONTINUE
      VR(2)=VR(1)+DT/6.0*(R(1)+2.0*R(2)+2.0*R(3)+R(4))
      ZR(2)=ZR(1)+DT/6.0*(Q(1)+2.0*Q(2)+2.0*Q(3)+Q(4))
      T=I*DT
      IF(N1.NE.10)GO TO 2
      N1=N1-10
      VRTH=VE*ALOG(MO/(MO-MED*T))-G*T
      ZRTH=VE*(T*(ALOG(MO/(MO-MED*T))+1.0)-MO/MED*
     1ALOG(MO/(MO-MED*T)))-5*G*T**2
      WRITE(108,51)T,VR(2),ZR(2),VRTH,ZRTH
    2 CONTINUE
      CALL EXIT
      END
```

The solution obtained from this program (for 100 time intervals) is as follows.

```
ROCKET PROBLEM
MASS FLOW RATE= 276.0    MASS OF ROCKET LOADED= 28000.0
      EXHAUST VELOCITY= 6540.0  CUT-OFF TIME= 66.00
NUMBER OF INTERVALS= 100
              NUMERICAL SOLUTION        CLOSED FORM SOLUTION
   TIME    VELOCITY      ALTITUDE       VELOCITY       ALTITUDE
    .00         .00   0000000E 00            .00    .0000000E 00
   6.60      227.62   .7348779E 03        227.62    .7349836E 03
  13.20      486.99   .3074381E 04        486.99    .3074371E 04
  19.80      783.04   .7243715E 04        783.03    .7244172E 04
  26.40     1121.96   .1350462E 05       1121.96    .1350481E 05
  33.00     1511.65   .2216502E 05       1511.65    .2216514E 05
  39.60     1962.43   .3359250E 05       1962.44    .3359275E 05
  46.20     2488.12   .4823349E 05       2488.13    .4823394E 05
  52.80     3107.80   .6664175E 05       3107.80    .6664244E 05
  59.40     3848.97   .8952238E 05       3848.99    .8952325E 05
  66.00     4753.27   .1178046E 06       4753.29    .1178063E 06
  *EXIT*
```

4.7.2 THE MOMENT-OF-MOMENTUM EQUATION

At the beginning of Section 4.7 the reader was introduced to the integral form of Newton's second law. Now we introduce the *moment-of-momentum equation* and its application to several rotating-flow configurations. The moment-of-momentum equation is an extension of Newton's second law.

For the arbitrary system within the continuum (see Fig. 4-24), the law is expressed as follows.

> The net moment developed by the external forces acting on the system equals the rate of increase in the moment of momentum in the system
>
> Stm. 4.7-3

The mathematical expression of the law is

$$\oiiint_{V_{sys}} \mathbf{r} \times \mathbf{F}\, d\mathcal{V} \oiint_{S_{sys}} \mathbf{r} \times \boldsymbol{\tau}\, dS = \frac{d}{dt} \oiiint_{V_{sys}} \frac{\rho}{g_c} \mathbf{r} \times \mathbf{v}\, d\mathcal{V} \qquad \text{Eq. 4.7-38}$$

Here the time derivative and velocity are to be taken with respect to an inertial reference frame, and **r** is the displacement vector about some fixed point in that reference frame. When a particular problem can be analyzed by considering a control volume fixed in an inertial reference frame, the application of Statement 4.2-2 gives the proper equation for the control-volume viewpoint. That is,

$$\oiiint_{\text{C.V.}} \mathbf{r} \times \mathbf{F}\, d\forall + \oiint_{\text{C.S.}} \mathbf{r} \times \boldsymbol{\tau}\, dS = \frac{d}{dt} \oiiint_{\text{C.V.}} \left(\frac{\rho}{g_c} \mathbf{r} \times \mathbf{v}\right) d\forall + \oiint_{\text{C.S.}} \frac{\rho}{g_c} (\mathbf{r} \times \mathbf{v})(\mathbf{v} \cdot d\mathbf{S})$$

Eq. 4.7-39

The moment-of-momentum equation finds its major application in the field of turbomachinery. In these devices, a rotating element either adds or extracts energy from the fluid passing through the machine. Pumps add energy, while turbines extract energy from the fluid. In a pump the rotating element is called an *impeller*. However, in a hydraulic turbine the rotating element is called a *runner*; in both the gas turbine and compressor it is a *rotor*. Turbines are classified as either *reaction* turbines or *impulse* turbines. In the reaction turbine, fluid enters the runner region over its entire outer circumference and therefore all runner blades react with the fluid simultaneously. In the impulse turbine, several jet streams impinge on rotating vanes attached to a wheel, as shown in Fig. 4-33. For each nozzle, only one blade reacts with the fluid at any instant of time. In the reaction turbine, the fluid velocity changes in magnitude and direction as it passes through the runner region. In the impulse turbine, the magnitude of the fluid velocity relative to the runner does *not* change as the fluid passes through the runner region.

Pumps and reaction turbines are further classified according to the direction of flow through the rotating element; these being either *radial*, *axial*, or *mixed*. In this textbook, we will only consider radial flow devices.

Suppose we consider a radial flow device (pump or turbine) with the rotating element rotating at a constant angular velocity (see Figs. 4-31 and 4-32). Let us select a control volume enclosing the rotating element (runner or impeller). The control surface is to contain the two end sections, thus cutting the shaft of the rotating element. The dashed lines in Figs. 4-31 and 4-32 represent the boundaries of our control volume. Also observe that our control volume contains both fluid and solid material. By this selection, the force that the blades exert on the fluid becomes an internal force inside the control volume. However, if the rotational speed, ω, of the runner or impeller is constant, the torque resulting from this force must be balanced by the torque developed by the internal stresses in the shaft. We thus evaluate the torque resulting from the fluid on the vanes indirectly; that is, from the torque developed by the internal shear stresses in the shaft at the section cut by our control surface.

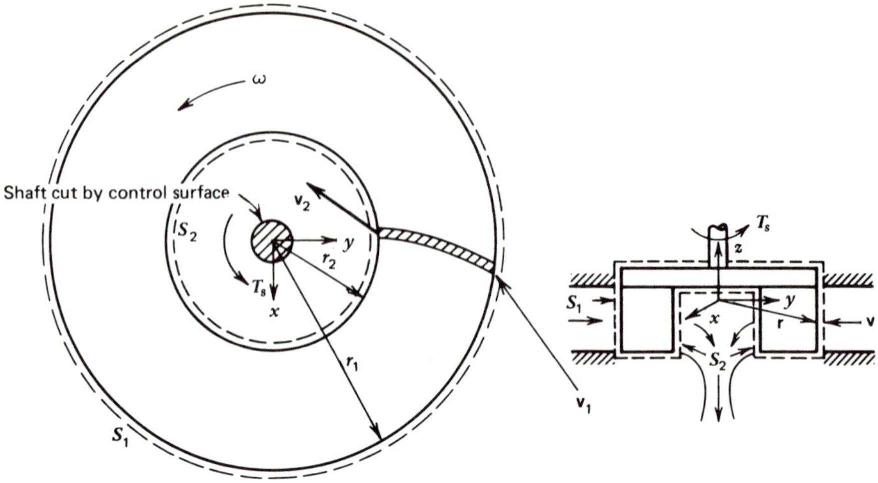

Fig. 4-31. Control volume for radial flow reaction turbine. Control volume is the region inside the dashed lines. Fluid enters at S_1 and leaves at S_2.

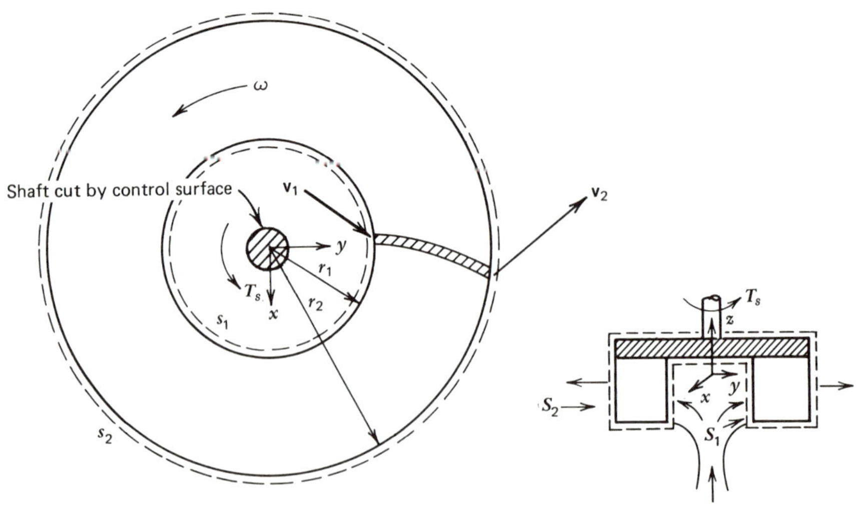

Fig. 4-32. Control volume for radial flow pump. Control volume is the region inside the dashed lines. Fluid enters at S_1 and leaves at S_2.

Suppose we select a coordinate system with the Z-axis coinciding with the axis of rotation. Our interest here is the **k** component of Eq. 4.7-39. Let us examine each term in this equation. Proceeding from left to right, we observe that: (1) if the axis of rotation is vertical, then

$$\mathbf{k} \cdot (\mathbf{r} \times \mathbf{F}) = -\mathbf{k} \cdot (\mathbf{r} \times \gamma \mathbf{k}) = -(\mathbf{r} \times \gamma \mathbf{k}) \cdot \mathbf{k} = -\mathbf{r} \cdot (\gamma \mathbf{k} \times \mathbf{k}) = 0$$

and (2) if the axis of rotation is horizontal, then, due to symmetry, the net torque caused by the gravitational field is zero. In either case

$$\mathbf{k} \cdot \underset{\text{C.V.}}{\oiiint} (\mathbf{r} \times \mathbf{F})\, d\mathcal{V} = 0$$

Next, we observe that the pressure field on all control surfaces does not produce any torque, since the pressure forces are radial or axial. However, there is a torque developed by the internal shear stresses at the section of the shaft cut by our control surface. We will designate this torque as T_S. Since, T_S, is considered as an unknown, we take it to be in the positive **k** direction. Then,

$$\mathbf{k} \cdot \underset{\text{C.S.}}{\oiint} \mathbf{r} \times \boldsymbol{\tau}\, dS = T_S$$

In evaluating the first term on the right-hand side of Eq. 4.7-39, we should realize that our control volume contains both fluid and the rotating element. Since the positions of the blades (those attached to the rotating element) change with time, the flow cannot be truly classified as a steady flow. That is, the velocity and density[8] at points within the control volume change with time. However, if (for the constant angular speed case) we consider the blades to be infinitesimally thin and the fluid velocity only a function of r, then the nonsteady aspects of the problem can be neglected. That is,

$$\frac{d}{dt} \underset{\text{C.V.}}{\oiiint} \frac{\rho}{g_c} (\mathbf{r} \times \mathbf{v})\, d\mathcal{V} = 0$$

In evaluating the last term, take

$$\mathbf{r} \times \mathbf{v} = (z\mathbf{k} + r\mathbf{e}_r) \times (v_r \mathbf{e}_r + v_\theta \mathbf{e}_\theta) = r v_\theta \mathbf{k} + z v_r \mathbf{e}_\theta - z v_\theta \mathbf{e}_r$$

[8] The material located at a particular point in space alternates between being fluid and solid (blade material).

Also observe that fluid only flows through surfaces S_1 and S_2; therefore,

$$\mathbf{k} \cdot \underset{\text{C.S.}}{\oiint} \frac{\rho}{g_c} (\mathbf{r} \times \mathbf{v})(\mathbf{v} \cdot d\mathbf{S}) = \underset{S_1}{\oiint} \frac{\rho}{g_c} r v_\theta (\mathbf{v} \cdot d\mathbf{S}) + \underset{S_2}{\oiint} \frac{\rho}{g_c} r v_\theta (\mathbf{v} \cdot d\mathbf{S})$$

For the turbine, fluid flows radially inward and thus

$$\underset{S_1}{\oiint} (\mathbf{v} \cdot d\mathbf{S}) = -Q \qquad \text{and} \qquad \underset{S_2}{\oiint} (\mathbf{v} \cdot d\mathbf{S}) = Q$$

Thus,

$$\mathbf{k} \cdot \underset{\text{C.S.}}{\oiint} \frac{\rho}{g_c} (\mathbf{r} \times \mathbf{v})(\mathbf{v} \cdot d\mathbf{S}) = \frac{\rho Q}{g_c} (r_2 v_{\theta,2} - r_1 v_{\theta,1})$$

For the turbine, the **k**-component of Eq. 4.7-39 becomes

$$T_S = \frac{\rho Q}{g_c} (r_2 v_{\theta,2} - r_1 v_{\theta,1})$$

In a turbine, v_θ, which is designated as the whirl, decreases as the fluid flows radially inward. Therefore, T_S is negative (the opposite direction to the rotation). Since the runner rotates at a constant angular velocity,

$$\sum_j \text{torques on the runner} = 0$$

Therefore, if T is the torque exerted *by the fluid on the runner*, then

$$T = -T_S = \frac{\rho Q}{g_c} (r_1 v_{\theta,1} - r_2 v_{\theta,2}) > 0 \qquad \text{Eq. 4.7-40}$$

The power, P, developed by the *turbine* is given by

$$P = \omega T = \frac{\rho Q \omega}{g_c} (r_1 v_{\theta,1} - r_2 v_{\theta,2}) \qquad \text{Eq. 4.7-41}$$

For the pump, fluid flows radially outward and thus

$$\oiint_{S_1} (\mathbf{v} \cdot d\mathbf{S}) = -Q \qquad \text{and} \qquad \oiint_{S_2} (\mathbf{v} \cdot d\mathbf{S}) = Q$$

For the pump, the **k**-component of Eq. 4.7-39 becomes

$$T_S = \frac{\rho Q}{g_c}(r_2 v_{\theta,2} - r_1 v_{\theta,1}) \qquad \text{Eq. 4.7-42}$$

The power, P, required to *drive the pump impeller* is

$$P = T\omega = \frac{\rho Q \omega}{g_c}(r_2 v_{\theta,2} - r_1 v_{\theta,1}) \qquad \text{Eq. 4.7-43}$$

But the power required to drive the pump impeller equals the power transmitted to the fluid, $(\delta W/\delta t)_p$. By the definition of the head developed by the pump, h_p, we can obtain the following relationship

$$P = \left(\frac{\delta W}{\delta t}\right)_p = \gamma Q h = \frac{\rho Q \omega}{g_c}(r_2 v_{\theta,2} - r_1 v_{\theta,1})$$

Thus,

$$h_p = \frac{\omega}{g}(r_2 v_{\theta,2} - r_1 v_{\theta,1}) = \frac{1}{g}(u_2 v_{\theta,2} - u_1 v_{\theta,1})$$

where $u =$ velocity of the impeller. It should be remembered that the velocity terms in Eqs. 4.7-40 to 4.7-43 are with respect to a fixed coordinate system, (X, Y, Z).

It should also be pointed out that the elementary angular momentum relationships developed here apply only to devices with pure radial flows. These expressions cannot be used in the case of mixed-flow turbines and pumps where both radial and axial flows exist. Treatment of these latter situations is beyond the intended scope of this book.

In a similar manner we can analyze the *Pelton wheel*-type turbine shown in Fig. 4-33. In this device a high-velocity jet stream impinges on a series of moving, hemispherical-shaped cups mounted on the circumference of the wheel, thus providing the torque required for rotation. Suppose we select a control volume

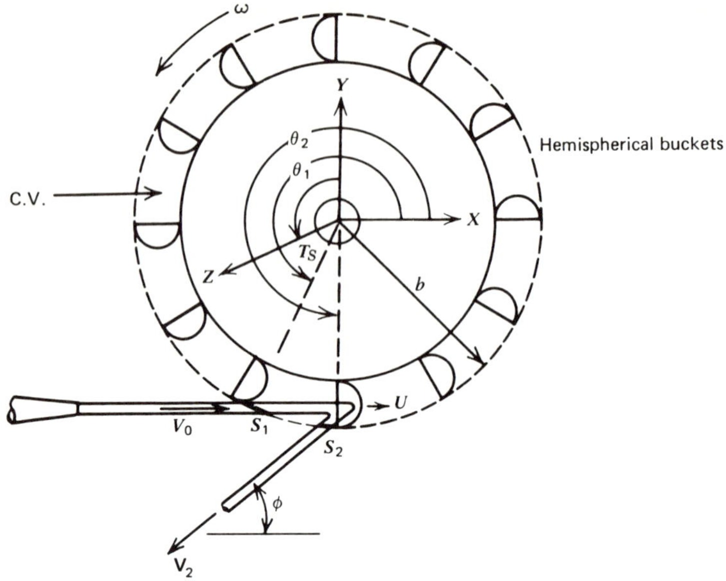

Fig. 4-33. The Pelton wheel turbine.

fixed in space, as shown by the dashed line in Fig. 4-33. Again consider the control surface to contain the two end sections cutting the wheel's shaft. We will consider the rotational speed, ω, to be a constant. By similar arguments to those used in analyzing the radial flow hydraulic turbine, the **k** component of Eq. 4.7-39 becomes

$$T_S = \mathbf{k} \cdot \oiint_{\text{C.S.}} \frac{\rho}{g_c} (\mathbf{r} \times \mathbf{v})(\mathbf{v} \cdot d\mathbf{S})$$

Here, we neglected the weight of the fluid inside the control volume. Since fluid crosses the control surface only at S_1 and S_2 (see Fig. 4-33), the above equation becomes

$$T_S = \mathbf{k} \cdot \left[\oiint_{S_1} \frac{\rho}{g_c} (b\mathbf{e}_r \times \mathbf{v})(\mathbf{v} \cdot d\mathbf{S}) + \oiint_{S_2} \frac{\rho}{g_c} (b\mathbf{e}_r \times \mathbf{v})(\mathbf{v} \cdot d\mathbf{S}) \right] \quad \text{Eq. 4.7-44}$$

We observe that on S_1, $\mathbf{v} = V_0\mathbf{i}$. Since the wheel radius is very large compared to the bucket, S_1 occurs at $\theta \approx \frac{3}{2}\pi$ and, therefore, $\mathbf{i} \approx \mathbf{e}_\theta(S_1)$. Therefore,

$$(b\mathbf{e}_r \times \mathbf{v})_{S_1} \approx b\mathbf{e}_r \times V_0\mathbf{e}_\theta = bV_0\mathbf{k}$$

Since the flow is into the control volume at S_1

$$\oiint_{S_1} \mathbf{v} \cdot d\mathbf{S} = -Q$$

where Q is the volume flow rate issuing from the nozzle. Thus,

$$\oiint_{S_1} \frac{\rho}{g_c}(b\mathbf{e}_r \times \mathbf{v})(\mathbf{v} \cdot d\mathbf{S}) = -\frac{\rho}{g_c} bV_0Q\mathbf{k}$$

To evaluate the second integral in Eq. 4.7-44 we note that $\mathbf{v}(S_2)$ is to be taken relative to a fixed coordinate system (X, Y, Z). Since the fluid moves with the bucket, the velocity of the fluid relative to the bucket can readily be determined. This was done in Example 4-11. There we determined that if (x, y, z) is a coordinate system fixed to the bucket, then

$$|\mathbf{v}_{xyz}(S_2)| = (V_0 - U)$$

where U is the velocity of the bucket relative to (X, Y, Z). Then

$$\begin{aligned}\mathbf{v}_{XYZ}(S_2) &= \mathbf{v}_{xyz}(S_1) + U\mathbf{i} = (V_0 - U)(-\cos\phi\mathbf{i} - \sin\phi\mathbf{j}) + U\mathbf{i} \\ &= [-(V_0 - U)\cos\phi + U]\mathbf{i} - (V_0 - U)\sin\phi\mathbf{j}\end{aligned}$$

On S_2, $\mathbf{i} \approx \mathbf{e}_\theta$ and $\mathbf{j} \approx -\mathbf{e}_r$. Therefore,

$$\begin{aligned}\mathbf{e}_r \times \mathbf{v}_{XYZ}(S_2) &= \mathbf{e}_r \times [-(V_0 - U)\cos\phi + U]\mathbf{e}_\theta - \mathbf{e}_r \times (V_0 - U)\sin\phi\mathbf{e}_r \\ &= [-(V_0 - U)\cos\phi + U]\mathbf{k}\end{aligned}$$

Then

$$\oiint_{S_2} \frac{\rho}{g_c}[b\mathbf{e}_r \times \mathbf{v}(S_1)](\mathbf{v} \cdot d\mathbf{S}) = \frac{\rho}{g_c} bQ[-(V_0 - U)\cos\phi + U]\mathbf{k}$$

Collecting terms, we obtain

$$T_S = \frac{\rho}{g_c} bV_0(-Q) + \frac{\rho}{g_c} bQ[-(V_0 - U)\cos\phi + U]$$

$$= \frac{\rho}{g_c} bQ[(V_0 - U) + (V_0 - U)\cos\phi] = -\frac{\rho}{g_c} bQ(V_0 - U)(1 + \cos\phi)$$

Since the bucket rotates at a constant angular velocity, $U = \omega b$. In the above equation, T_S represents the torque exerted on the shaft by the internal stresses at the section cut by our control surface. Since the wheel is in equilibrium, the sum of the torques acting on the wheel is zero; that is,

$$\sum_j \text{torques on the wheel} = 0$$

If T *is the torque exerted by the fluid on the buckets*, then

$$T = -T_S = \frac{\rho b Q}{g_c}(V_0 - \omega b)(1 + \cos\phi) \qquad \text{Eq. 4.7-45}$$

Equation 4.7-45 gives the torque developed by a single jet impinging on the buckets. If the turbine contains several nozzles, then, obviously, the torque given by Eq. 4.7-45 must be multiplied by the number of working nozzles. The power developed by the turbine is given by

$$P = T\omega = \frac{n\rho b Q\omega}{g_c}(V_0 - b\omega)(1 + \cos\phi) \qquad \text{Eq. 4.7-46}$$

where n = the number of working nozzles.

4.8 THE FIELD EQUATION BASED ON NEWTON'S SECOND LAW

In Section 4.5 the field equation based on the law of conservation of mass, denoted as the continuity equation, was derived. In this section, the field equation based on Newton's second law is derived from Eq. 4.7-3 by using concepts of vector calculus. Recall that this equation was derived for an arbitrary system within the continuum. Applying Eq. 2.2-17 to the second term on the left-hand side of Eq. 4.7-3, it becomes

$$\oiiint_{V} \mathbf{F}\, dV - \oiiint_{V} \nabla p\, dV = \oiiint_{V} \frac{\partial}{\partial t}\left(\frac{\rho}{g_c}\mathbf{v}\right) dV + \oiint_{S} \frac{\rho}{g_c}\mathbf{v}(\mathbf{v}\cdot d\mathbf{S}) \quad \text{Eq. 4.8-1}$$

This vector equation decomposes into the three scalar equations

$$\oiiint_{V} \left[\frac{\partial}{\partial t}\left(\frac{\rho}{g_c}v_1\right) + \frac{\partial p}{\partial x} - F_1\right] dV + \oiint_{S} \frac{\rho}{g_c} v_1 \mathbf{v}\cdot d\mathbf{S} = 0$$

$$\oiiint_{V} \left[\frac{\partial}{\partial t}\left(\frac{\rho}{g_c}v_2\right) + \frac{\partial p}{\partial y} - F_2\right] dV + \oiint_{S} \frac{\rho}{g_c} v_2 \mathbf{v}\cdot d\mathbf{S} = 0 \quad \text{Eqs. 4.8-2}$$

$$\oiiint_{V} \left[\frac{\partial}{\partial t}\left(\frac{\rho}{g_c}v_3\right) + \frac{\partial p}{\partial z} - F_3\right] dV + \oiint_{S} \frac{\rho}{g_c} v_3 \mathbf{v}\cdot d\mathbf{S} = 0$$

The last term on the left-hand side of each equation can now be converted to a volume integral by the divergence theorem. For example, in the first equation

$$\oiint_{S} \frac{\rho}{g_c} v_1 \mathbf{v}\cdot d\mathbf{S} = \oiiint_{V} \nabla\cdot\left(\frac{\rho}{g_c} v_1 \mathbf{v}\right) dV$$

Thus, the first equation becomes

$$\oiiint_{V} \left[\frac{\partial}{\partial t}\left(\frac{\rho}{g_c}v_1\right) + \frac{\partial p}{\partial x} - F_1 + \nabla\cdot\left(\frac{\rho}{g_c} v_1 \mathbf{v}\right)\right] dV = 0 \quad \text{Eq. 4.8-3}$$

Since, the volume of integration is arbitrary, the integrand must be identically equal to zero. Expanding the first and fourth terms (the fourth by Eq. 2.2-16), it becomes

$$\frac{\rho}{g_c}\frac{\partial v_1}{\partial t} + \frac{v_1}{g_c}\frac{\partial \rho}{\partial t} + \frac{\partial p}{\partial x} - F_1 + \frac{\rho}{g_c}\mathbf{v}\cdot\nabla v_1 + \frac{v_1}{g_c}\nabla\cdot(\rho\mathbf{v}) = 0 \quad \text{Eq. 4.8-4}$$

The sum of the second and last term is zero by the continuity equation, Eq. 4.5-2. Then the above equation reduces to

$$\frac{\rho}{g_c}\left(\frac{\partial v_1}{\partial t} + \mathbf{v}\cdot\nabla v_1\right) = F_1 - \frac{\partial p}{\partial x}$$

In a similar fashion, the last two scalar equations in the following set can be derived

$$\frac{\rho}{g_c}\left(\frac{\partial v_1}{\partial t} + \mathbf{v}\cdot\nabla v_1\right) = F_1 - \frac{\partial p}{\partial x}$$

$$\frac{\rho}{g_c}\left(\frac{\partial v_2}{\partial t} + \mathbf{v}\cdot\nabla v_2\right) = F_2 - \frac{\partial p}{\partial z} \qquad \text{Eqs. 4.8-5}$$

$$\frac{\rho}{g_c}\left(\frac{\partial v_3}{\partial t} + \mathbf{v}\cdot\nabla v_3\right) = F_3 - \frac{\partial p}{\partial z}$$

This set is called *Euler's equations of motion*. They are applicable when viscous effects are negligible. These three equations can be combined into a single vector equation

$$\frac{\rho}{g_c}\left(\frac{\partial \mathbf{v}}{\partial t} + (\mathbf{v}\cdot\nabla)\mathbf{v}\right) = \mathbf{F} - \nabla p \qquad \text{Eq. 4.8-6}$$

If the body force, **F**, results only from the gravitational field, and if the gravitational field points in the $-z$ direction, then Eq. 4.8-6 reduces to

$$\frac{\rho}{g_c}\left(\frac{\partial \mathbf{v}}{\partial t} + (\mathbf{v}\cdot\nabla)\mathbf{v}\right) = -\nabla p - \gamma\mathbf{k} \qquad \text{Eq. 4.8-7}$$

If the equation of state is such that $\rho = \rho(p)$, then Euler's equation can be integrated along a streamline. Streamlines are of interest mainly in steady-flow situations. In unsteady flow the streamline pattern changes from instant to instant, thus losing its usefulness. In the following development we will take the fluid properties to be time-independent. Under these circumstances, Eq. 4.8-7 becomes

$$\frac{\rho}{g_c}(\mathbf{v}\cdot\nabla)\mathbf{v} = -\nabla p - \gamma\mathbf{k} \qquad \text{Eq. 4.8-8}$$

Now $(\mathbf{v} \cdot \nabla)\mathbf{v}$ can be replaced by its vector identity, given by Eq. 2.2-21. Also dividing both sides by ρ/g_c, Eq. 4.8-8 becomes

$$\tfrac{1}{2}\nabla(v^2) - \mathbf{v} \times (\nabla \times \mathbf{v}) = -\frac{g_c}{\rho}\nabla p - g\mathbf{k} \qquad \text{Eq. 4.8-9}$$

since $\gamma = \rho g/g_c$. Let $d\mathbf{r}$ be a displacement vector *along a streamline*. Take the dot product of both sides of Eq. 4.8-9 with $d\mathbf{r}$; that is,

$$\tfrac{1}{2}\nabla(v^2)\cdot d\mathbf{r} - \mathbf{v} \times (\nabla \times \mathbf{v})\cdot d\mathbf{r} = -\frac{g_c}{\rho}\nabla p \cdot d\mathbf{r} - g\mathbf{k}\cdot d\mathbf{r} \qquad \text{Eq. 4.8-10}$$

Now the vector $\mathbf{v} \times (\nabla \times \mathbf{v})$ is perpendicular to the plane made by the vectors $\mathbf{v}$ and $(\nabla \times \mathbf{v})$. Therefore, the vector $\mathbf{v} \times (\nabla \times \mathbf{v})$ is perpendicular to both $\mathbf{v}$ and to $(\nabla \times \mathbf{v})$. But if $d\mathbf{r}$ is a vector along the streamline, it is parallel to $\mathbf{v}$. Therefore, $\mathbf{v} \times (\nabla \times \mathbf{v})\cdot d\mathbf{r} = 0$. Now if $\Phi(x, y, z)$ is any scalar function, then

$$d\Phi = \nabla\Phi \cdot d\mathbf{r} = \frac{\partial\Phi}{\partial x}dx + \frac{\partial\Phi}{\partial y}dy + \frac{\partial\Phi}{\partial z}dz \qquad \text{Eq. 4.8-11}$$

Equation 4.8-11 gives the change in Φ resulting from a displacement $d\mathbf{r} = dx\,\mathbf{i} + dy\,\mathbf{j} + dz\,\mathbf{k}$. If $d\mathbf{r}$ is along a streamline, then Eq. 4.8-11 gives the change in Φ resulting from a displacement along the streamline. Integrating Eq. 4.8-11 along a streamline from point A to point B, we obtain

$$\frac{1}{2}\int_A^B \nabla(v^2)\cdot d\mathbf{r} = -g_c\int_A^B \frac{1}{\rho}\nabla p \cdot d\mathbf{r} - g\int_A^B \mathbf{k}\cdot d\mathbf{r} \qquad \text{Eq. 4.8-12}$$

and applying Eq. 4.8-11 to Eq. 4.8-12 we obtain

$$\frac{1}{2}\int_{v_A}^{v_B} d(v^2) = -g_c\int_{p_A}^{p_B}\frac{1}{\rho}\,dp - g\int_{z_A}^{z_B} dz \qquad \text{Eq. 4.8-13}$$

or

$$\int_{p_A}^{p_B}\frac{1}{\rho}\,dp = \frac{1}{2g_c}({v_A}^2 - {v_B}^2) + \frac{g}{g_c}(z_A - z_B) \qquad \text{Eq. 4.8-14}$$

If ρ is a function of pressure alone, the fluid motion is described as being barotropic. Barotropic flows include all cases in which density changes result from compressive effects. For example, the isentropic case mentioned at the end of this section fits into this category. However, cases in which ρ changes as a result of temperature effects are not barotropic.

If ρ is a constant then the terms on the left-hand side of Eq. 4.8-14 can readily be integrated, giving, after some rearranging

$$\frac{p_B}{\rho} + \frac{v_B^2}{2g_c} + \frac{g}{g_c} z_B = \frac{p_A}{\rho} + \frac{v_A^2}{2g_c} + \frac{g}{g_c} z_A \qquad \text{Eq. 4.8-15}$$

If we multiply both sides of 4.8-15 by g_c/g and substitute γ for $\rho g/g_c$, which is sometimes more convenient, then

$$\frac{p_B}{\gamma} + \frac{v_B^2}{2g} + z_B = \frac{p_A}{\gamma} + \frac{v_A^2}{2g} + z_A \qquad \text{Eq. 4.8-16}$$

Although Eq. 4.8-16 appears to be exactly the same as Eq. 4.6-18 (except for the designation of the endpoints), there is a significant difference. In Eq. 4.6-18, the velocity terms, V_1 and V_2, represent average velocities at their respective cross sections; whereas the velocity terms in Eq. 4.8-16, v_A and v_B, represent the actual fluid velocity at their respective locations. Equation 4.8-16 is commonly known as the *Bernoulli equation.* This equation is valid when (1) the flow is steady, (2) viscous effects are negligible, and (3) points A and B lie along the same streamline.

The requirement that points A and B be along the same streamline may be dropped if the flow is irrotational. By definition, the flow is irrotational when $\nabla \times \mathbf{v} = 0$. Since under these circumstances, the second term on the left-hand side of Eq. 4.8-9 is zero; and Eqs. 4.8-15 and 4.8-16 become valid for any two points A and B in the flow field when the flow is steady. If the flow is irrotational but nonsteady then the development is slightly more complicated.

In dealing with flows around bodies, such as airfoils, pitot tubes, and the like, where streamlines can be identified, or where the flow is known to be irrotational, Eq. 4.8-15 or Eq. 4.8-16 becomes very useful. This is discussed in much greater detail in Chapter 9. However, a discussion of the pitot tube is given in the following example.

Example 4-14

A pitot tube is a device that enables one to determine the local velocity in a flow (by use of Eq. 4.8-16). A sketch of the device is shown in Fig. 4-34. Determine an expression for the local velocity in terms of the manometer reading, h.

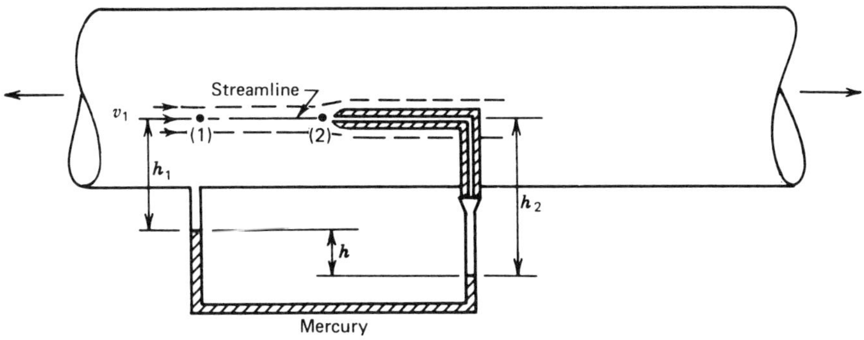

Fig. 4-34. Pitot tube used for local velocity measurement.

SOLUTION

Observe that points 1 and 2 are on the same streamline and that point 2 is a stagnation point.[9] Since points 1 and 2 are both on the same streamline, Eq. 4.8-16 is applicable. Then

$$\frac{p_1}{\gamma} + \frac{v_1^2}{2g} + Z_1 = \frac{p_2}{\gamma} + \frac{v_2^2}{2g} + Z_2$$

Since $Z_1 = Z_2$, and v_2 is zero

$$v_1^2 = \frac{2g}{\gamma}(p_2 - p_1)$$

From the manometer reading, we have that

$$p_2 = p_1 + \gamma h_1 + \gamma_{hg} h - \gamma h_2$$

Also

$$h_1 + h = h_2$$

or

$$p_2 - p_1 = (\gamma_{hg} - \gamma)h$$

[9] By definition, a stagnation point is a point where the velocity is zero.

Therefore,

$$v_1 = \sqrt{2gh\left(\frac{\gamma_{hg}}{\gamma} - 1\right)} \qquad \text{Eq. 4.8-17}$$

It should be noted that in order to obtain an accurate reading for v_1, the longitudinal axis of the pitot tube must be parallel to the flow direction.

If the flow under consideration is a subsonic airflow, then the assumption that the flow is isentropic is more appropriate. In addition, it is also appropriate to assume that the air behaves as an ideal gas. The isentropic flow assumption gives

$$p = K\rho^k \qquad \text{or} \qquad \frac{1}{\rho} = \left(\frac{K}{p}\right)^{1/k} \qquad \text{Eq. 4.8-18}$$

where $k = c_p/c_v$. The ideal gas behavior gives

$$p = \rho RT \qquad \text{Eq. 4.8-19}$$

Then the integral on the left-hand side of Eq. 4.8-14 becomes

$$\begin{aligned}\int_{p_A}^{p_B} \frac{1}{\rho}\,dp &= \int_{p_A}^{p_B} \frac{K^{1/k}}{p^{1/k}}\,dp = K^{1/k}\left[\frac{p^{1-(1/k)}}{1-(1/k)}\right]_{p_A}^{p_B} \\ &= \frac{k}{k-1} K^{1/k}(p_B{}^{(k-1)/k} - p_A{}^{(k-1)/k}) \\ &= \frac{k}{k-1} K^{1/k} p_B{}^{(k-1)/k}\left[1 - \left(\frac{p_A}{p_B}\right)^{(k-1)/k}\right] \qquad \text{Eq. 4.8-20}\end{aligned}$$

But

$$K = \frac{p_B}{\rho_B{}^k} \qquad \text{or} \qquad K^{1/k} = \frac{p_B{}^{1/k}}{\rho_B} \qquad \text{Eq. 4.8-21}$$

Substituting Eqs. 4.8-21 and 4.8-19 into Eq. 4.8-20, we obtain

$$\begin{aligned}\int_{p_A}^{p_B} \frac{1}{\rho}\,dp &= \frac{k}{k-1}\frac{p_B}{\rho_B}\left[1 - \left(\frac{p_A}{p_B}\right)^{(k-1)/k}\right] \\ &= \frac{k}{k-1} RT_B\left[1 - \left(\frac{p_A}{p_B}\right)^{(k-1)/k}\right] \qquad \text{Eq. 4.8-22}\end{aligned}$$

In dealing with gas flows, gravitational effects are usually negligible. Under these circumstances, Eq. 4.8-14 reduces to

$$\frac{1}{2g_c}(v_A^2 - v_B^2) = \frac{k}{k-1} RT_B\left[1 - \left(\frac{p_A}{p_B}\right)^{(k-1)/k}\right] \qquad \text{Eq. 4.8-23}$$

If the pitot tube in Example 4.14 was being used to measure the local velocity of an airflow, then by Eq. 4.8-23

$$v_1 = \sqrt{\frac{2k}{k-1} g_c RT_2\left[1 - \left(\frac{p_1}{p_2}\right)^{(k-1)/k}\right]} \qquad \text{Eq. 4.8-24}$$

*4.9 MORE ON THE LAGRANGIAN AND EULERIAN VIEWPOINTS—ACCELERATION OF A FLUID PARTICLE AS DESCRIBED FROM AN EULERIAN VIEWPOINT

It was mentioned in Section 4.1.1 that there exist two different viewpoints for describing a fluid-flow field. In the Lagrangian viewpoint, a single fluid particle or element is observed as it moves through the flow field. On the other hand, in the Eulerian viewpoint, fluid particles are observed as they move into a particular point in space. In this section, the relationship between these two viewpoints is discussed and an expression for the acceleration of a fluid particle in the Eulerian viewpoint is obtained.

*4.9.1 THE LAGRANGIAN VIEWPOINT

Consider a fluid occupying region R_0 at $t = 0$. Suppose at time t it occupies region R, as shown in Fig. 4-35. Suppose the fluid element located at (x', y', z') in R_0 at $t = 0$ moves to (x, y, z) in R at t. If the motion was known, then it should be possible to obtain the relationship between the position of the fluid element (x, y, z) at t and its initial position (x', y', z') at $t = 0$; that is,

$$\begin{aligned} x &= x(t; x', y', z') \\ y &= y(t; x', y', z') \\ z &= z(t; x', y', z') \end{aligned} \qquad \text{Eqs. 4.9-1}$$

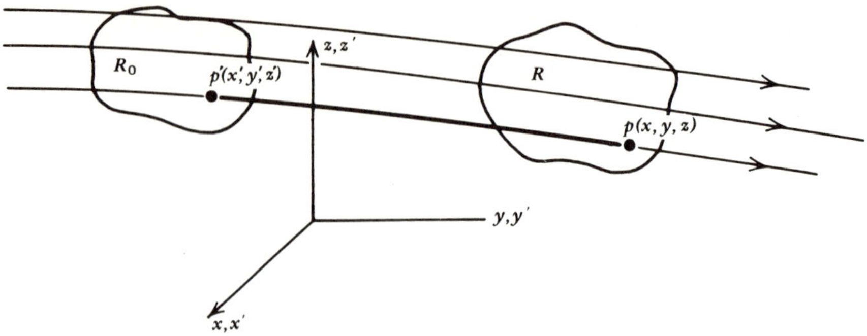

Fig. 4-35. System located in R_0 at $t = 0$ moves to region R at t.

It should be noted that (x', y', z') identifies the element under consideration. This approach is used in particle dynamics. If the velocity vector, **V**, in terms of its (x, y, z) components is

$$\mathbf{V} = V_1\mathbf{i} + V_2\mathbf{j} + V_3\mathbf{k} \qquad \text{Eq. 4.9-2}$$

then

$$V_1 = \frac{dx}{dt}(t; x', y', z') \qquad \text{Eq. 4.9-3}$$

Here (x', y', z') are considered to be constants during the differentiation process. If the acceleration vector, **A**, in terms of its (x, y, z) components is

$$\mathbf{A} = A_1\mathbf{i} + A_2\mathbf{j} + A_3\mathbf{k} \qquad \text{Eq. 4.9-4}$$

then

$$A_1 = \frac{d^2x}{dt^2}(t; x', y', z') = \frac{dV_1}{dt}(t; x', y', z') \qquad \text{Eq. 4.9-5}$$

Here, again, (x', y', z') are considered to be constants during the differentiation process.

*4.9.2 THE EULERIAN VIEWPOINT

Suppose that $V_1(t; x', y', z')$ is known and that the acceleration a_1 of the material located at (x, y, z) at time t is desired. Assume that we can obtain the inverse transformation between the coordinates; that is,

$$\begin{aligned} x' &= x'(x, y, z, t) \\ y' &= y'(x, y, z, t) \\ z' &= z'(x, y, z, t) \end{aligned} \qquad \text{Eqs. 4.9-6}$$

then

$$\begin{aligned} V_1(t; x', y', z') &= V_1(t; x'(x, y, z, t), y'(x, y, z, t), z'(x, y, z, t)) \\ &= v_1(x, y, z, t) \end{aligned} \qquad \text{Eq. 4.9-7}$$

and

$$a_1 = A_1 = \frac{dv_1}{dt} = \frac{\partial v_1}{\partial x}\frac{dx}{dt} + \frac{\partial v_1}{\partial y}\frac{dy}{dt} + \frac{\partial v_1}{\partial z}\frac{dz}{dt} + \frac{\partial v_1}{\partial t} \qquad \text{Eq. 4.9-8}$$

since x, y, and z are all functions of t. But

$$\frac{dx}{dt} = V_1 = v_1, \qquad \frac{dy}{dt} = V_2 = v_2, \qquad \frac{dz}{dt} = V_3 = v_3 \qquad \text{Eqs. 4.9-9}$$

Thus,

$$a_1(x, y, z, t) = \frac{\partial v_1}{\partial t} + v_1\frac{\partial v_1}{\partial x} + v_2\frac{\partial v_1}{\partial y} + v_3\frac{\partial v_1}{\partial z} \qquad \text{Eq. 4.9-10}$$

The right-hand side of Eq. 4.9-10 is called the *material derivative* or *substantial derivative* of v_1. It represents the x-component of the acceleration of a fluid element in the Eulerian viewpoint. Since Eq. 4.8-6 is the vector form of the three scalar equations, Eqs. 4.8-5, we can interpret the expression inside the brackets in Eq. 4.8-6 as the acceleration, **a**, of a fluid element in the Eulerian viewpoint. We can further conclude that **a** is the material derivative of the fluid velocity; that is,

$$\mathbf{a} = \frac{d\mathbf{v}}{dt} = \frac{\partial \mathbf{v}}{\partial t} + (\mathbf{v} \cdot \nabla)\mathbf{v} \qquad \text{Eq. 4.9-11}$$

Then Eq. 4.8-7 can be rewritten as

$$\frac{\rho}{g_c}\frac{d\mathbf{v}}{dt} = -\nabla p - \gamma \mathbf{k} \qquad \text{Eq. 4.9-12}$$

We can now interpret the meaning of the various terms in Eq. 4.9-12. The left-hand side of the equation is mass/unit volume divided by g_c, times the acceleration. The right-hand side represents the net forces acting on the fluid element per unit volume, that is,

$$\frac{m}{\Delta \forall}\frac{\mathbf{a}}{g_c} = \frac{\text{net forces}}{\Delta \forall}$$

If we multiply both sides by $\Delta \forall$, the equation becomes the familiar form of Newton's second law applied to a particle.

Although our discussion was concerned with the development of the material derivative of v_1, the concept is applicable to any physical variable. For example, in the Eulerian viewpoint, the time rate of change of the temperature of a fluid element moving through the flow field is

$$\frac{dT}{dt}(x, y, z, t) = \frac{\partial T}{\partial t} + v_1 \frac{\partial T}{\partial x} + v_2 \frac{\partial T}{\partial y} + v_3 \frac{\partial T}{\partial z} \qquad \text{Eq. 4.9-13}$$

The following example illustrates the difference between expressing the acceleration of a fluid element in the Eulerian description from that in the Lagrangian description.

Example 4-15

Suppose a rod fixed at one end was made from a particular material which behaves in such a manner that if the free end moves to position $(L + \varepsilon_L)$, then an element of material, initially at x', moves to position x governed by

$$\frac{x}{x'} = \frac{L + \varepsilon_L}{L} \qquad \text{Eq. 4.9-14}$$

See Fig. 4-36.[10] Now suppose that $\varepsilon_L = A \sin \omega t$. Determine expressions for the acceleration at an arbitrary point on the rod from both the Lagrangian and Eulerian viewpoints.

[10] It should be noted that a typical elastic body would not behave in this manner. The example is presented simply to illustrate the difference in the Eulerian and Lagrangian viewpoints.

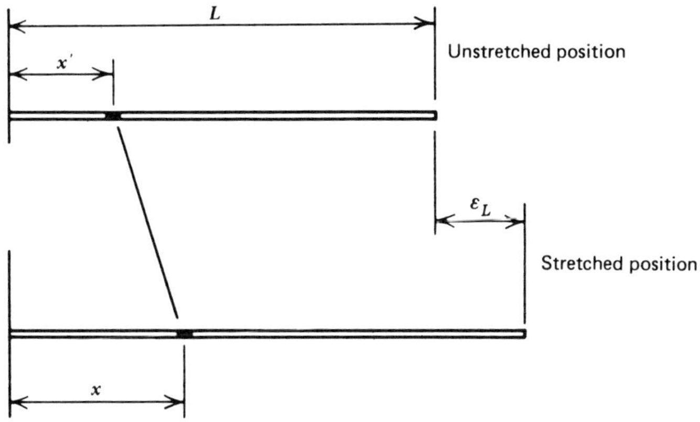

Fig. 4-36. Rod with end stretched to position $(L + \varepsilon_L)$.

SOLUTION

By Eq. 4.9-14, the position on the rod can be expressed by

$$x = x'\left(1 + \frac{A}{L}\sin\omega t\right) = x(t; x') \qquad \text{Eq. 4.9-15}$$

The velocity and acceleration of the material initially located at x' described by the Lagrangian viewpoint are

$$V_1 = \frac{dx}{dt} = x'\frac{A\omega}{L}\cos\omega t \qquad \text{Eq. 4.9-16}$$

$$A_1 = \frac{dV_1}{dt} = -x'\frac{A\omega^2}{L}\sin\omega t \qquad \text{Eq. 4.9-17}$$

To obtain a_1 from the Eulerian description, let $v_1(x, t) = V_1(t', x')$; that is,

$$v_1 = \frac{x}{1 + \frac{A}{L}\sin\omega t}\left(\frac{A\omega}{L}\cos\omega t\right) = \left(\frac{A\omega x}{L}\right)\left(\frac{\cos\omega t}{1 + \frac{A}{L}\sin\omega t}\right) \qquad \text{Eq. 4.9-18}$$

Then by Eq. 4.9-10,

$$
\begin{aligned}
a_1 &= \frac{\partial v_1}{\partial t} + v_1 \frac{\partial v_1}{\partial x} \\
&= \left(\frac{A\omega x}{L}\right)\left[\frac{-\omega \sin \omega t}{1 + \dfrac{A}{L}\sin \omega t} - \left(\frac{\cos \omega t}{\left(1 + \dfrac{A}{L}\sin \omega t\right)^2}\right)\left(\frac{A\omega}{L}\cos \omega t\right)\right] \\
&\quad + \left(\frac{A\omega x}{L}\right)\left(\frac{\cos \omega t}{1 + \dfrac{A}{L}\sin \omega t}\right)\left(\frac{A\omega}{L}\right)\left(\frac{\cos \omega t}{1 + \dfrac{A}{L}\sin \omega t}\right) \\
&= -\left(\frac{A\omega^2 x}{L}\right)\left(\frac{\sin \omega t}{1 + \dfrac{A}{L}\sin \omega t}\right) = -x' \frac{A\omega^2}{L}\sin \omega t
\end{aligned}
\qquad \text{Eq. 4.9-19}
$$

Comparing Eqs. 4.9-17 and 4.9-19, we see they are the same; that is, $a_1 = A_1$. No matter which description is used, the end result must be the same.

*4.10 THE SECOND LAW OF THERMODYNAMICS[11]

There is a tendency in natural processes for various forms of energy to be converted into thermal energy. Thermal energy tends to redistribute itself so that the temperature equalizes throughout the system. This behavior on the microscopic level is interpreted as a tendency for ordered phenomena to become random in character. The conversion of various forms of energy into thermal energy is associated with the generation of the thermodynamic variable called entropy. This behavior is expressed in the second law of thermodynamics, which states that, for an isolated system, the entropy in the system never decreases. Thus, if an isolated system undergoes an irreversible process, there will be entropy generation. We will designate this entropy generation per unit volume as $\dot{\Sigma}$. It should be noted that $\dot{\Sigma}$ is never negative. The law can readily be extended for a nonisolated system. Select an arbitrary system within the continuum, as shown in Fig. 4-37. As a consequence of the second law, the following statement can be formulated.

[11] For a review of thermodynamics, see Appendix A.3.

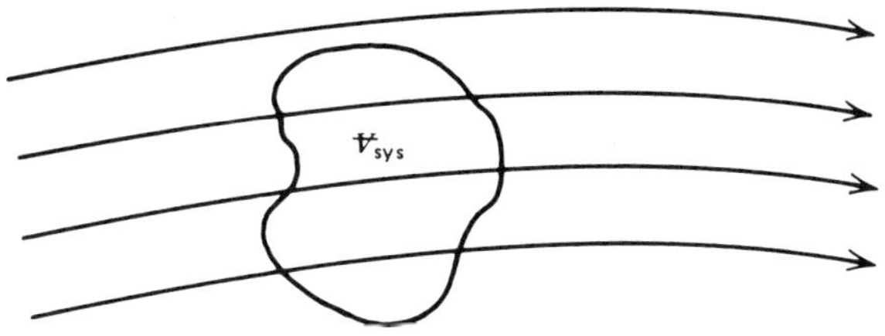

Fig. 4-37. Arbitrary system in the continuum.

> The rate of increase in the total entropy in the system equals the rate at which entropy is generated within the system minus the rate entropy flows out of the system. Stm. 4.10-1

If $\mathcal{s}$ is the entropy/unit mass and $\dot{\Sigma}$ is the rate entropy is generated per unit volume, then the mathematical equation expressing this law is

$$\frac{d}{dt}\iiint_{\mathcal{V}} \rho \mathcal{s}\, d\mathcal{V} = \iiint_{\mathcal{V}} \dot{\Sigma}\, d\mathcal{V} - \oiint_{S} \frac{1}{T}\mathbf{q}\cdot d\mathbf{S} \qquad \text{Eq. 4.10-1}$$

Note that $\dot{\Sigma} \geq 0$. Applying Eq. 4.2-9 to the left-hand side of Eq. 4.10-1, we obtain

$$\iiint_{\mathcal{V}} \frac{\partial(\rho \mathcal{s})}{\partial t}\, d\mathcal{V} + \oiint_{S} \rho \mathcal{s}\mathbf{v}\cdot d\mathbf{S} = \iiint_{\mathcal{V}} \dot{\Sigma}\, d\mathcal{V} - \oiint_{S} \frac{1}{T}\mathbf{q}\cdot d\mathbf{S} \qquad \text{Eq. 4.10-2}$$

Now apply the divergence theorem, Eq. 2.2-13, to all surface integral terms and combine all terms under one integral sign

$$\iiint \left[\frac{\partial(\rho \mathcal{s})}{\partial t} + \nabla\cdot(\rho \mathcal{s}\mathbf{v}) - \dot{\Sigma} + \nabla\cdot\left(\frac{1}{T}\mathbf{q}\right)\right] d\mathcal{V} = 0 \qquad \text{Eq. 4.10-3}$$

Since this equation is valid for an arbitrary system within the continuum, the integrand must be identically zero; that is,

$$\frac{\partial(\rho \mathcal{s})}{\partial t} + \nabla\cdot(\rho \mathcal{s}\mathbf{v}) - \dot{\Sigma} + \nabla\cdot\left(\frac{1}{T}\mathbf{q}\right) = 0 \qquad \text{Eq. 4.10-4}$$

Expanding the terms, gives

$$\rho \frac{\partial s}{\partial t} + \rho \mathbf{v} \cdot \nabla s + s \underbrace{\left[\frac{\partial \rho}{\partial t} + \nabla \cdot (\rho \mathbf{v})\right]}_{\text{zero}} - \dot{\Sigma} + \frac{1}{T} \nabla \cdot \mathbf{q} + \mathbf{q} \cdot \nabla\left(\frac{1}{T}\right) = 0 \quad \text{Eq. 4.10-5}$$

The third term on the left-hand side is zero by the continuity equation, Eq. 4.5-2. The first and second terms together can be expressed as ρ times the material derivative of s. With the expansion of the last term on the left-hand side, Eq. 4.10-5 reduces to

$$\rho \frac{ds}{dt} - \dot{\Sigma} + \frac{1}{T} \nabla \cdot \mathbf{q} - \mathbf{q} \cdot \frac{1}{T^2} \nabla T = 0$$

or

$$\rho T \frac{ds}{dt} = T\dot{\Sigma} + \mathbf{q} \cdot \frac{1}{T} \nabla T - \nabla \cdot \mathbf{q} \qquad \text{Eq. 4.10-6}$$

Finally for a reversible process, $\dot{\Sigma} = 0$. Students should recall from their courses in thermodynamics that a reversible process involving heat transfer can only be approached when the heat transfer takes place across a small temperature differential. For this condition $\nabla T \to 0$. Thus, for a reversible process

$$\rho \frac{ds}{dt} = -\nabla \cdot \mathbf{q} \qquad \text{Eq. 4.10-7}$$

4.11 THE FIELD EQUATION BASED ON THE FIRST LAW OF THERMODYNAMICS

In Sections 4.5 and 4.8 the field equations based on the law of conservation of mass and Newton's second law, respectively, were derived. In this section the field equation based on the first law of thermodynamics is derived. First, we will rewrite the rate of work term $(\delta W/\delta t)_{\text{shaft}}$, appearing in Eq. 4.6-8, as a volume integral representing the rate of work per unit volume done by non-conservative forces on the system; that is,

$$\left(\frac{\delta W}{\delta t}\right)_{\text{shaft}} = \oiiint_{\mathcal{V}} \Phi \, d\mathcal{V} \qquad \text{Eq. 4.11-1}$$

Although it is beyond the scope of this text, an exact expression for Φ can be derived when the nonconservative forces result strictly from viscous effects. We now apply the divergence theorem, Eq. 2.2-13, to the surface integral terms in Eq. 4.6-8, obtaining

$$\iiint_{\mathcal{V}} \frac{\partial}{\partial t}\left[\rho\left(Ju + \frac{v^2}{2g_c} + \frac{gz}{g_c}\right)\right] d\mathcal{V} + \iiint_{\mathcal{V}} \nabla \cdot \left[\rho\left(Ju + \frac{v^2}{2g_c} + \frac{gz}{g_c}\right)\mathbf{v}\right] d\mathcal{V}$$

$$= -\iiint_{\mathcal{V}} \nabla \cdot (p\mathbf{v})\, d\mathcal{V} + \iiint_{\mathcal{V}} \Phi\, d\mathcal{V} - \iiint_{\mathcal{V}} J(\nabla \cdot \mathbf{q})\, d\mathcal{V} \qquad \text{Eq. 4.11-2}$$

or

$$\iiint_{\mathcal{V}} \left\{\frac{\partial}{\partial t}\left[\rho\left(Ju + \frac{v^2}{2g_c} + \frac{gz}{g_c}\right)\right] + \nabla \cdot \left[\rho\left(Ju + \frac{v^2}{2g_c} + \frac{gz}{g_c}\right)\mathbf{v}\right]\right.$$

$$\left. + \nabla \cdot (p\mathbf{v}) + \Phi + J\nabla \cdot \mathbf{q}\right\} d\mathcal{V} = 0 \qquad \text{Eq. 4.11-3}$$

Since this equation is valid for an arbitrary system in the continuum, the integrand must be identically equal to zero

$$\frac{\partial}{\partial t}\left[\rho\left(Ju + \frac{v^2}{2g_c} + \frac{gz}{g_c}\right)\right] + \nabla \cdot \left[\rho\left(Ju + \frac{v^2}{2g_c} + \frac{gz}{g_c}\right)\mathbf{v}\right] + \nabla \cdot (p\mathbf{v})$$

$$+ J(\nabla \cdot \mathbf{q}) + \Phi = 0 \qquad \text{Eq. 4.11-4}$$

Expanding the first, second, and third terms, we obtain

$$\rho \frac{\partial}{\partial t}\left(Ju + \frac{v^2}{2g_c} + \frac{gz}{g_c}\right) + \left(Ju + \frac{v^2}{2g_c} + \frac{gz}{g_c}\right)\left(\frac{\partial \rho}{\partial t}\right) + \left(Ju + \frac{v^2}{2g_c} + \frac{gz}{g_c}\right)(\nabla \cdot \rho\mathbf{v})$$

$$+ \rho\mathbf{v} \cdot \nabla\left(Ju + \frac{v^2}{2g_c} + \frac{gz}{g_c}\right) + p(\nabla \cdot \mathbf{v}) + \mathbf{v} \cdot \nabla p + J(\nabla \cdot \mathbf{q}) + \Phi = 0$$

$$\text{Eq. 4.11-5}$$

The second and third terms can be combined; and, by applying the continuity equation, Eq. 4.5-2, the combination can be set to zero; that is,

$$\left(Ju + \frac{v^2}{2g_c} + \frac{gz}{g_c}\right)\overbrace{\left(\frac{\partial \rho}{\partial t} + \mathbf{\nabla} \cdot \rho \mathbf{v}\right)}^{\text{zero}} = 0 \qquad \text{Eq. 4.11-6}$$

The first and the fourth terms in Eq. 4.11-5 can also be combined, and by the definition of the material derivative they can be expressed, as given in Eq. 4.11-7,

$$\rho\left[\frac{\partial}{\partial t}\left(Ju + \frac{v^2}{2g_c} + \frac{gz}{g_c}\right) + \mathbf{v} \cdot \mathbf{\nabla}\left(Ju + \frac{v^2}{2g_c} + \frac{gz}{g_c}\right)\right] = \rho \frac{d}{dt}\left(Ju + \frac{v^2}{2g_c} + \frac{gz}{g_c}\right)$$

Eq. 4.11-7

Applying Eqs. 4.11-6 and 4.11-7 to Eq. 4.11-5, we obtain

$$\rho \frac{d}{dt}\left(Ju + \frac{v^2}{2g_c} + \frac{gz}{g_c}\right) + p(\mathbf{\nabla} \cdot \mathbf{v}) + \mathbf{v} \cdot \mathbf{\nabla} p + J(\mathbf{\nabla} \cdot \mathbf{q}) + \Phi = 0 \qquad \text{Eq. 4.11-8}$$

The first term represents the time rate of change in the energy of the fluid element as it moves through the flow field per unit volume; the second term represents the negative of the rate of work done by pressure forces in dilating the fluid element per unit volume; the third term represents the negative of net rate of work done by the pressure forces in moving the fluid element per unit volume; the fourth term represents the rate at which heat flows out of the fluid element per unit volume; and, finally, the fifth term represents the rate of work done on the fluid element by the nonconservative forces per unit volume. Equation 4.11-8 can be further developed in two different directions. The first direction focuses on the thermodynamic conditions necessary for arriving at the Bernoulli equation. These conditions are then compared with the mechanical conditions used in Section 4.8 for arriving at the Bernoulli equation. The second direction enables us to eliminate the kinetic and potential energy terms from the field equation and arrive at an equation that is applicable to some problems in heat transfer.

Two conditions used previously to arrive at the Bernoulli equation are (1) the flow is steady and (2) the fluid is incompressible. We will again impose these conditions. The latter condition, by Eq. 4.5-4, gives

$$\mathbf{\nabla} \cdot \mathbf{v} = 0 \qquad \text{Eq. 4.11-9}$$

This condition also allows us to rewrite the third term in Eq. 4.11-8 as follows

$$\mathbf{v} \cdot \nabla p = \rho \mathbf{v} \cdot \frac{1}{\rho} \nabla p = \rho \mathbf{v} \cdot \nabla \left(\frac{p}{\rho} \right) \qquad \text{Eq. 4.11-10}$$

In the Eulerian viewpoint, the following equality can be written when the flow is steady

$$\rho \frac{d}{dt} \left(\frac{p}{\rho} \right) = \rho \mathbf{v} \cdot \nabla \left(\frac{p}{\rho} \right) \qquad \text{Eq. 4.11-11}$$

Applying Eqs. 4.11-9, 4.11-10, and 4.11-11 to Eq. 4.11-8, and rearranging, we obtain

$$\rho \frac{d}{dt} \left(Ju + \frac{p}{\rho} + \frac{v^2}{2g_c} + \frac{gz}{g_c} \right) + J(\nabla \cdot \mathbf{q}) + \Phi = 0 \qquad \text{Eq. 4.11-12}$$

or

$$J \frac{du}{dt} = - \frac{d}{dt} \left(\frac{p}{\rho} + \frac{v^2}{2g_c} + \frac{gz}{g_c} \right) - \frac{J}{\rho} (\nabla \cdot \mathbf{q}) - \frac{\Phi}{\rho} \qquad \text{Eq. 4.11-13}$$

In Section A.3 an equation that combines both the first and second laws of thermodynamics is given, that is, Eq. A.3-13. Expressing this equation as a rate equation, it becomes

$$\frac{du}{dt} = T \frac{ds}{dt} - \frac{p}{J} \frac{d\tilde{v}}{dt} = T \frac{ds}{dt} - \frac{1}{J} p \frac{d}{dt} \left(\frac{1}{\rho} \right) \qquad \text{Eq. 4.11-14}$$

Although this equation was derived for a reversible process, the equation is valid for any process, since the equation only involves quantities that are properties. Recall that changes in thermodynamic properties are independent of path. For an incompressible flow, the last term on the right-hand side of Eq. 4.11-14 is zero, then

$$\frac{du}{dt} = T \frac{ds}{dt} \qquad \text{Eq. 4.11-15}$$

Substituting Eq. 4.11-15 into Eq. 4.11-13, we obtain

$$JT \frac{ds}{dt} = - \frac{d}{dt} \left(\frac{p}{\rho} + \frac{v^2}{2g_c} + \frac{gz}{g_c} \right) - \frac{J}{\rho} (\nabla \cdot \mathbf{q}) - \frac{\Phi}{\rho} \qquad \text{Eq. 4.11-16}$$

If the process is reversible, then by Eq. 4.10-7

$$\rho T \frac{ds}{dt} = -(\nabla \cdot \mathbf{q}) \qquad \text{Eq. 4.11-17}$$

If, in addition, there are no nonconservative forces present, then Φ can be set to zero. Applying this condition and Eq. 4.11-17 to Eq. 4.11-16 and canceling like terms, we obtain

$$\frac{d}{dt}\left(\frac{p}{\rho} + \frac{v^2}{2g_c} + \frac{gz}{g_c}\right) = 0 \qquad \text{Eq. 4.11-18}$$

Since the derivative in Eq. 4.11-18 is a material derivative, Eq. 4.11-18 implies that the properties of a fluid element as it moves along its path satisfy the equation

$$\frac{p}{\rho} + \frac{v^2}{2g_c} + \frac{gz}{g_c} = \text{constant} \qquad \text{Eq. 4.11-19}$$

Since in steady flow, the path trajectory is the same as a streamline, Eq. 4.11-19 is valid along a streamline. Thus, once again, we have arrived at the Bernoulli equation, but from a different starting point. Before, we arrived at the Bernoulli equation by the integration of Euler's equation along a streamline. This time, we arrived at the Bernoulli equation by the use of the energy field equation. Let us now compare the conditions in each case which allowed us to arrive at the same result. In the case of the energy equation the conditions are: (1) steady flow, (2) incompressible fluid, (3) no work done by nonconservative forces, and (4) the fluid element undergoes a reversible process. In the case of the integration of Euler's equation along a streamline, the conditions are: (1) steady flow, (2) incompressible fluid, (3) no nonconservative forces acting on the fluid, and (4) no viscous effects. By comparing the two lists of conditions, we may conclude that the "no viscous effects condition" required for the integration of Euler's equation is equivalent to the "reversible process condition" required for the integration of the energy equation.

We now return to the second direction for the further development of Eq. 4.11-8. We will consider the case when there are no nonconservative forces acting within the continuum. Under these circumstances, Φ in Eq. 4.11-8 can be set to zero, giving

$$\rho \frac{d}{dt}\left(Ju + \frac{v^2}{2g_c} + \frac{g_z}{g_c}\right) + p(\nabla \cdot \mathbf{v}) + \mathbf{v} \cdot \nabla p + J(\nabla \cdot \mathbf{q}) = 0 \qquad \text{Eq. 4.11-20}$$

The second term and third term within the first parentheses can be rewritten as follows:

$$\rho \frac{d}{dt}\left(\frac{v^2}{2g_c}\right) = \frac{\rho}{2g_c}\frac{d}{dt}(\mathbf{v}\cdot\mathbf{v}) = \frac{\rho}{g_c}\mathbf{v}\cdot\left(\frac{d\mathbf{v}}{dt}\right) = \mathbf{v}\cdot\left(\frac{\rho}{g_c}\frac{d\mathbf{v}}{dt}\right) \qquad \text{Eq. 4.11-21}$$

and

$$\rho \frac{d}{dt}\left(\frac{gz}{g_c}\right) = \rho\left[\frac{\partial}{\partial t}\left(\frac{gz}{g_c}\right) + \mathbf{v}\cdot\nabla\left(\frac{gz}{g_c}\right)\right] \qquad \text{Eq. 4.11-22}$$

In the Eulerian viewpoint, time and position coordinates are independent. Thus, the first term within the bracket on the right-hand side of Eq. 4.11-22 is zero. Since g/g_c is a constant, $\nabla z = \mathbf{k}$ and $\rho g/g_c = \gamma$, Eq. 4.11-22 becomes

$$\rho \frac{d}{dt}\left(\frac{gz}{g_c}\right) = \mathbf{v}\cdot\left(\frac{\rho g}{g_c}\mathbf{k}\right) = \mathbf{v}\cdot\gamma\mathbf{k} \qquad \text{Eq. 4.11-23}$$

Substituting Eqs. 4.11-21 and 4.11-23 into Eq. 4.11-20, and rearranging terms, Eq. 4.11-20 becomes

$$J\rho\frac{du}{dt} + \mathbf{v}\cdot\left(\frac{\rho}{g_c}\frac{d\mathbf{v}}{dt} + \gamma\mathbf{k} + \nabla p\right) + p(\nabla\cdot\mathbf{v}) + J(\nabla\cdot\mathbf{q}) = 0 \qquad \text{Eq. 4.11-24}$$

But by Eq. 4.8-7

$$\frac{\rho}{g_c}\frac{d\mathbf{v}}{dt} + \gamma\mathbf{k} + \nabla p = 0$$

Thus Eq. 4.11-24 reduces to

$$\rho\frac{du}{dt} + \frac{p}{J}(\nabla\cdot\mathbf{v}) + \nabla\cdot\mathbf{q} = 0 \qquad \text{Eq. 4.11-25}$$

If we limit our discussion to liquids, with no phase changes taking place within the continuum, then we may take ρ = constant, $u = c_v T$, and $c_v = c_p = c$. By the continuity equation, Eq. 4.5-4, $\nabla\cdot\mathbf{v} = 0$. Thus, Eq. 4.11-25 reduces to

$$\rho c\frac{dT}{dt} = -\nabla\cdot\mathbf{q} \qquad \text{Eq. 4.11-26}$$

If the fluid is isotropic, then Fourier's heat-conduction law is applicable; that is,

$$\mathbf{q} = -k_t \nabla T \qquad \text{Eq. 4.11-27}$$

where k_t is the thermal conductivity of the fluid. Fourier's heat-conduction law is to be regarded as a constitutive equation, since not all materials behave in this manner. Substituting Eq. 4.11-27 into Eq. 4.11-26 and expanding the material derivative, we obtain

$$\rho c\left(\frac{\partial T}{\partial t} + \mathbf{v} \cdot \nabla T\right) = \nabla \cdot (k_t \nabla T) \qquad \text{Eq. 4.11-28}$$

If we are dealing with temperature variations, which are not too large, we may take k_t to be a constant; then Eq. 4.11-28 reduces to

$$\rho c\left(\frac{\partial T}{\partial t} + \mathbf{v} \cdot \nabla T\right) = k_t(\nabla^2 T) \qquad \text{Eq. 4.11-29}$$

When the boundaries containing a fluid are subjected to either a heat source or a heat sink, Eq. 4.11-29 becomes relevant. Problems of this type are covered in a course concerned with heat transfer.

Problems

4-1 Determine the streamline pattern and the equation of the streamline passing through the point (1, 3) for the following two dimensional velocity fields:

(a) $v_1 = -\dfrac{\Gamma}{2\pi}\dfrac{y}{x^2 + y^2} \qquad v_2 = \dfrac{\Gamma}{2\pi}\dfrac{x}{x^2 + y^2}$

(b) $v_1 = \dfrac{Q}{2\pi}\dfrac{x}{x^2 + y^2} \qquad v_2 = \dfrac{Q}{2\pi}\dfrac{y}{x^2 + y^2}$

(c) $v_1 = \dfrac{Q(y^2 - x^2)}{(x^2 + y^2)^2} \qquad v_2 = -\dfrac{2Qxy}{(x^2 + y^2)^2}, \qquad$ where $y \geq 0$

Hint. Look for a differential equation made exact by an integrating factor involving one variable.

4-2 A cylindrical tank 3 m in diameter has an orifice 2 cm in diameter. If a plug from the orifice is removed when the water level in the tank is 6 m above the orifice, how long will it take for the water level in the tank to drop 3 m? Take the discharge coefficient for the orifice to be equal to 0.75.

4-3 The side walls of a holding tank have the shape of a hyperboloid of revolution, as shown in Fig. 4-38. The equation of the hyperboloid is given by

$$r = 10\sqrt{1 + \frac{3}{400} z^2} \text{ ft}$$

The tank has a $\frac{3}{4}$ in. diameter orifice, located at $z = 0$, which may be plugged or unplugged. If the orifice is unplugged when the water level in the tank is 20 ft above the orifice, how long will it take the water level in the tank to drop 8 ft, if $c_D = 0.8$?

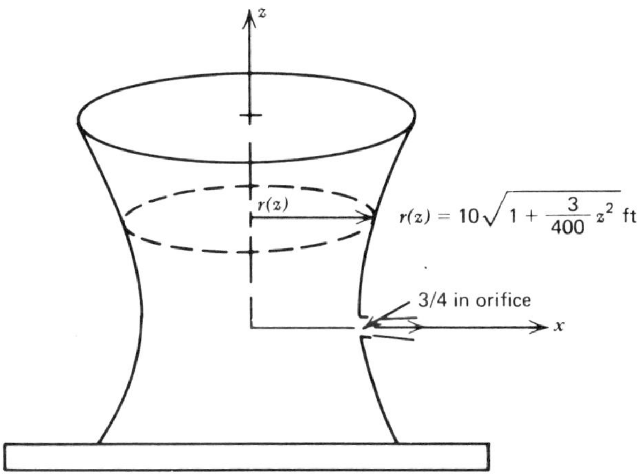

Fig. 4-38. Tank shaped as a hyperboloid of revolution.

4-4 A device used to measure the viscosity of oils is the Saybolt viscometer. It consists of a brass tube that discharges through a capillary tube, as shown in Fig. 4-39. With the cork in place, the test oil is slowly poured into the larger tube until it overflows. Excess oil is then removed from the overflow region by a pipette. Since the tube rests inside a liquid bath whose temperature can be controlled by a heater, the temperature of the test oil can also be varied. After thermal equilibrium between bath liquid and test oil is obtained, the cork is pulled out and the tube is discharged. The time it takes to collect 60 cm^3 of oil depends on the oil's viscosity. Although the relationship is empirically determined, we can obtain a good estimate of this relationship by considering (1) the flow inside the tube to be an ideal flow (viscous

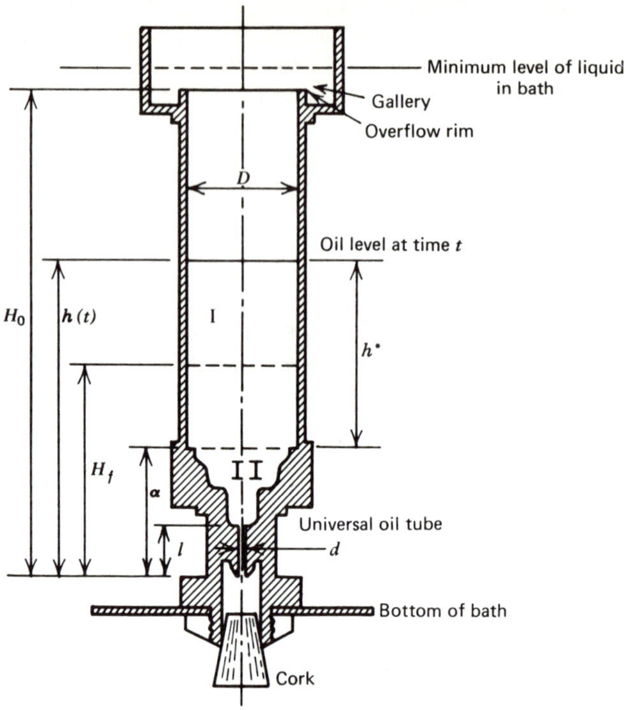

Fig. 4-39. The Saybolt viscometer. (From Method of Test for Saybolt Viscosity, ASTM D88-56 by R. W. Fox and A. T. McDonald. Courtesy of ASTM.)

effects neglected) and (2) the flow to be a fully developed laminar flow[12] in the capillary tube. Under these conditions, it is known that the average velocity, V_2, at the exit of the capillary tube is given by

$$V_2 = c_1(\sqrt{1 + c_2 h} - 1), \qquad c_1 = \frac{16l\nu}{1.14d^2}, \qquad \text{and} \qquad c_2 = \frac{1.14gd^4}{256l^2\nu^2}$$

(*a*) Determine a relationship between t, H_0, H_f, c_1, c_2, d, and D.
(*b*) If the test time, t_f, to collect 60 cm^3 of a particular oil is 45.0 s, estimate the kinematic viscosity. Take $D = 2.976$ cm, $H_0 = 12.50$ cm, $d = 0.1766$ cm, and $l = 1.225$ cm.

Hint. Plot t_f versus ν on semilog paper.

[12] A laminar flow is a smooth flow where the fluid flows in layers.

(*c*) The empirical relationship between ν and t_f is

$$\nu = 0.0022 t_f - \frac{1.80}{t_f} \text{ cm}^2/\text{s}$$

Determine ν from this empirical relationship.
(*d*) Assuming that ν determined in part (*c*) is correct, what is the percent error of the ν value determined in part (*b*).

4-5 When a flow through a circular pipe is turbulent[13] the velocity distribution may be approximated by

$$v_1(r) = V_0\left(1 - \frac{r}{R}\right)^{1/7}$$

where R is the radius of the pipe and V_0 is the fluid velocity at the centerline. Determine the bulk velocity in terms of V_0. Assume that the flow is incompressible.

4-6 The velocity profile in the boundary layer[14] (see Fig. 9-2) of a two-dimensional laminar flow over a flat plate is given by

$$v_1 = v_\infty \frac{df}{d\bar{y}}(\bar{y}) \qquad \text{and} \qquad v_2 = \frac{1}{2}\sqrt{\frac{V_\infty \nu}{x}}\left[\bar{y}\frac{df}{d\bar{y}}(\bar{y}) - f(\bar{y})\right]$$

with

$$f(\bar{y}) = \sum_{n=0}^{\infty}\left(-\frac{1}{2}\right)^n \frac{c_n \alpha^{n+1}}{(3n+2)!}(\bar{y})^{3n+2}$$

where c_1, c_2, and so on, are known constants (the first three are given below), $\alpha = 0.332$, and $\bar{y} = \sqrt{V_\infty/(\nu x)}\,y$. Approximate the bulk velocity in the boundary layer ($0 \le y \le \delta$) at a section normal to the plate at $x = 0.5$ ft, by using the first three terms in the series for $f(\bar{y})$. Take $V_\infty = 50$ ft/s, $c_0 = 1$, $c_1 = 1$, $c_2 = 11$, and $\delta = 5\sqrt{\nu x/V_\infty}$. Assume that the fluid is air at 50 F.

4-7 In the water jet pump, a jet stream is introduced into the mainstream, as shown in Fig. 4-40. At section 2, a short distance downstream from section 1, the fluid is

[13] Turbulence is an unsteady, chaotic flow with strong transverse mixing within the fluid. This topic is discussed in Section 5.1.
[14] The region immediately adjacent to a solid boundary is referred to as boundary layer. Velocity gradients in the direction perpendicular to the solid boundary are relatively large. Boundary layers are discussed in Sections 9.1.1 and 9.10.

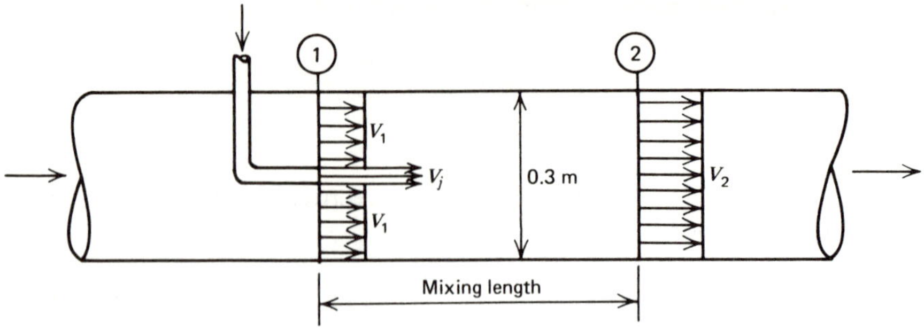

Fig. 4-40. The water jet pump.

thoroughly mixed. If the bulk velocities of the mainstream and jet at section 1 are 3 m/s and 25 m/s, respectively, determine the bulk velocity at section 2. Take the area of the jet, A_j, to be 0.006 m^2.

4-8 A fluid flows through a pipe that has an annulus cross section, as shown in Fig. 4-41. The inner radius of the annulus is R_1 and the outer radius is R_2. The flow is

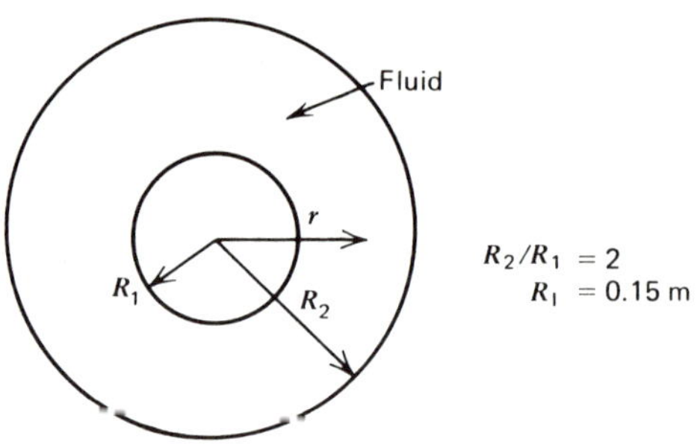

Fig. 4-41. Flow through an annulus.

steady and fully developed (velocity profile does not change from section to section). A pitot tube traverse in the radial direction indicates that the velocity vector along a radial line is given by

$$\mathbf{v}(r) = 3\left\{1 - \left(\frac{r}{R_1}\right)^2 + \left[\left(\frac{R_2}{R_1}\right)^2 - 1\right]\frac{\ln(r/R_1)}{\ln(R_2/R_1)}\right\}\mathbf{i}\ \text{m/s}$$

Determine: (*a*) the maximum velocity, v_{max}, at a section; (*b*) the bulk (or average) velocity, V_A, at a section; (*c*) the volume flow rate, Q.

4-9 An ideal (nonviscous) incompressible fluid has velocity components, v_1 and v_2, in the x and y directions as follows

$$v_1(x, y, z, t) = C_1 V_0(x^2 - y^2 + xy)$$

$$v_2(x, y, z, t) = C_1 V_0(-2xy - 2yz + 2z^2)$$

where C_1 and V_0 are constants. Find the z-component of velocity that satisfies the continuity equation. Is this solution unique? Why?

4-10 An ideal incompressible fluid (such as water) enters a circular passage with a diameter of 0.25 m with a volume rate of flow of 2.0 m^3/s. What is the volume rate of flow at a second section of the passage where the diameter is 0.15 m if the flow is steady? What is the velocity at the second point?

4-11 Water flows at the rate of 200 gal/min in a 2-in. ID (inside diameter) pipe. Consider the fluid incompressible and the flow steady. The flow is subsequently split into two 1-in. ID pipes that exhaust into open tanks at different elevations. If the velocity in one of the small pipes is determined to be 40 ft/s, what is the velocity at the exit of the other small pipe.

4-12 Air (assumed to be an ideal gas) flows isentropically through a duct of varying area in one-dimensional steady flow. Show that the velocity, V, at section x is given by the expression

$$V = V_0\left[\frac{D_0}{D}\left(\frac{p_0}{p}\right)^{1/(2k)}\right]^2$$

where p and D are the pressure and diameter at section x, p_0 and D_0 are the pressure and diameter at reference section 0, and k is the ratio of specific heats.

4-13 Air flows reversibly and adiabatically in a pipe having a diameter of 0.15 m. At state 1 the static temperature and pressure are 40 C and 27×10^4 N/m^2, respectively. The velocity at this initial state is 10 m/s. At a second state the pressure has been reduced to 15×10^4 N/m^2. What is the pipe diameter at this second point? What is the static temperature at this second state? Find the mass flow rate in the pipe.

4-14 Water flows at the rate of 7×10^{-4} m^3/s in a 2.0-cm ID garden hose. Assuming steady flow, find the exit velocity if the nozzle diameter is 0.5 cm.

4-15 During a fire, a fire fighter directs a nozzle at the end of a fire hose at an angle of 45° with the horizontal, toward a flaming building 60 ft away (horizontally). The diameters of the fire hose and nozzle are 4 in. and 1 in., respectively. (*a*) What pressure is required in the hose at 1 ft away from nozzle exit for the water stream to reach a section of the building 50 ft above the elevation of the nozzle? Assume that the water stream behaves as a free-falling body after it leaves the nozzle. (*b*) Does the water-stream trajectory reach its maximum height before it is intercepted by the building?

4-16 The pressure distribution, $p(x)$, in an axially symmetric channel of radius, $r(x)$, is given by

$$p(x) = p_0 + \beta x$$

If the fluid flowing through the channel is a liquid, determine an expression for $r(x)$. Take $r(0) = r_0$. Assume that the channel axis is horizontal.

4-17 Water at 15 C flows in a 5.0-cm ID smooth pipe. A venturi meter, similar to that shown in Fig. 4-19, with a 2.5-cm throat diameter is placed in the system and a mercury manometer is used to measure the pressure differential. Find the velocity and the mass flow rate in the 5.0-cm pipe if c_D equals 0.94 and the differential height, h is 5.0 cm.

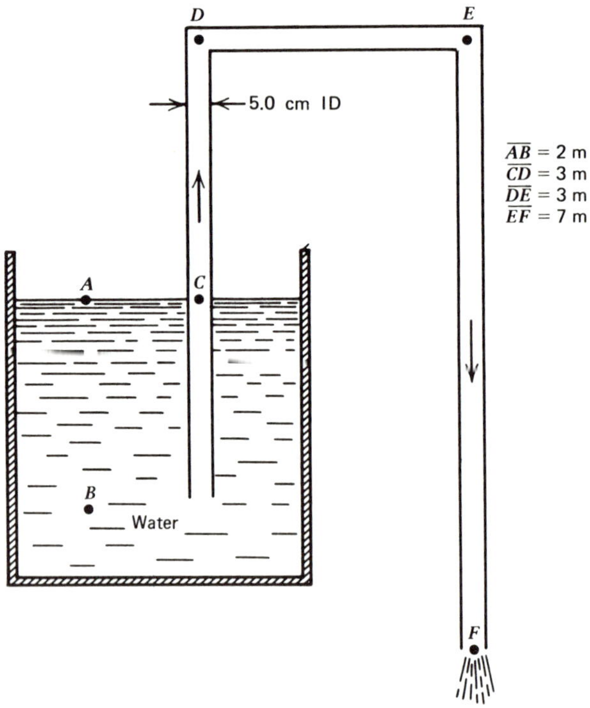

Fig. 4-42. A siphon.

4-18 Find the mass flow rate of water from the system shown in Fig. 4-42. The pipe has an inside diameter (ID) of 5.0 cm and the flow is assumed to be ideal in all respects. Can the flow rate be increased? How? What is the maximum length of the pipe section labeled $\overline{CD}$?

4-19 The insulated water storage tank shown in Fig. 4-43 is used to supply water to a scientific mission based on the moon. The water is maintained at a pressure of 27 psia through helium pressurization and at a temperature of 40 F by heaters. Water flows from the tank to an administration building located in a pressurized air lock maintained at 13 psia. The water pipe has an inside diameter of 2 in. as shown. If the system is assumed to be ideal, and if the gravitational acceleration on the moon is one-sixth of that on the surface of the earth, find the discharge from the pipe in cubic feet per minute.

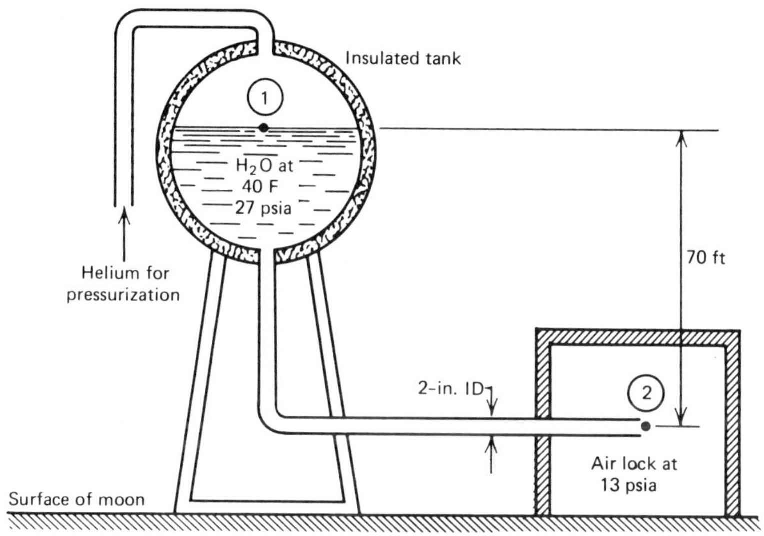

Fig. 4-43. Flow from a water storage tank.

4-20 Water from a reservoir feeds the nozzle of a Pelton wheel turbine (see Fig. 4-33). The nozzle is located 50 m below the reservoir surface. The nozzle diameter is 0.05 m and the flow from the reservoir through the nozzle is assumed to be ideal. After exiting from the buckets of the Pelton wheel, the spent water drops with negligible velocity into the tailrace (the collecting area for the spent water) immediately below. If the Pelton wheel extracts 85% of the available energy in the stream, compute the power output from the system.

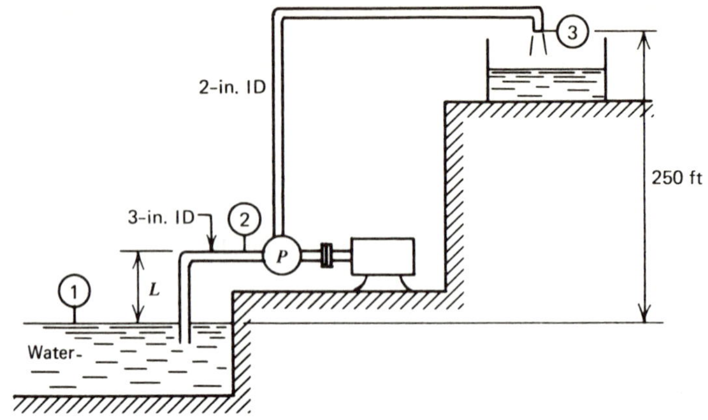

Fig. 4-44. A pump delivering water to a tank.

4-21 A pump is used to deliver water from a lake to a tank, as shown in Fig. 4-44. If the pump delivers 0.8 ft^3/s to the tank under the conditions shown in the figure, determine (*a*) the head, h_p, developed by the pump, (*b*) the horsepower that the electric motor delivers to the pump, if the efficiency of the pump is 30%, (*c*) the maximum height, *L*, above the free surface of the lake, at which the pump may be located if the pump specifications require a minimum suction pressure of 2 psia (at pump entrance).

4-22 Find the power developed by the turbine shown in Fig. 4-45.

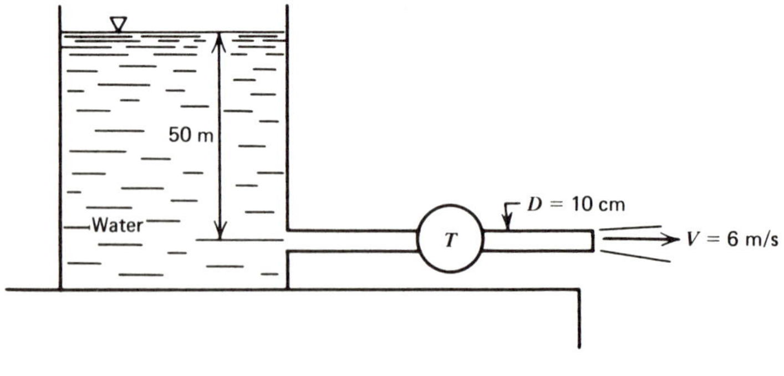

Fig. 4-45. A reservoir discharging through a turbine.

4-23 Suppose that in a particular laboratory test on the equipment similar to that shown in Fig. 4-22, the flow rate was measured to be 0.14 m^3/min, the pressure gage on the discharge side read 3.0 m of water and the vacuum gage on the suction side read 25 mm of Hg. Assume that the pressure gage is 0.5 m above the vacuum gage. Take the pipe diameters on the discharge and suction sides to be 1.3 cm and 2.0 cm, respectively. Determine: (*a*) the head, h_p, developed by the pump and (*b*) the horsepower delivered to the water by the pump.

4-24 Suppose in a particular wind-tunnel test, with instrumentation similar to that shown in Fig. 4-23, the following data were taken: barometric pressure = 745.9 mm of Hg, $z_1 = 28.5$ cm, $z_2 = 7.1$ cm, and $T_0 = 40$ C. Determine: (*a*) the free-stream velocity by using both Eq. 4.6-38 and Eq. 4.6-40 and (*b*) the temperature in the test section.

4-25 Water, flowing in a 2-in. ID horizontal pipe at the rate of 50 gal/min. enters a 90° elbow. Find the magnitude and direction of the force exerted by the elbow on the water. The flow is assumed to be frictionless, and the pressure in the pipe is 50 psig. Neglect the weight of water in the elbow.

4-26 Water, flowing at the rate of 0.0015 m^3/s, in a pipe of inside diameter equal to 0.05 m, enters a 45° reducing elbow. The exit diameter of the elbow is 0.025 m, and the water pressure at the entrance state is 4×10^4 N/m^2 (see Fig. 4-46). Find the com-

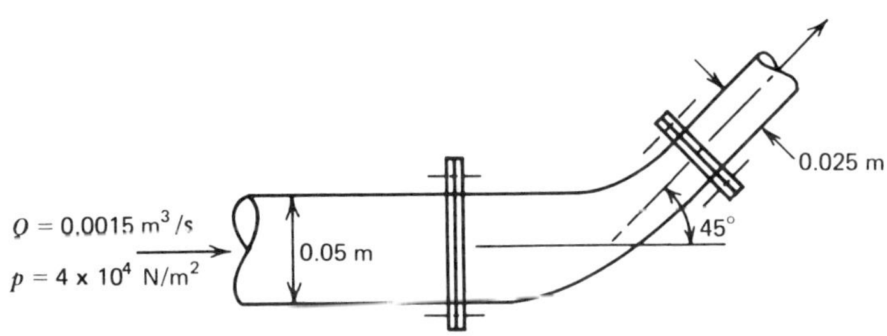

Fig. 4-46. A 45° reducing elbow.

ponents of the force exerted by the elbow on the interior fluid, if the volume contained inside the elbow is 3×10^{-4} m^3. Assume that the elbow lies in the horizontal plane.

4-27 If the flow rate through a venturi tube (see Fig. 4-19) is gradually increased until the pressure at the throat reaches the vapor pressure of the liquid, bubbles begin to appear in the liquid just downstream from the throat. It is believed that these bubbles form from nuclei of undissolved gas trapped in the crevices of minute solid particles suspended in the liquid. This phenomenon is known as *cavitation*. As the bubbles move into the higher pressure regions, they collapse and on rebounding cause momentary high-pressure waves that frequently result in damage to nearby solid surfaces. Suppose that methyl alcohol at 68 F flowed through the venturi described in Example 4-4. At this temperature the density of methyl alcohol is 49.3 lb_m/ft^3. If the volume flow rate is 0.85 ft^3/s, the ratio of the diameters D_2/D_1 is 0.5, and diameter D_1 is 4 in., what pressure at section 1 will just cause cavitation to occur?

4-28 A fire truck pumps water through a 2-in. ID fire hose under a total pressure of 150 psig. Two fire fighters at ground level are holding a nozzle with a 1-in. ID opening. Find the force exerted by the nozzle on the fire fighter and state its direction. Assume an ideal (frictionless) flow.

4-29 Does the answer to Problem 4-28 somehow seem counter to your own experience with a garden hose? With the momentum principle as a basis, attempt to explain this seeming paradox.

4-30 A stream of water, 2.5 cm in diameter, having a velocity of 30 m/s, strikes the vane shown in the Fig. 4-47 and is deflected through an angle of 30°. Find the magnitude of the force components R_1 and R_2.

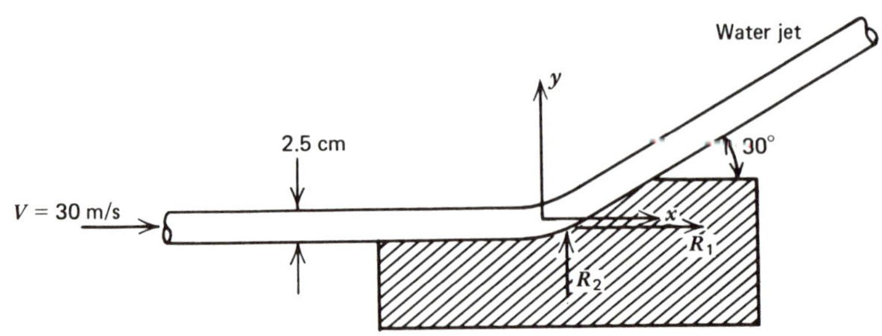

Fig. 4-47. Jet impinging on vane.

4-31 A stream of water, 0.05 m in diameter, strikes the U-block shown in Fig. 4-48 at a velocity of 15 m/s. Find the resultant force R_1.

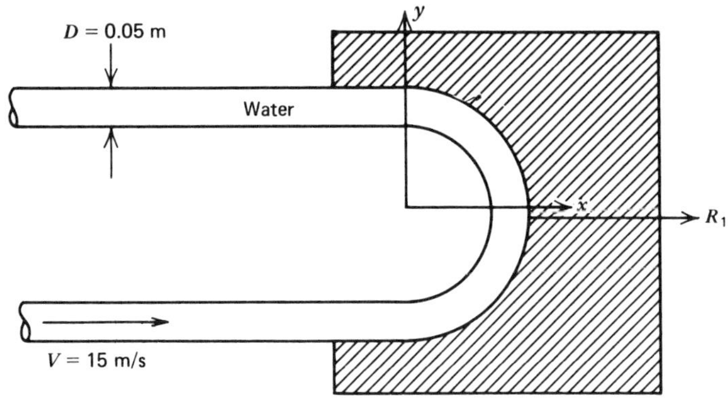

Fig. 4-48. Jet impinging on U-block.

4-32 A vertical jet of water, 2 in. in diameter, strikes a horizontal circular flat plate 3 ft in diameter. Assume that the centerline of the jet intersects the center of the plate. Assume the plate is fixed and the jet velocity is 300 ft/s. Find the force exerted by the water on the plate.

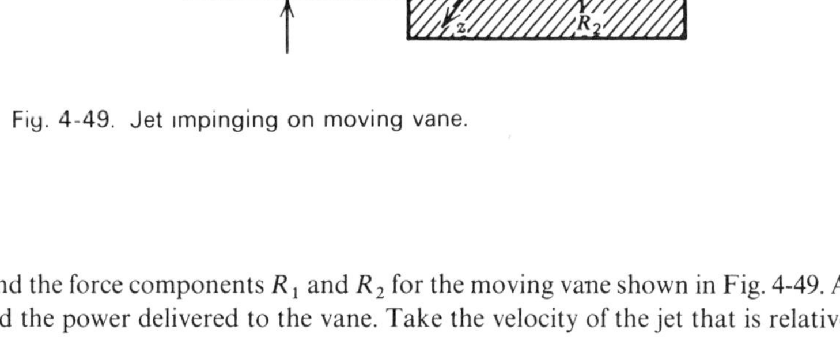

Fig. 4-49. Jet impinging on moving vane.

4-33 Find the force components R_1 and R_2 for the moving vane shown in Fig. 4-49. Also find the power delivered to the vane. Take the velocity of the jet that is relative to the ground to be 60 m/s. The fluid is water.

4-34 For the moving vane shown in Fig. 4-49, draw the complete velocity-vector diagram for the exiting stream, showing and labeling the following velocity components:

Absolute exit velocity

Vane velocity

Relative exit velocity

The x-component, absolute exit velocity

The x-component, relative exit velocity

The y-component, absolute exit velocity

The y-component, relative exit velocity

4-35 A crude propulsion system consists of a tank and a compressor. The tank sits on top of a boat, as shown in Fig. 4-50. Water is discharged through a nozzle having a 2.5-cm diameter. The compressor pressurizes the system, causing the water in the tank to discharge at a relatively high velocity. The boat experiences a drag of

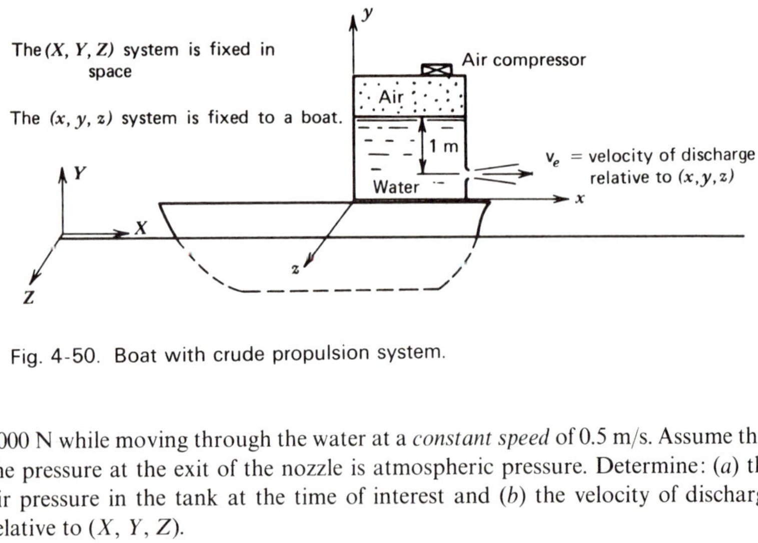

Fig. 4-50. Boat with crude propulsion system.

4000 N while moving through the water at a *constant speed* of 0.5 m/s. Assume that the pressure at the exit of the nozzle is atmospheric pressure. Determine: (*a*) the air pressure in the tank at the time of interest and (*b*) the velocity of discharge relative to (X, Y, Z).

4-36 An elbow–nozzle combination is supported by four bolts connecting the flanges, as shown in Fig. 4-51. If the elbow–nozzle combination lies in the horizontal plane and has an interior volume of 3 ft^3, find the total force on the bolts.

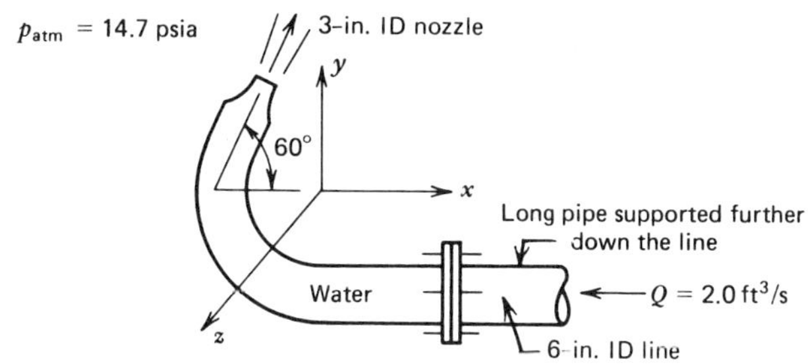

Fig. 4-51. Elbow–nozzle combination.

4-37 A horizontal jet impinges on a vertical flat plate. The plate is equipped with stream guides that keep the stream moving in a horizontal direction until it leaves the plate, as shown in the Fig. 4-52. Assume that the flow is steady and neglect gravitational

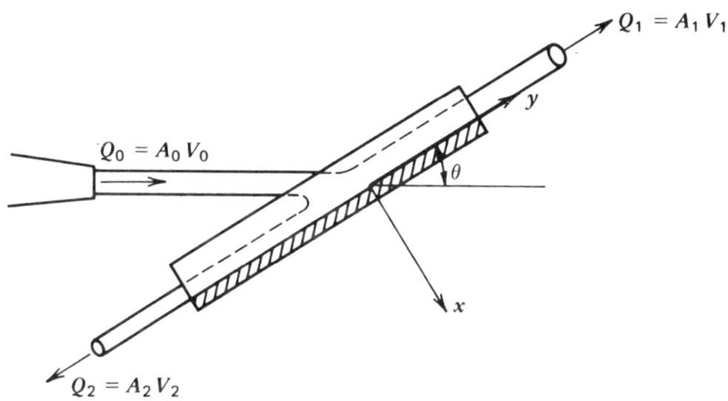

Fig. 4-52. Jet impinging on vertical flat plane.

and viscous effects. Observe that by the use of Bernoulli's equation one can readily show that $V_2 = V_1 = V_0$. Determine: (*a*) the resultant force that the jet exerts on the plate and (*b*) Q_1 and Q_2. Answers should be in terms of ρ, Q_0, V_0, and θ.

Hint. Since viscous effects are negligible, the resultant force found in part (*a*) must be normal to the plate.

4-38 (*a*) Compute each of the velocity components called for in Problem 4-34 in terms of the inlet absolute velocity, V_0; the vane velocity, U; and the relative discharge angle, θ.

(*b*) Show that if (x, y, z) moves linearly at a constant velocity, $\mathbf{U}$, with respect to (X, Y, Z), and a control volume fixed in (x, y, z) is selected, then Newton's second law may also be written as

$$\underset{\text{C.V.}}{\oiint} \mathbf{F}\, d\forall + \underset{\text{C.S.}}{\oiint} \boldsymbol{\tau}\, dS = \frac{\partial}{\partial t_{xyz}} \underset{\text{C.V.}}{\iiint} (\rho \mathbf{v}_{XYZ})\, d\forall + \underset{\text{C.S.}}{\oiint} \rho \mathbf{v}_{XYZ}(\mathbf{v}_{xyz} \cdot d\mathbf{S})$$

Hint. Use the law of mass conservation.

(*c*) Show that the force R_1 determined by the equation given in part (*b*) gives the same answer as that found in Problem 4-33.

4-39 The doubly directed vane shown in Fig. 4-53 moves in the x direction at 15 m/s. Water discharging from a nozzle at an exit velocity of 90 m/s relative to the ground impinges on the vane. Determine the force the jet exerts on the vane.

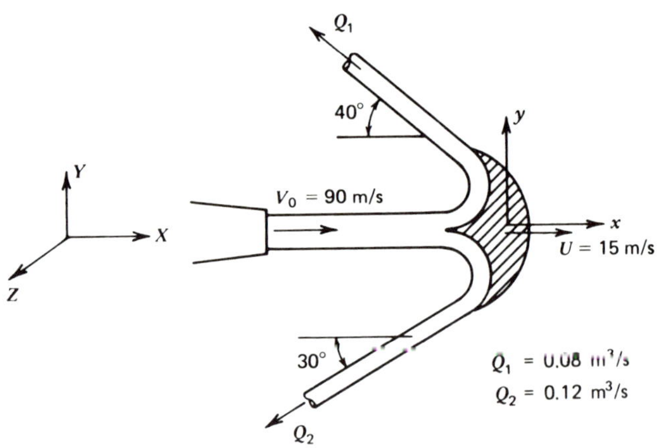

Fig. 4-53. Jet impinging on moving vane.

4-40 Suppose the vane shown in Problem 4-33 was free to move on a level surface without air or frictional resistance. If the jet stream is turned on when the vane is at rest, determine an expression for the velocity of the vane as a function of time.

4-41 A gas turbine jet engine, similar to that shown in Fig. 4-28, has the following specifications:

Inlet diameter	3 ft
Exhaust diameter	1.5 ft
Length	12 ft
Airflow	10,000 lb_m/min
Fuel flow	200 lb_m/min
Atmospheric pressure	14.7 psia
Static pressure, inlet	14.7 psia
Static pressure, exhaust	20.0 psia
Temperature, inlet	60 F
Temperature, exhaust	800 F

Assume that both inlet air and exhaust gases behave as ideal gases. Take R, for the exhaust gas, to be 53.4 (lb_f-ft)/(lb_m-R) and compute the thrust developed by the engine.

4-42 Repeat Problem 4-41 for a jet engine with the following specifications:

Inlet diameter	0.8 m
Exhaust diameter	0.5 m
Length	3.0 m
Airflow	65 kg/s
Fuel flow	1.39 kg/s
Atmospheric pressure	4.9×10^4 N/m^2 abs.
Static pressure, inlet	5.1×10^4 N/m^2 abs.
Static pressure, exhaust	7.0×10^4 N/m^2 abs.
Temperature, inlet	293 K
Temperature, exhaust	700 K

4-43 Find the horsepower of the jet engine of Problem 4-41 if the airspeed of the plane (and engine) is 200 mi/h.

4-44 Hot gases stream from the exhaust of a rocket engine with a velocity of 1000 m/s at a temperature of 1700 K. The pressure in the exhaust stream is 6.0×10^4 N/m^2, and the exhaust nozzle has a diameter of 0.2 m. Find the mass flow rate of the fuel plus oxidizer, and the thrust developed by the rocket. The exhaust may be assumed to be an ideal gas with a molecular weight of 27; assume the entire system operates outside the earth's atmosphere.

4-45 A Pelton wheel (a type of water wheel) is shown in Fig. 4-54. A single stream of water, 2.5 in. in diameter with a velocity of 100 ft/s, continuously strikes the buckets of the Pelton wheel. The angular velocity is 120 rpm and the mean radius of the buckets is 2 ft. Find the power delivered by the wheel if the flow is assumed to be frictionless.

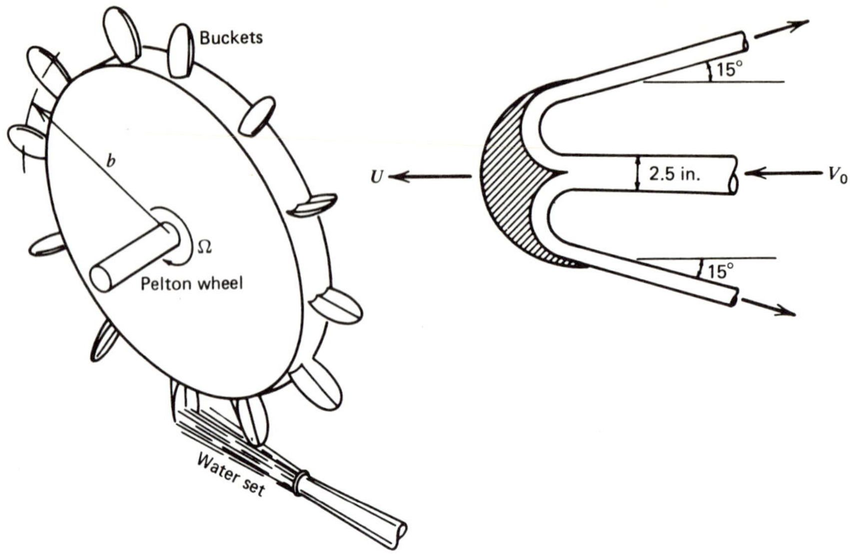

Fig. 4-54. Pelton wheel.

4-46 A water turbine with radial flow has the dimensions shown in Fig. 4-55. The absolute entering velocity is 50 ft/s, and it makes an angle of 30°, with the tangent to the rotor. The absolute exit velocity is directed radially inward. The angular speed of the rotor is 120 rpm. Find the power delivered to the shaft of the turbine if the flow is assumed to be ideal.

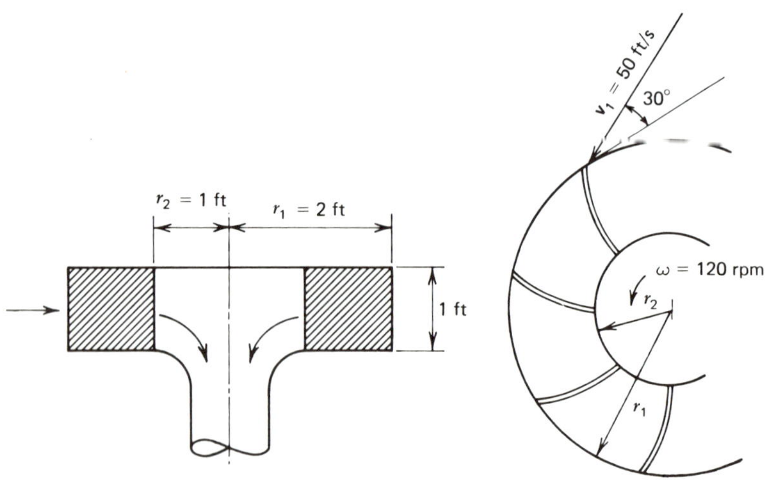

Fig. 4-55. A water turbine.

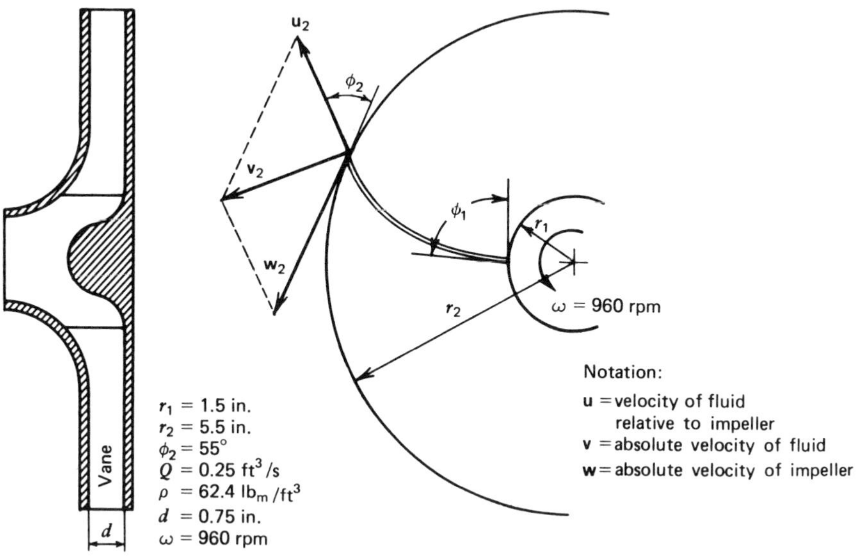

Fig. 4-56. A centrifugal water pump.

4-47 A centrifugal radial water pump has the dimensions shown in Fig. 4-56. The volume rate of flow is 0.25 ft^3/s, and the absolute inlet velocity is directed radially outward. The angular velocity of the impeller is 960 rpm. The exit velocity as seen from a coordinate system attached to the impeller can be assumed to be tangent to the vane at its trailing edge. (*a*) Find the power required to drive the pump. (*b*) It is conventional to design the vane on the impeller so that the fluid velocity relative to the impeller is tangential to the vane at r_1. Find angle ϕ_1.

4-48 A pitot tube is to be used for determining the velocity profile in a pipeline containing an oil flow. The pitot tube is connected to a mechanism that allows it to be traversed across a diameter of the pipe. The pitot tube is connected to a mercury manometer, as shown in Fig. 4-34. The pipe radius, R, is 0.375 in., γ_{oil} is 54.8 lb_f/ft^3, and γ_{hg} is 847.7 lb_f/ft^3. The following data have been taken:

r (in.)	h (in.)
0.075	0.45
0.225	0.20
0.325	0.10

Determine the oil velocities in meters per second at the three radial positions.

4-49 A differential pitot tube differs from that shown in Fig. 4-34 by the fact that it has a second tube passage, which leads to an opening in the side walls of the tube, as shown in Fig. 4-57. Since the pitot tube is a very small device, the stream velocity passing over the side opening can be assumed to be the same as the free-stream velocity. Therefore, by Bernoulli's equation, we can assume that the pressure at the side opening is the same as the free-stream pressure. Suppose that such a device

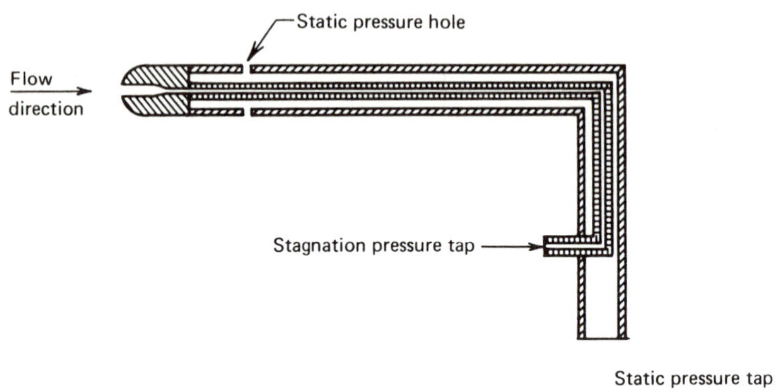

Fig. 4-57. Differential pitot tube.

is used to measure the flight speed of an airplane. We will assume that the pitot tube is projected sufficiently far in front of the airplane that it lies in the free stream. Take both pressure taps on the pitot tube to be connected to a mercury manometer, with a reading $h = 0.924$ ft. A barometer aboard the airplane indicates that the free-stream pressure is 10.1 psia. A thermocouple whose temperature measuring junction is located at the stagnation temperature is 85.3 F. (*a*) Determine the flight speed of the airplane using Eq. 4.8-24. (*b*) Determine the flight speed of the airplane if the air is assumed to be incompressible; that is, by using Eq. 4.8-17. (*c*) What percent error results from the incompressible assumption?

4-50 Air flows through a circular duct, as shown in Fig. 4-58. Suppose we wish to make an estimate of the mass flow rate through the duct by a single measurement with a pitot tube. Let's assume that the pressure and temperature are uniform at a section and that the flow is laminar. However, assume that the flow around the differential pitot tube (see Fig. 4-57) is compressible and is positioned at the duct centerline. Determine the mass flow rate.

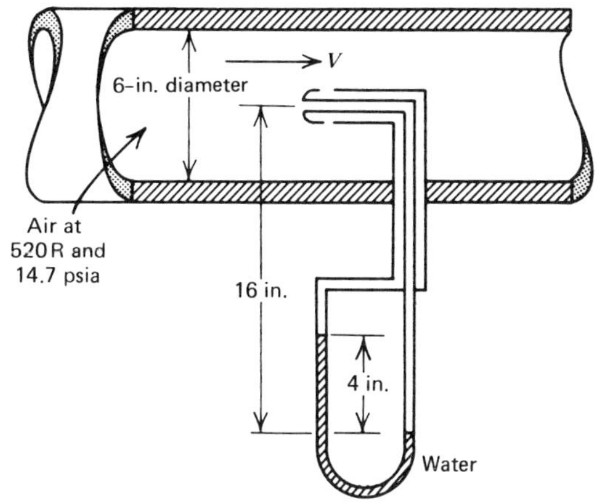

Fig. 4-58. Differential pitot tube positioned at center-line of duct.

4-51 (*a*) Beginning with Eq. 4.8-7 show that consideration of the unsteady case alters Eq. 4.8-14 so that it becomes

$$\frac{1}{g_c}\oint_A^B \frac{\partial \mathbf{v}}{\partial t}\cdot d\mathbf{r} + \oint_A^B \frac{dp}{\rho} = \frac{1}{2g_c}(v_A{}^2 - v_B{}^2) + \frac{g}{g_c}(z_A - z_B)$$

where points A and B lie on the same instantaneous streamline at time t, and where $d\mathbf{r}$ is a displacement vector along that same streamline.

(*b*) Now apply this equation to show that the natural frequency of oscillation in the thin U-tube is, $\omega = \sqrt{2g/L}$, where L is the total length of the fluid column. That

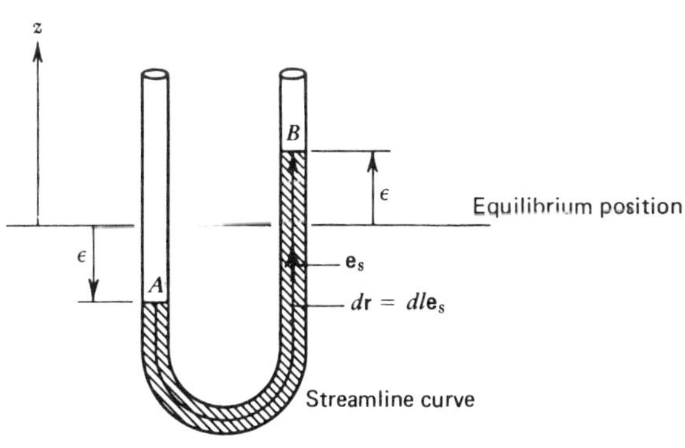

Fig. 4-59. Oscillating liquid column.

is, show that $\varepsilon = c_1 \cos \sqrt{2g/L}t + c_2 \sin \sqrt{2g/L}t$, where ε is the position of the free surface of the right-hand liquid column as measured from its equilibrium position, and where c_1 and c_2 are constants that are determined by initial conditions (see Fig. 4-59). You may consider the fluid column to be a single instantaneous streamline. Note that $\mathbf{v} = (d\varepsilon/dt)\mathbf{e}_s$, where $d\varepsilon/dt$ is uniform along the streamline and $\mathbf{e}_s$ is a vector tangent to the streamline and pointing in the direction of integration along the curve. Also note that $d\mathbf{r} = d\ell\,\mathbf{e}_s$. Write the above equation between points A and B and evaluate the terms, thereby obtaining a differential equation.

4-52 Consider a temperature probe that responds instantaneously to its local temperature. Assume that its output is a voltage, $e(t)$, which is proportional to the local temperature; that is, $e(t) = aT$. Suppose the probe moves through a steady temperature field given by

$$T(x) = \frac{T_2 - T_1}{L} x + T_1$$

If the probe moves in the x direction from $x = 0$ to $x = L$ at a uniform speed of 5 ft/s, determine $e(t)$.

4-53 Let (x, y, v_1, v_2) be the position and velocity-vector components in Eulerian coordinates, and (x, y, V_1, V_2) be the position and velocity-vector components in Lagrangian coordinates. Suppose $v_1(x, y, t) = axt$ and $v_2(x, y, t) = -ayt$. Determine the Lagrangian position relations: $x = x(t; x_0, y_0)$ and $y = y(t; x_0, y_0)$.

4-54 Suppose in Problem 4-53, the Eulerian velocity components were

$$v_1 = A \sin \omega(y - ct) \quad \text{and} \quad v_2 = B,$$

where A and B are constants. Determine the Lagrangian position relations: $x = x(t; x_0, y_0)$ and $y = y(t; x_0, y_0)$.

Chapter Five

Flow of a Real Fluid, Dimensional Analysis, and Similitude

5.1 INTRODUCTION—FLOW OF A REAL FLUID

In nearly all of the problems discussed in the previous chapters, viscous effects were neglected. However, in many fluid-flow problems viscous effects are of sufficient importance that they cannot be neglected. In fact, in some cases, viscous effects become dominant. For example, under certain conditions, the drag on a streamline body, such as an airfoil, results predominantly from viscous effects. The pressure drop in a flow through a pipe of constant cross-sectional area, held at a fixed elevation, also results from viscous effects. Many other examples illustrating this point can be cited.

When viscous effects are included in the momentum and the energy equations, a set of partial differential equations is obtained which are very complicated. In addition, a stress rate-of-strain relationship is needed to obtain a complete set of equations. Viscosity results in the conversion of kinetic energy into thermal energy and this process is irreversible. Because of the complexity of the governing equations, solutions are obtained, except for a few very special problems, by making approximations. Furthermore, there are many problems that can only be solved by extensive experimentation.

An additional complication exists because real flows also exhibit instabilities, which result in a turbulent flow regime. *Turbulence* is an unsteady, chaotic flow with strong transverse mixing within the fluid; whereas a laminar flow is a smooth flow where the fluid flows in layers. If a flow discharging from a pipe is a laminar flow, it will have a glossy appearance. Some of the differences in behavior between a laminar flow and a turbulent flow can be seen in Reynolds' experiment.

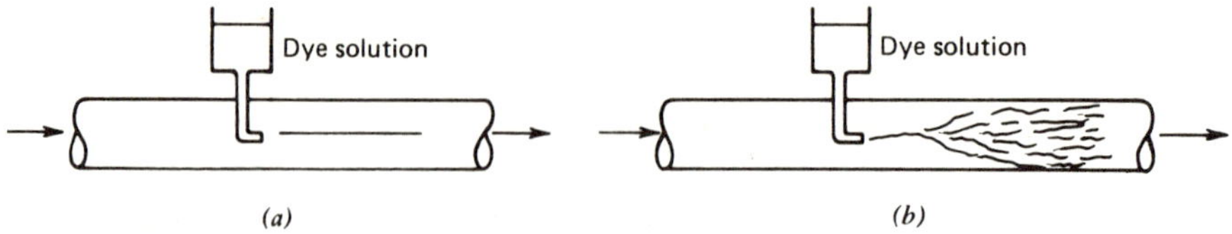

Fig. 5-1. Apparatus for Reynolds' pipe flow experiment. (*a*) Laminar flow. (*b*) Turbulent flow.

In Reynolds' experiment, a dye solution is injected into an established water flow through a transparent pipe. The flow rate through the pipe can be controlled by a discharge valve (see Fig. 5-1). A dye solution having the same specific weight as water is used. If the flow is started at a low volume flow rate and a stream of dye is released into the flow, the dye will flow as a thin line with the fluid when the flow is laminar. As the velocity is increased, a point is reached where the line breaks up and the dye diffuses throughout the liquid. This occurs when the flow is turbulent. Furthermore, if the velocity, v_1 is measured at a particular point in the flow with a fast responding instrument, such as a hot-wire anemometer, and a plot made of v_1 versus t, for some fixed valve setting, plots are obtained as shown in Fig. 5-2.

We see that for the laminar case, the flow can be regarded as a truly steady flow. But for the turbulent case, the flow is unsteady. Hence, when a velocity magnitude is referred to in turbulent flow, it is frequently a time-averaged value. Methods of treating turbulent flows are not nearly as well established as those treating laminar flows. For laminar flows there exists a complete set of equations to describe the phenomenon. However, this is not the case for a general turbulent flow.

The velocity at which a flow turns turbulent depends in part on the existence of small disturbances within the flow region, and the frequency of those dis-

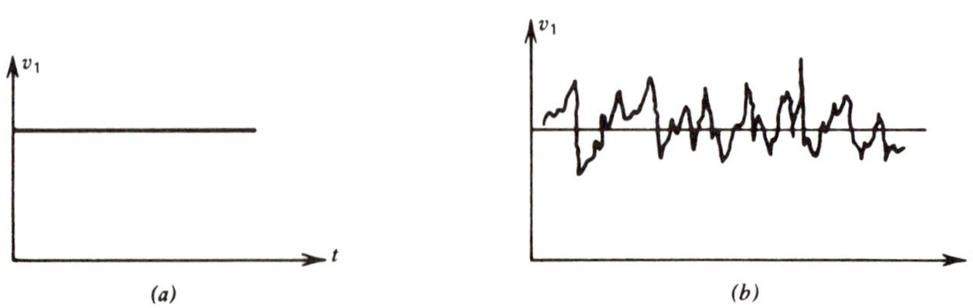

Fig. 5-2. Plots of v_1 versus t. (*a*) Laminar flow. (*b*) Turbulent flow.

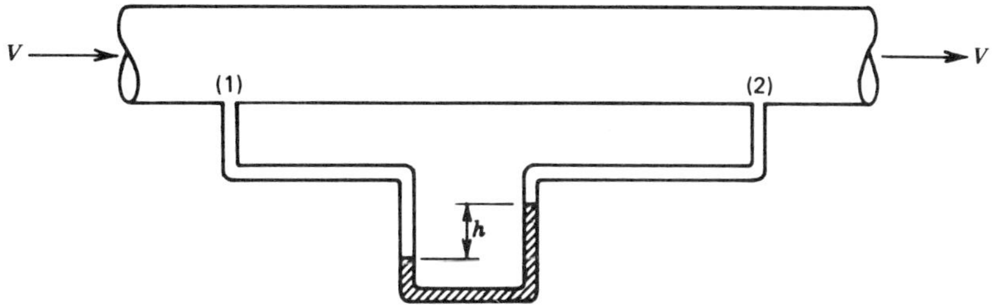

Fig. 5-3. Apparatus for demonstrating pressure drop in a pipe flow.

turbances. The minimum velocity at which a flow will turn turbulent is called the *critical velocity.*

In the opening paragraph we mentioned the fact that a pressure drop occurs in a flow through a pipe of constant cross-sectional area, which is held at a fixed elevation. When the flow turns turbulent we find that the pressure drop increases significantly. For example, consider a flow through a pipeline having a uniform cross section. If a manometer is connected to pressure taps placed in the line, as shown in Fig. 5-3, and sufficient data are taken to construct a plot of h_f versus V, where h_f is defined by

$$h_f = \frac{p_1 - p_2}{\gamma} \qquad \text{Eq. 5.1-1}$$

then one obtains the type of curve shown in Fig. 5-4. We see that the curve labeled turbulent lies above the curve labeled laminar. This indicates that if two flows existed, each having the same average velocity, but one being in the turbulent

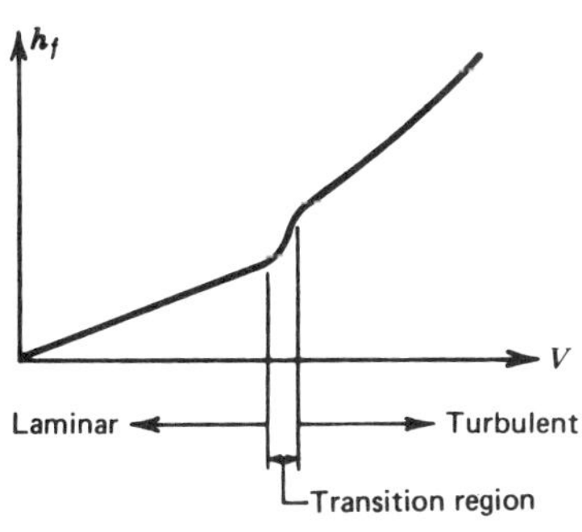

Fig. 5-4. Plot of h_f versus V.

regime and the other in the laminar regime, the one being in the turbulent regime would have a larger pressure drop.

Pipe and duct flows are frequently treated by adding a head-loss term to the energy equation that is obtained for an ideal fluid (see Eq. 4.6-18). The modified equation becomes

$$\frac{p_1}{\gamma} + z_1 + \frac{V_1^2}{2g} = \frac{p_2}{\gamma} + z_2 + \frac{V_2^2}{2g} + h_f \qquad \text{Eq. 5.1-2}$$

The term h_f is determined empirically when the flow is turbulent. This is discussed in much greater detail in the next chapter. For laminar flow h_f can be determined analytically.

The important parameter that indicates whether a flow will behave in a viscous manner is the Reynolds number, Re, which is defined by the following equation

$$\text{Re} = \frac{Vd\rho}{\mu} = \frac{Vd}{\nu} \qquad \text{Eq. 5.1-3}$$

where d is the diameter of the pipe or some other physical dimension, such as size of body in a flow field; V is the mean velocity or free-stream velocity; ρ is the mass density; μ is the absolute viscosity; and ν is the kinematic viscosity. The Reynolds number also indicates when a flow will turn turbulent. The critical Reynolds number $(\text{Re})_c$ is defined to be the minimum Reynolds number at which a flow will turn turbulent. $(\text{Re})_c$ depends on the type of flow. For example,

1. For a flow between parallel walls, $(\text{Re})_c \approx 1000$. Here V is the average velocity and d is the distance between the walls.

2. For a flow around a sphere, $(\text{Re})_c \approx 1$. Here V is the free-stream velocity and d is the diameter of the sphere.

3. For a flow through a cylindrical pipe, $(\text{Re})_c \approx 2100$. Here V is the average velocity and d is the diameter of the pipe.

Example 5-1

Water flows through a 1-in. diameter pipe. The kinematic viscosity of water is 0.1×10^{-4} ft^2/s. Calculate the largest flow rate for which the flow will *definitely* be laminar.

SOLUTION

The largest flow rate for which a flow will definitely be laminar occurs when $\text{Re} = (\text{Re})_c$. For a flow through a pipe $(\text{Re})_c \approx 2100$, then

$$(\text{Re})_c = 2100 = \frac{V_c d}{\nu}$$

$$V_c = \frac{2100\nu}{d} = \frac{2100(0.1 \times 10^{-4})}{\frac{1}{12}} = 0.252 \text{ ft/s}$$

$$Q_c = AV_c = \frac{\pi}{4}\left(\frac{1}{12}\right)^2 \times 0.252 = 1.38 \times 10^{-3} \text{ ft}^3\text{/s}$$

5.2 WHY DIMENSIONAL ANALYSIS?

In the previous section, it was pointed out that real fluids behave in a complex manner and that certain fluid-flow phenomena can only be understood by extensive experimentation. The first step in the process is to determine the variables involved in the phenomenon of interest. The second step is to set up the experimental program in such a manner that the data can readily be interpreted. This requires that some empirical relationship between the variables be determined. By a procedure called *dimensional analysis*, the variables may be grouped in a manner that facilitates the determination of the desired relationship. The student should not be left with the impression that having observed a phenomenon, one can immediately apply the concepts of dimensional analysis. This is just not so. Notice that before the method can be applied, a knowledge of the variables involved in the phenomenon is required. This may be obtained either by extensive experimentation or by a thorough analysis that provides the mathematical equations which govern the phenomenon. In the latter case, even if a mathematical solution is available, there are still many advantages in applying some of the concepts of dimensional analysis.

The following example illustrates the complexity in obtaining a functional relationship between the variables involved in a particular phenomenon *without* the use of dimensional analysis. Suppose a smooth sphere moves through a fluid at a low speed. Assume that the drag[1] on the sphere, D, is a function of (1)

[1] The integration of the surface traction, τ, over the exterior surface of a body gives the resultant aerodynamic force acting on the body. The component of this force in the direction of the free stream is defined as the *drag*, while the component normal to the free stream (in the upward direction) is defined as the *lift*.

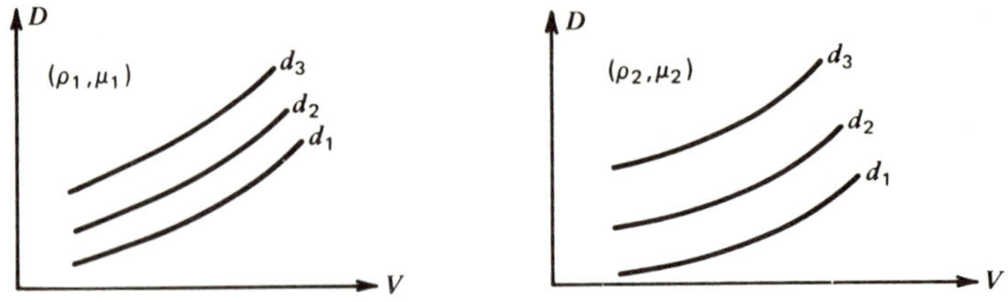

Fig. 5-5. Drag versus V for spheres of various diameters.

the diameter, d; (2) the velocity, V; (3) the density, ρ; and (4) the viscosity, μ; that is,

$$D = f(d, V, \rho, \mu)$$

Without dimensional analysis, one could develop an experimental program giving D versus V for several values of d, μ, and ρ. These may be plotted on a series of graphs (see Fig. 5-5). Clearly, it would be very difficult to establish a relationship between the variables. However, as we will see later, the use of dimensional analysis facilitates the means by which a relationship between the variables can be established empirically. The basic concepts involved in dimensional analysis are discussed in the following section.

5.3 DIMENSIONAL HOMOGENEITY AND DIMENSIONAL ANALYSIS

Equations that are in a form that do not depend on the fundamental units of measurement are called *dimensionally homogeneous* equations. Fundamental equations of physics and engineering are dimensionally homogeneous equations. To illustrate this point observe the equation

$$\dot{m} = \rho A V \qquad \text{Eq. 5.3-1}$$

which gives the mass flow rate through a pipe in terms of the mass density of the fluid, ρ; the cross-sectional area of the pipe, A; and the average velocity of the fluid, V. The formula is valid for any basic set of units of mass, length, and time. For example, if the basic set of units is (lb_m, ft, h), then the dimensions of $\dot{m}$

are (lb_m/h). On the other hand, if the basic set of units is (kg, m, s), then the dimensions of $\dot{m}$ are (kg/s). However, for each set of basic dimensions, the formula is correct. On the other hand, if we substitute 1 ft^2 for A in the equation, then the equation is valid only if the units of density are in mass/ft^3 and the units of velocity are in ft/unit time. Thus, the equation

$$\dot{m} = \rho V$$

is not dimensionally homogeneous.

We can develop this concept more fully by further considering Eq. 5.3-1. Let $(\dot{m}, \rho, A, V)$ be the variables in the (lb_m, ft, h), system of units and $(\bar{\dot{m}}, \bar{\rho}, \bar{A}, \bar{V})$ be the corresponding variables in the (kg, m, s) system of units. If 1 kg $= P$ lb_m, 1 m $=$ Q ft and 1 s $=$ R h, then

$$\bar{\rho}\,\frac{\text{kg}}{m^3} = \rho\,\frac{\text{lb}_\text{m}}{\text{ft}^3}\cdot\frac{1}{P}\frac{\text{kg}}{\text{lb}_\text{m}}\cdot\left(Q\,\frac{\text{ft}}{m}\right)^3$$

or

$$\bar{\rho} = \rho\left(\frac{Q^3}{P}\right)$$

Similarly

$$\bar{A} = A\left(\frac{1}{Q^2}\right)$$

$$\bar{V} = V\left(\frac{R}{Q}\right)$$

$$\bar{\dot{m}} = \dot{m}\left(\frac{R}{P}\right) \qquad \text{Eq. 5.3-2}$$

If Eq. 5.3-1 is homogeneous, then by definition

$$\bar{\dot{m}} = \bar{\rho}\bar{A}\bar{V} \qquad \text{Eq. 5.3-3}$$

That is, the equation in each set of units must have the same functional relationship. Equation 5.3-1 is homogeneous, if and only if, Eq. 5.3-3 is an identity in the variables (ρ, A, V, P, Q, R). To check to see that this is so, substitute Eqs. 5.3-2 into Eq. 5.3-3.

$$\dot{m} \cdot \left(\frac{R}{P}\right) = \rho\left(\frac{Q^3}{P}\right) \cdot A\left(\frac{1}{Q^2}\right) \cdot V\left(\frac{R}{Q}\right) = \rho A V\left(\frac{R}{P}\right)$$

Then by Eq. 5.3-1, we see that

$$\rho A V\left(\frac{R}{P}\right) = \rho A V\left(\frac{R}{P}\right)$$

Thus, Eq. 5.3-3 is an identity and Eq. 5.3-1 is homogeneous.

For a generalization of these concepts, the reader is referred to Chapter 4 of *Dimensional Analysis and Theory of Models*, by H. L. Langharr.

A special form of the dimensionally homogeneous equation is one that involves only *dimensionless groups* of terms. This type of equation plays a very important role in the dimensional analysis. For example, the equation

$$\frac{g_c \, \Delta p}{\rho V^2} = F\left(\text{Re}, \frac{\ell}{d}\right) \qquad \text{Eq. 5.3-4}$$

is an equation involving only dimensionless groups. In this equation Δp is a pressure drop; l is the length of a pipe; d is the diameter of the pipe, and Re is the Reynolds number as defined by Eq. 5.1-3. Now let us examine the dimensions of $(g_c \, \Delta p)/(\rho V^2)$ and Re. It is obvious that for any consistent set of length units, ℓ/d is dimensionless. If we desire to write the dimensions of a variable, u, we write $\mathscr{D}(u)$. For example, the dimensions of velocity are written

$$\mathscr{D}(V) = \text{ft/s}$$

If u is dimensionless, we write $\mathscr{D}(u) = 1$. Then

$$\mathscr{D}(\text{Re}) = \mathscr{D}\left(\frac{\rho V d}{\mu}\right) = \left(\frac{\text{lb}_\text{m}}{\text{ft}^3}\right)\left(\frac{\text{ft}}{\text{s}}\right)(\text{ft})\left(\frac{\text{ft-s}}{\text{lb}_\text{m}}\right) = 1$$

and

$$\mathscr{D}\left(\frac{g_c \, \Delta p}{\rho V^2}\right) = \left(\frac{\text{lb}_\text{m}\text{-ft}}{\text{lb}_\text{f}\text{-s}^2}\right)\left(\frac{\text{lb}_\text{f}}{\text{ft}^2}\right)\left(\frac{\text{ft}^3}{\text{lb}_\text{m}}\right)\left(\frac{\text{s}^2}{\text{ft}^2}\right) = 1$$

Thus, Eq. 5.3-4 is indeed an equation involving only dimensionless groups.

Now let us return to the problem involving the drag or sphere. We will consider units of mass, length, and time as the basic set of units. To simplify the notation, we will represent this set of dimensions by the symbol (M, L, T). Force is not included as a basic dimension since $\mathscr{D}(g_c \cdot \text{force})$ is equivalent to $\mathscr{D}(\text{mass} \cdot \text{acceleration})$ by Newton's second law. Thus, whenever the dimensions of a variable contain lb_f, we use $(g_c \cdot \text{variable})$ instead of the variable itself. Suppose we assume that *the phenomenon can be described by a functional relationship between a set of dimensionless products*, π_1, π_2, and so on; that is,

$$F(\pi_1, \pi_2, \ldots) = 0$$

where each π-group is a product of the variables raised to power: thus,

$$\pi = (g_c D)^{x_1}(\ell)^{x_2}(V)^{x_3}(\rho)^{x_4}(\mu)^{x_5} \qquad \text{Eqs. 5.3-5}$$

Each π-group must satisfy the conditions of dimensional homogeneity. Suppose the variables appearing in Eq. 5.3-5 are in the units (lb_m, ft, min). Now suppose those variables are replaced by a corresponding set, say $\bar{\pi}$, $\overline{g_c D}$, $\bar{\ell}$, $\bar{\rho}$, $\bar{\mu}$, expressed in a second set of units, say (kg, m, s). If Eq. 5.3-5 is dimensionally homogeneous, then

$$\bar{\pi} = (\overline{g_c D})^{x_1}(\bar{\ell})^{x_2}(\bar{V})^{x_3}(\bar{\rho})^{x_4}(\bar{\mu})^{x_5} \qquad \text{Eqs. 5.3-6}$$

If, 1 kg $= P$ lb_m, 1 m $= Q$ ft, and 1 s $= R$ h, then the relationship between a variable in the first set of units and the same variable in the second set of units can be obtained. For the variables under consideration, these relationships are

$$\bar{\pi} = \pi$$

$$\bar{V}\,\frac{\text{m}}{\text{s}} \cdot Q\,\frac{\text{ft}}{\text{m}} \cdot \frac{1}{R}\,\frac{\text{s}}{\text{h}} = V\,\frac{\text{ft}}{\text{h}}$$

$$(\overline{g_c D})\,\frac{\text{kg-m}}{\text{s}^2} \cdot P\,\frac{\text{lb}_m}{\text{kg}} \cdot Q\,\frac{\text{ft}}{\text{m}} \cdot \left(\frac{1}{R}\,\frac{\text{s}}{\text{h}}\right)^2 = (g_c D)\,\frac{\text{lb}_m\text{-ft}}{\text{h}^2}$$

$$\bar{\rho}\,\frac{\text{kg}}{\text{m}^3} \cdot P\,\frac{\text{lb}_m}{\text{kg}} \left(\frac{1}{Q}\,\frac{\text{m}}{\text{ft}}\right)^3 = \rho\,\frac{\text{lb}_m}{\text{ft}^3}$$

$$\bar{\ell}\,\text{m} \cdot Q\,\frac{\text{ft}}{\text{m}} = \ell\,\text{ft}$$

$$\bar{\mu}\,\frac{\text{kg}}{\text{m-s}} \cdot P\,\frac{\text{lb}_m}{\text{kg}} \cdot \frac{1}{Q}\,\frac{\text{m}}{\text{ft}} \cdot \frac{1}{R}\,\frac{\text{s}}{\text{h}} = \mu\,\frac{\text{lb}_m}{\text{ft-h}}$$

Solving the above set of equations for $\bar{\pi}$, $(\overline{g_c D})$, $\bar{\ell}$, $\bar{\rho}$, and $\bar{\mu}$,

$$\bar{\pi} = \pi \qquad \bar{V} = \frac{R}{Q} V$$

$$\overline{g_c D} = \frac{R^2}{PQ}(g_c D) \qquad \bar{\rho} = \frac{Q^3}{P}\rho$$

$$\bar{\ell} = \frac{1}{Q}\ell \qquad \bar{\mu} = \frac{QR}{P}\mu \qquad \text{Eq. 5.3-7}$$

Substituting Eqs. 5.3-7 into Eq. 5.3-6, gives

$$\pi = \left(\frac{R^2}{PQ}\right)^{x_1}\left(\frac{1}{Q}\right)^{x_2}\left(\frac{R}{Q}\right)^{x_3}\left(\frac{Q^3}{P}\right)^{x_4}\left(\frac{QR}{P}\right)^{x_5}(g_c D)^{x_1}(\ell)^{x_2}(V)^{x_3}(\rho)^{x_4}(\mu)^{x_5} \qquad \text{Eqs. 5.3-8}$$

Substituting for π, as given by Eq. 5.3-5, and canceling like terms on both sides of the equation, we obtain

$$1 = (P)^{-x_1 - x_4 - x_5}(Q)^{-x_1 - x_2 - x_3 + 3x_4 + x_5}(R)^{2x_1 + x_3 + x_5}$$

Equation 5.3-5 will be dimensionally homogeneous if the above equation is an identity. This will be true if the exponents of P, Q, and R are all zero. Setting these exponents to zero gives the following set of linear equations

$$\begin{aligned} -x_1 - x_4 - x_5 &= 0 \\ -x_1 - x_2 - x_3 + 3x_4 + x_5 &= 0 \\ 2x_1 + x_3 + x_5 &= 0 \end{aligned} \qquad \text{Eqs. 5.3-9}$$

We have three homogeneous[2] equations in five unknowns. There is a theorem in linear algebra that states that the most general solution for a system of m linear, homogeneous equations in n unknowns involves $(n - r)$ independent arbitrary parameters, where r is the rank of the coefficient matrix.[3] The set of equations, Eqs. 5.3-9, written in matrix notation is $AX = 0$, where

$$A = \begin{bmatrix} -1 & 0 & 0 & -1 & -1 \\ -1 & -1 & -1 & 3 & 1 \\ 2 & 0 & 1 & 0 & 1 \end{bmatrix} \quad \text{and} \quad X = \begin{bmatrix} x_1 \\ x_2 \\ x_3 \\ x_4 \\ x_5 \end{bmatrix} \qquad \text{Eq. 5.3-10}$$

[2] Here homogeneous refers to the fact that all the coefficients on the right-hand side of the equations are zero.

[3] In this type of problem, the rank of the coefficient matrix, in 99 out of 100 cases, is equal to the number of basic dimensions needed to express the variables dimensionally. This leads to Buckingham's π theorem, which is discussed later in this section. Some matrix concepts and the rank of a matrix are discussed in Appendix A.4.

Matrix A is the coefficient matrix. The highest rank, in this case, that A may have is three. The three corresponds to *the number of basic dimensions that describe the phenomenon*—mass, length, and time. If we take the determinant of the first three columns of A, we obtain

$$|\tilde{A}| = \begin{vmatrix} -1 & 0 & 0 \\ -1 & -1 & -1 \\ 2 & 0 & 1 \end{vmatrix} = (-1)[-1 - 0] = 1$$

Thus, the rank of A is three. The number of independent arbitrary parameters, $(n - r)$, is two. If we were to select two arbitrary parameters, such as $x_1 = \lambda_1$ and $x_5 = \lambda_2$, and solve the system in terms of λ_1 and λ_2, we would obtain a solution involving two dimensionless groups, one raised to the λ_1 power and the other raised to the λ_2 power. It should be realized that the x values can serve as the arbitrary parameters. In this example, we would be solving for x_2, x_3, and x_4 in terms of x_1 and x_5. A single dimensionless group can be obtained by setting $x_1 = 1$ and $x_5 = 0$. Under these circumstances, Eqs. 5.3-9 reduce to

$$\begin{aligned} x_4 &= -1 \\ x_3 &= -2 \\ x_2 &= -1 - x_3 + 3x_4 = -1 + 2 - 3 = -2 \end{aligned} \qquad \text{Eqs. 5.3-11}$$

Substituting Eqs. 5.3-11 into Eq. 5.3-5, we obtain the first of the two dimensionless groups; that is,

$$\pi_1 = (g_c D)(\ell)^{-2}(V)^{-2}(\rho)^{-1}(\mu)^0 = \frac{g_c D}{\rho \ell^2 V^2} \qquad \text{Eq. 5.3-12}$$

If we now select $x_1 = 0$ and $x_5 = -1$, then Eqs. 5.3-9 reduce to

$$\begin{aligned} x_4 &= 1 \\ x_3 &= 1 \\ x_2 &= -x_3 + 3x_4 - 1 = -1 + 3 - 1 = 1 \end{aligned} \qquad \text{Eqs. 5.3-13}$$

Substituting Eqs. 5.3-13 into Eq. 5.3-5, we obtain the second of the dimensionless groupings

$$\pi_2 = (g_c D)^0 \ell V \rho (\mu)^{-1} = \frac{\rho V \ell}{\mu} \qquad \text{Eq. 5.3-14}$$

TABLE 5-1 Powers of (M, L, T)

	$(g_c D)$	ℓ	V	ρ	μ
M	1	0	0	1	1
L	1	1	1	−3	−1
T	−2	0	−1	0	−1

It should be observed that $\pi_2 = \text{Re}$ as defined by Eq. 5.1-3, with ℓ replacing d as the characteristic length.

It is interesting to note that matrix A can be obtained by tabulating the powers of (M, L, T) that are associated with each variable in the set (see Table (5-1). If we take the tabulated numbers to be a matrix, H, then

$$H = \begin{bmatrix} 1 & 0 & 0 & 1 & 1 \\ 1 & 1 & 1 & -3 & -1 \\ -2 & 0 & -1 & 0 & -1 \end{bmatrix}$$

We can then see that $H = -A$. This is equivalent to multiplying each equation in Eqs. 5.3-9 by (-1). Obviously, this would not affect the end result. Thus, the set of Eqs. 5.3-9 can be obtained immediately by writing the Table 5-1. By selecting two different parameters than those chosen above, we could establish other π-groups. For example, suppose we select $x_4 = 0$ and $x_1 = 1$; then Eqs. 5.3-9 reduce to

$$\begin{aligned} x_5 &= -1 \\ x_3 &= -2 - x_5 = -2 + 1 = -1 \\ x_2 &= -1 - x_3 + x_5 = -1 + 1 - 1 = -1 \end{aligned} \qquad \text{Eqs. 5.3-15}$$

Substituting Eqs. 5.3-15 into Eq. 5.3-5, we obtain

$$\pi_3 = (g_c D)(\ell)^{-1}(V)^{-1}(\rho)^0(\mu)^{-1} = \frac{(g_c D)}{\ell V \mu} \qquad \text{Eq. 5.3-16}$$

Now it can be shown that π_3 is a product of π_1 and π_2 each raised to a power; that is,

$$\pi_3 = (\pi_1)^{b_1}(\pi_2)^{b_2} \qquad \text{Eq. 5.3-17}$$

where b_1 and b_2 are constants. When such a relationship exists, then π_3 is said to be *dependent* on the other dimensionless groups. To evaluate these constants, substitute Eqs. 5.3-12, 5.3-14, and 5.3-16 into Eq. 5.3-17, giving

$$\frac{g_c D}{\ell V \mu} = \left(\frac{g_c D}{\rho \ell^2 V^2}\right)^{b_1} \left(\frac{\rho V \ell}{\mu}\right)^{b_2} \qquad \text{Eq. 5.3-18}$$

Obviously $b_1 = 1$ and $b_2 = 1$. On the other hand, π_1 and π_2 are *independent dimensionless groups.* That is, there does not exist a constant b such that $\pi_1 = (\pi_2)^b$. This is apparent, since π_2 contains the variable μ and π_1 does not.

Suppose we had assumed that the phenomenon is describable by a functional relationship between a set of dimensionless products. That is

$$G(\pi_1, \pi_2, \pi_3, \pi_4, \ldots) = 0 \qquad \text{Eq. 5.3-19}$$

where π_1 and π_2 are independent groups and π_3, π_4, and so on, are all dependent on π_1 and π_2. Since π_3, π_4, ..., can all be expressed in terms of π_1 and π_2, Eq. 5.3-19 reduces to

$$F(\pi_1, \pi_2) = 0 \qquad \text{or} \qquad \pi_1 = f(\pi_2) \qquad \text{Eq. 5.3-20}$$

Now the number of independent groups equals the number of independent parameters involved in the solution of Eqs. 5.3-9, which turns out to be the rank of the matrix established by Table 5-1. Furthermore, if we assume that the rank of this matrix is equal to the number of basic dimensions needed to express the variables dimensionally (which, as mentioned in the footnote on page 256, is valid in 99 out of 100 cases), then we have established, for this special case, the theorem known as Buckingham's π theorem. We can state this theorem in the following way.

> *The number of independent dimensionless groups* that may be employed to describe a phenomenon known to involve n variables is equal to the number $(n - \bar{r})$, where $\bar{r}$ is the number of basic dimensions needed to express the variables dimensionally. Stm. 5.3-1

In this example, the variables are: $(g_c D)$, ℓ, ρ, V, μ. Thus, $n = 5$. The basic dimensions are: (M, L, T) = (mass, length, time). Thus, $\bar{r} = 3$. The number of independent groups is $(n - \bar{r}) = 2$. Thus, the phenomenon is describable by the two dimensionless groups found in Eqs. 5.3-12 and 5.3-14; that is,

$$\frac{g_c D}{\rho \ell^2 V^2} = f(\text{Re}) \qquad \text{Eq. 5.3-21}$$

In the above example, two of the six most commonly used dimensionless products in fluid mechanics were developed. We can summarize the method and simultaneously develop the remaining four dimensionless products:

1. Select the variables involved in the phenomenon.

The most common variables encountered in fluid mechanics are mass density, ρ; length, ℓ; velocity, V; force, F; absolute viscosity, μ; acceleration of gravity, g; speed of sound, c; surface tension, σ; and a second length, ℓ^*.

2. Determine the basic dimensions needed to express the variables dimensionally.

The variables listed under step 1 can be described dimensionally by an (M, L, T) system. However, recall that any variable, Z, containing the dimensions of force is to be written as $(g_c \cdot Z)$.

3. Write the dimensions of each variable.

The dimensions of the variable listed under step 1 are as follows: $\mathscr{D}(\rho) = M/L^3$, $\mathscr{D}(\ell) = L$, $\mathscr{D}(V) = L/T$, $\mathscr{D}(g_c \cdot F) = ML/T^2$, $\mathscr{D}(\mu) = M/LT$, $\mathscr{D}(g) = L/T^2$, $\mathscr{D}(c) = L/T$, $\mathscr{D}(g_c\sigma) = M/T^2$, and $\mathscr{D}(\ell^*) = L$.

4. Select the dimensionless group to be a product of each variable raised to an unknown power.

For the variables of interest,

$$\pi = (\rho)^{x_1}(\ell)^{x_2}(V)^{x_3}(g_c F)^{x_4}(\mu)^{x_5}(g)^{x_6}(c)^{x_7}(g_c \sigma)^{x_8}(\ell^*)^{x_9} \qquad \text{Eq. 5.3-22}$$

5. Construct a table listing the powers of (M, L, T) that correspond to the dimensions of each variable. The elements in the table become the elements in the coefficient matrix, A.

For the variables of interest the table is as follows:

	ρ	ℓ	V	$(g_c F)$	μ	g	c	$(g_c\sigma)$	ℓ^*
M	1	0	0	1	1	0	0	1	0
L	−3	1	1	1	−1	1	1	0	1
T	0	0	−1	−2	−1	−2	−1	−2	0

Then the coefficient matrix A is

$$A = \begin{bmatrix} 1 & 0 & 0 & 1 & 1 & 0 & 0 & 1 & 0 \\ -3 & 1 & 1 & 1 & -1 & 1 & 1 & 0 & 1 \\ 0 & 0 & -1 & -2 & -1 & -2 & -1 & -2 & 0 \end{bmatrix}$$

6. Construct the set of linear homogeneous equations of the form $AX = 0$, where A is the coefficient matrix described in step 5, and

$$X = \begin{bmatrix} x_1 \\ x_2 \\ x_3 \\ x_4 \\ x_5 \\ x_6 \\ x_7 \\ x_8 \\ x_9 \end{bmatrix}$$

For the variables of interest, the set of equations is

$$x_1 + x_4 + x_5 + x_8 = 0$$

$$-3x_1 + x_2 + x_3 + x_4 - x_5 + x_6 + x_7 + x_9 = 0$$

$$-x_3 - 2x_4 - x_5 - 2x_6 - x_7 - 2x_8 = 0 \qquad \text{Eqs. 5.3-23}$$

This set of linear equations comes about by requiring the equation expressing the π-group to be dimensionally homogeneous and π to be dimensionless. That is, in a second set of units, the Eq. 5.3-22 in step 4 must still be valid. Thus,

$$\bar{\pi} = (\bar{\rho})^{x_1}(\bar{\ell})^{x_2}(\bar{V})^{x_3}(\overline{g_c F})^{x_4}(\bar{\mu})^{x_5}(\bar{g})^{x_6}(\bar{c})^{x_7}(\overline{g_c \sigma})^{x_8}(\bar{\ell}^*)^{x_9} \qquad \text{Eq. 5.3-24}$$

If

$$\text{One unit of} \begin{Bmatrix} \text{mass} \\ \text{length} \\ \text{time} \end{Bmatrix} \text{in the second set}$$

$$= \begin{Bmatrix} P \\ Q \\ R \end{Bmatrix} \text{units of} \begin{Bmatrix} \text{mass} \\ \text{length} \\ \text{time} \end{Bmatrix} \text{in the first set}$$

then

$$\bar{\pi} = \pi[(P)^{x_1 + x_4 + x_5 + x_8}][(Q)^{-3x_1 + x_2 + x_3 + x_4 - x_5 + x_6 + x_7 + x_9}] \times [(R)^{-x_3 - 2x_4 - x_5 - 2x_6 - x_7 - 2x_8}]$$

Since π is dimensionless, $\bar{\pi} = \pi$ and the powers of P, Q, and R must be zero, giving Eqs. 5.3-23.

7. Determine the number of arbitrary parameters, q, contained in a solution to Eqs. 5.3-23, according to the formula

$$q = n - r$$

where n is the number of variables and r is the number of basic dimensions needed to express the variables dimensionally. Here we have assumed that the rank of matrix A is equal to r. In the problem of interest, $q = 9 - 3 = 6$.

8. Let $\lambda_1, \lambda_2, \ldots, \lambda_q$ be the arbitrary parameters mentioned in step 7. Assign a set of x values to be equal to the λ values. *However, be certain that this selection is consistent with the set of equations found in step 6.*[4] Wherever

[4] For example, if one of the equations obtained in step 6 was

$$x_1 + x_6 = 0$$

one could not select $x_1 = \lambda_1$ and $x_6 = \lambda_2$; then in step 9 set $\lambda_2 = 0$ and $\lambda_2 = 1$.

possible the set of x values selected should be the exponents of those variables that may not be important in certain types of problems associated with the set of variables.

In the problem of interest we set $x_4 = \lambda_1, x_5 = \lambda_2, x_6 = \lambda_3, x_7 = \lambda_4, x_8 = \lambda_5$, and $x_9 = \lambda_6$. We have selected this subset of x values because in many fluid-flow problems any one of the following effects may be negligible: (a) viscous, (b) gravity, (c) speed of sound, and (d) surface tension. In addition, in kinematic problems, force measurements may not be involved. Finally, many problems may not involve two characteristic lengths.

9. Successively set $(q - 1)$ of the parameters to zero and the remaining parameter to 1 or (-1) and solve Eqs. 5.3-23. Then substitute the established set of x values into Eq. 5.3-22, obtaining a π-group.

Suppose in our problem of interest, we set

$$\lambda_1 = \lambda_2 = \lambda_3 = \lambda_4 = \lambda_6 = 0$$

and

$$\lambda_5 = -1$$

Then Eqs. 5.3-23 reduce to

$$\begin{aligned} x_1 &= 1 \\ -3x_1 + x_2 + x_3 &= 0 \\ -x_3 &= -2 \end{aligned}$$

giving $x_1 = 1$, $x_2 = 1$, and $x_3 = 2$. Then a dimensionless group, say π_4, is

$$\pi_4 = (\rho)^1(\ell)^1(V)^2(g_c\sigma)^{-1} = \frac{\rho \ell V^2}{g_c \sigma}$$

The value π_4 is commonly designated as the Weber number.

In a similar manner to obtaining π_5 in step (9), the following set of dimensionless products can be established.

π_1 = Force (or pressure) coefficient, $P = \dfrac{g_c F}{\rho \ell^2 V^2}$ (also called the Euler number)

π_2 = Reynolds number, $\text{Re} = \dfrac{\rho \ell V}{\mu}$

π_3 = Froude number $\mathscr{F} = \dfrac{V^2}{\ell g}$

π_4 = Weber number, $W = \dfrac{\rho \ell V^2}{g_c \sigma}$

π_5 = Mach number, $M = \dfrac{V}{c}$

π_6 = geometric ratio, $G = \ell/\ell^*$ Eqs. 5.3-25

Although many other dimensionless products can be formed from the variables used in the example just discussed, all are dependent on six dimensionless groups, given in Eqs. 5.3-25. On the other hand, it can be seen that the six dimensionless groups given in Eqs. 5.3-25 are independent, since each contains a variable not present in the others. This set of dimensionless products is a *complete set* for the group of variables from which they were derived. Finally, any fluid-flow problem involving all of the variables specified in item (1) of this section can be described by some functional relationship between this complete set of dimensionless products; that is,

$$f(P, \text{Re}, \mathscr{F}, W, M, G) = 0 \qquad \text{Eqs. 5.3-26}$$

The exact *form of the functional relationship is to be determined by experimentation.*

5.4 PHYSICAL INTERPRETATION OF THE DIMENSIONLESS PRODUCTS

It is now possible, with closer examination, to assess the physical significance of each of the groupings given by Eqs. 5.3-25. Since the interpretation of the

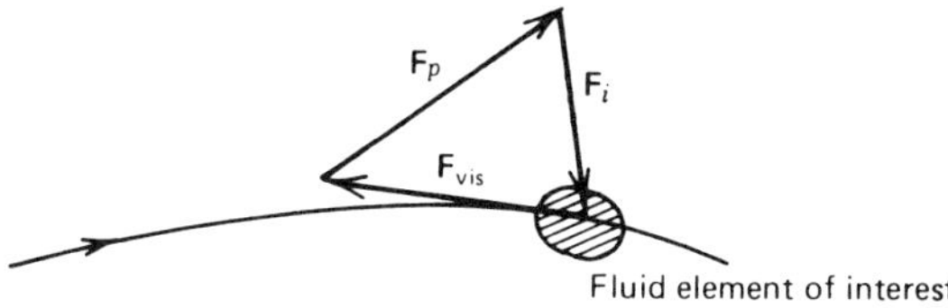

Fig. 5-6. Forces acting on a fluid element.

geometric ratio falls into a different category than the others, it will be discussed first.

As the name implies, the geometric ratio is merely an expression regarding the geometry or shape of some characteristic portion of the fluid system under consideration. Examples of the geometric ratio are such quantities as the length/diameter, (ℓ/D), ratio for a pipe; the roughness factor/diameter, (e/D), ratio of a pipe, where e is a measure of the average height of the interior ridges on the pipe surface; and the thickness/chord length, (ℓ_t/ℓ_c), ratio of an airfoil. More will be said of the geometric ratio in the discussion on similitude, Section 5.5.

By considering the variables as local quantities, each of the first four π-groups (the pressure coefficient, Reynolds number, Froude number, and Weber number) can be shown to represent a ratio between inertia forces and forces of pressure, viscosity, gravity, and surface tension. The product of fluid mass, acceleration, and $(1/g_c)$ is considered as the negative of an *inertia force*, $\mathbf{F}_i$. To demonstrate this concept, consider the forces that act on a fluid element as it moves along a streamline. Suppose for the moment, we limit our discussion to the forces resulting from the pressure field, say $\mathbf{F}_p$, and from viscous effects, say $\mathbf{F}_{vis}$. If pressure and viscous forces are the only true forces acting on a fluid element, then according to Newton's second law, $\mathbf{F}_p$, $\mathbf{F}_{vis}$, and $\mathbf{F}_i$ must form a closed triangle, as shown in Fig. 5-6. We now wish to obtain an order of magnitude estimate for these various forces. Recall that for a steady-flow situation, the acceleration of a fluid particle in the Eulerian description for a steady, parallel-flow situation is given by

$$a_1 = v_1 \frac{\partial v_1}{\partial x}$$

Let b be a measure of the length of the fluid element. As a measure of v_1 and $\partial v_1/\partial x$, we take $v_1 \sim V$ and $\partial v_1/\partial x \sim V/b$, then

$$a_1 \sim \frac{V^2}{b}$$

A measure of the volume of the fluid element is b^3, then

$$F_i \sim \left(\frac{\rho b^3}{g_c}\right)\left(\frac{V^2}{b}\right) = \frac{\rho}{g_c} V^2 b^2$$

We can obtain an order of magnitude estimate of the viscous force, F_{vis}, by recalling that for a parallel-flow situation,

$$F_{\text{vis}} = \frac{\mu}{g_c}\frac{dv_1}{dy} A_S$$

As a measure of dv_1/dy and A_S, we take $dv_1/dy \sim V/b$ and $A_S \sim b^2$, then

$$F_{\text{vis}} \sim \frac{\mu}{g_c} Vb$$

Let us now examine the order of magnitude of the ratio of F_i and F_{vis}, then

$$\frac{F_i}{F_{\text{vis}}} \sim \left(\frac{\rho V^2 b^2}{g_c}\right)\left(\frac{g_c}{\mu V b}\right) = \frac{\rho V b}{\mu}$$

We see that (F_i/F_{vis}) *is essentially a local Reynolds number*. Problems where viscous and inertia forces are significant are likely to involve the Reynolds number. The Reynolds number perhaps appears more often in fluid-flow problems than any of the other five dimensionless products. The value of the Reynolds number determines whether a given flow is laminar (smooth) or turbulent (confused). The Reynolds number is important in the study of both liquids and gases and finds application in the flow through pipes, channels, pumps, and turbines, as well as in the flow of fluids about airfoils and other submerged bodies.

If the pressure on the upper half of our fluid element of interest differs from that on the lower half (the same is true with respect to the front half and rear half), then there would be a net pressure force on the element. If Δp is a measure of this pressure difference and b_2 is a measure of the surface area of the fluid element, then a measure of the pressure force, F_p, acting on the element is

$$F_p \sim \Delta p b^2$$

A measure of the ratio of pressure and inertia forces is

$$\frac{F_p}{F_i} \sim (\Delta p\, b^2)\left(\frac{g_c}{\rho V^2 b^2}\right) = \frac{g_c\, \Delta p}{\rho V^2}$$

We see that (F_p/F_i) *is essentially a local pressure coefficient.* Problems where pressure and inertia forces are significant are likely to involve the pressure coefficient. The force coefficient is likely to be involved whenever the total force acting on a body affects the phenomenon of interest, even though the forces arise from effects other than pressure.

A measure of the gravitational force, F_g, acting on the fluid element can similarly be established. Taking b^3 to be a measure of the volume of the fluid element and γ to be its specific weight, then

$$F_g \sim \gamma b^3$$

A measure of the ratio of inertia and gravitational forces becomes

$$\frac{F_i}{F_g} \sim \left(\frac{\rho V^2 b^2}{g_c}\right)\left(\frac{1}{\gamma b^3}\right) = \frac{V^2}{bg}$$

Here we have used the fact that $\gamma = \rho g/g_c$. We see that (F_i/F_g) *is essentially a local Froude number.* Problems where gravitational and inertia effects are important are likely to involve the Froude number. This occurs in the study of flow around ships and certain other problems involving wave action that is governed to a large extent by gravity forces. In other types of fluid-flow situations, the Froude number is generally of little importance.

If our fluid element happened to be at a curved interface of two different fluids, then it would also experience a surface tension force, F_{st}. If σ is the surface-tension constant, then a measure of the surface-tension force becomes

$$F_{st} = \sigma b$$

A measure of the ratio of inertia and surface-tension forces becomes

$$\frac{F_i}{F_{st}} \sim \left(\frac{\rho V^2 b^2}{g_c}\right)\left(\frac{1}{\sigma b}\right) = \frac{\rho V^2 b}{g_c \sigma}$$

We see that the ratio, (F_i/F_{st}), *is essentially a local Weber number.* The Weber number may play an important role in those fluid situations where a curved interface between two fluids exists and the surface-to-volume ratios are large. Such a situation occurs in the study of liquid droplets, for example, with a colloidal suspension of liquid droplets in a gas environment.

As the fluid element moves from one region to another it may experience a pressure change. This will cause the fluid element to compress. Recall that the fluid-bulk modulus of elasticity, K, is defined by

$$K = -\frac{dp}{d\tilde{v}/\tilde{v}} = \frac{dp}{d\rho/\rho}$$

Since the dimensions of K are force/unit area, we take the elastic force, F_e, which is defined as a product of E and the surface area of the fluid element, to be a measure of this compressibility effect

$$F_e \sim Kb^2$$

A measure of the ratio of inertial and elastic forces becomes

$$\frac{F_i}{F_e} \sim \left(\frac{pV^2b^2}{g_c}\right)\left(\frac{1}{Kb^2}\right) = \frac{\rho V^2}{g_c}\,\frac{1}{\rho\dfrac{dp}{d\rho}} = \frac{V^2}{g_c\left(\dfrac{dp}{d\rho}\right)}$$

Recall that the acoustic velocity is given by

$$c = \sqrt{g_c\left(\frac{dp}{d\rho}\right)_S}$$

For an ideal gas the above equation reduces to

$$c = \sqrt{g_c kRT} \qquad \text{Eq 5.4-1}$$

where k = ratio of specific heats, R is the gas constant and T is the gas temperature. If we assume that the above compression process is isentropic, then

$$\frac{F_i}{F_e} \sim \frac{V^2}{c^2}$$

We see that (F_i/F_e) *is essentially the square of the local Mach number*. The Mach number plays an important role in the study of high-speed gas flows where compressibility effects become of critical importance. The Mach number also indicates whether boundary irregularities will be propagated throughout the flow field. This is further discussed in Section 5.5.

From the preceding discussion, it is apparent that a knowledge of which forces predominate in a given fluid-flow situation provides the clue as to which of the dimensionless parameters should be involved in an experimental program designed to produce empirical relationships. These same dimensionless groupings, when used with care, can help predict the performance of full-scale equipment through the study of models, as we will now discover.

5.5 SIMILITUDE—A STUDY OF MODEL AND PROTOTYPE

One extremely important application of the art of dimensional analysis is the study of models and their relationship to full-scale equipment. The full-scale equipment is designated as the prototype. In cases where analytical analysis is impossible or inconclusive, the experimental study of proposed designs becomes a necessity. Prohibitive cost or practical impossibility often make the test of full-scale mock-ups of each proposed design an economic impossibility. In such cases the behavior of the prototype can frequently be predicted from the study of a small-scale model of the proposed design, provided certain restrictions are met.

The airplane designer tests small-scale models in a wind tunnel; the naval architect tests new hull designs in a towing basin, using a model; and the civil engineer plans flood control and power generation systems by making use of small models of the land formations and the proposed dams and canals. In order to use a model to simulate the conditions of the prototype, the model system must be geometrically, kinematically, and dynamically similar to the prototype system.

For geometric similarity, the model must have the same shape as the prototype. That is, the model must be a scaled-down version of the prototype. Then the ratios between corresponding linear dimensions in the model and in the prototype will be everywhere the same. Thus, if ℓ_1 and ℓ_2 are two characteristic lengths of a particular body, then

$$\left(\frac{\ell_1}{\ell_2}\right)_m = \left(\frac{\ell_1}{\ell_2}\right)_p$$

Here the subscripts m and p denote the variables associated with the model and prototype respectively. Area and volume ratios between model and prototype

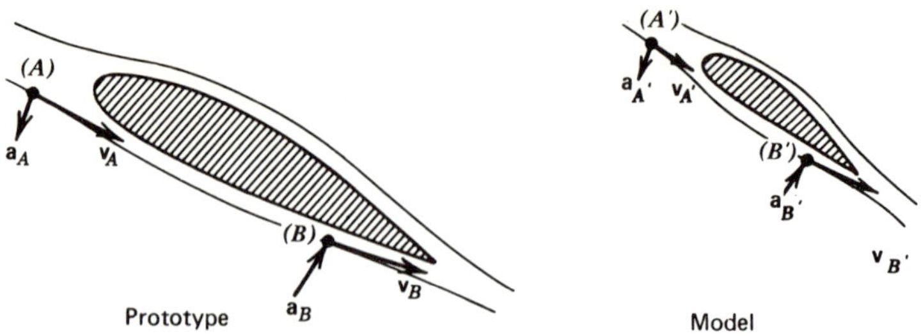

Fig. 5-7. Kinematic similarity between geometrically similar model and prototype.

will then respectively equal the square and cube of the ratio of their characteristic lengths; that is,

$$\frac{A_m}{A_p} = \left[\frac{(\ell_1)_m}{(\ell_1)_p}\right]^2 \quad \text{and} \quad \frac{\forall_m}{\forall_p} = \left[\frac{(\ell_1)_m}{(\ell_1)_p}\right]^3$$

Thus, the geometric ratio, Π_6, of prototype and model will be the same when geometric similarity is preserved.

When the streamline patterns of two flows are geometrically similar everywhere, the flows are said to be *kinematically similar*. When this occurs, the ratio of velocities at any two corresponding points in both flows will be equal. The same can be said with regard to accelerations at any two corresponding points in both flows (see Fig. 5-7); therefore,

$$\left(\frac{v_A}{v_B}\right)_m = \left(\frac{v_{A'}}{v_{B'}}\right)_p \quad \text{and} \quad \left(\frac{a_A}{a_B}\right)_m = \left(\frac{a_{A'}}{a_{B'}}\right)_p$$

where points (A', B') in one flow correspond to points (A, B) in the other. Since solid boundaries are streamlines, two flows can only be kinematically similar, if the flows are around geometrically similar bodies. The converse of this statement is not necessarily true. That is, it is possible to have flows around geometrically similar bodies without the two being kinematically similar. An example of this situation is the flows over two geometrically similar double-wedge wings, as shown in Fig. 5-8.

In the flow on the left, the Mach number is less than one, $M < 1$, while the flow on the right, $M > 1$. The streamline pattern indicates that when $M < 1$, boundary irregularities are smoothed out in the flow. On the other hand, when $M > 1$, boundary irregularities are propagated through the flow.

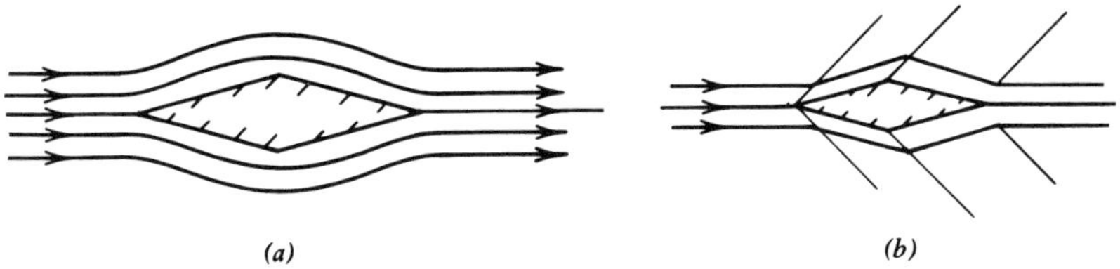

Fig. 5-8. Geometrically similar bodies in nonkinematically similar flow fields. (a) $M < 1$. Boundary irregularities are smoothed out. (b) $M > 1$. Boundary irregularities are propagated throughot flow field.

We have previously indicated that the forces that act on a fluid can be broken down into a viscous force, $\mathbf{F}_{\text{vis}}$, an inertia force, $\mathbf{F}_i$, a gravitational force, $\mathbf{F}_g$, a pressure force, $\mathbf{F}_p$, and a surface-tension force, $\mathbf{F}_{st}$. When the force polygons for every pair of corresponding points in two flows are similar, the flows are said to be *dynamically similar* (see Fig. 5-9). The similar force polygons at corresponding points in two flows imply that the ratio of any two force types (pressure, viscous, etc.) must be the same in both flows. This means that the local dimensionless groups (pressure coefficient, Reynolds number, etc.) must be the same everywhere. Under these circumstances, the resultant aerodynamic or hydrodynamic forces acting on two geometrically similar bodies in dynamically similar flows will be related in a very special way. The question immediately arises: How do we obtain dynamic similarity between two flows? It turns out that the situation is not quite as complicated as one might first expect. First, experimentation and experience indicate that we may use characteristic values for the variables in the dimensionless groupings. For example, when considering the problem of a body in a flow field, the free-stream velocity may be used as the velocity variable in all of the dimensionless groups. In a similar fashion, the chord length or the diameter of a cylinder may be used as the

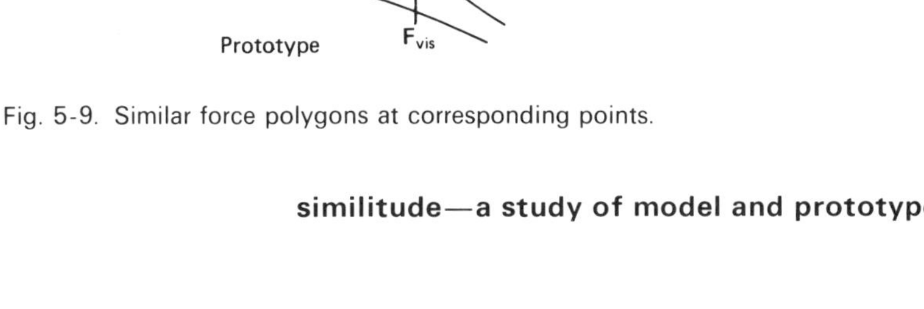

Fig. 5-9. Similar force polygons at corresponding points.

length dimension. Under these circumstances, it becomes feasible, in many cases, to match Reynolds numbers, Mach numbers, and so on, for two different flows. However, even with this simplification, it is virtually impossible or impractical to match all of the previously mentioned dimensionless groups. Fortunately, in most problems of interest, some of the previously mentioned forces do not play a significant role in the phenomenon. Hence, the associated dimensionless group can be deleted from the analysis. Several examples are discussed below.

5.5.1 WIND-TUNNEL TESTS

Most wind-tunnel tests are carried out to predict the forces on a prototype from force measurements on the model. This can only be accomplished when the two flows are dynamically similar. However, in gas flows, neither surface tension nor gravitational effects are significant factors. However, compressibility effects may be significant. Thus, dynamic similarity can be attained by making the model geometrically similar to the prototype $[(\ell/\ell^*)_m = (\ell/\ell^*)_p]$ and making

$$\left(\frac{\rho \ell V}{\mu}\right)_m = \left(\frac{\rho \ell V}{\mu}\right)_p \quad \text{and} \quad \left(\frac{V}{c}\right)_m = \left(\frac{V}{c}\right)_p \qquad \text{Eqs. 5.5-1}$$

Then

$$\left(\frac{g_c F}{\rho \ell^2 V^2}\right)_m = \left(\frac{g_c F}{\rho \ell^2 V^2}\right)_p \qquad \text{Eqs. 5.5-2}$$

It is unfortunate that in most real flow situations the property-control capacity of existing wind tunnels does not permit both conditions (described by Eqs. 5.5-1) to be carried out. It therefore often becomes necessary to carry out tests under incomplete similarity conditions. At moderate speeds, ($\mathrm{Re} > 50$, $M < 0.5$) compressibility effects are not very significant and thus dynamic similarity between model and prototype can be approximately obtained by making $(\mathrm{Re})_m = (\mathrm{Re})_p$. At the higher speeds ($M > 1.5$), compressibility effects become dominant; dynamic similarity is then approximately obtained by making $M_m = M_p$. This is further discussed in the next paragraph.

The total drag force on a body can usually be broken down into two parts: pressure drag and skin-friction drag. Pressure drag, as the name implies, results from the pressure field on the body, whereas skin-friction drag results from viscous shear forces. In moderate-speed flows around blunt bodies (cylinders, spheres, discs, and boxlike shapes), the pressure field on the front half of the body is greater than the pressure on the rear half. This results from a phenomenon known as a *boundary-layer separation*, which is discussed in Section 9.1.1. The point where the boundary layer separates depends on the Reynolds number of the flow. Thus, the Reynolds number affects both the pressure drag and the skin-friction drag on blunt bodies in moderate-speed flows. In the case of a blunt body, the pressure drag is the dominant drag force on the body. However, well-designed streamlined bodies (tear drop, wing shapes, etc.), not severely inclined to the free-stream direction, do not experience boundary-layer separation; thus, the skin-friction drag becomes dominant. But, in either case (blunt body or streamlined body in a moderate-speed flow), the Reynolds number essentially determines the drag.[5] However, in a high-speed flow (Mach number greater than 1.0), a phenomenon known as *shock waves* (see Sections 8.1 and 8.2) develops in the flow field. This causes the pressure field in the front part of the body to be much larger than the pressure on the rear portion of the body. This phenomenon is known as *wave drag*.[6] It turns out that in a high-speed flow the wave drag (or pressure drag) is the dominant type of drag. Since the wave drag is strongly influenced by the Mach number, dynamic similarity can be approached by making the Mach numbers for model and prototype the same.

Perhaps at this point it would be appropriate to point out that it is common for aerodynamicists to replace the force coefficient by either the "drag coefficient," C_D, or the "lift coefficient," C_L. These coefficients are defined by

$$D = C_D A \frac{\rho V^2}{2g_c} \qquad \text{and} \qquad L = C_L A \frac{\rho V^2}{2g_c} \qquad \text{Eqs. 5.5-3}$$

where A is some reference area (such as average chord length times the span length, or the frontal area of a sphere, etc.). Suppose we solve for C_D using the frontal area of a sphere (the frontal area is that area which results from projecting a body onto a plane perpendicular to the free stream). If ℓ is the diameter of the sphere, the frontal area equals $\pi\ell^2/4$ and

$$C_D = \left(\frac{Dg_c}{\rho V^2 \ell^2}\right)\left(\frac{8}{\pi}\right)$$

We see that C_D is a constant times the force coefficient. Therefore, C_D or C_L may replace the force coefficient in the analysis.

[5] The whole topic of "drag" is beautifully covered in the film entitled *The Fluid Dynamics of Drag* by Ascher H. Shapiro. The film is in four parts and is available through ASME.

[6] See Section 4-16 in H. W. Liepman and A. Roshko, *Elements of Gasdynamics*, Wiley, New York, 1957.

In all cases, lift is determined primarily from the pressure field. It is interesting to note that at moderate speeds and small angles of attack the lift coefficient, C_L, is mainly a function of the angle of attack. The Reynolds number becomes important only when the stall angle is approached (see Fig. 9-29).

5.5.2 TOWING BASIN TESTS

In the study of ship hulls, surface tension and compressibility effects are not significant. Therefore, for geometrically similar bodies, dynamic similarity occurs when

$$\left(\frac{V^2}{\ell g}\right)_m = \left(\frac{V^2}{\ell g}\right)_p \quad \text{and} \quad \left(\frac{\rho V \ell}{\mu}\right)_m = \left(\frac{\rho V \ell}{\mu}\right)_p$$

For a fixed value of the gravitational acceleration, g, the simultaneous solution of the above set of equations gives

$$\left(\frac{\nu_m}{\nu_p}\right) = \left(\frac{\ell_m}{\ell_p}\right)^{3/2} \qquad \text{Eq. 5.5-4}$$

where $\nu = \mu/\rho$. Equation 5.5-4 indicates that once the scale of the model is decided on, the kinematic viscosity of the fluid in the towing tank is fixed. This precludes, of course, the use of water in the towing basin for any model scale other than 1:1. Experience has shown, however, that the Froude number is of greater significance than the Reynolds number in this particular application. Thus, the fluid used in the towing tank is generally water; the Froude number alone is maintained between model and prototype; and empirical corrections are made to compensate for the differences that exist between the Reynolds numbers.

When using model studies for quantitative analysis, experience alone will show which of the dimensionless parameters are of importance, and which may be neglected in preserving dynamic similitude between model and prototype.

Example 5-2

A marine research facility uses a towing basin to test models of proposed ship hull configurations. A new hull shape utilizing a bulbous underwater bow is proposed for a nuclear-powered aircraft carrier that is to be 300 m long. A 3-m model has been tested in the towing tank and found to have a maximum practical hull speed of 1.4 m/s. What is the anticipated hull speed for the prototype?

SOLUTION

We will assume, as already noted, that sufficient similitude is maintained by preserving constant Froude numbers for the model and the prototype. Hence we ignore the viscous effects, as measured by the Reynolds number, and concentrate on the hull's wave-making characteristics, as measured by the Froude number.

$$\left(\frac{V^2}{\ell g}\right)_m = \left(\frac{V^2}{\ell g}\right)_p$$

Since the gravitational acceleration is the same for model and prototype, the anticipated prototype velocity becomes

$$V_p = V_m\left(\frac{\ell_p}{\ell_m}\right)^{1/2}$$

or

$$V_p = 1.4(10.0) = 14 \text{ m/s}$$

As a matter of interest, a speed of 14 m/s translates into 27.2 knots, where 1 knot = 1 nautical mile per hour and 1 nautical mile = 6080 feet.

Example 5-3

A model of a proposed low-speed crop-dusting plane is to be tested in a wind tunnel capable of maintaining velocities of 264 ft/s, using atmospheric air. The prototype plane has a design wing span of 27 ft and a proposed speed of 88 ft/s. In order to preserve dynamic similarity, what should be the wing span of the model if the wind tunnel is to operate at full speed? Assume the air at these low velocities to be essentially incompressible. In addition, assume that the temperature of atmospheric air and the temperature of the wind-tunnel air are the same.

SOLUTION

For low-speed flows, where viscous and inertia forces predominate, dynamic similarity requires that the Reynolds numbers be the same for both the model and the prototype, and that they be geometrically similar. Thus,

$$\left(\frac{\rho V \ell}{\mu}\right)_p = \left(\frac{\rho V \ell}{\mu}\right)_m \quad \text{and} \quad \left(\frac{b}{\ell}\right)_m = \left(\frac{b}{\ell}\right)_p$$

Here, ℓ is the average chord length and b is the span length. Now we may assume that $p_m \approx p_p$; then we may take $\rho_m \approx \rho_p$, since $T_m = T_p$ (this follows from the ideal gas law). Similarly, $\mu_m = \mu_p$, since $T_m = T_p$. Therefore,

$$b_m = \frac{b_p \ell_m}{\ell_p} = b_p \left(\frac{\rho V}{\mu}\right)_p \left(\frac{\mu}{\rho V}\right)_m = b_p \left(\frac{V_p}{V_m}\right) = \frac{27 \times 88}{264} = 9 \text{ ft}$$

Example 5-4

A gravity fed lock in a proposed shipping canal is to be studied with a 1/100-scale geometrically similar model. If the model lock fills in 0.8 min, estimate the time for the prototype to fill.

SOLUTION

Here again, dynamic similarity is assumed to be based on equality of the Froude number in model and prototype. Let the geometric ratio ℓ_p/ℓ_m be designated as λ (lambda). From Example 5-2, we see that

$$V_p = V_m(\lambda)^{1/2}$$

Fill time in the model studies is proportional to the ratio $(\ell^3/Q)_m = (\ell^3/AV)_m$, while time in the prototype is proportional to $(\ell^3/AV)_p$. Since $A_m/A_p = \ell_m^2/\ell_p^2$, the ratio of the fill times, t_p/t_m, is given by

$$\frac{t_p}{t_m} = \left(\frac{\ell}{V}\right)_p \left(\frac{V}{\ell}\right)_m = \frac{\lambda}{\sqrt{\lambda}} = \sqrt{\lambda}$$

Therefore, $t_p = t_m\sqrt{\lambda}$, or for this specific case,

$$t_p = 0.80\sqrt{100} = 8 \text{ min}$$

Example 5-5

Find the ratio of corresponding mass flow rates in the model and prototype locks of Example 5-4.

SOLUTION

The mass flow rates can be represented as the product of appropriate densities, velocities, and areas. Thus,

$$\frac{\dot{m}_p}{\dot{m}_m} = \frac{(\rho V A)_p}{(\rho V A)_m}$$

or

$$\frac{\dot{m}_p}{\dot{m}_m} = \frac{\rho_p}{\rho_m} \times \sqrt{\lambda} \times \lambda^2 = \lambda^{5/2}$$

since $\rho_p/\rho_m = 1.0$. (The fluid is water in each case.)

Problems

5-1 Suppose that in Example 5-1 the fluid flowing through the 1-in. diameter pipe was engine oil at 68 F. Calculate the largest flow rate for which the flow will definitely be laminar.

5-2 Which of the following combinations of fluid variables are indeed dimensionless groupings? (*a*) $\rho V^3/\mu g$; (*b*) $\ell^2 g \rho^2 / g_c \sigma$; (*c*) $g_c \sigma / c \rho \ell$; (*d*) $g_c \ell \, \Delta p / \mu V$; (*e*) $\ell V^2 \, \Delta \rho / \mu$; (*f*) $\rho \ell^2 / c V$.

5-3 The pressure loss in a straight pipe due to friction is given by the Darcy equation

$$\Delta p = f\left(\frac{L}{D}\right)\left(\frac{\rho V^2}{2g_c}\right)$$

where $f = 8\tau_0 g_c / \rho V^2$, Δp is the pressure drop, τ_0 is the wall shear stress, and L and D, respectively, are the length and diameter of the pipe. Show that the same general result, except for the numerical constants, can be obtained by dimensional analysis, if we assume that $\pi_1 \propto \pi_2 \pi_3$.

5-4 The flow of water about a ship's hull results in a drag, D, which depends on the following variables: a geometric dimension of the ship, b; the velocity of the ship relative to the water, V; the density of the water, ρ; the absolute viscosity of the water, μ; and the local acceleration due to gravity, g. By using dimensional analysis, determine the π-groups for this case.

5-5 The thrust force $\mathscr{T}$ of an airplane propeller is assumed to be dependent on the following variables: the angular velocity, ω; the diameter of the propeller, D; the air density, ρ; the absolute viscosity of the air, μ; and the flight velocity, V. By using dimensional analysis, determine the π-groups for this case.

5-6 The velocity at which a given steel sphere falls through various viscous liquids (after having reached a uniform velocity) is assumed to depend on the diameter of the ball, d; the density of the fluid, ρ; the absolute viscosity of the fluid, μ; and the acceleration due to gravity, g. Determine the dimensionless groups pertaining to this situation and assess their physical significance.

5-7 By assuming the velocity of propagation of a surface wave, V_s, in a liquid to be a function of the depth, ℓ, and the gravitational acceleration, g, find a dimensionless parameter relating to this situation. Show that this parameter is equivalent to the Froude number.

5-8 A wing model is to be tested in a wind tunnel for the purpose of predicting the drag on the prototype. Assume that the wind tunnel characteristics are such that both the Mach number and the Reynolds number for the test can be made to match those of the prototype. Determine the chord length of the model for the following conditions:

Prototype	Model
Altitude of flight = 6000 m	Free-stream pressure = 4.1×10^5 N/m abs.
V = 180 m/s	Free-stream temperature = 5 C
Chord length = 2.4 m	

5-9 In dealing with problems involving heat transfer from solid walls to surrounding fluids, it is common practice to introduce the convective heat transfer coefficient, $\bar{h}$, which is defined by the following empirical relation

$$\left(\frac{\delta Q}{\delta t}\right)_{\text{wall}\to\text{fluid}} = \bar{h}A_s(T_W - T_f)$$

where $(\delta Q/\delta t)_{\text{wall}\to\text{fluid}}$ is the rate at which heat flows out of the solid wall and is carried away by convection; A_s is the wall surface area; T_W is the temperature of the wall on the outside surface; and T_f is the fluid temperature a short distance from the wall. (See the paragraph following Eq. 8.6-5, as well as Fig. 8-22 and the paragraph above Eq. 8.6-14.) (*a*) In experiments dealing with a forced cross flow over a heated cylinder it has been determined that $\bar{h}$ is a function of the following variables: V, ρ, μ, c_p, k_t, and d, where ρ, μ, c_p, and k_t are, respectively, the mass density, absolute viscosity, specific heat at constant pressure, and thermal conductivity of the fluid; V is the free-stream velocity and d is the diameter of the cylinder. Show that the phenomenon should be describable by the following π-groups: $\mathrm{Nu} = \bar{h}d/k_t$, $\mathrm{Re} = \rho V d/\mu$, and $\mathrm{Pr} = c_p \mu/k_t$. The respective names given to these dimensionless groups are: the Nussult number, the Reynolds number, and the Prandtl number. (*b*) In experiments dealing with the natural convection from a heated vertical wall, it has been experimentally determined that the convective heat transfer coefficient, $\bar{h}$, is a function of ρ, μ, c_p, k_t, β, g, $(T_W - T_f)$, and L, where ρ, μ, c_p, k_t are the same as those defined in part (*a*), β is the coefficient of thermal expansion of the fluid (see Section 1.4.4); g is the acceleration of gravity; $(T_W - T_f)$ is the temperature difference between the wall surface and the fluid a short distance from the wall; and L is wall length. Show that the phenomenon should be describable by the following π-groups: $\mathrm{Nu} = \bar{h}L/k_t$, $\mathrm{Pr} = c_p\mu/k_t$, and $\mathrm{Gr} = \rho^2 g\beta(T_w - T_f)L^3/\mu^2$. The respective names given to these dimensionless groups are: the Nussult number, the Prandtl number, and the Grashof number.

5-10 Demonstrate the physical significance of the following dimensionless group

$$\pi = \frac{\Delta p \ell}{\sigma}$$

5-11 Repeat Problem 5-10 for the grouping

$$\pi = \frac{\rho \ell^2 g}{\mu V}$$

5-12 Assess the physical significance of the π-groups found in Problems 5-4 and 5-5.

5-13 An airplane flies just above sea level in a standard atmosphere with a speed of 500 mi/h. At what speed must the airplane fly at an altitude of 30,000 ft in order to preserve viscous similarity? (*b*) Repeat part (*a*) for an altitude of 60,000 ft.

5-14 An airplane flies just above sea level in a standard atmosphere with a speed of 450 m/s. (*a*) At what speed must the airplane fly in order to maintain Mach number similarity at an altitude of 9000 m? (*b*) Repeat part (*a*) for an altitude of 18,000 m.

5-15 A proposed Great Lakes ore carrier, 200 m long, is to be studied in a towing tank from a model 3 m long. Assuming the mean temperature of the Great Lakes to be 10 C, what would be the required kinematic viscosity for the fluid used in the towing tank if dynamic similarity is maintained between the model and the prototype?

5-16 The proposed ore carrier of Problem 5-15 is expected to have a speed of approximately 6 m/s. If water is used in the towing tank and the Froude number alone is maintained between model and prototype, at what velocity must the 3 m model be towed?

5-17 A 1:10-scale model of an ICBM is to be tested in a supersonic, blowdown wind tunnel, using nitrogen as the atmosphere. In atmospheric air the missile is designed to reach a Mach number of 2.0 at 15,000 ft. The nitrogen is stored in a large tank at a pressure of 2000 psia and at a temperature of 100 F for use in the blowdown tunnel. If the flow of the nitrogen is assumed isentropic, to what pressure must it be expanded at the nose of the model to maintain dynamic similarity?

5-18 What volume in the lock model of Examples 5-4 and 5-5 represents a flow rate of 35 m^3/min in the prototype?

5-19 The hydrostatic equation tells us that pressure in a submerged liquid is proportional to the fluid's specific weight and the depth below the free surface. If the filled prototype lock of Example 5-4 exerts a force of 1.35×10^5 N on one of the two downstream gates, what is the equivalent force in the model?

5-20 Suppose in Problem 5-8, the measured drag on the model during a particular run was 175 N. What drag force would you expect from the prototype?

5-21 A flat-plate wing model has a span length of 5 ft and a chord length of 1 ft. The model is towed in a water basin at 20 ft/s at zero angle of attack (see Section 9.1.1). The water temperature is 68 F and the measured drag is 12 lb_f. (*a*) Find the dimensions of the prototype plate that will move through air at 14.7 psia and 59 F, with a speed of 120 ft/s at zero angle of attack. (*b*) What drag force is expected from the prototype?

A Computer Problem

5-22 Develop a computer program for determining the velocity, $V_R(t)$, and the altitude, $Z_R(t)$, of the rocket described in Example 4-13. But this time account for aerodynamic drag and the variation of both g and the free-stream properties with altitude. The governing equations for the rocket were given by Eqs. 4.7-30 and 4.7-33. These are, after some rearranging of Eq. 4.7-30,

$$\frac{dZ_R}{dt} = V_R \qquad \text{Eq. 4.7-33}$$

and

$$\frac{dV_R}{dt} = \frac{g_c}{M_0 - \dot{m}_e t}\left\{-\text{Drag}_{\text{engine off}} + (p_e - p_\infty)A_e - \frac{\rho V_\infty^{\ 2}}{2g_c}\tilde{c}_{p,b}A_e + \frac{\dot{m}_e V_e}{g_c}\right\}$$
$$- g = f_1(t, V_R, Z_R) \qquad \text{Eq. 4.7-30}$$

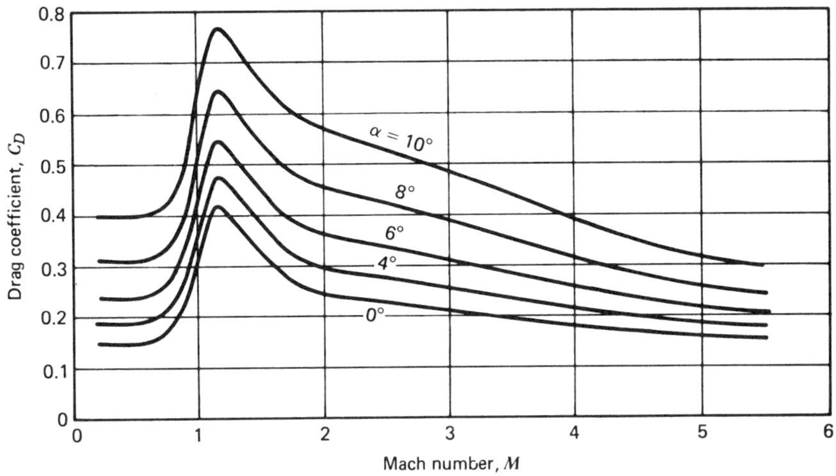

Fig. 5-10. Variation of drag coefficient with Mach number of rocket, based on body-cross-sectional area with jet off. (From *Rocket Propulsion Elements* by G. P. Sutton, Copyright © 1956 by John Wiley & Sons. Reprinted by permission of the publisher.)

where $V_\infty = V_R$ and where the right-hand side of the above equation is an implicit function of V_R and Z_R through the Drag, $\tilde{c}_{p,b}$ and g. Take $\dot{m} = 276\ \text{lb}_\text{m}/\text{s}$; $v_e = 6540$ ft/s; $p_e = 14.696$ psia; and $M_0 = 20{,}000\ \text{lb}_\text{m}$. Assume that the diameters of the nozzle's exit area (diameter of surface, S_e) and the rocket's frontal area are 29.1 in. and 40.0 in., respectively. Assume that a wind tunnel test on the rocket of interest provided values for C_D (with engine off), as shown in Fig. 5-10. Take the rocket to be moving upward at a zero angle of attack. Approximate the C_D curve by a series of straight lines as tabulated below.

M_∞ range	C_D
$0.0 \rightarrow 0.2$	0.0
$0.2 \rightarrow 0.7$	0.15
$0.7 \rightarrow 1.18$	$0.15 + 0.5625\,(M - 0.7)$
$1.18 \rightarrow 1.8$	$0.42 - 0.2742\,(M - 1.18)$
$1.8 \rightarrow 4.4$	$0.25 - 0.03077\,(M - 1.8)$
$4.4 \rightarrow 6.0$	$0.17 - 0.01875\,(M - 4.4)$
$6.0 \rightarrow \infty$	0.14

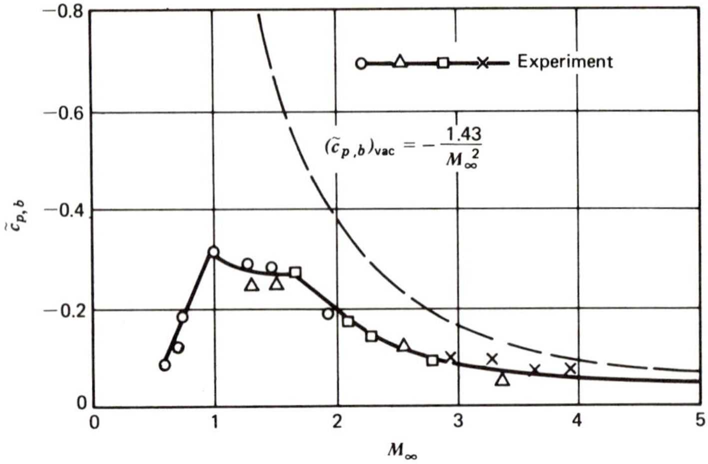

Fig. 5-11. Experimental and theoretical data for the base-pressure coefficient. (From *Aerodynamics of Bodies of Revolution* by N. F. Krasnov, Copyright © 1970 by Elsevier North Holland, Inc. Reprinted by permission of the publishers.)

Take the gravitational acceleration to vary with altitude according to

$$g = g_0\left(1 - \frac{2Z}{r_0}\right)$$

where $g_0 = 32.17$ ft/s^2 and $r_0 = 2.090 \times 10^7$ ft. Take the free-stream pressure and temperature (p_∞, T_∞) to be that of the U.S. Standard Atmosphere at the altitude of the rocket. Evaluate these quantities from Eqs. 3.8-8, 3.8-9 (or 3.8-12), and Table 3-1. Assume that wind-tunnel testing on the rocket of interest provided values for $\tilde{c}_{p,b}$, as plotted in Fig. 5-11. Represent this graph by a series of straight lines as tabulated below.

M_∞ range	$\tilde{c}_{p,b}$
$0.0 \rightarrow 0.2$	0.0
$0.2 \rightarrow 0.4$	$-0.02 - 0.35\,(M_\infty - 0.2)$
$0.4 \rightarrow 1.0$	$-0.09 - 0.22\,(M_\infty - 0.4)/0.6$
$1.0 \rightarrow 1.7$	$-0.31 - 0.03\,(M_\infty - 1.0)/0.7$
$1.7 \rightarrow 2.8$	$-2.28 + 0.185\,(M_\infty - 1.7)/1.1$
$2.8 \rightarrow 4.0$	$-0.095 + 0.045\,(M_\infty - 2.8)/1.2$
$4.0 \rightarrow \infty$	-0.05

Use the Runge-Kutta method for integrating Eqs. 4.7-30 and 4.7-33, as described in Example 4-13. Use Eqs. 4.7-35 and 4.7-37 and replace Eqs. 4.7-36 with the following set of equations

$$r_1 = f_1(t_i, (V_R)_i, (Z_R)_i)$$

$$r_2 = f_1\left(t_i + \frac{\Delta t}{2}, (V_R)_i + \frac{\Delta t}{2} r_1, (Z_R)_i + \frac{\Delta t}{2} q_1\right)$$

$$r_3 = f_1\left(t_i + \frac{\Delta t}{2}, (V_R)_i + \frac{\Delta t}{2} r_2, (Z_R)_i + \frac{\Delta t}{2} q_2\right)$$

$$r_4 = f_1(t_i + \Delta t, (V_R)_i + \Delta t\, r_3, (Z_R)_i + \Delta t\, q_3)$$

Print out values of t, V_R, and Z_R for every tenth interval on a time span of 66 s (the time of power cutoff). Compare your results to those given in Example 4-13.

Chapter Six

Steady, Incompressible Flow in Conduits and Pipes

6.1 INTRODUCTION

The concepts developed in the study of incompressible fluid flow in conduits and pipes have numerous applications in engineering. For example, the proper design of water supply, drainage, and liquid fuel and lubricating systems depends on a thorough understanding of this topic. Our first concern is with the pressure drop that occurs in a fully wetted closed conduit or pipe. The two flow regimes discussed in Section 5.1, laminar and turbulent, are exhibited in the flow through pipes. Early investigators in the field of laminar pipe flow were Hagen and Poiseuille, whose work we will discuss later in this chapter. Our understanding of the transition from laminar to turbulent flow is based on the work of Osborne Reynolds. Detailed study of turbulent flow is of more recent origin, and much of our present knowledge of this topic is due to the work of Reynolds, Prandtl, von Kármán, and Nikuradse. Although the pressure drop in a pipe can be determined either analytically or empirically when the flow is laminar, it must be determined empirically when the flow is turbulent. Thus, we first treat the pressure drop in a pipe from the viewpoint of dimensional analysis. Other topics treated in this chapter are: pipe entrance conditions, single pipeline problems, and simple networks.

6.2 PIPE ENTRANCE CONDITIONS AND FULLY DEVELOPED FLOW

As an incompressible fluid such as water flows from a tank into a circular, horizontal pipe, the velocity distribution in the pipe is noted to change through a small length of pipe. Near the entrance the velocity distribution is nearly uniform, except near the pipe walls, as shown in Fig. 6-1. This region of nearly uniform velocity, called the *core*, decreases in size as the distance from the entrançe section increases. Since the shear stress, τ_t, is proportional to dv_1/dr, we see that viscous effects are important only outside the core ($dv_1/dr \approx 0$ inside the core). In this respect, the entrance section of a pipe behaves in a similar manner to the flow over a flat plate, which is discussed in Section 9.1.1. That is, there is a *boundary-layer* buildup at the entrance of the pipe. A boundary layer is a region adjacent to the wall where viscous effects are important in developing the velocity profile, and this region grows in size as the distance from the leading edge increases. After some finite distance from the entrance, the core region disappears and the flow becomes *fully developed.*

The velocity profile in a fully developed flow is parabolic for the case of laminar flow and somewhat more blunt near the centerline of the pipe for the case of turbulent flow, as shown in Fig. 6-2. The velocity at the walls for all real flows is zero, owing to the no-slip boundary condition at a solid boundary. For laminar flow, the length, L', required to reach a fully developed flow con-

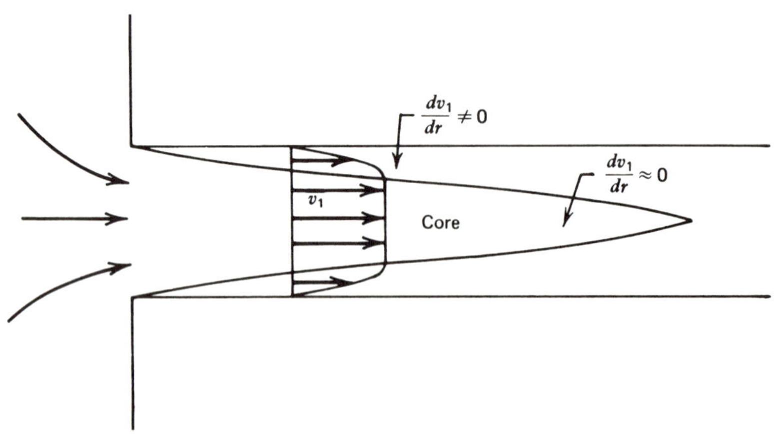

Fig. 6-1. Velocity distribution at pipe entrance.

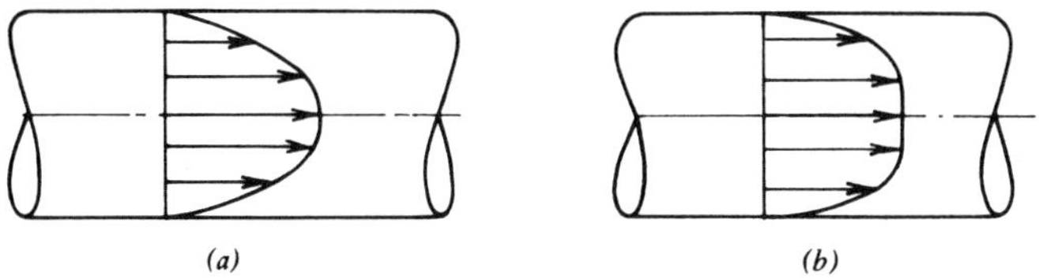

Fig. 6-2. Velocity profile in fully developed pipe flow. (*a*) Laminar flow. (*b*) Turbulent flow.

dition can be found from the following expression, which is attributed to Langhaar

$$\frac{L'}{D} = 0.058 \text{ Re} \qquad \text{Eq. 6.2-1}$$

where D is the pipe diameter and Re is the Reynolds number. For turbulent flow, Nikuradse states that L'/D is of the order of 25 to 40.

6.3 PRESSURE DROP IN FULLY DEVELOPED PIPE FLOW—THE FRICTION FACTOR AND THE MOODY DIAGRAM

In Section 5.1, it was mentioned that pipe flows are usually treated by adding a head-loss term to the energy equation obtained for an ideal flow; the modified equation being

$$\frac{p_1}{\gamma} + z_1 + \frac{V_1^2}{2g} = \frac{p_2}{\gamma} + z_2 + \frac{V_2^2}{2g} + h_f \qquad \text{Eq. 6.3-1}$$

where h_f is the head loss, which must be determined empirically for the case of turbulent flow. If the cross-sectional area and elevation of the pipe are the same at both sections (see Fig. 5-3), then

$$z_1 = z_2 \qquad \text{and} \qquad V_1 = V_2$$

The latter equation results from the conservation of mass, Eq. 4.4-6. Then we see that the pressure drop, $p_1 - p_2$, is given by

$$p_1 - p_2 = H_f = \gamma h_f \qquad \text{Eq. 6.3-2}$$

The pressure drop in the pipe, H_f, results from viscous shear stresses in the case of laminar flow and an *apparent shear stress* (also called Reynolds stresses) in the case of turbulent flow. Experiments indicate that

$$H_f = F(\rho, V, D, L, \mu, e) \qquad \text{Eq. 6.3-3}$$

where D is the diameter of the pipe; L is the length over which the pressure drop takes place; and e is a *roughness factor* for the pipe. All other symbols have the same meaning as previously indicated.

The roughness factor, e, is important only in the case of turbulent flow and is based on the experiments carried out by Nikuradse in the early 1930s. He cemented sand grains of a uniform size to the interior surface of smooth pipes in an effort to determine the effect of surface-roughness conditions normally found on commercial pipes. Thus, e represents the average diameter of the sand grains. Applying the concepts of dimensional analysis presented in Chapter 5, we see that the number of variables, n, equals 7. The number of basic dimensions, $\bar{r}$, needed to describe the variables dimensionally equals 3. Thus, the number of independent dimensionless groups that may be employed to describe the phenomenon, as given by Statement 5.3-1, equals 4.

Two dimensionless groups may be immediately chosen by observation; that is, it is obvious that L/D and e/D are dimensionless and may thus be selected as independent dimensionless groups, say π_3 and π_4, which may be important in describing the phenomenon. We can also see that the variables contained in the Reynolds number are present and thus, $\text{Re} = \rho V D/\mu$ may also be selected as the third independent dimensionless group, say π_2, which may be important in describing the phenomenon. To obtain the last dimensionless group, observe that the term (F/ℓ^2) appearing in the expression for the pressure coefficient, P, in Section 5.3 has the same dimensions as H_f. Thus, the last independent dimensionless group, say π_1, is a modified form of the pressure coefficient; that is,

$$\pi_1 = \frac{g_c H_f}{\rho V^2} \qquad \text{Eq. 6.3-4}$$

Of course, π_1 can be directly derived from the procedure outlined in Section 5.3. This is left to the student as an exercise. Then, by Eq. 5.3-26, we may write

$$\frac{g_c H_f}{\rho V^2} = G_1\left(\text{Re}, \frac{L}{D}, \frac{e}{D}\right) \qquad \text{Eq. 6.3-5}$$

The exact form of the functional relationship is to be determined by experimentation. The first bit of information obtained from experimentation is that G_1 is linear in (L/D),

$$\frac{g_c H_f}{\rho V^2} = \frac{L}{D} G\left(\text{Re}, \frac{e}{D}\right)$$

or

$$H_f = \frac{\rho V^2}{g_c} \frac{L}{D} G\left(\text{Re}, \frac{e}{D}\right)$$

Conventionally, a factor of 1/2 has been incorporated into the expression, giving

$$H_f = \frac{\rho V^2}{2g_c} \frac{L}{D} (2G) = \frac{\rho V^2}{2g_c} \cdot \frac{L}{D} f\left(\text{Re}, \frac{e}{D}\right) \qquad \text{Eq. 6.3-6}$$

The factor $2G$ is generally designated as the pipe *friction factor*, f. Equation 6.3-6 is known as the Darcy formula. This equation can also be written in terms of the head loss, h_f, by substituting Eqs. 6.3-2 and 1.4-3 into Eq. 6.3-6 and canceling out terms where appropriate; this gives us

$$h_f = \frac{V^2}{2g} \frac{L}{D} f \qquad \text{Eq. 6.3-7}$$

The dimensions of h_f are either in feet or meters.

A chart that gives f versus Re for various values of e/D is known as a *Stanton diagram*. The most commonly used Stanton diagram was constructed by Moody using data collected by Nikuradse and Colebrook. A small version of the Moody diagram, which will be used for discussion, is shown in Fig. 6-3. A larger version, one for use in problem solving, is shown in Fig. 6-4. A table of recommended values for the roughness factor, e, for standard commercial pipes, is used in conjunction with the *Moody diagram*. For our convenience this table has been printed directly onto the Moody diagram (see Fig. 6-4).

Examination of the Moody diagram indicates that

1. The f value is not a function of e/D in the laminar region. In fact, in the laminar region, $f = 64/\text{Re}$. This fact will be derived later in this chapter.

2. When Re $\approx$ 2100, a transition region occurs (from laminar to turbulent). Points in this region may lie either on the laminar curve, or on one of the turbulent curves, or somewhere between.

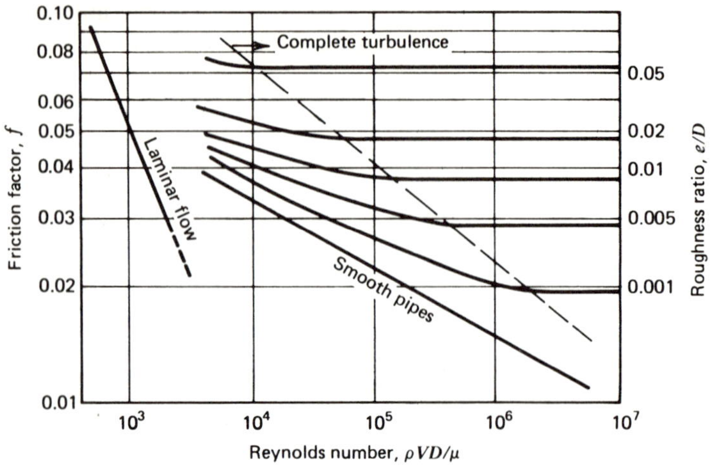

Fig. 6-3. The Moody diagram. Relationship between friction factor and Reynolds number for various values of roughness ratio. (From L. F. Moody, "Friction Factors for Pipe Flow," *Transactions*, ASME, Nov. 1944.)

3. For all roughness ratios, e/D, there exists a region, designated as the complete turbulence region, where f is only a function of (e/D).

4. All curves of $e/D < 0.001$ blend into the smooth pipe curve at sufficiently low Reynolds numbers.

Example 6-1

Water at 80 F flows through a horizontal, rough pipe at an average velocity of 9.30 ft/s. If the pipe diameter is 1.20 in. and the roughness ratio of the pipe is 0.02, find the pressure loss in pounds per square inch through 10 ft of pipe.

SOLUTION

The pressure loss, which is a result of viscous effects, is given by the Darcy equation as

$$H_f = f\frac{L}{D}\frac{\rho V^2}{2g_c}$$

In order to find the friction factor, f, from Fig. 6-4, it is necessary to know the Reynolds number, Re. To calculate the Reynolds number, we need the kinematic

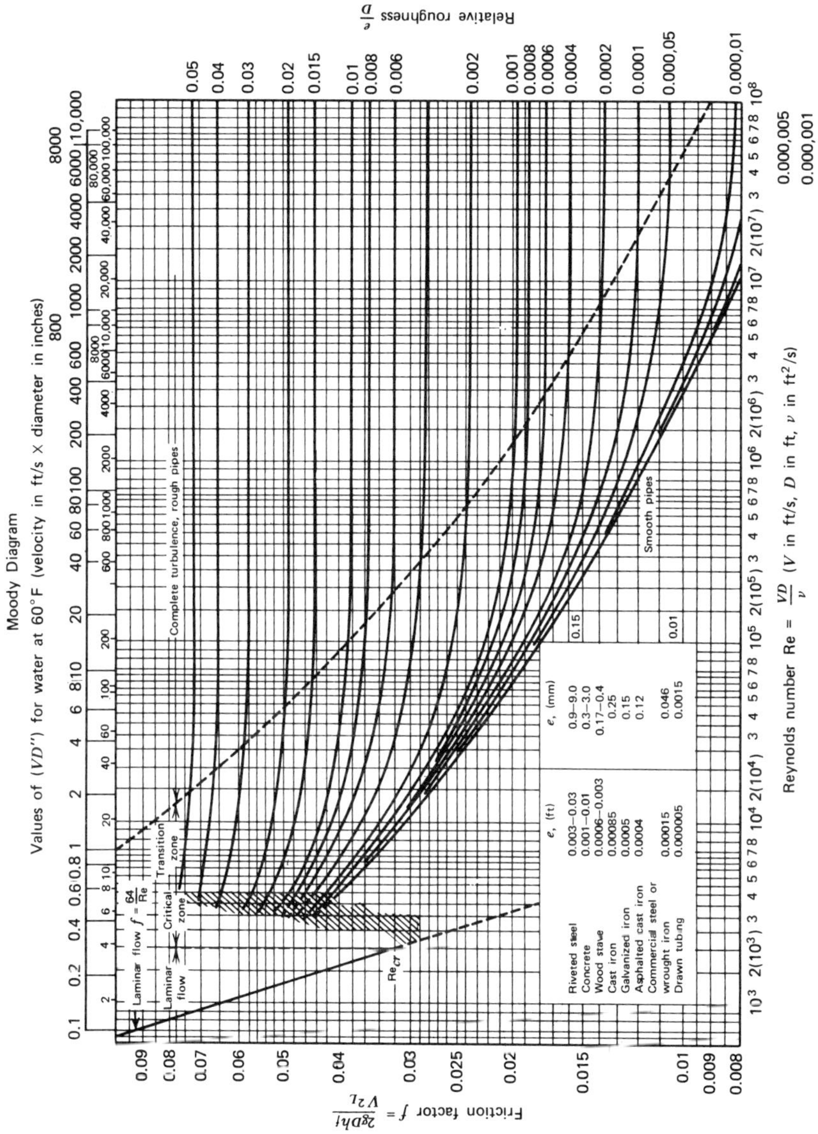

Fig. 6-4. The Moody diagram. (From L. F. Moody, "Friction Factors for Pipe Flow," *Transactions*, ASME, November 1944.)

viscosity, ν. From Table A.6-1, for water at 80 F, $\nu = 0.930 \times 10^{-5}$ ft²/s. Thus,

$$\text{Re} = \frac{VD}{\nu} = \frac{(9.3)(0.10)}{0.930 \times 10^{-5}} = 10^5$$

From Fig. 6-4, $f = 0.048$ for a roughness ratio of 0.02 and a Reynolds number of 10^5. Therefore,

$$H_f = (0.048)\left(\frac{10}{0.10}\right)\frac{62.2(9.3)^2}{64.4} = 402\ \text{lb}_f/\text{ft}^2$$

$$= 2.78\ \text{lb}_f/\text{in.}^2$$

The existence of valves, bends, sudden contractions, and so on, in pipelines also causes minor head losses. These losses are not included in the Darcy equation and therefore must be added to the pipe friction factor loss of a straight pipe. This is discussed in greater detail in the following section.

6.3.1 FLOW IN NONCIRCULAR CONDUITS

In considering a flow through a conduit having a noncircular cross section, the question arises as to what characteristic length, D_{ch}, is to be used in evaluating the Reynolds number for the flow. In practice, D_{ch} is based on the *hydraulic radius*, R_H. The hydraulic radius is defined to be the cross-sectional area containing the fluid divided by the wetted perimeter[1] (the perimeter of the fluid region in contact with the solid boundary). The relationship between D_{ch} and R_H can be obtained by considering a flow through a conduit having a circular cross section; for then $D_{ch} = D$. For a circular pipe

$$R_H = \frac{\text{cross-sectional area}}{\text{wetted perimeter}} = \frac{\pi D^2}{\pi D} = \frac{D}{4} = \frac{D_{ch}}{4}$$

Therefore,

$$D_{ch} = 4R_H$$

[1] When a liquid flows through a conduit, the definition allows for the possibility that the liquid does not completely fill the conduit. Such flows are classified as open-channel flows, which are treated in Chapter 7.

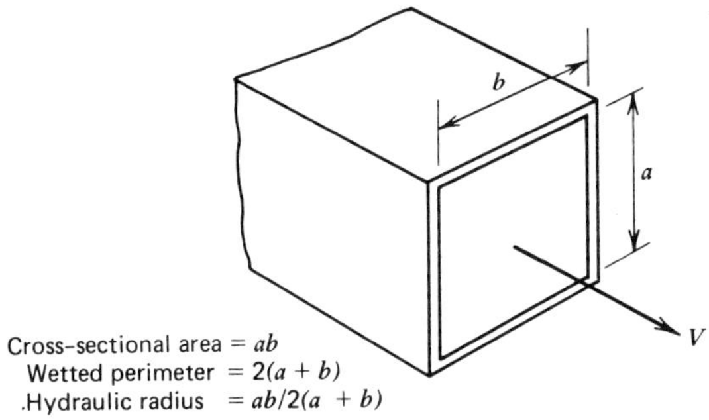

Fig. 6-5. A rectangular cross-sectional conduit.

Then, for a flow through a noncircular conduit,

$$\text{Re} = \frac{VD_{ch}}{\nu}$$

We will examine two cases.

(1) The Rectangular Cross-Sectional Conduit

For conduits of rectangular cross section, as shown in Fig. 6-5,

$$R_H = \frac{ab}{2(a+b)}$$

Then

$$D_{ch} = 4R_H = \frac{2ab}{a+b} \qquad \text{Eq. 6.3-8}$$

(2) The Concentric Annulus Conduit

For the concentric annulus conduit, as shown in Fig. 6-6,

$$R_H = \frac{(\pi/4)(D^2 - d^2)}{\pi(D+d)} = \frac{D-d}{4}$$

Then

$$D_{ch} = 4R_H = D - d \qquad \text{Eq. 6.3-9}$$

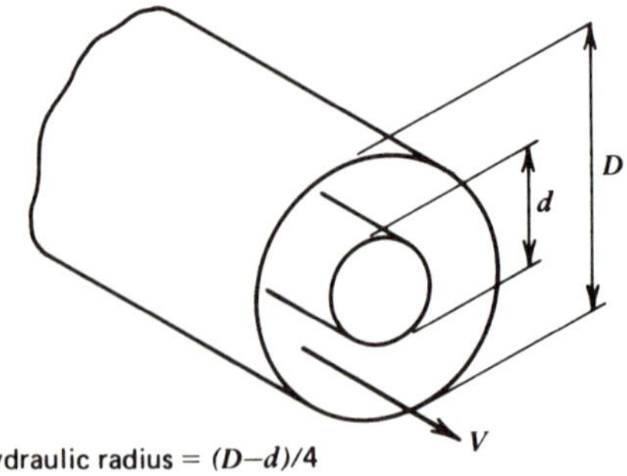

Hydraulic radius = $(D-d)/4$
Equivalent diameter = $(D-d)$

Fig. 6-6. A concentric annulus conduit.

Experimentation indicates that the use of D_{ch}, as defined here for evaluating the friction loss in a flow through a noncircular conduit, is reasonably accurate when the flow is *turbulent*. For noncircular conduits, D_{ch} based on the hydraulic radius should *not* be used when the flow is in the laminar region.

6.4 MINOR LOSSES AS A RESULT OF PIPE FITTINGS AND SUDDEN PIPE CONTRACTIONS AND EXPANSIONS

It has been observed that pipe fittings (such as elbows, valves, couplings, etc.) and sudden pipe contractions and expansions all cause a pressure drop to occur in a pipeline system. Experiments indicate that this pressure drop or head loss is a function of the square of the velocity. Therefore, if a pipe fitting lies between sections 1 and 2, then Eq. 6.3-1 is modified to include a minor head loss, $h_{L,\text{fitting}}$, on the right-hand side of the equation; that is

$$\frac{p_1}{\gamma} + z_1 + \frac{V_1^2}{2g} = \frac{p^2}{\gamma} + z_2 + \frac{V_2^2}{2g} + h_f + h_{L,\text{fitting}} \qquad \text{Eq. 6.4-1}$$

where

$$h_{L,\text{fitting}} = C\frac{V^2}{2g} \qquad \text{Eq. 6.4-2}$$

The constant, C, is found by experimentation. Table 6-1 gives values of C for a number of common pipe fittings.

TABLE 6-1 Loss Coefficients for Fittings and Valves

Type of Fitting or Valve	Loss Coefficient (C)
45° elbow, standard	0.35
45° elbow, long radius	0.2
90° elbow, standard	0.75
Long radius	0.45
Square or miter	1.3
180° bend, close return	1.5
Tee, standard, along run, branch blanked off	0.4
Used as elbow, entering run	1.3
Used as elbow, entering branch	1.5
Branching flow	1
Coupling	0.04
Union	0.04
Gate valve, open	0.17
$\frac{3}{4}$ open	0.9
$\frac{1}{2}$ open	4.5
$\frac{1}{4}$ open	24.0
Diaphragm valve, open	2.3
$\frac{3}{4}$ open	2.6
$\frac{1}{2}$ open	4.3
$\frac{1}{4}$ open	21.0
Globe valve, bevel seat, open	6.4
$\frac{1}{2}$ open	9.5
Composition seat, open	6.0
$\frac{1}{2}$ open	8.5
Plug disk, open	9.0
$\frac{3}{4}$ open	13.0
$\frac{1}{2}$ open	36.0
$\frac{1}{4}$ open	112.0
Angle valve, open	3.0
Y or blowoff valve, open	3.0
Plug cock, $\theta = 5°$	0.05
10°	0.29
20°	1.56
40°	17.3
60°	206.0
Butterfly valve, $\theta = 5°$	0.24
10°	0.52
20°	1.54
40°	10.8
60°	118.0
Check valve, swing	2.0
Disk	10.0
Ball	70.0
Foot valve	15.0
Water meter, disk	7.0
Piston	15.0
Rotary (star-shaped disk)	10.0
Turbine wheel	6.0

Source. From *Chemical Engineers' Handbook* by Perry & Chilton, copyright © 1963 by McGraw-Hill, Inc., used with permission of McGraw-Hill Book Company.

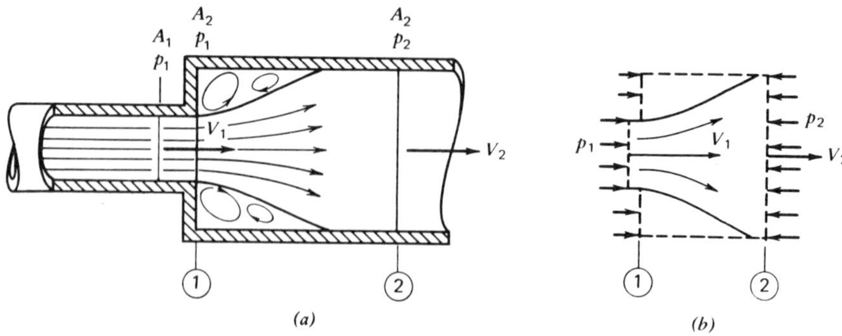

Fig. 6-7. Sudden expansion in a pipe. (a) Pipe section, (b) Control volume.

Head losses also occur when there exists either a sudden expansion or a sudden contraction in a pipeline (see Figs. 6-7*a* and 6-8). Again the head loss, h_L, is given by Eq. 6.4-2. For a sudden expansion, it can be shown (see Section 6.6.1) that

$$C = \left(1 - \frac{A_1}{A_2}\right)^2 \qquad \text{Eq. 6.4-3}$$

For this case, the V in Eq. 6.4-2 is to be taken as V_1.

The C value for a sudden contraction is plotted in Fig. 6-9, as a function of D_2/D_1. For this case, the V in Eq. 6.4-2 is to be taken as V_2.

Finally, the head loss at either an entrance or exit of a pipe from or to a reservoir is also given by Eq. 6.4-2. Values for C for the various cases are shown in Fig. 6-10.

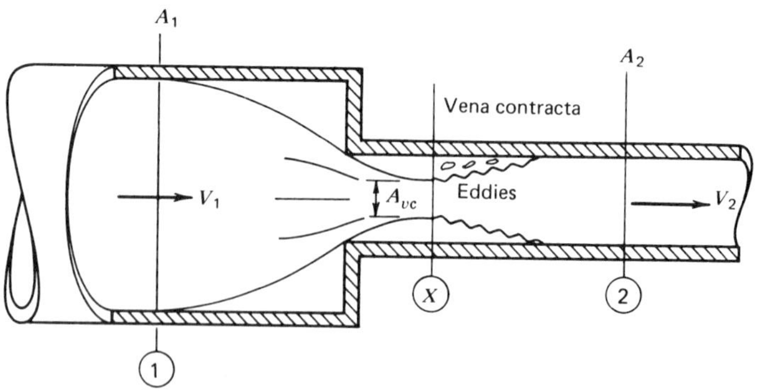

Fig. 6-8. Sudden contraction in a conduit.

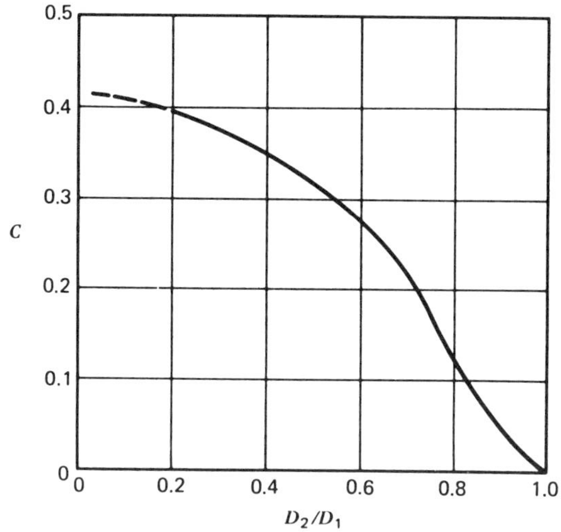

Fig. 6-9. Head-loss coefficient for a sudden contraction. (*Source.* R. A. Kenyon, *Principles of Fluid Mechanics,* 1960.)

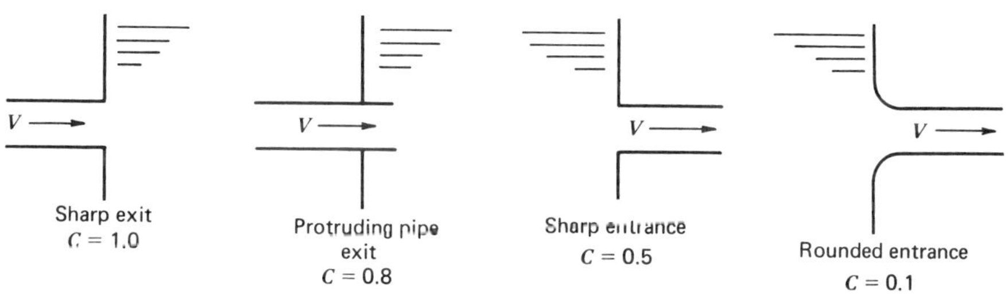

Fig. 6-10. Head-loss coefficient at a pipe entrance or exit.

6.5 SINGLE-PATH PIPELINE PROBLEMS

In many single-path pipeline problems we find that there exist several different types of head losses between the end sections of the pipe system being analyzed. A useful equation which can readily be demonstrated is as follows. If the flow takes place from section α to section β, then the modified energy equation becomes

> The total head at section α equals the total head at section β plus the sum of the head losses which occur in the path between sections α and β.
>
> Stm. 6.5-1

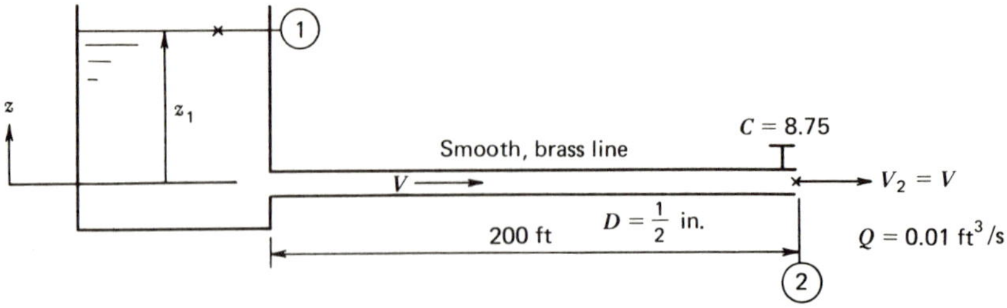

Fig. 6-11. The discharge of a reservoir through a pipeline.

This statement expressed mathematically is

$$\left(\frac{p}{\gamma} + z + \frac{V^2}{2g}\right)_A = \left(\frac{p}{\gamma} + z + \frac{V^2}{2g}\right)_\beta + \sum_i h_{L,i} \qquad \text{Eq. 6.5-1}$$

where i is an index enumerating the number of different head losses existing between sections α and β.

Single-path pipeline problems can usually be solved in a direct fashion, if V is either given or can be determined from Q/A. The unknown in this type of problem is usually either the sum of the head losses or the difference in elevation between the end sections. When V *cannot* readily be determined, because either Q or D is the unknown, then the problem must be solved by an iteration technique. Methods of solution are illustrated in the following examples.

Example 6-2

A reservoir discharges water through a 200 ft long, horizontal, smooth brass pipeline open to the atmosphere at one end, as shown in Fig. 6-11. The flow rate is controlled by an angle valve near the exit of the pipe. If the pipe entrance is square edged and the water temperature is 50 F, determine the elevation of the water in the open reservoir if the inside diameter (ID) of the pipe is $\frac{1}{2}$ in. and the flow rate is 0.01 ft^3/s when the angle valve is partly open. Assume that the loss coefficient, C, is 8.75.

SOLUTION

Applying Eq. 6.5-1 to the problem, we obtain

$$\frac{p_1}{\gamma} + z_1 + \frac{V_1^2}{2g} = \frac{p_2}{\gamma} + z_2 + \frac{V_2^2}{2g} + h_f + h_{L,V} + h_{L,\text{entrance}} \qquad \text{Eq. 6.5-2.}$$

From the specified conditions of the problem, we have

$$z_2 = 0, \qquad V_2 = V, \qquad \text{and} \qquad A_2 = A = \frac{\pi}{4}\frac{(0.5)^2}{144} = 0.1364 \times 10^{-2}\ \text{ft}^2$$

Since the cross-sectional area of the reservoir is very large compared to the cross-sectional area of the pipe, we may take $V_1 = 0$. Finally, since sections 1 and 2 are open to the atmosphere, $p_1 = p_2 = p_{atm}$. Then Eq. 6.5-2 reduces to

$$z_1 = \frac{V^2}{2g} + h_f + h_{L,V} + h_{L,\text{entrance}} \qquad \text{Eq. 6.5-3}$$

By Eqs. 6.3-7, 6.4-2, the specified value for C, and Fig. 6-10, we have

$$z_1 = \frac{V^2}{2g}\left(1 + \frac{L}{D}f + 8.75 + 0.5\right) = \frac{V^2}{2g}\left(\frac{L}{D}f + 10.25\right) \qquad \text{Eq. 6.5-4}$$

Knowing the volume flow rate, the average fluid velocity in the pipe, V, can be determined; that is,

$$V = \frac{Q}{A} = \frac{0.01}{0.1364 \times 10^{-2}} = 7.33\ \text{ft/s}$$

The friction factor, f, can be determined from the Moody diagram. To enter the Moody diagram, we need the Reynolds number. To calculate Re, we need the kinematic viscosity, ν. From Table A.6-1, $\nu = 1.4 \times 10^{-5}$ ft^2/s. Then from Eq. 5.1-3,

$$\text{Re} = \frac{VD}{\nu} = \frac{7.33(0.5)}{12(1.4 \times 10^{-5})} = 2.2 \times 10^4$$

From Fig. 6-4, $f = 0.0255$. Then

$$z_1 = \frac{(7.33)^2}{64.4}\left[\frac{200(12)(0.0255)}{0.5} + 10.25\right]$$

$$= 0.8343[122.4 + 10.25] = 110.7\ \text{ft}$$

Example 6-3

Suppose in Example 6-2, z_1 is known to be 50 ft, but the flow rate, Q, is unknown. The problem is to find Q.

SOLUTION

The solution to this problem is obtained by an iterative technique. Again, Eq. 6.5-4 is valid. Substituting for z_1 and multiplying both sides of Eq. 6.5-4 by $2g$, we obtain

$$50(64.4) = V^2\left[\frac{200(12)}{0.5}f + 10.25\right] \qquad \text{Eq. 6.5-5}$$

Since Q is unknown, we cannot immediately determine either V or f. Thus, we must solve the problem by iteration. Here are the steps we may follow.

1. Assume a value for f. (For the case of a smooth pipe, a good starting value for f is 0.03. For the case of a rough pipe, a good starting value is the value determined for f in the completely turbulent region of the Moody diagram.)

2. Solve Eq. 6.5-5 for V.

3. Solve Eq. 5.1-3 for Re.

4. Use the Moody diagram, knowing e/D and Re from step 3, and determine f.

5. Compare the value of f obtained in step 4 with the assumed f. If different, take the value of f obtained in step 4 as the assumed value and repeat steps 2 to 5 until agreement is reached. This iterative process is usually a fast converging process. We can now carry out this procedure for this example.

We assume $f = 0.03$ and determine V from Eq. 6.5-5, giving $V = 4.57$ ft/s. We can then determine that $\text{Re} = 1.36 \times 10^4$. Since in this case the pipe is smooth, we may use the smooth curve on the Moody diagram. From the Moody diagram, $f = 0.028$. We repeat the process assuming $f = 0.028$; then $V = 4.72$ ft/s and $\text{Re} = 1.4 \times 10^4$. From the Moody diagram, $f = 0.028$. Thus, the assumed f agrees with the f obtained from the Moody diagram. These results are shown in the Table 6-2. Therefore, the correct V is 4.72 ft/s, giving

$$Q = AV = \frac{\pi(0.5)^2(4.72)}{4(144)} = 0.00643 \text{ ft}^3/\text{s}$$

TABLE 6-2 Determination of f by Iteration

f (Assumed)	V (ft/s)	Re	f (Determined from Moody Diagram)
0.030	4.57	1.36×10^4	0.028
0.028	4.72	1.40×10^4	0.028

Example 6-4

Suppose that in Example 6-2, z_1 is known to be 50 ft. The problem is to determine the diameter of the smooth, brass pipe that will result in a flow rate of 0.01 ft^3/s.

SOLUTION

This problem may be considered as a pipe design problem. This is so because, in many instances, the engineer will be faced with designing a piping system that will provide a specific flow rate to a certain part of a system at a specific pressure. The problem is to determine the pipe sizes that will satisfy the specified conditions.

Although Eq. 6.5-4 is still valid in this problem, the velocity, V, cannot immediately be determined, since the cross-sectional area, A, is unknown. Without knowing V, we can neither determine Re nor the friction factor, f. The solution is obtained by either an iterative process or by plotting two curves and obtaining the point of intersection. We will use the point-of-intersection method.

First let us solve for V in terms of Q and D; that is,

$$V = \frac{Q}{A} = \frac{4(0.01)}{\pi D^2} = \frac{0.01273}{D^2} \qquad \text{Eq. 6.5-6}$$

Substituting the values of z_1 and L and Eq. 6.5-6 into Eq. 6.5-4, we obtain

$$50 = \frac{(0.01273)^2}{64.4D^4}\left(10.25 + \frac{200f}{D}\right)$$

Solving for f in terms of D gives

$$f = 9.935 \times 10^4 D^5 - 0.05125D \qquad \text{Eq. 6.5-7}$$

We can also express Re in terms of D; that is,

$$\mathrm{Re} = \frac{VD}{\nu} = \frac{0.01273}{1.4 \times 10^{-5} D} = \frac{9.093 \times 10^2}{D} \qquad \text{Eq. 6.5-8}$$

A method of solution is as follows.

1. Assume a value of D. (A good starting value can be obtained by neglecting the second term on the right-hand side of Eq. 6.5-7. We then assume that $f = 0.03$ and solve for D. This gives the assumed D.)

2. Determine f from Eq. 6.5-7.

3. Determine Re from Eq. 6.5-8 and e/D for those problems where the pipe is not smooth.

4. Determine f from the Moody diagram.

By assuming other values of D, two curves may be plotted: one being f versus D, where f is determined from Eq. 6.5-7, the other being f versus D, where f is determined from the Moody diagram. The solution is the value of D at the intersection of the two curves. Table 6-3, shown below, gives points on each curve. The first assumed D, obtained according to step 1, was $D \approx 0.05$ ft. It should be observed that a small change in D causes: (1) a large change in f when it is determined from Eq. 6.5-7 and (2) only a small change in f when it is determined from the Moody diagram. This fact can be employed in deciding whether to use a smaller or larger D for the next trial. That is, if in some particular trial f as determined from Eq. 6.5-7 is smaller than f as determined from the Moody diagram, then use a larger D in the next trial. A plot of the two curves is shown in

TABLE 6-3 Determination of f by Point-of-Intersection Method

D (ft)	f (Determined from Eq. 6.5-8)	Re	f (Determined from Moody Diagram)
0.035	0.00342	2.60×10^4	0.0240
0.040	0.00812	2.30×10^4	0.0250
0.050	0.02848	1.80×10^4	0.0265
0.055	0.04718	1.65×10^4	0.0270

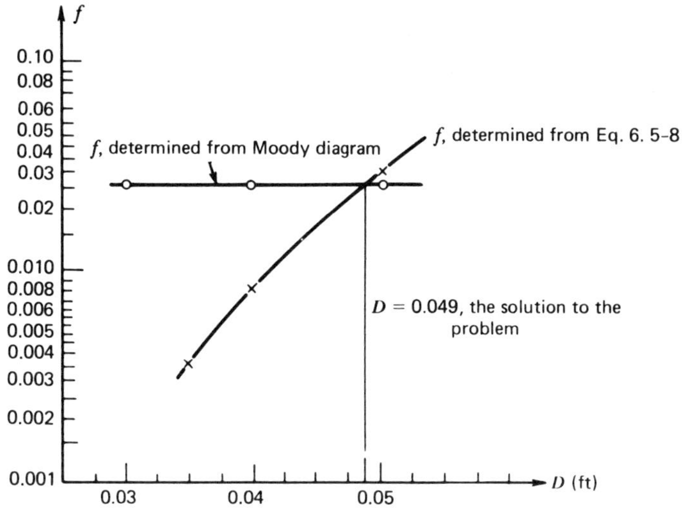

Fig. 6-12. Intersecting curves give the correct value for D.

Fig. 6-12. The solution is $D = 0.049$ ft $= 0.588$ in. In most design problems, it is sufficient to select the nearest standard pipe size.

Finally, it should be noted that when the pipeline arrangement for the type of problem described in Example 6-4 results in no D term in Eq. 6.5-7, then the problem may be more readily solved by the iterative method described in Example 6-3. This is so because the D term is the term that prevents an easy solution for D when an f value is assumed. The D term will be present whenever there exist minor head losses between the two end sections or when a bulk-velocity difference exists at the two end sections.

6.6 ANALYTICAL DETERMINATION OF THE FRICTION FACTOR FOR LAMINAR FLOW

Consider a laminar flow through a cylindrical pipe, as shown in Fig. 6-13. First, we will assume that the flow is a steady, parallel flow with $\mathbf{v} = v_3\mathbf{k}$. We can show that v_3 is only a function of r. The continuity equation, Eq. 4.5-4, for the case of a steady, incompressible flow, reduces to

$$\nabla \cdot \mathbf{v} = 0$$

The continuity equation is valid even when viscous effects are included.

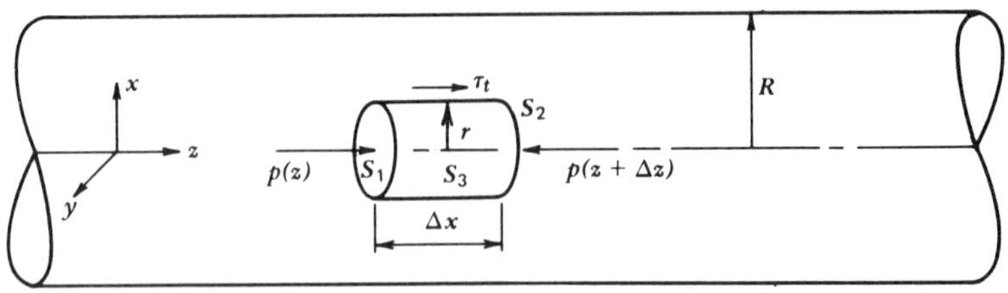

Fig. 6-13. Forces on fluid within control volume.

In cylindrical coordinates,[2] with $\mathbf{v} = v_r \mathbf{e}_r + v_\theta \mathbf{e}_\theta + v_3 \mathbf{k}$,

$$\nabla \cdot \mathbf{v} = \frac{\partial v_r}{\partial r} + \frac{1}{r}\frac{\partial v_\theta}{\partial \theta} + \frac{\partial v_3}{\partial z} = 0$$

But for parallel flow, $v_r = v_\theta = 0$; therefore, $\partial v_3/\partial z = 0$. Thus, $v_3 \neq f(z)$. From symmetry considerations, we can also conclude that $v_3 \neq f(\theta)$, leaving $\mathbf{v} = v_3(r)\mathbf{k}$. We now apply Newton's second law, Eq. 4.7-5, to the control volume shown in the Fig. 6-13, neglecting the body force, $\mathbf{F}$, to obtain

$$\oint\!\!\oint_S \boldsymbol{\tau}\, dS = \oint\!\!\oint_S \frac{\rho}{g_c} v_3 \mathbf{k}(v_3 \mathbf{k} \cdot d\mathbf{S}) \qquad \text{Eq. 6.6-1}$$

In the paragraph just following Eq. 4.6-28, it was noted that experimentation indicates that in a pipe or channel flow, the pressure does not significantly vary in the transverse direction to the flow. Thus we may take[3] $p = p(z)$. Thus p is a constant on both surface S_1 and S_2. Also, noting that $v_3 = v_3(r)$ and

$$\tau_t = \frac{\mu}{g_c}\frac{dv_3}{dr} \qquad \text{Eq. 6.6-2}$$

we can see that τ_t is a constant on surface S_3. Although the surface traction, $\boldsymbol{\tau}$, on each surface is, in general, given by

$$\boldsymbol{\tau} = -p\mathbf{n} + \tau_t \mathbf{e}_t$$

[2] See Eq. 2.2-28 for the expression of $\nabla \cdot \mathbf{v}$ in cylindrical coordinates.
[3] If we were to approach the problem completely from the field equations (field equations which include viscous effects are known as the Navier–Stokes equations), then we would be able to directly show that $p = p(z)$ only. However, the derivation of the Navier–Stokes equations is very involved and beyond the scope of this text. Students interested in this topic should see *Boundary-Layer Theory* by H. Schlichting.

our interest lies only in the **k**-component of Eq. 6.6-1. To evaluate the **k**-component of $\oiint_S \boldsymbol{\tau}\, dS$, note that

$$\mathbf{k} \cdot \boldsymbol{\tau}(S_1) = p(z), \qquad \mathbf{k} \cdot \boldsymbol{\tau}(S_2) = -p(z + \Delta z), \qquad \text{and} \qquad \mathbf{k} \cdot \boldsymbol{\tau}(S_3) = \tau_t(r)$$

Thus, the **k**-component of Eq. 6.6-1 becomes

$$p(z)\pi r^2 - p(z + \Delta z)\pi r^2 + 2\pi r \tau_t\, \Delta z = \int_0^{2\pi} \int_0^r \frac{\rho}{g_c} \underbrace{[v_3^2(r') - v_3^2(r')]}_{0} r'\, dr'\, d\theta$$

or

$$\frac{p(z + \Delta z) - p(z)}{\Delta z} = \frac{2}{r} \tau_t$$

Taking the $\lim_{\Delta z \to 0}$ of both sides of the equation, we find that

$$\frac{dp}{dz} = \frac{2}{r} \tau_t \qquad \text{Eq. 6.6-3}$$

Substituting Eq. 6.6-2 into Eq. 6.6-3, we obtain

$$\frac{dp}{dz} = \frac{2}{r} \frac{\mu}{g_c} \frac{dv_3}{dr} \qquad \text{Eq. 6.6-4}$$

Close examination of Eq. 6.6-4 reveals a subtle mathematical point that might at first appear as an inconsistency in the equation. Since $p = p(z)$ and $v_3 = v_3(r)$, observe that the left-hand side of the equation is a pure function of z and the right-hand side of the equation is a pure function of r. Yet, z and r are independent variables. This means that we may change the value of r without changing the value of z, or vice versa. But by changing the value of r and holding z fixed, we would expect the value of the right-hand side of the equation to change, but not the left side. Thus Eq. 6.6-4 would appear to be inconsistent. However, if each side of the equation is equal to the same constant, say $(-\lambda)$, then the inconsistency is eliminated, giving

$$\frac{dp}{dz} = -\lambda \qquad \text{Eq. 6.6-5}$$

and

$$\frac{2\mu}{g_c} \frac{1}{r} \frac{dv_3}{dr} = -\lambda \qquad \text{Eq. 6.6-6}$$

Since dp/dz is negative, λ is positive. Integrating the above equation gives

$$v_3 = -\frac{\lambda g_c}{2\mu}\frac{r^2}{2} + C$$

By the no-slip boundary condition, $v_3(R) = 0$, then

$$v_3(R) = 0 = -\frac{\lambda g_c}{4\mu} R^2 + C$$

or

$$C = \frac{\lambda g_c R^2}{4\mu}$$

giving

$$v_3 = \frac{\lambda g_c}{4\mu}(R^2 - r^2) \qquad \text{Eq. 6.6-7}$$

We see that the velocity profile is parabolic. Integration of Eq. 6.6-7 gives

$$\int_{p_1}^{p_2} dp = p_2 - p_1 = -H_f = -\lambda \int_0^L dz = -\lambda L$$

or

$$p_1 - p_2 = H_f = \lambda L \qquad \text{Eq. 6.6-8}$$

We can determine the volume flow rate, Q, by Eq. 4.4-7; that is,

$$Q = \int_0^R \int_0^{2\pi} v_3(r) r \, dr \, d\theta = \int_0^R \int_0^{2\pi} \frac{\lambda g_c}{4\mu}(R^2 - r^2) r \, dr \, d\theta$$

or

$$Q = 2\pi \frac{\lambda g_c}{4\mu}\left[R^2 \frac{r^2}{2} - \frac{r^4}{4}\right]_0^R = \frac{\pi \lambda g_c R^4}{8\mu} \qquad \text{Eq. 6.6-9}$$

The average velocity, V, can then be determined from $Q = AV$, giving

$$V = \frac{\lambda g_c}{8\mu} R^2 \qquad \text{or} \qquad \lambda = \frac{8\mu V}{g_c R^2}$$

Substituting into Eq. 6.6-8 the expression for H_f as given by Eq. 6.3-6 and the above expression for λ, we obtain

$$\frac{\rho V^2}{2g_c}\frac{L}{D} f = \frac{8\mu V L}{g_c R^2}$$

Solving for the friction factor, f, we find that

$$f = \frac{16\mu D}{\rho V R^2} = \frac{16\,\mu D}{\rho V \left(\frac{D}{2}\right)^2} = \frac{64}{\frac{\rho V D}{\mu}} = \frac{64}{\text{Re}} \qquad \text{Eq. 6.6-10}$$

Thus, Eq. 6.6-10 gives, for the case of laminar pipe flow, the functional relationship of the friction factor in terms of the Reynolds number.

A relationship for Q in terms of the pressure drop, H_f, can be obtained by solving Eq. 6.6-8 for λ and substituting the resulting equation into Eq. 6.6-9, giving

$$Q = \frac{\pi g_c R^4 H_f}{8\mu L} = \frac{\pi g_c D^4 H_f}{128\mu L} \qquad \text{Eq. 6.6-11}$$

Equation 6.6-11 is known as the *Hagen–Poiseuille equation* after the men who first derived it, independently of one another. This relation is valid only for flow in the laminar region when the velocity distribution is parabolic. The flow rate Q is seen to be directly proportional to the pressure drop per unit length of pipe. Note that the flow of liquid in laminar flow through a pipe is analogous to the flow of electric current in a conductor and the flow of heat by conduction through a plane wall. In the first analogy, the electric current, I, corresponds to the volume flow rate, Q, and the voltage drop, E, corresponds to the pressure drop, H_f. Thus, Eq. 6.6-11 corresponds to the equation $I = E/R$. In the second analogy, the rate of heat flow per unit surface area, q/A, corresponds to the volume flow rate, Q, and the temperature drop, ΔT, corresponds to the pressure drop, H_f. Then Eq. 6.6-11 corresponds to the equation $q/A = k(\Delta T/\Delta x)$.

6.6.1 ANALYTICAL DETERMINATION OF THE HEAD-LOSS COEFFICIENT FOR A SUDDEN EXPANSION IN A PIPELINE

In Section 6.4 the head loss coefficient, C, was given for a sudden expansion in a pipeline (see Eq. 6.4-3). This coefficient can be analytically derived, and we do so here.

The head loss resulting from a sudden expansion is caused by the formation of a pocket, containing turbulent eddies, just inside the larger pipe. The pressure in this pocket tends to remain at the same pressure as in the upstream section. Furthermore, there is a loss of energy due to dissipation in the eddies. Following a sudden enlargement, the flow may not return to its fully developed state until it reaches a distance as far as 50 pipe diameters downstream from the enlargement. If we apply the energy and momentum equations to the control volume shown in Fig. 6-7*b*, we can obtain an expression for the head loss. For such small lengths, viscous forces can be neglected. However, in the energy equation we account for the dissipation in the eddies by adding a head loss term to the ideal flow energy equation; that is,

$$\frac{p_1}{\gamma} + \frac{V_1^2}{2g} = \frac{p_2}{\gamma} + \frac{V_2^2}{2g} + h_L \qquad \text{Eq. 6.6-12}$$

The momentum equation in the x direction gives

$$p_1 A_2 - p_2 A_2 = \frac{\rho}{g_c} V_1(-V_1 A_1) + \frac{\rho}{g_c} V_2(V_2 A_2) \qquad \text{Eq. 6.6-13}$$

Conservation of mass yields $A_1 V_1 = A_2 V_2$. Substituting for V_1 in Eq. 6.6-13, we obtain

$$(p_1 - p_2)A_2 = \frac{\rho}{g_c} V_2^2\left[A_2 - \left(\frac{A_2}{A_1}\right)^2 A_1\right] = \frac{\rho}{g_c} V_2^2 A_2\left(1 - \frac{A_2}{A_1}\right)$$

or

$$(p_1 - p_2) = \frac{\rho}{g_c} V_2^2\left(1 - \frac{A_2}{A_1}\right) \qquad \text{Eq. 6.6-14}$$

By substituting Eq. 6.6-14 into Eq. 6.6-12 and solving for h_L we find that

$$h_L = \frac{V_1^2 - V_2^2}{2g} + \frac{\rho V_2^2}{g_c \gamma}\left(1 - \frac{A_2}{A_1}\right)$$

Noting that $\rho/\gamma g_c = 1/g$ and substituting for V_2, we obtain, after some rearranging,

$$h_L = \frac{V_1^2}{2g}\left[1 - 2\left(\frac{A_1}{A_2}\right) + \left(\frac{A_1}{A_2}\right)^2\right] = \frac{V_1^2}{2g}\left(1 - \frac{A_1}{A_2}\right)^2 \qquad \text{Eq. 6.6-15}$$

Comparing Eqs. 6.6-15 and 6.4-2, we see that the V in Eq. 6.4-2 is to be taken as V_1 and that

$$C = \left(1 - \frac{A_1}{A_2}\right)^2 \qquad \text{Eq. 6.4-3}$$

The problem of an abrupt contraction, in which the fluid suddenly enters a pipe of smaller diameter, is shown in Fig. 6-8. This situation differs somewhat from being merely the reverse of sudden enlargement. With an abrupt contraction, the fluid inertia causes the fluid stream to contract after entering the small-diameter pipe, reaching a minimum area or *vena contracta* at station X, well inside the entrance to the smaller pipe. This situation is the same as that of the orifice discussed in Chapter 4. An analysis similar to that used for the case of sudden expansion would require the knowledge of the vena contracta area in relation to areas A_1 and A_2, and this information can be determined only by experiment. For this reason no analytical development is given for the head loss in a sudden contraction. Experimental results were given in Fig. 6-9.

6.7 MULTIPATH PIPELINE PROBLEMS AND THE HARDY–CROSS METHOD

In the previous section we analyzed three types of single-line pipe flow problems commonly encountered by engineers. In this section, we consider problems associated with multipath systems, which, if extensive, would be classified as a *flow network*. In general, water supply systems are considered as flow networks, since nearly all such systems consist of many interconnecting pipes. Distribution systems of electric power are similarly classified as electrical networks. In treating flow networks the Darcy equation and the Moody diagram are replaced by a single empirical relationship. For water supply systems the Hazen-Williams formula—which relates Q, h_f, D, and L for any single pipeline—is most often used. To see that this can be done note that

$$h_f = \frac{V^2}{2g}\frac{L}{D}f$$

where $f = F(\text{Re}, e/D)$. Observe that (1) V can be expressed in terms of Q and D and that (2) Re can also be expressed in terms of Q and D when the problem is confined to a single fluid at a particular temperature, because then ν becomes a constant. Under these circumstances

$$h_f = F(Q, D, L, C)$$

where C is a constant that depends on the pipe material and the condition of the pipe. Here C takes the place of the roughness factor, e. The Hazen-Williams formula for water flow through pipes at customary temperatures in a water supply system is

$$h_f = \frac{k_1 L}{C^{1.852} D^{4.8704}} Q^{1.852} \qquad \text{Eq. 6.7-1}$$

or

$$Q = k_2 C D^{2.63} (h_f/L)^{0.54} \qquad \text{Eq. 6.7-2}$$

In these formulas Q and h_f are taken as positive numbers with h_f, L, and D in feet. The C value is considered to be dimensionless; then,

$$k_1 = 4.727 \qquad \text{and} \qquad k_2 = 0.432$$

when Q is in cubic feet per second, and

$$k_1 = 10.63 \qquad \text{and} \qquad k_2 = 0.279$$

when Q is in million gallons per day. If h_f, L, and D are expressed in meters, then

$$k_1 = 10.466 \qquad \text{and} \qquad k_2 = 0.2814$$

when Q is in cubic meters per second. Here "million gallons per day" can be written in an abbreviated form, "mgd" (a set of units commonly used by engineers who are associated with water supply systems). Values of C for various materials and pipe conditions are listed in Table 6-4. To simplify the notation, we will designate

$$K = \frac{k_1 L}{C^{1.852} D^{4.87}} \qquad \text{Eq. 6.7-3}$$

TABLE 6-4 Hazen–Williams Coefficient

Types of Pipe	C
Extremely smooth pipes	140
New steel or cast iron	130
Wood, average concrete	120
New riveted steel, clay	110
Old cast iron, brick	100
Old riveted steel	95
Badly corroded cast iron	80
Very badly corroded iron or steel	60

Source. From *Essentials of Engineering Fluid Mechanics* by R. M. Olson, Copyright © 1973 by Harper & Row. Reprinted by permission of the publisher.

Then

$$h_f = KQ^{1.852} \qquad \text{Eq. 6.7-4}$$

A nomograph, which is a graph relating Q, D, h_f, and V through the Hazen–Williams formula, with $C = 100$ and $L = 1000$, is shown in Fig. 6-14.

The design of a water supply system is based on: (1) the anticipated local maximum rates of water consumption at various points within the network and (2) the need to maintain the pressure within a certain range at various points within the network. In residential areas, pressures from 40 to 50 psig are adequate but, for commercial areas, pressures above 60 psig are necessary. Main feeders should be designed for pressures close to 75 psig. In the design of a system, pipe sizes are selected to provide average fluid velocities from 3 to 5 ft/s. Designs must also meet the requirements of the National Board of Fire Underwriters. System configurations are either grid or branch or a combination of both. A branch configuration consists of a main supply line, with branches that meet at a dead end, as shown in Fig. 6-15*a*. The grid configuration, shown in Fig. 6-15*b*, consists of a series of closed loops and, thus, any point in the system may be fed from at least two different paths. It is this feature that makes a grid system more desirable than a branch-type system. Figure 6-15*c* shows a configuration that is a combination of the two. Flow analyses of established networks are usually made by the use of the Hardy–Cross method. This is described in the next section.

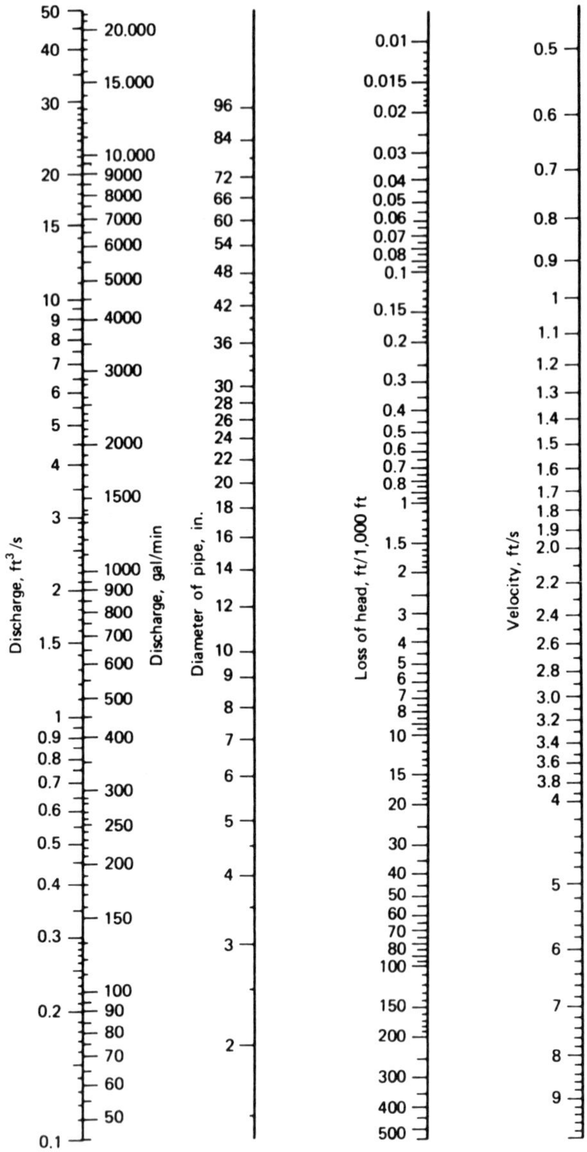

Fig. 6-14. Nomograph for Hazen–Williams formula in which $C = 100$. (From *Water Supply and Pollution Control,* 2nd ed., by J. W. Clark, W. Viessman, Jr., and M. J. Hammer: (International Textbook Co.), Fig. 5.9 (p. 149), Copyright © 1971 by Harper & Row. Reprinted by permission of the publisher.)

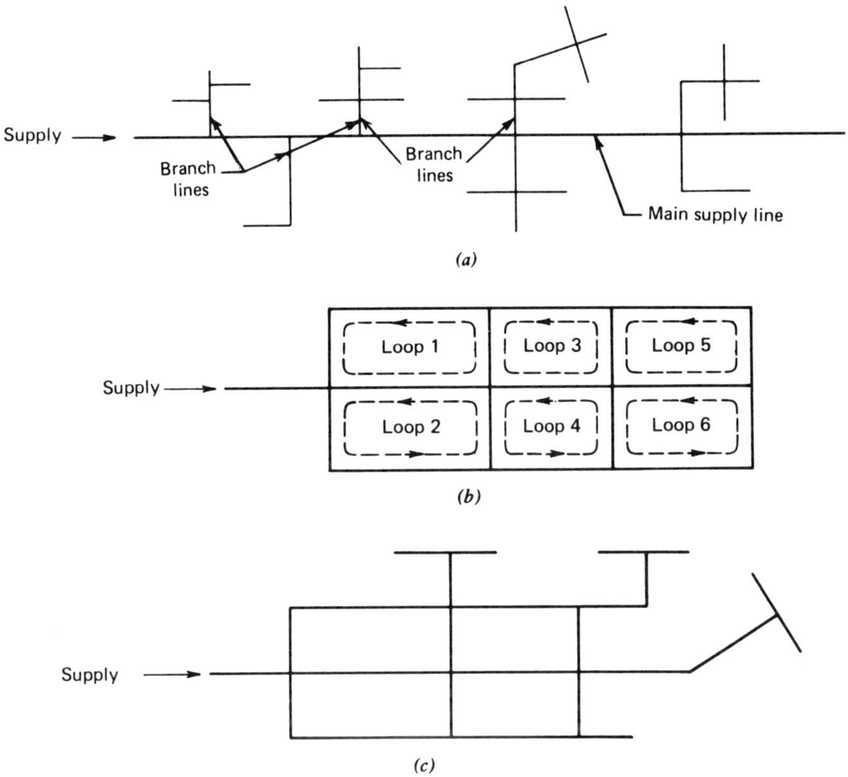

Fig. 6-15. Three different supply system configurations. (*a*) A branch configuration. (*b*) A grid configuration (*c*) A combination of grid and branch configuration.

6.7.1 FLOW ANALYSIS BY THE HARDY–CROSS METHOD

The Hardy–Cross method provides a means for determining the distribution of the flow rates and head losses throughout a network, if the pipe sizes, lengths, and roughnesses are known. To outline the method we need to define two important terms.

1. A *loop* is a series of pipes forming a closed path, as shown in Fig. 6-15*b*. When applying certain loop rules, which will shortly be discussed, a sign convention is needed. We will take the flow rate, Q, and the head loss, h_f, to have positive values, if the flow is in the counterclockwise direction around that loop. It should be realized that a line, which appears in two adjacent loops, may have a positive Q value for one loop and a negative Q value for the other.

2. A *node* is a point where two or more lines are joined. A sign convention also exists for rules applied to a node. For a node rule, we will use the sign convention that a flow rate has a positive value if the flow direction is toward the node.

It should be realized that a flow rate in a line may be considered positive when the loop rule is used, and negative when a node rule is used, or visa versa. The sign convention is strictly arbitrary, but once selected, it should be adhered to. It should also be noted that a similar situation exists for the analysis of electrical networks.

The Hardy–Cross method is an iterative method and is based on two concepts: (1) the law of mass conservation and (2) the fact that the total head at any node is single valued. Concept (1) leads to a node rule that may be stated as follows

$$\sum_{\beta} Q_{\alpha\beta} = 0 \qquad \text{Eq. 6.7-5}$$

In this equation, the first subscript, α, indicates that the flow is either into or out of node α, and the second subscript, β, indicates that the flow is from or to node β. The sign convention, mentioned above, tells us the direction of flow. Concept 2 leads to the loop rule, which may be stated as follows

$$\sum_{j} h_{ij} = 0 \qquad \text{Eq. 6.7-6}$$

where

$$h_{ij} = \text{head loss in the } j\text{th line in the } i\text{th loop}$$

In this equation, the subscript f, which is usually written with h to indicate a head loss due to viscous effects (or apparent viscous effects), has been dropped in order to minimize the number of subscripts. If we apply Eq. 6.5-1 to a closed loop, the validity of the above equation becomes obvious; then, section β is the same as section α, and Eq. 6.7-6 is obtained. It should be noted that minor head losses are usually neglected in network analysis. On the rare occasions when they are included, they are usually added as an additional equivalent length. It should also be observed that elevation differences along a loop cancel and, therefore, these differences do not enter the calculations. Finally, it should be noted that these rules are completely analogous to Kirchoff's rules for electrical circuits. In the analogy, flow rate corresponds to electrical current and pressure drop corresponds to voltage drop.

The procedure for carrying out the Hardy–Cross method is as follows.

1. Subdivide the network into a number of loops. Be sure that all pipes are included in at least one loop.

2. Determine the zeroth estimate for the flow rate, $Q_{\alpha\beta}^{(0)}$, for each line according to the following procedure. Let s equal the total number of nodes in the system and r equal the total number of lines. Invariably, r will be greater than s. Write a node equation, Eq. 6.7-5, for each node, using the sign convention for the node rule. Since $r > s$, there will be more unknowns than equations. If we assume $(r - s)$ $Q_{\alpha\beta}^{(0)}$ values, then the system should reduce to s linear algebraic equations in s unknowns. If the resulting set is linearly independent, then a solution for all other $Q_{\alpha\beta}^{(0)}$ values can be determined. This would give the set of values $\{Q_{\alpha\beta}^{(0)}\}$. Linear independence can be tested by taking the determinant of the coefficient matrix.[4] In many problems, the system of equations turns out to be linearly dependent, and one additional $Q_{\alpha\beta}^{(0)}$ value must be assumed before a set of values $\{Q_{\alpha\beta}^{(0)}\}$ can be established. Note that if the established $Q_{\alpha\beta}^{(0)}$ values are reasonably accurate, the convergence will be quick.

3. Relabel the set of $Q_{\alpha\beta}^{(0)}$ values obtained in step 2 according to the loop rule, giving a set of values $\{Q_{ij}^{(0)}\}$. The index (ij) designates the jth line in the ith loop, where the positive direction is counterclockwise around the loop.

4. Now determine a correction factor, ΔQ_i, for each loop according to the equation[5]

$$\Delta Q_i = -\frac{\sum_{j=1}^{J} K_{ij} Q_{ij}^{(0)} |Q_{ij}^{(0)}|^{n-1}}{n \sum_{j=1}^{J} K_{ij} |Q_{ij}^{(0)}|^{n-1}} \qquad \text{Eq. 6.7-7}$$

   where J is the total number of lines in the loop and K_{ij} equals the constant appearing in Eq. 6.7-3 for the jth line in the ith loop.

5. After obtaining a (ΔQ) for each loop, obtain algebraically a new value for the flow rate in each line; that is,

$$Q_{ij}^{(1)} = Q_{ij}^{(0)} + \Delta Q_i \qquad \text{Eq. 6.7-8}$$

It should be noted that the procedure automatically preserves the node rule. Having new values for the flow rate in each line, we may now repeat the process until each (ΔQ) is less than a certain specified value.

The development of Eq. 6.7-7 for a simple network now follows. Consider the network shown in Fig. 6-16. The Hardy–Cross method requires that we

[4] See Section 5.3 for definition of coefficient matrix. The system is linearly independent if the determinant of the coefficient matrix is nonzero.

[5] This formula is derived for a three-line loop later in this section.

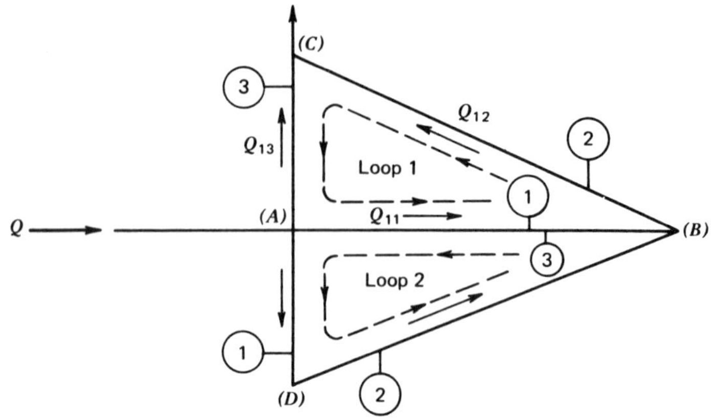

Fig. 6-16. Two three-line loops.

assume a value for the flow rate in each line in each loop. For loop 1, we would have three values, $Q_{11}{}^{(0)}$, $Q_{12}{}^{(0)}$, and $Q_{13}{}^{(0)}$. Since the assumed values for the flow rates will in all likelihood be incorrect for most lines, the sum of the head losses around the loop based on $Q_{ij}{}^{(0)}$ and Eq. 6.7-4 will not be zero; that is,

$$\sum_j h_{ij}{}^{(0)} \neq 0$$

which is a violation of Eq. 6.7-6. However, if a correction, ΔQ_i, is applied to each $Q_{ij}{}^{(0)}$, then a new value for the sum of the head losses, $h_{ij}{}^{(1)}$, around the loop is obtained. Since the sum of the head losses around the loop must be zero for a correct solution, we select ΔQ_i to satisfy this condition; thus,

$$\sum_j h_{ij}{}^{(1)} = 0 \qquad \text{Eq. 6.7-9}$$

Since the Hazen–Williams formula as given by Eq. 6.7-2 does not allow for negative values of Q and our sign convention requires the existence of negative Q values, the formula is restated for both positive and negative Q values; that is,

$$h = \begin{cases} KQ^n, & \text{if } Q \geq 0 \\ -K(-Q)^n, & \text{if } Q < 0 \end{cases}$$

and

$$\frac{dh}{dQ} = \begin{cases} nKQ^{n-1}, & \text{if } Q \geq 0 \\ -nK(-Q)^{n-1}(-1) = nk(-Q)^{n-1}, & \text{if } Q < 0 \end{cases} \qquad \text{Eq. 6.7-10}$$

where, if the fluid is water, $n = 1.852$ and K is given by Eq. 6.7-3. Taking $h = h(Q)$, we can write a Taylor series expansion for h

$$h(Q + \Delta Q) = h(Q) + h'(Q)(\Delta Q) + \frac{h''(Q)}{2!}(\Delta Q)^2 + \cdots \qquad \text{Eq. 6.7-11}$$

For $Q > 0$,

$$h^{(1)} = h(Q + \Delta Q) = KQ^n + nKQ^{n-1}(\Delta Q) + \frac{n(n-1)}{2}KQ^{n-2}(\Delta Q)^2 + \cdots$$

For $Q < 0$,

$$h^{(1)} = h(Q + \Delta Q) = -K(-Q)^n + nK(-Q)^{n-1}(\Delta Q) - \frac{n(n-1)}{2}K(-Q)^{n-2}(\Delta Q)^2 + - \cdots \qquad \text{Eq. 6.7-12}$$

If ΔQ is small, then terms containing (ΔQ) raised to a power of two or greater may be neglected. Now suppose we apply Eqs. 6.7-9, 6.7-10, 6.7-11, and 6.7-12 to loop 1 in Fig. 6-16 and assume that ΔQ is small. Also suppose that the assumed values for Q result in the flow directions as indicated by the arrows around the loop; that is, $Q_{11}{}^{(0)} > 0$, $Q_{12}{}^{(0)} > 0$, and $Q_{13}{}^{(0)} < 0$. Then

$$h_{11}{}^{(1)} + h_{12}{}^{(1)} + h_{13}{}^{(1)} = 0$$

or

$$K_{11}[(Q_{11}{}^{(0)})^n + n(Q_{11}{}^{(0)})^{n-1}\Delta Q_1] + K_{12}[(Q_{12}{}^{(0)})^n + n(Q_{12}{}^{(0)})^{n-1}\Delta Q_1] - K_{13}[(-Q_{13}{}^{(0)})^n - n(-Q_{13}{}^{(0)})^{n-1}\Delta Q_1] = 0$$

Solving for ΔQ_1, we obtain

$$\Delta Q_1 = -\frac{K_{11}(Q_{11}{}^{(0)})^n + K_{12}(Q_{12}{}^{(0)})^n - K_{13}(-Q_{13}{}^{(0)})^n}{n[K_{11}(Q_{11}{}^{(0)})^{n-1} + K_{12}(Q_{12}{}^{(0)})^{n-1} + K_{13}(-Q_{13}{}^{(0)})^{n-1}]}$$

or

$$\Delta Q_1 = -\frac{K_{11}Q_{11}{}^{(0)}|Q_{11}{}^{(0)}|^{n-1} + K_{12}Q_{12}{}^{(0)}|Q_{12}{}^{(0)}|^{n-1} + K_{13}Q_{13}{}^{(0)}|Q_{13}{}^{(0)}|^{n-1}}{n[K_{11}|Q_{11}{}^{(0)}|^{n-1} + K_{12}|Q_{12}{}^{(0)}|^{n-1} + K_{13}|Q_{13}{}^{(0)}|^{n-1}]} \qquad \text{Eq. 6.7-13}$$

We can generalize Eq. 6.7-13 for the ith loop containing J lines, establishing Eq. 6.7-7; that is,

$$\Delta Q_i = -\frac{\sum_{j=1}^{J} K_{ij} Q_{ij}^{(0)} |Q_{ij}^{(0)}|^{n-1}}{n \sum_{j=1}^{J} K_{ij} |Q_{ij}^{(0)}|^{n-1}} \qquad \text{Eq. 6.7-7}$$

The following example illustrates the method.

Example 6-5

A network consisting of two loops is shown in Fig. 6-17. (*a*) If the system is new and made from cast iron, determine the flow rate distribution through the system. (*b*) If the total head at node (A) is 140 ft, determine the total head at the other nodes.

SOLUTION

First, we calculate the K value for each line and tabulate as shown in Table 6-5. Note that for new cast iron, $C = 130$. We can observe that there are seven unknown Q values ($r = 7$) and six nodes ($s = 6$). If we assume one Q value, then the number of unknowns is reduced to six. Furthermore, if the resulting set of equations is nonsingular,[6] then a unique solution for the six unknowns exists and their values can be determined. Suppose, we assume that $Q_{FE}^{(0)} = 1.0$ mgd. The notation used here is the one designated for the node rule; this notation can be defined as follows. The first subscript (in this case F) indicates that the node rule is being applied at node F; the second subscript (in this case E) indicates that the flow is either from or to node E. The sign of Q indicates the direction of flow. The sign convention is that Q is a positive quantity when the flow is into the node (the one at which the rule is being applied), whereas Q is negative if the flow is out of that node. Thus, $Q_{FE}^{(0)} = 1.0$ mgd implies that 1.0 mgd flows through line EF from node E to node F. We may then write the following set of node equations (unknown flow rates are always taken to be positive).

At node F:

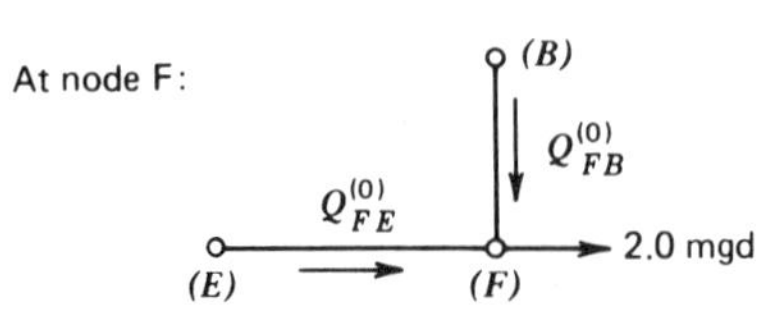

$$\overbrace{Q_{FE}^{(0)}}^{1.0} + Q_{FB}^{(0)} - 2.0 = 0$$

or

$$Q_{FB}^{(0)} = 1.0 \text{ mgd}$$

[6] If the determinant of the coefficient matrix is zero, the set of equations is singular, and no unique solution exists.

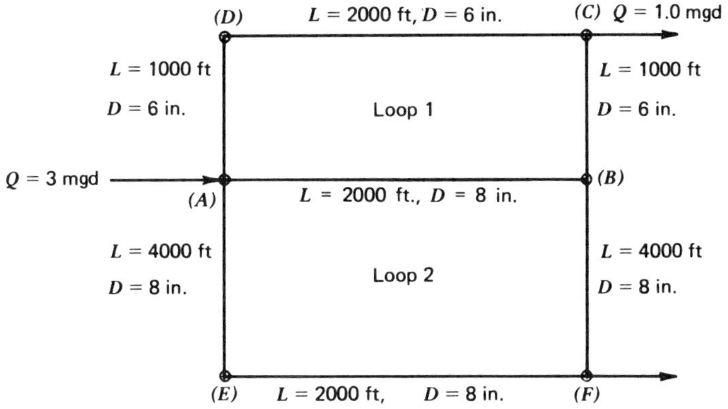

(a)

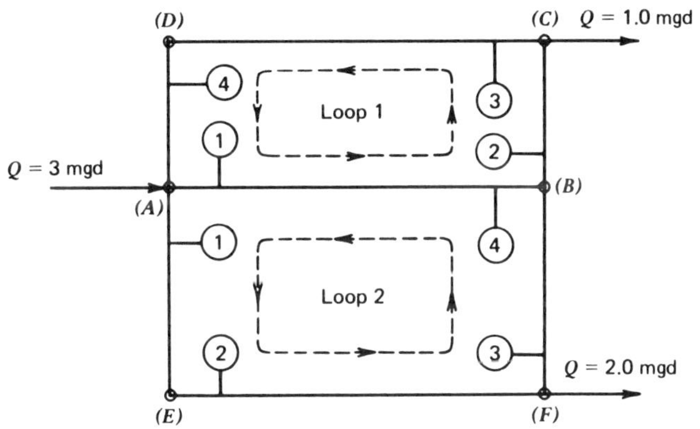

(b)

Fig. 6-17. A two-loop network. (a) Network dimensions. (b) Network line and loop numbers.

TABLE 6-5 Hardy–Cross Computation

Loop No.	Pipe No.	Pipe Diameter (in.)	Pipe Length (ft.)	K	Assumed Q (mgd)	$K\|Q\|^{0.852}$	$KQ\|Q\|^{0.852}$	ΔQ (mgd)
Trial No. 1								
1	1	8	2000	18.625	0.8	15.401	12.321	0.509
	2	6	1000	37.803	−0.2	9.594	−1.919	
	3	6	2000	75.606	−1.2	88.312	−105.974	
	4	6	1000	37.803	−1.2	44.156	−52.987	
					Sums:	157.462	−148.559	
2	1	8	4000	37.251	1.0	37.251	37.251	−0.031
	2	8	2000	18.625	1.0	18.625	18.625	
	3	8	4000	37.251	−1.0	37.251	−37.251	
	4	8	2000	18.625	−0.8	15.401	−12.321	
					Sums:	108.527	6.305	

Loop No.	Pipe No.	K	Assumed Q (mgd)	$K\|Q\|^{0.852}$	$KQ\|Q\|^{0.852}$	ΔQ (mgd)	Assumed Q (mgd)	$K\|Q\|^{0.852}$	$KQ\|Q\|^{0.852}$	ΔQ (mgd)
Trial No. 2							**Trial No. 3**			
1	1	18.625	1.340	23.912	32.061	0.093	1.346	23.999	32.316	0.021
	2	37.803	0.309	13.915	4.306		0.402	17.407	7.005	
	3	75.606	−0.691	55.152	−38.086		−0.598	48.758	−29.136	
	4	37.803	−0.691	27.576	−19.043		−0.598	24.379	−14.568	
			Sums:	120.555	−20.763		*Sums*:	114.543	−4.384	
2	1	37.251	0.969	36.253	35.116	0.087	1.056	39.017	41.198	0.018
	2	18.625	0.969	18.126	17.558		1.056	19.508	20.599	
	3	37.251	−1.031	38.244	−39.444		−0.944	35.470	−33.487	
	4	18.625	−1.340	23.912	−32.061		−1.346	23.999	−32.316	
			Sums:	116.535	−18.831		*Sums*:	117.994	−4.007	
Trial No. 4							**Trial No. 5**			
1	1	18.625	1.349	24.035	32.420	0.004	1.349	24.032	32.412	0.001
	2	37.803	0.423	18.165	7.686		0.427	18.312	7.821	
	3	75.606	−0.577	47.317	−27.298		−0.573	47.037	−26.947	
	4	37.803	−0.577	23.659	−13.649		−0.573	23.518	−13.474	
			Sums:	113.176	−0.842		*Sums*:	112.899	−0.187	
2	1	37.251	1.074	39.594	42.532	0.004	1.078	39.724	42.839	0.001
	2	18.625	1.074	19.797	21.266		1.078	19.862	21.419	
	3	37.251	−0.926	34.382	−32.293		−0.922	34.748	−32.024	
	4	18.625	−1.349	24.035	−32.420		−1.349	24.032	−32.412	
			Sums:	118.307	−0.914		*Sums*:	118.367	−0.179	

At node E:

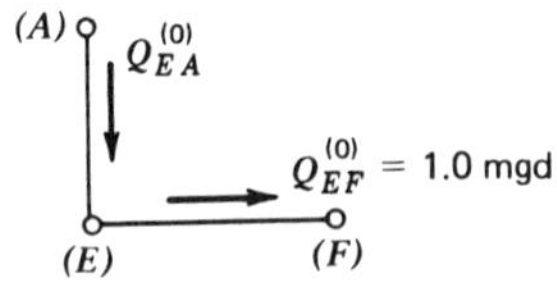

$$Q_{EA}{}^{(0)} + \overbrace{Q_{EF}{}^{(0)}}^{-1.0} = 0$$

or

$$Q_{EA}{}^{(0)} = 1.0 \text{ mgd}$$

At node B:

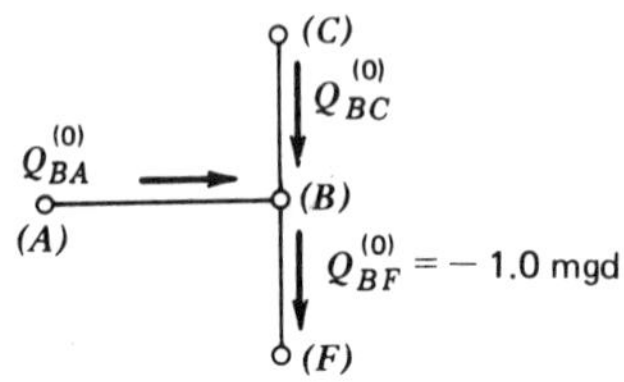

$$Q_{BA}{}^{(0)} + Q_{BC}{}^{(0)} + \overbrace{Q_{BF}{}^{(0)}}^{-1.0} = 0$$

or

$$-Q_{AB}{}^{(0)} + Q_{BC}{}^{(0)} = 1.0 \text{ mgd} \qquad \text{Eq. 6.7-14}$$

At node C:

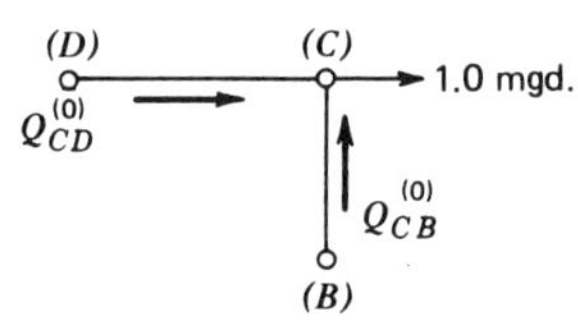

$$Q_{CB}{}^{(0)} + Q_{CD}{}^{(0)} - 1.0 = 0$$

or

$$-Q_{BC}{}^{(0)} + Q_{CD}{}^{(0)} = 1.0 \text{ mgd} \qquad \text{Eq. 6.7-15}$$

At node D:

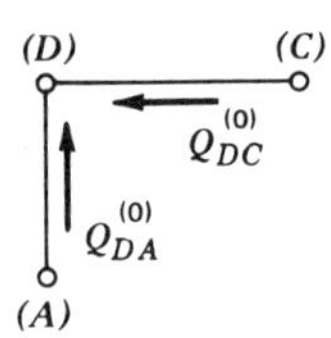

$$Q_{DC}{}^{(0)} + Q_{DA}{}^{(0)} = 0$$

or

$$Q_{AD}{}^{(0)} + Q_{CD}{}^{(0)} = 0 \qquad \text{Eq. 6.7-16}$$

At node A:

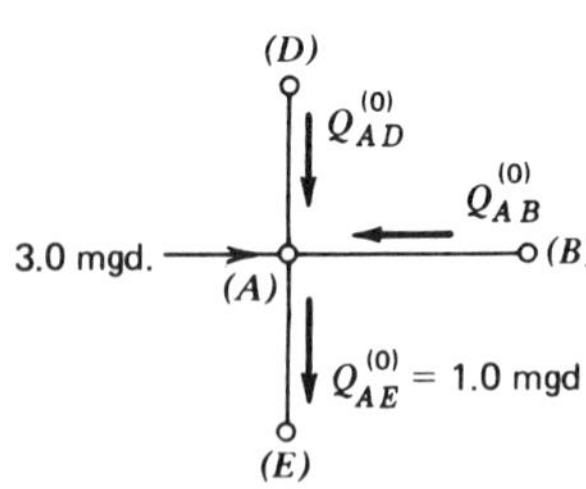

$$Q_{AB}{}^{(0)} + Q_{AD}{}^{(0)} + \overbrace{Q_{AE}{}^{(0)}}^{-1.0} + 3.0 = 0$$

or

$$Q_{AB}{}^{(0)} + Q_{AD}{}^{(0)} = -2.0 \text{ mgd} \qquad \text{Eq. 6.7-17}$$

Adding Eqs. 6.7-14 and 6.7-15 gives

$$-Q_{AB}{}^{(0)} + Q_{CD}{}^{(0)} = 2.0 \text{ mgd} \qquad \text{Eq. 6.7-18}$$

Equations 6.7-16, 6.7-17, and 6.7-18 involve the three variables $Q_{AB}{}^{(0)}$, $Q_{AD}{}^{(0)}$, and $Q_{CD}{}^{(0)}$. That is,

$$0 \cdot Q_{AB}{}^{(0)} + Q_{AD}{}^{(0)} + Q_{CD}{}^{(0)} = 0$$

$$Q_{AB}{}^{(0)} + Q_{AD}{}^{(0)} + 0 \cdot Q_{CD}{}^{(0)} = -2.0$$

$$-Q_{AB}{}^{(0)} + 0 \cdot Q_{AD}{}^{(0)} + Q_{CD}{}^{(0)} = 2.0$$

From linear algebra we know that a unique solution exists, if the determinant of the coefficient matrix is not zero. Let's evaluate this determinant, which is defined

$$\begin{vmatrix} 0 & 1 & 1 \\ 1 & 1 & 0 \\ -1 & 0 & 1 \end{vmatrix} = 0 + (-1)(1 - 0) + 1(0 - 1) = 0$$

Since the determinant is zero, the solution is not unique and the value of one additional variable must be selected.[7] Suppose we select, $Q_{AB}{}^{(0)} = -0.8$ mgd. Then $Q_{AD}{}^{(0)} = -1.2$ mgd, $Q_{CD}{}^{(0)} = 1.2$ mgd, and $Q_{BC}{}^{(0)} = 0.2$ mgd. These established Q values could have been obtained quicker, if we a priori accepted the fact that we needed to assume $(s - r + 1)$ Q values to establish the set. In fact, under this assumption, the Q values could almost have been obtained by observation. That is, having assumed a value for $Q_{FE}{}^{(0)}$, we immediately established $Q_{FB}{}^{(0)}$ and $Q_{EA}{}^{(0)}$. Then, having proceeded to either node (A) or node (B), we would have observed that there were two unknown Q values at these nodes. If we had then assumed that $Q_{AB}{}^{(0)} = -0.8$ mgd, then all other Q values would immediately have been established.

The established $Q_{\alpha\beta}{}^{(0)}$ values are tabulated in Table 6-6, under the heading of Trial No. 1. In this table the loop sign convention was adhered to. Line and loop numbers are shown in Fig. 6-17*b*. Also observe that line 1 in loop 1 is the same line as line 4 in loop 2. Therefore, this line has two corrections applied to it, one from loop 1 and the other from loop 2. Then

$$Q^{(1)}_{\text{line 1}} = Q^{(0)}_{\text{line 1}} + \Delta Q_{\text{loop 1}} - \Delta Q_{\text{loop 2}} \qquad \text{Eq. 6.7-19}$$

The minus sign in front of $\Delta Q_{\text{loop 2}}$ results from the fact that

$$Q_{\text{line 4, loop 2}} = -Q_{\text{line 1, loop 1}} \qquad \text{Eq. 6.7-22}$$

[7] It should be observed that the rank of the coefficient matrix is two, indicating that two of the variables can be expressed in terms of the other.

Examination of Table 6-5 indicates that after 5 trials, the correction to the flow rates in each loop is sufficiently small; therefore, we may take the flow rates, after applying the correction to the assumed Q values in trial No. 5, to be the correct values. We may then determine the head losses in each line using Eq. 6.7-4.

The final Q values in each line and the total head at each node are tabulated in Table 6-6. In this table the total head at node β was obtained by applying Eq. 6.7-4 around a loop in the positive direction. That is, if the loop i starts at node α and node β is another node in the same loop, then

$$(\text{total head})_\beta = (\text{total head})_\alpha - \sum_{j=1}^{\bar{J}} h_{ij}$$

where h_{ij} is the pressure drop in feet in the jth line in the ith loop, and $\bar{J}$ equals the number of lines between node α and node β. Values for h_{ij} are obtained from Eq. 6.7-4. If the Q values are absolutely correct, then the total head at the end of the loop would be the same as at the beginning. Notice that in both loops the error of closure to two significant figures, is 0.00. By using Eq. 6.5-1, the pressures at the nodes can also be determined.

TABLE 6-6 Pressures at the Nodes

Loop No.	Node	Line No.	Q_{ij} (mgd)	K_{ij}	h_{ij} (ft)	Total Head (ft)
	A					140.00
		1	1.349	18.625	32.416	
1	*B*					107.58
		2	0.428	37.803	7.859	
	C					99.73
		3	−0.572	75.606	−26.851	
	D					126.58
		4	−0.572	37.803	−13.425	
	A					140.00
	A					140.00
		1	1.079	37.251	42.914	
	E					97.09
2		2	1.079	18.625	21.457	
	F					75.63
		3	−0.921	37.251	−31.958	
	B					107.59
		4	−1.349	18.625	−32.416	
	A					140.00

Problems

6-1 Water enters a pipeline from a large reservoir. Estimate the length of pipe needed to attain fully developed laminar flow under the following conditions: (*a*) Bulk velocity is 40 ft/min, water temperature is 50 F, and pipe ID is $\frac{1}{2}$ in. (*b*) Bulk velocity is 0.2 m/s, water temperature is 10 C, and pipe ID is 1.25 cm.

6-2 Consider the Saybolt viscometer described in Problem 4-4. In that problem, the bulk velocity, V_2, inside the capillary tube was expressed by the equation

$$V_2(t) = c_1(\sqrt{1 + c_2 h(t)} - 1)$$

where $c_1 = 16\ell\nu/1.14d^2$ and $c_2 = 1.14gd^4/256\ell^2\nu^2$. This expression was based on the assumption that the flow in the capillary tube was a fully developed laminar flow. If the length, L', required to reach a fully developed flow is to be less than or equal to 0.1ℓ when $h(t) = H_0$, find the minimum value for ν to satisfy this condition.

6-3 Using the procedure outlined in Section 5.3 and Eq. 6.3-3, show that $\pi_1 = (g_c H_f)/(\rho V^2)$.

6-4 Find the pressure loss, in pounds per square inch, in a pipe for the flow conditions described in Problem 6-1*a*, if the pipe length is 100 ft.

6-5 Find the pressure loss, in newtons per square meter, for the pipe and flow conditions described in Problem 6-1*b*, if the pipe length is 30 m.

6-6 Water at 50 F flows in a 2-in. ID smooth pipe with an average velocity of 4 ft/s. Find the head loss per 1000 ft of pipe.

6-7 Crude oil, having a specific gravity of 0.86 and a kinematic viscosity of 6×10^{-6} m/s^2, flows through a commercial steel pipeline at a volume flow rate of 5×10^{-4} m^3/s. Find the pressure drop, in newtons per square meter, for a line length of 1 m, if (*a*) the line has a circular cross section, with an ID of 2 cm, (*b*) the line has a square cross section, with 2 cm sides.

6-8 Water flowing in a 1-in. ID pipe, with an average velocity of 10 ft/s, suddenly enters a 3-in. ID pipe. Estimate the head loss for this sudden enlargement.

6-9 Oil flows at a bulk velocity of 3 m/s in a 2-cm ID pipe. If the pipe has a sudden enlargement to 4-cm ID, estimate the resulting head loss.

6-10 Suppose in Problem 6-8, the flow direction was reversed. Find the head loss resulting from the sudden contraction.

6-11 Suppose in Problem 6-9, the flow direction was reversed. Find the head loss resulting from the sudden contraction.

6-12 Suppose that in Example 6-2 (see Fig. 6-11) the water temperature was 10 C, the pipe length was 60 m, and had an ID of 1.25 cm. Determine the elevation of the free surface, z_1, in the tank, if the volume flow rate through the pipe was 285 cm^3/s.

6-13 For the system shown in Fig. 6-18, (*a*) find the sum of the head losses if the discharge through the line is 0.25 m^3/s. (*b*) Determine the flow rate, Q, if the pipe line is cast iron and the valve is a plug disk, fully open globe valve.

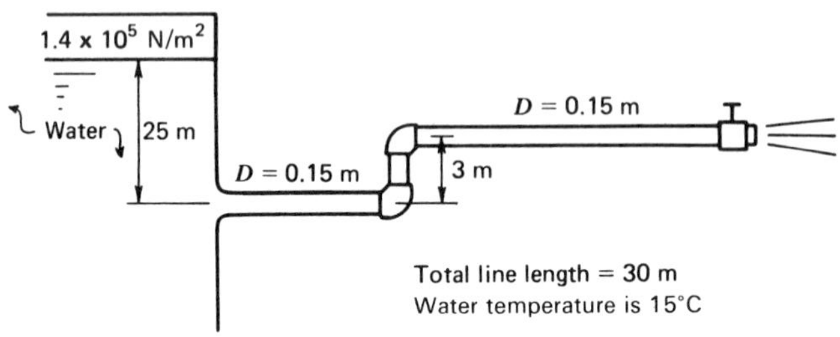

Fig. 6-18. A discharging line.

6-14 A pump moves 0.2 ft^3/s of light oil through a 2-in. ID, cast-iron line, 500 ft long, as shown in Fig. 6-19. Determine the horsepower developed by the pump.

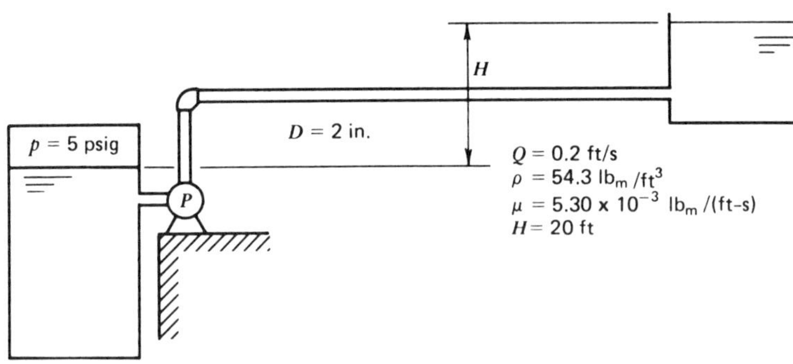

Fig. 6-19. A pump and discharging line.

6-15 Water at 15 C, flows through a 2.5-cm ID pipe, 30 m long, at an average velocity of 4 m/s; then the flow passes through a standard 90° elbow and through a sudden enlargement into a 5-cm ID pipe, 15 m long. The flow then suddenly re-enters a 2.5-cm ID pipe, 15 m long, and then exits from the system through a gate valve, which is half open. If the pipes are all wrought iron, with e equal to 0.046 mm, estimate the pressure drop through the entire system.

6-16 A certain machine, requiring 0.30×10^{-3} m^3/s of water, is to be installed in a building. Suppose a water supply source is located at a distance of 180 m from the machine site. If the pressures at the beginning and end of the connecting line are to be 3.10×10^5 and 1.70×10^5 N/m^2, respectively, what size line should be used to connect the machine to the water supply? Take the water temperature to be 50 F. Assume that the line is to be smooth copper tubing.

6-17 Repeat Problem 6-16, but now the connecting line is to be cast iron.

6-18 Suppose that in Example 6-2 (see Fig. 6-11) z_1 and the pipe length were, respectively, 15 m and 60 m, but the pipe ID was unknown. Find the pipe ID that would result in a flow rate equal to 285 cm^3/s, if the water temperature was 10 C.

6-19 Suppose that in Example 6-2 (see Fig. 6-11) z_1, the pipe ID, and the pipe length were, respectively, 15 m, 1.25 cm, and 60 m. Find the flow rate, Q, if the water temperature is 10 C.

6-20 Suppose the machine described in Problem 6-16 used a smaller volume flow rate when it was operated at less than full load. At this condition the line pressure taps located at the beginning and end of the connecting line indicate pressures of 3.10×10^5 N/m^2 and 2.06×10^5 N/m^2, respectively, If the connecting line is a smooth, copper 2-cm ID tube, determine the volume flow rate through the line. Take the water temperature to be 15 C.

6-21 Use the nomograph, Fig. 6-14, to solve Problem 6-6. Compare this answer with the one obtained using the Darcy formula, as given by Eq. 6.3-7.

6-22 Suppose in Fig. 6-16, the flow rate, $Q_{A,i}$, entering node (A) is 0.056 m^3/s. If the total head at node (A) is 30 m of water, and the lengths and diameters of the lines are as defined in the table that follows, determine the volume flow rate in each line and the total head at each node.

Loop No.	Line No.	Length (m)	Diameter (cm)
	1	1200	15
1	2	1500	15
	3	900	10
	1	900	10
2	2	1500	15
	3	1200	15

6-23 Suppose that in Example 6-5 the pipe network consisted of badly corroded cast iron, with an inflow at node (*A*) of 5 mgd, an outflow at node (*C*) of 2.0 mgd, and an outflow at node (*F*) of 3.0 mgd. Determine the flow-rate distribution and the total head at each node. Take the total head at node (*A*) to be 140 ft.

6-24 Determine the volume flow rate in each line and the pressure at each node for the system shown in Fig. 6-20. Pipe lengths, diameters, and elevations at the nodes are tabulated in the table that follows. The pressure at node (*B*) is 60 psig. Take the Hazen-Williams coefficient, *C*, to be equal to 100. Assume that each line starts and ends at nodes of sufficient volume capacity so that the fluid velocity head at each node may be considered as negligible. Start the iteration by taking $Q_{BA} = -0.833\ \text{ft}^3/\text{s}$.

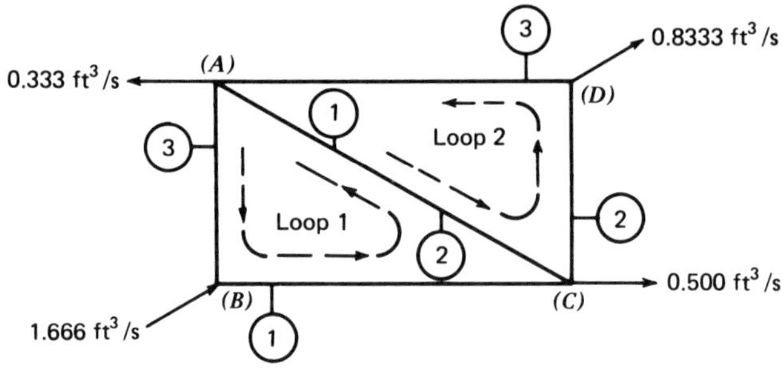

Fig. 6-20. A two-loop network.

Loop No.	Line No.	Length (ft)	Diameter (in.)	Node	Elevation (ft)
	1	4000	8	*A*	30.0
1	2	5000	8	*B*	50.0
	3	3000	6	C	20.0
				D	0.0
	1	5000	8		
2	2	3000	8		
	3	4000	6		

6-25 Determine the volume flow rate in each line and the total head at each node for the system shown in Fig. 6-21. Pipe lengths and diameters are in the table that follows. Take C to be 100 and the total head at node (A) to be 180 ft of water.

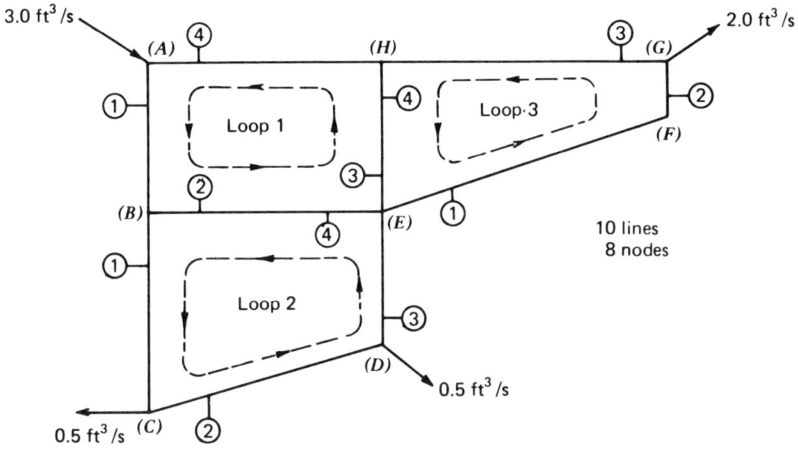

Fig. 6-21. A three-loop network.

Loop No.	Line No.	Length (ft)	Diameter (in.)
1	1	3000	10
	2	6000	6
	3	3000	6
	4	6000	16
2	1	2500	8
	2	6020.8	8
	3	2000	8
	4	6000	6
3	1	6082.8	8
	2	2000	8
	3	6000	12
	4	3000	6

6-26 An experiment consisting of three parts was carried out on the equipment shown in Fig. 6-22. The equipment consists of a circulating pump, a long, smooth circular test pipe, a weighing tank, a reservoir, a pitot tube, and a mercury manometer board. The test pipe has 17 pressure taps that are connected to the manometer

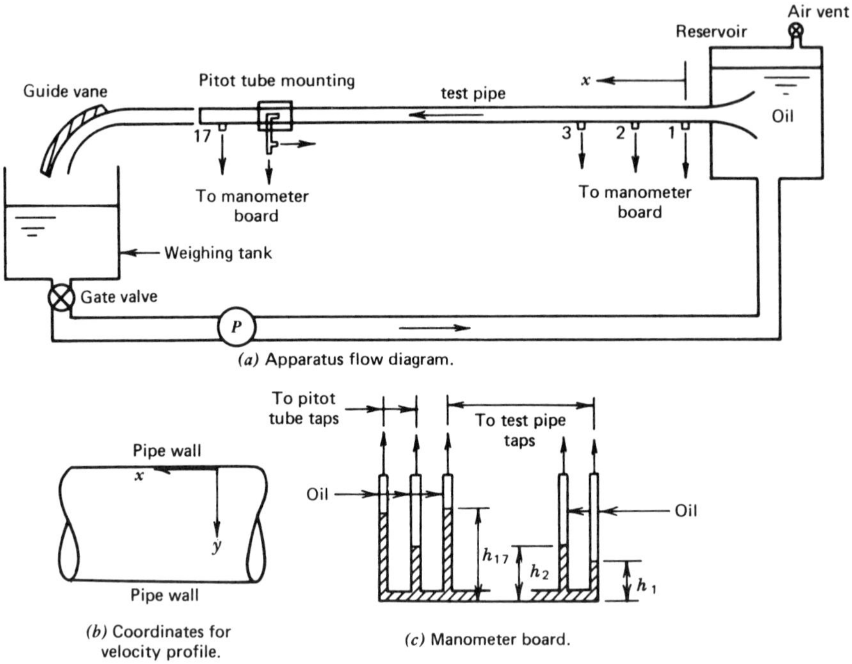

Fig. 6-22. A pipe flow apparatus.

board. The pitot tube is mounted on a device that will permit the pitot tube to traverse the diameter of the test pipe. Pressure taps from the pitot tube are also connected to the manometer board.

The data given below are applicable to all three parts of the experiment.

Fluid is oil at 72 F with the following properties:

Density = 54.8 lb_m/ft^3

Kinematic viscosity = 2.35×10^{-4} ft^2/s.

Pipe diameter = 0.75 in.

PART I: Determination of the pressure drop along the pipe [$(p_1 - p_n)$ versus x, where p_n is the pressure at tap number n and p_1 is the pressure at tap number 1].

Data: Weight of oil collected = 50 lb_f.
Time of collection = 1 min −39 s for laminar flow.
Time of collection = 32.5 s for turbulent flow.

		Manometer Reading	
Tap No.	Position x (in.)	Laminar h (in.)	Turbulent h (in.)
1	0	25.30	12.60
2	6	25.40	13.00
3	12	25.50	13.40
4	18	25.55	13.85
5	24	25.60	14.25
6	30	25.70	14.60
7	36	25.75	14.95
8	42	25.85	15.35
9	48	25.95	15.80
10	54	26.00	16.20
11	66	26.15	17.10
12	78	26.30	17.90
13	90	26.40	18.75
14	102	26.55	19.55
15	132	26.90	21.70
16	162	27.30	23.80
17	192	27.65	24.80

a. Write a computer program that will

1. Read in the data.
2. Calculate and write out the flow rate, Q, and the Reynolds number, Re.
3. Calculate $(p_1 - p_n)$ in pounds per square inch.
4. Tabulate: the tap No., the x position, and $(p_1 - p_n)$ for both the laminar and turbulent flow cases.

b. Plot results (this may be done by hand).

PART II: Determination of velocity profile (local velocity versus y, where y is the distance across a pipe diameter). See Fig. 6-22*b*.

Data: Weight of oil collected = 50 lb_f.
Time of collection = 1 min – 39 s for laminar flow.
Time of collection = 32.5 s for turbulent flow.

Position y (in.)	Manometer Readings: Laminar Δh (in.)	Manometer Readings: Turbulent Δh (in.)
0.05	0.10	1.60
0.15	0.20	2.15
0.25	0.40	2.40
0.30	0.45	2.40
0.40	0.50	2.20
0.50	0.45	2.10
0.60	0.20	1.65
0.70	0.10	0.90

a. Write a computer program that will

1. Read in the data.
2. Calculate and write out the flow rate, Q, and the Reynolds number, Re.
3. Compute the local velocity.
4. Tabulate: the y position and the local velocities for both the laminar and turbulent flow cases.

b. Plot results (this may be done by hand).

PART III: Determination of friction factor, f, versus Re for the laminar flow case.

Data: Weight of oil collected = 50 lb_f
$H_f = p_{10} - p_{17}$.
$\Delta h = h_{17} - h_{10}$.
The flow is laminar.

Trial No.	Time of Collection (s)	Manometer Reading Δh (in.)
1	149	1.10
2	123	1.40
3	95	1.60
4	82	1.90
5	72	2.15

a. Write a computer program that will

1. Read in the data.
2. Calculate: the Reynolds number, Re, the experimental friction factor, f_{exp}, and the theoretical friction factor, f_{th}.
3. Tabulate: the trial number, time of collection, Re, f_{exp}, and f_{th}.

b. Plot f_{th} and f_{exp} versus Re on log-log paper. Both curves should be on the same graph. This may be done by hand.

6-27 Write a *computer program* that will determine: (*a*) the flow-rate distribution in a loop network system by the Hardy–Cross method and (*b*) the total head at each node. Assume that starting values for Q_{ij} will be read into the program. Allow for a maximum of six loops with a maximum of six lines in a loop. Assume that all lines are made of the same material and are in the same condition. In writing your program, assume that the starting Q values for the Hardy–Cross iteration scheme will be read into the program. Write your program so that part (*a*) is first completed before it proceeds to part (*b*). Using the following notation list, try the program out on Problem 6-25. Take the maximum allowable difference in flow rates between iterations to be 0.0001.

List of Symbols for Computer Program

NL	=	Number of loops.
NJ(I)	=	Number of lines in Ith loop.
NMAXL	=	Maximum number of lines in a single loop.
ZL(I,J)	=	Length of Jth line in Ith loop.
EPSLN	=	Maximum allowable difference in flow rates between iterations (use 0.0001).
C	=	Hazen-Williams constant for a pipe (all pipes assumed to be made of same material).
ZK(I,J)	=	K value for Jth line in Ith loop (see Eq. 6.7-3).
DI(I,J)	=	Diameter of Jth line in Ith loop, in inches.
D(I,J)	=	Diameter of Jth line in Ith loop, in feet.
Q(I,J)	=	Volume-flow rate in Jth line in Ith loop.

HF(I,J) = Pressure drop in Jth line in Ith loop.
H(I,K) = Pressure at Kth node in Ith loop.
DELQ(I) = Flow-rate correction applied to each line in Ith loop.
ID(I,J) = Index for Jth line in Ith loop. (Its value equals zero if line (I,J) is *not* a common line, and equals K if line (I,J) is in common with a line in loop K.)
L2(I) = The node number in loop (I − 1) in common with node (I,1)

The determination of the total head at each node requires that the calculations proceed in consecutive order from node to node and from loop to loop. The following rules for labeling loops and nodes is helpful in this regard.

Rules for Labeling Loops and Nodes

Select loop 1 to contain the node where the total head is given. Label this node as node (1, 1). Select line (1, 1) to contain node (1, 1) on its upstream end, according to the positive loop direction (see Fig. 6-23). Nodes are then numbered consecutively around a loop in the

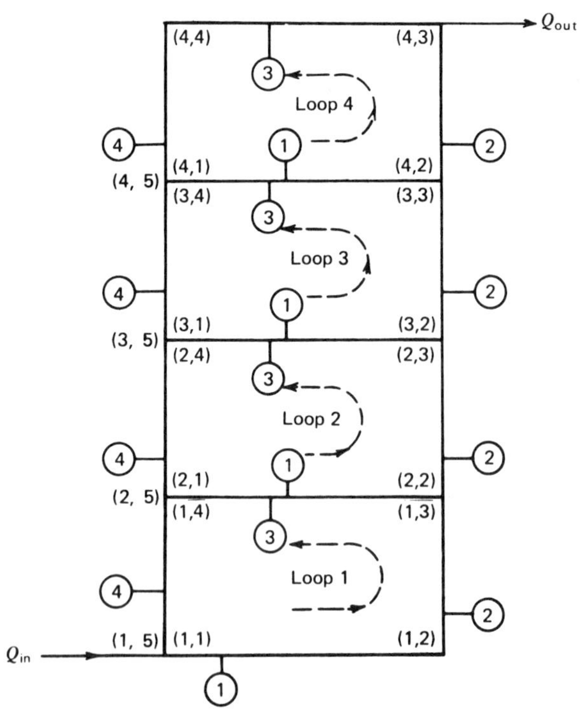

Fig. 6-23. A four-loop network. Total head given at node (1,1).

positive direction. The first node in the loop will have two numbers 1 and [NJ(1) + 1]; use 1 when considering line 1 and use [NJ(1) + 1] when considering line numbered NJ(1). Loop 2 is selected to be adjacent to loop 1 and is to contain at least one common node with loop 1. This common node for loop 2 calculations is to be designated as node (2, 1). The lines and nodes in this loop are then designated in the same manner as that described for loop 1 (see Fig. 6-23). Index LL(2) equals the node number in loop 1 that is in common with node (2, 1). The remaining loop and node numbers are selected in the same manner. It should be noted that, if the node where the total head is given is an interior node point, then it may be necessary to slightly modify the progression.

Chapter Seven

An Introduction to Open-Channel Flow

7.1 INTRODUCTION

A flow is classified as an open-channel flow when the liquid is only partially surrounded by a solid boundary, its top surface being open to the atmosphere. The flow in rivers, canals, drainage and irrigation channels, partially filled culverts, and the like, are examples of open-channel flow (see Fig. 7-1). In almost all cases of interest, the liquid in question is water and the flow is turbulent.

In open-channel flow, the forces causing the flow are gravitational; whereas viscous shear tends to retard the flow. In channels of uniform cross section, the slope of the channel flow strongly influences the depth and the flow rate through the channel. Flows may be classified as steady or unsteady, uniform or nonuniform. Uniform flow can be achieved only when the cross-sectional area of the channel and the slope of the floor are both constant over a sufficiently long distance. Suppose we consider a channel of uniform cross section. If the slope of the channel floor of a particular section is steeper than the slope of an adjacent upstream section, then the flow between the sections accelerates, causing the depth to decrease, as shown in Fig. 7-2*a*. For the reverse situation, the flow decelerates, causing the depth to increase, as shown in Fig. 7-2*b*. Obstructions in a channel, such as barriers, dams, weirs, and spillways, also cause nonuniform flow conditions. A nonuniform flow may be gradually varying or rapidly varying. For example, in a flow over a spillway, the flow upstream from the spillway is usually a gradually varying flow, while the flow over the spillway is a rapidly varying flow. In a uniform and a gradually varying

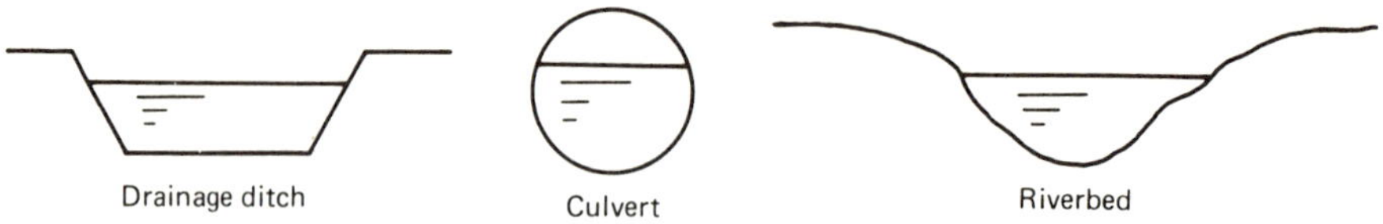

Fig. 7-1. Examples of open-channel flow.

flow, the pressure field within the flow is essentially hydrostatic. In a rapidly varying flow the streamlines have significant curvature and the pressure field is more complicated. When the slope of the floor changes from a steep to a less steep condition, a phenomenon known as a *hydraulic jump* may occur (see Fig. 7-2*d*). The conditions that cause a hydraulic jump to take place are discussed later in this section. The hydraulic jump is somewhat analogous to a shock wave, which occurs in a gaseous, compressible flow. Shock waves are discussed in Chapter 8. We will find that the Froude number plays the same type of role in open-channel flow as the Mach number plays in compressible gas flows.

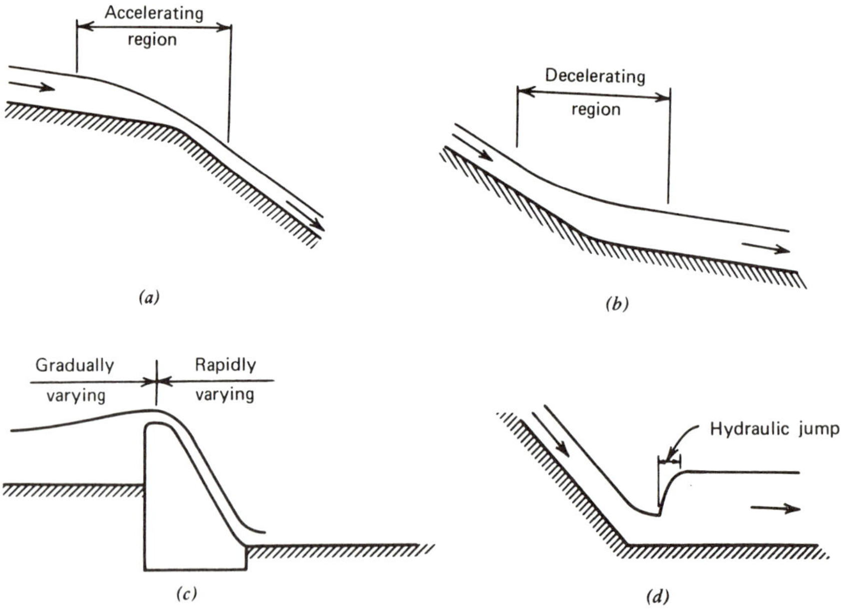

Fig. 7-2. Several types of open-channel flow. (*a*) Accelerating flow. (*b*) Decelerating flow. (*c*) Flow over a spillway. (*d*) Hydraulic jump.

7.2 ONE-DIMENSIONAL ANALYSIS IN A NONVISCOUS OPEN-CHANNEL FLOW

Because of the complex nature of the real velocity field found in open-channel flow, it is convenient (as was done in pipe flow) to make a one-dimensional analysis of the problem. In our analysis we will assume that the fluid velocity is uniform at each section. Velocity values are thus bulk or average values. Recall that in pipe flow (Chapters 4 and 6), we assumed in our one-dimensional model that all fluid properties were uniform at each section. In open-channel flow we account for the vertical (z) variation both in the pressure field and in the potential energy term appearing in the first law of thermodynamics. We do this because of the large depths that frequently occur in open-channel flow, and because the inclusion of these effects does not overly complicate the analysis.

The flow rate through a channel can be measured by inserting a *venturi flume* in the system. We will assume that the flume has vertical side walls and a horizontal floor. As in a pipeline venturi, the flume venturi has a reduced cross-sectional area at the throat. This is accomplished by reducing the width at the throat section, which then affects the liquid depth and the fluid velocity in the flume (see Fig. 7-3).

We can analyze the flume by assuming the conditions described in the previous paragraph and by neglecting viscous effects. Select a control volume

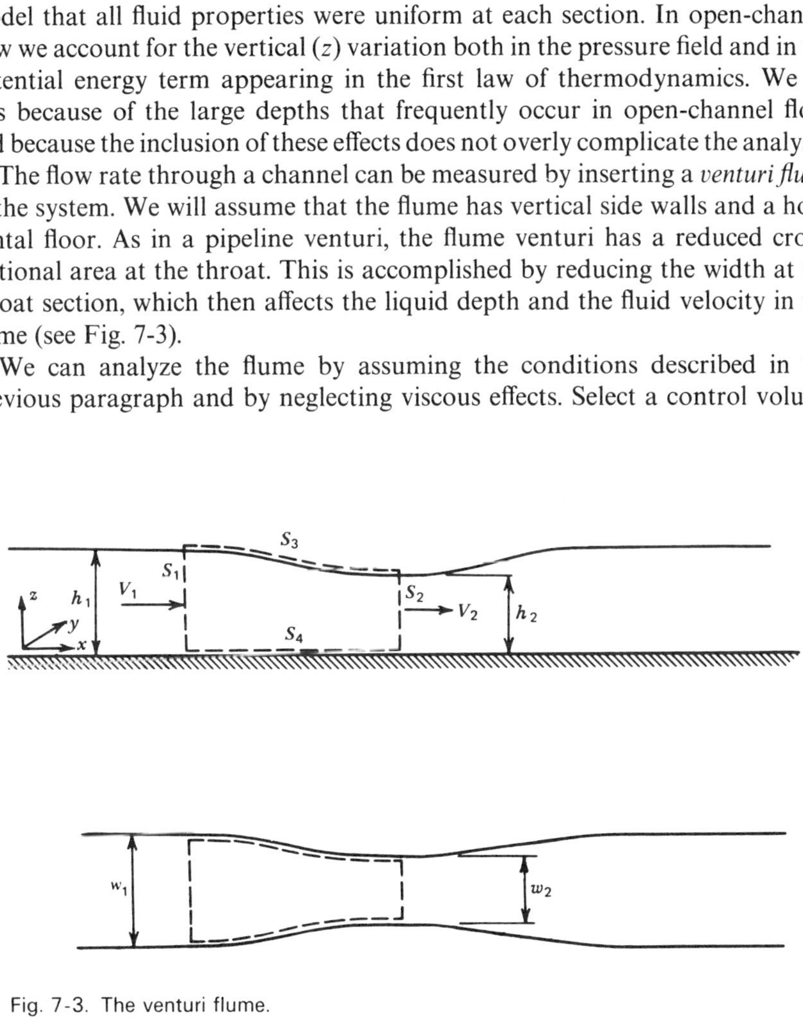

Fig. 7-3. The venturi flume.

as shown by the dashed line in Fig. 7-3, and apply the first law of thermodynamics (Eq. 4.6-11) to the control volume. For steady flow this gives

$$-\oiint_{S_1} \rho\left[\frac{V_1^2}{2g_c} + \frac{gz}{g_c}\right] V_1 \mathbf{i}\cdot\mathbf{i}\, dS_1 + \oiint_{S_2} \rho\left[\frac{V_2^2}{2g_c} + \frac{gz}{g_c}\right] V_2 \mathbf{i}\cdot\mathbf{i}\, dS_2$$

$$= \oiint_{S_1} p\mathbf{i}\cdot\mathbf{i}V_1\, dS_1 - \oiint_{S_2} p\mathbf{i}\cdot\mathbf{i}V_2\, dS_2$$

Taking the pressure at sections S_1 and S_2 to be hydrostatic, $dS_1 = w_1 dz$ and $dS_2 = w_2 dz$, the above equation becomes

$$-\frac{\rho V_1^2 V_1 w_1}{2g_c}\int_0^{h_1} dz - \frac{\rho V_1 w_1 g}{g_c}\int_0^{h_1} z\, dz + \frac{\rho V_2^2 V_2 w_2}{2g_c}\int_0^{h_2} dz + \frac{\rho V_2 w_2 g}{g_c}\int_0^{h_2} z\, dz$$

$$= V_1 w_1 \int_0^{h_1} \gamma(h_1 - z)\, dz - V_2 w_2 \int_0^{h_2} \gamma(h_2 - z)\, dz$$

Then

$$-\frac{\rho V_1^2 V_1 w_1 h_1}{2g_c} - \frac{\rho g V_1 w_1 h_1^2}{2g_c} + \frac{\rho V_2^2 V_2 w_2 h_2}{2g_c} + \frac{\rho g V_2 w_2 h_2^2}{2g_c}$$

$$= \frac{V_1 w_1 \rho g}{g_c}\left(h_1^2 - \frac{h_1^2}{2}\right) - \frac{V_2 w_2 \rho g}{g_c}\left(h_2^2 - \frac{h_2^2}{2}\right) \qquad \text{Eq. 7.2-1}$$

Also, from conservation of mass (Eq. 4.4-4 is also applicable here), we have

$$\rho V_1 h_1 w_1 = \rho V_2 h_2 w_2 = \dot{m} \qquad \text{Eq. 7.2-2}$$

Since ρ is a constant, Eq. 7.2-2 reduces to

$$V_1 h_1 w_1 = V_2 h_2 w_2 = Q \qquad \text{Eq. 7.2-3}$$

Dividing both sides of Eq. 7.2-1 by $\dot{m}$, we obtain

$$-\frac{V_1^2}{2g_c} - \frac{g}{g_c}\frac{h_1}{2} + \frac{V_2^2}{2g_c} + \frac{g}{g_c}\frac{h_2}{2} = \frac{gh_1}{2g_c} - \frac{gh_2}{2g_c}$$

By combining and rearranging, we find that

$$\frac{V_1^{\,2}}{2} + gh_1 = \frac{V_2^{\,2}}{2} + gh_2 \qquad \text{Eq. 7.2-4}$$

Solving Eq. 7.2-3 for V_2 and substituting the resulting relation into Eq. 7.2-4, we obtain

$$V_1^{\,2} = \frac{2gh_2^{\,2}w_2^{\,2}(h_2 - h_1)}{h_2^{\,2}w_2^{\,2} - h_1^{\,2}w_1^{\,2}} \qquad \text{Eq. 7.2-5}$$

Thus, by knowing w_1 and w_2 and measuring h_1 and h_2, we can determine V_1. Then Q can be determined from Eq. 7.2-3.

A more general analysis than that given above can be made for a variable width channel. Suppose we consider a variable width channel that has a horizontal floor and vertical walls. We will consider section 1 to be fixed and section 2 to represent some arbitrary section, located at position x. Then Eqs. 7.2-3 and 7.2-4 will still be valid. To simplify the notation, we will take

$$h = h_2, \qquad V = V_2, \qquad \text{and} \qquad w = w_2$$

Then Eqs. 7.2-3 and 7.2-4, respectively, become

$$Vhw = \text{constant}$$

$$\frac{V^2}{2} + gh = \text{constant}$$

In differential form, these equations reduce to

$$Vh\,dw + Vw\,dh + hw\,dV = 0 \qquad \text{Eq. 7.2-6}$$

$$V\,dV + g\,dh = 0 \qquad \text{Eq. 7.2-7}$$

Dividing each term in Eq. 7.2-6 by Vhw, we obtain

$$\frac{dw}{w} + \frac{dh}{h} + \frac{dV}{V} = 0 \qquad \text{Eq. 7.2-8}$$

Rearranging Eq. 7.2-7, we find that

$$\frac{dh}{dV} = -\frac{V}{g} \qquad \text{Eq. 7.2-9}$$

We see that as V increases in a channel, h decreases, and vice versa. To obtain a similar relationship between dV and dw, substitute Eq. 7.2-8 into Eq. 7.2-7 and rearrange

$$\frac{V}{gh}dV - \frac{dw}{d} - \frac{dV}{V} = 0$$

or

$$\left(\frac{V^2}{gh} - 1\right)dV = \frac{V}{w}dw \qquad \text{Eq. 7.2-10}$$

Recall from Chapter 5 that the Froude number, $\mathscr{F}$, was defined by

$$\mathscr{F} = \frac{V^2}{gh}$$

Also, recall that the Froude number is an important parameter whenever gravitational forces are significant in determining the behavior of some particular phenomenon. That is exactly the case here. Then Eq. 7.2-10 becomes

$$dV = \frac{V}{w}\frac{1}{\mathscr{F} - 1}dw \qquad \text{Eq. 7.2-11}$$

We see that if $\mathscr{F} < 1$, V decreases as w increases, and V increases as w decreases. When $\mathscr{F} < 1$, the flow is said to be *subcritical.* If $\mathscr{F} > 1$, then V increases as w increases, and V decreases as w decreases. When $\mathscr{F} > 1$, the flow is said to be *supercritical.* The flow is *critical* when $\mathscr{F} = 1$. These conclusions are summarized in Fig. 7-4. Note that the Froude number plays the same role here that the Mach number plays in one-dimensional compressible flow (see Section 8.3).

Another example in which frictionless flow may be considered is in the flow established by a *sluice gate,* a gate used to control the discharge through a channel. We will assume that the channel downstream from the sluice gate is rectangular, having the same width as the sluice gate (see Fig. 7-5).

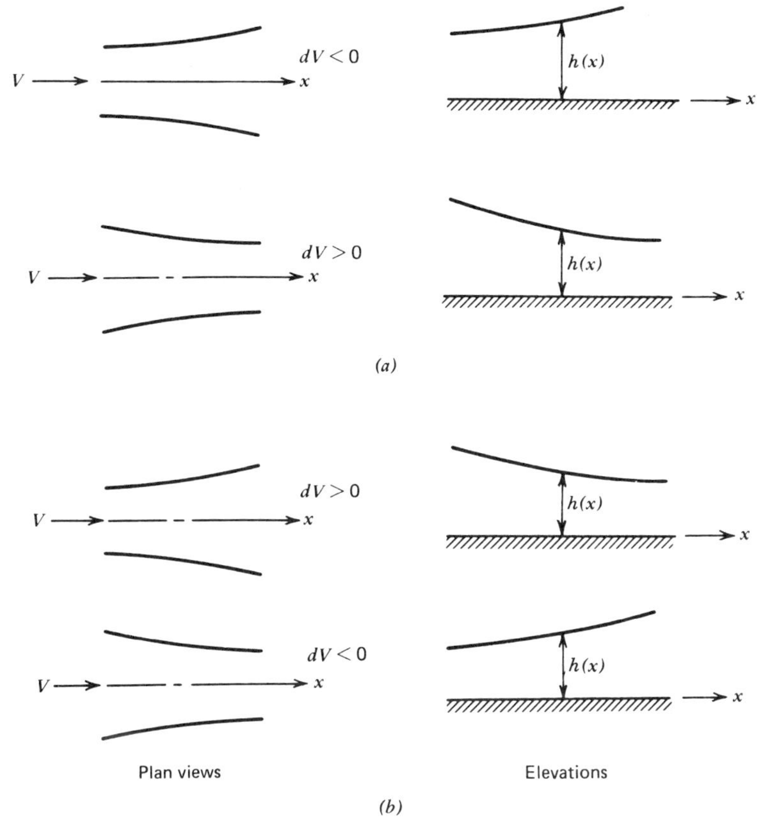

Fig. 7-4. Variation in depth versus variation in channel width (floor horizontal). (a) $\mathcal{F} < 1$. (b) $\mathcal{F} > 1$.

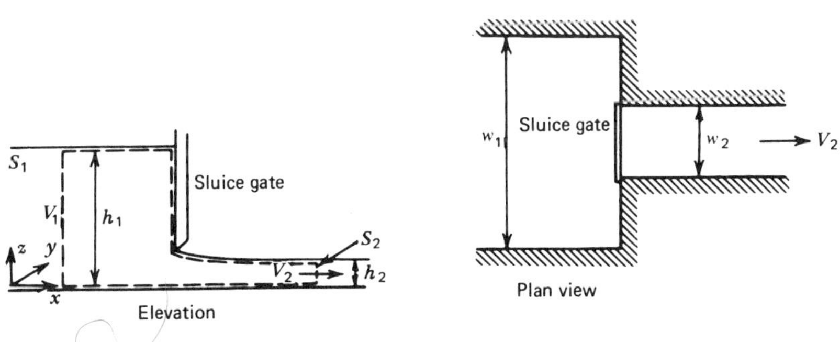

Fig. 7-5. Flow through a sluice gate.

Select a control volume as shown by the dashed line in Fig. 7-5, and apply the first law, of thermodynamics to the control volume (Eq. 4.6-11). If we again assume that the velocity field is uniform at S_1 and S_2 and take the pressure field to be hydrostatic, then Eq. 4.6-11 reduces to Eq. 7.2-4. In sluice gate problems, surface S_1 is so much larger than S_2 that V_1 becomes negligible compared to V_2. Thus, Eq. 7.2-4 becomes

$$gh_1 = \frac{V_2^2}{2} + gh_2$$

Solving Eq. 7.2-4 for V_2 in terms of Q and substituting the resulting equation into the one above, we obtain

$$gh_1 = \frac{Q^2}{2h_2^2 w_2^2} + gh_2$$

Rearranging gives

$$\left(\frac{Q}{w_2}\right)^2 = 2gh_2^2(h_1 - h_2)$$

To simplify the notation, let us set

$$h = h_2 \qquad \text{and} \qquad w = w_2$$

then

$$\left(\frac{Q}{w}\right)^2 = 2gh^2(h_1 - h) = 2gh_1^3\left(\frac{h}{h_1}\right)^2\left(1 - \frac{h}{h_1}\right) \qquad \text{Eq. 7.2-12}$$

Since raising or lowering the gate changes the volume flow rate through the channel, it is of interest to examine if there exists a relative maximum for Q when h_1 and w are held constant. Suppose we take our independent variable to be the nondimensionalized channel depth, ζ, where

$$\zeta = h/h_1$$

Then, Eq. 7.2-12 becomes

$$\left(\frac{Q}{w}\right)^2 = 2gh_1^3(\zeta^2 - \zeta^3) \qquad \text{Eq. 7.2-13}$$

To find a relative maximum for Q, we take the derivative of both sides of Eq. 7.2-13 with respect to ζ. That is,

$$2\left(\frac{Q}{w}\right)\frac{dQ}{d\zeta} = 2gh_1{}^3(2\zeta - 3\zeta^2) = 2gh_1{}^3\zeta(2 - 3\zeta)$$

The Q variable has a relative extremum where $dQ/d\zeta = 0$. Obviously, $\zeta = 2/3$ is the desired answer. Substituting $\zeta = 2/3$ into Eq. 7.2-13, we find that

$$\left(\frac{Q}{w}\right)^2_{\max} = \frac{8}{27}gh_1{}^3 \qquad \text{Eq. 7.2-14}$$

Let us now examine the Froude number when the flow through the sluice gate is a maximum. Recall that

$$\mathscr{F} = \frac{V^2}{gh} = \frac{Q^2}{hg(wh)^2} \qquad \text{Eq. 7.2-15}$$

Substituting Eq. 7.2-14 into the above equation, we obtain

$$(\mathscr{F})_{Q_{\max}} = \left(\frac{8gh_1{}^3}{27gh^3}\right)_{Q_{\max}} = \frac{8}{27}\left(\frac{h_1}{h}\right)^3_{Q_{\max}}$$

Since $\zeta = h/h_1 = 2/3$ when Q is a maximum, the above equation reduces to

$$(\mathscr{F})_{Q_{\max}} = \left(\frac{8}{27}\right)\left(\frac{3}{2}\right)^3 = 1$$

Thus, when Q is a relative maximum, the flow just downstream from the sluice gate is critical. A plot of $Q^2/(2gw^2h_1{}^3)$ versus h/h_1 is shown in Fig. 7-6. We see that as we raise the gate from the closed position (holding h_1 constant), the flow rate through the channel increases until h/h_1 reaches 2/3. Then, any further rise in the gate position causes the flow rate to decrease. Equation 7.2-12 is valid only for $h/h_1 < 1$. To see that this is so, note that if the bottom of the sluice gate lies above the free surface and $w_1 = w$, then $h/h_1 = 1$ and $V/V_1 = 1$. Since V_1 was neglected in developing Eq. 7.2-12, the equation is not valid for $h/h_1 = 1$. If, on the other hand, the bottom of the gate lies above the free surface and $w_1 \gg w$, then the channel acts as a flume and $h/h_1 < 1$, as shown in Fig. 7-3.

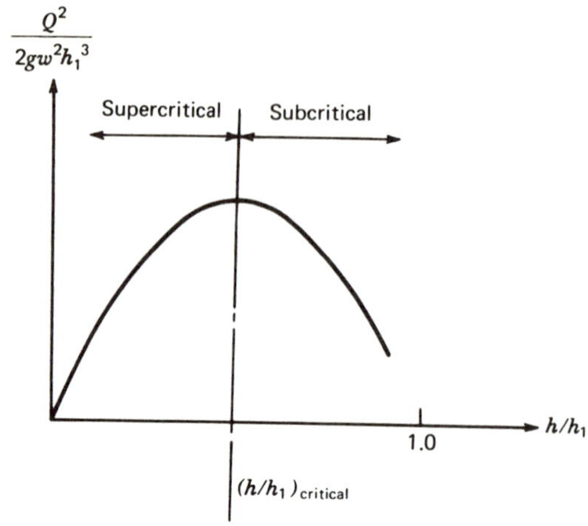

Fig. 7-6. A plot of $Q^2/(2gw^2h_1{}^3)$ versus h/h_1.

We can readily show that if $h/h_1 < 2/3$, then in the region just downstream from the sluice gate $\mathscr{F} > 1$. Thus, the flow there is supercritical. If $h/h_1 > 2/3$, then $\mathscr{F} < 1$ and the flow is subcritical. It is observed that when the flow is supercritical, a *hydraulic jump* takes place a short distance downstream from the sluice gate, as shown in Fig. 7-7*a*. We have assumed here that the channel floor under the sluice gate is horizontal. Conditions downstream of the jump, in terms of upstream conditions, can be obtained by applying the basic equations (conservation of mass, Newton's second law, and the first law of thermodynamics) to the idealized control volume, shown in Fig. 7-7*b*. Conservation of mass gives

$$V_1 h_1 w = V_2 h_2 w$$

or

$$V_1 h_1 = V_2 h_2 \qquad \text{Eq. 7.2-16}$$

Applying Eq. 4.7-4 to the control volume and dotting the resulting equation with the unit vector **i**, we obtain

$$\int_0^{h_1} pw\,dz - \int_0^{h_2} pw\,dz = -\int_0^{h_1} \frac{\rho}{g_c} V_1{}^2 w\,dz + \int_0^{h_2} \frac{\rho}{g_c} V_2{}^2 w\,dz$$

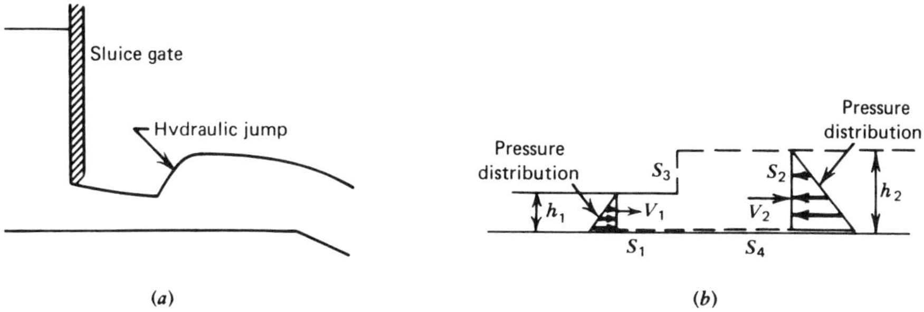

Fig. 7-7. Hydraulic jump. (*a*) A real case. (*b*) An idealized model.

Here, we have assumed that the x-component of the fluid-velocity vector on S_3 is zero. Noting that $p = \gamma(h_1 - z)$ on S_1 and $p = \gamma(h_2 - z)$ on S_2, we can integrate the above equation, giving

$$\frac{w\gamma h_1^2}{2} - \frac{w\gamma h_2^2}{2} = -\frac{\rho}{g_c} V_1^2 w h_1 + \frac{\rho}{g_c} V_2^2 w h_2 \qquad \text{Eq. 7.2-17}$$

Substituting Eq. 7.2-16 into the above equation, rearranging, and taking into account that $\gamma = \rho g/g_c$, we obtain

$$\frac{g}{2}(h_1^2 - h_2^2) = \frac{g}{2}(h_1 - h_2)(h_1 + h_2) = \frac{V_1^2 h_1(h_1 - h_2)}{h_2}$$

or

$$(h_1 - h_2)\left[\frac{g}{2}(h_1 + h_2) - \frac{V_1^2 h_1}{h_2}\right] = 0$$

Either

$$h_1 = h_2 \qquad \text{or} \qquad \frac{g}{2}(h_1 + h_2) - \frac{V_1^2 h_1}{h_2} = 0$$

The first solution is not of interest. The second gives

$$h_2^2 + h_1 h_2 - \frac{2}{g} h_1 V_1^2 = 0 \qquad \text{Eq. 7.2-18}$$

Solving for h_2 and taking the positive root, we find that

$$h_2 = \frac{h_1}{2}\left(\sqrt{1 + \frac{8V_1^{\,2}}{gh_1}} - 1\right) = \frac{h_1}{2}(\sqrt{1 + 8\mathscr{F}_1} - 1) \qquad \text{Eq. 7.2-19}$$

If $(V_1)^2/(gh_1) > 1$, then $h_2 > (3 - 1)h_1/2 = h_1$. If $(V_1)^2/(gh)_1 < 1$, then $h_2 < h_1$. There is nothing in either Newton's second law or the law of mass conservation that indicates that only the first situation is a possible one. That is, that

$$\frac{V_1^{\,2}}{gh_1} = \mathscr{F}_1 > 1$$

It is the second law of thermodynamics that gives us this information. Let us first apply the first law to the control volume shown in Fig. 7-7*b*. Since the hydraulic jump involves an irreversible process, the fluid does undergo a change in its internal energy. Application of Eq. 4.6-9 to the control volume, shown in Fig. 7-7*b*, gives

$$-\int_0^{h_1} \rho\left(u_1 + \frac{V_1^{\,2}}{2g_c} + \frac{gz}{g_c}\right)V_1\mathbf{i}\cdot\mathbf{i}w\,dz + \int_0^{h_2} \rho\left(u_2 + \frac{V_2^{\,2}}{2g_c} + \frac{gz}{g_c}\right)V_2\mathbf{i}\cdot\mathbf{i}w\,dz$$

$$= \int_0^{h_1} p\mathbf{i}\cdot\mathbf{i}V_1 w\,dz - \int_0^{h_2} p\mathbf{i}\cdot\mathbf{i}V_2 w\,dz$$

Here we have taken $v_1 = 0$ on S_3. By taking $p = \gamma(h_1 - z)$ on S_1 and $p = \gamma(h_2 - z)$ on S_2, carrying out the integration, dividing through by $\dot{m} = \rho V_1 w h_1 = \rho V_2 w h_2$, multiplying through by g_c/g, and rearranging, the above equation reduces to

$$\frac{Jg_c}{g}(u_2 - u_1) = \frac{V_1^{\,2} - V_2^{\,2}}{2g} + (h_1 - h_2)$$

The term $(Jg_c/g)(u_2 - u_1)$ is the term designated as the head loss, h_L. The head loss represents a conversion of mechanical energy to thermal energy. Thus

$$h_L = \frac{V_1^{\,2} - V_2^{\,2}}{2g} + (h_1 - h_2) \qquad \text{Eq. 7.2-20}$$

We can express h_L in terms of h_1 and h_2 exclusively, by substituting Eq. 7.2-16 into Eq. 7.2-20. This gives

$$h_L = \frac{V_1^2}{2gh_2^2}(h_2^2 - h_1^2) + (h_1 - h_2)$$

Solving Eq. 7.2-18 for V_1^2/g and substituting the resulting expression into the above equation, we obtain

$$h_L = \frac{h_2^2 + h_1h_2}{4h_1}\left(1 - \frac{h_1^2}{h_2^2}\right) + (h_1 - h_2)$$

The above equation can be simplified to give

$$h_L = \frac{(h_2 - h_1)^3}{4h_1h_2} \qquad \text{Eq. 7.2-21}$$

We now apply the second law of thermodynamics to show that $\mathscr{F}_1 > 1$. For a substance undergoing a process that only involves expansion work, we know from thermodynamics that

$$T\,ds - p\,d\tilde{v} = du \qquad \text{Eq. A.3-13}$$

For a liquid

$$u = cT \quad \text{and} \quad du = c\,dT$$

where c is its specific heat. Thus h_L can also be expressed by the equation[1]

$$h_L = \frac{Jg_c c}{g}(T_2 - T_1) \qquad \text{Eq. 7.2-22}$$

For an incompressible liquid, $d\tilde{v} = 0$; thus, Eq. A.3-13 reduces to

$$ds = \frac{du}{T} = \frac{c\,dT}{T}$$

[1] By examining Eq. 7.2-22, it can be seen that a head loss of 1 ft. would result in a temperature rise of 0.0013 F. Since head losses of 1 ft and less are typical, we find that the head loss is invariably determined from velocity and depth measurements.

By integrating between states 1 and 2, we find that

$$\mathscr{s}_2 - \mathscr{s}_1 = c \ln \frac{T_2}{T_1} \qquad \text{Eq. 7.2-23}$$

The second law of thermodynamics requires that

$$m \, d\mathscr{s} \geq \frac{\delta Q}{T} \qquad \text{Eq. A.3-8}$$

We may assume that a fluid element passing through a hydraulic jump undergoes an adiabatic process, giving $\delta Q = 0$ and

$$\int_1^2 d\mathscr{s} = \mathscr{s}_2 - \mathscr{s}_1 \geq 0 \qquad \text{Eq. 7.2-24}$$

Comparing Eq. 7.2-23 and 7.2-24, we see that the second law requires that $T_2 \geq T_1$. Applying this information to Eqs. 7.2-21 and 7.2-22, we find that $h_L > 0$, $h_2 > h_1$ and $\mathscr{F}_1 > 1$. Thus, the hydraulic jump can only take place in one direction. That is, *in the direction of flow, the fluid depth must increase.*

Although our discussion of the hydraulic jump originated from our discussion of flow under a sluice gate, a hydraulic jump can also occur as a result of a slope change in the channel floor. That is, when the slope of the channel floor changes from a steep slope to a milder slope, causing the flow to go from a supercritical to a subcritical, the transition takes place through a hydraulic jump. Since high-velocity flow causes erosion, supercritical flow should, whenever possible, be avoided.

Our last topic for discussion under the broad heading of one-dimensional, nonviscous open-channel flow is the sharp-crested rectangular weir. The sharp-crested weir may be used as a flow-rate measuring device as well as a simple overflow spillway for a water basin. A sketch of the weir is shown in Fig. 7-8*a*. The water stream downstream from the weir is called the *nappe*. The fluid particles in the nappe a short distance downstream from the weir behave like a free-falling body. Therefore, the equations for the upper and lower surfaces of the nappe are parabolas. The determination of the constants in the equation of the parabola would require that initial conditions be known. At present, no satisfactory theoretical means has been developed for establishing these initial conditions. However, empirical values have been established.[2] For high weirs, the approach velocities (stream velocities far upstream from the

[2] See Ven Te Chow, *Open Channel Hydraulics*, McGraw-Hill, New York, 1959.

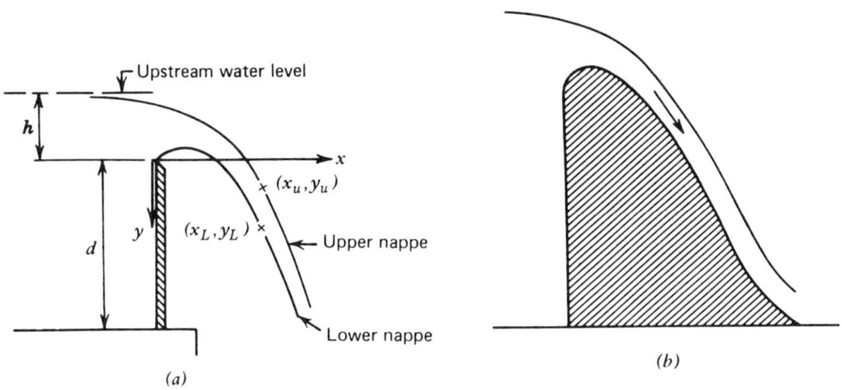

Fig. 7-8. The sharp-crested weir and the overflow spillway. (*a*) Sharp-crested weir. (*b*) Overflow spillway.

weir) are usually negligible. In these cases, the equations for the lower and upper surfaces of the nappe are

$$\frac{y_L}{h} = 0.150 + 0.055\frac{x_L}{h} - 0.425\left(\frac{x_L}{h}\right)^2 \qquad \text{Eq. 7.2-25}$$

and

$$y_u = y_L + 0.56h \qquad \text{Eq. 7.2-26}$$

These equations are valid only for $x_L/h \geq 0.5$. When $x_L/h < 0.5$, the pressure within the nappe is above that of the atmosphere, indicating that the fluid particles, in this region, do not behave as a free-falling particle.

A number of empirical formulas exist for determining the flow rate over a sharp-crested weir. These formulas can usually be expressed in the following form

$$Q = Cbh^{3/2} \qquad \text{Eq. 7.2-27}$$

where C is the discharge coefficient and b is the width of the weir. The dimensions of Q, C, b, and h are, respectively, ft^3/s, $ft^{1/2}/s$, ft, and ft in English units and m^3/s, $m^{1/2}/s$, m and m in SI units. A formula for the determination of C has been given by Rehbock and is defined

$$C = \begin{cases} 3.27 + 0.40\dfrac{h}{d}, & \text{in English units} \\ 1.81 + 0.221\dfrac{h}{d}, & \text{in SI units} \end{cases} \qquad \text{Eq. 7.2-28}$$

where d is the height of the weir. The equation is approximately valid for $h/d < 10$. Of course for a laboratory system, it would be best to calibrate the weir.

In the past the overflow spillway of a dam (see Fig. 7-8*b*) was designed to have the same shape as the lower surface of the nappe occurring in a flow over a sharp-crested weir. It was thought that such a profile would result in spillway-surface pressures being equal to atmospheric pressure everywhere. However, when actual flow conditions differed from design conditions, pressures below atmospheric pressure on the spillway surface could not be avoided. Fear of cavitation, with the resulting cavitation damage, instigated numerous investigations for a proper design for the spillway surface. Recommended spillway surface shapes have since been developed by the U.S. Army Corp. of Engineers. For further information on this topic see Chow's *Open Channel Hydraulics*.

7.3 EMPIRICAL FORMULAS IN CHANNEL FLOW

So far we have analyzed a number of one-dimensional, open-channel flow problems by neglecting losses due to viscous effects and turbulence. We now want to introduce such losses into our formulation. One approach that is frequently used is to modify the Bernoulli equation, Eq. 4.8-16, in order to include a head-loss term.

Consider the channel flow shown in Fig. 7-9. If we initially assume that viscous effects are negligible, then the Bernoulli equation is valid along a

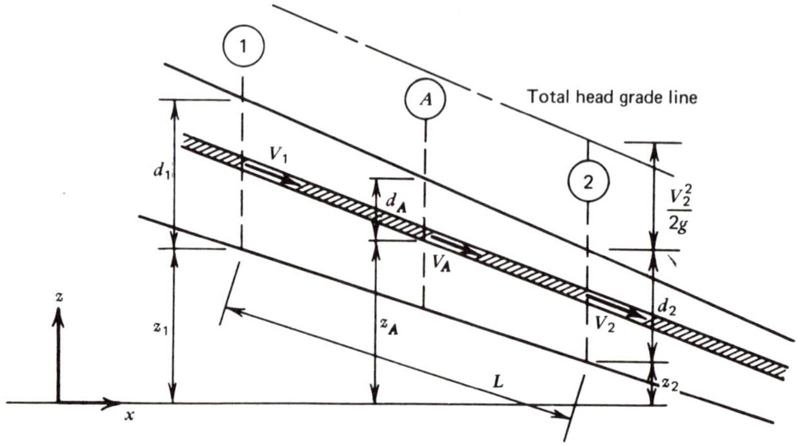

Fig. 7-9. A general channel flow.

streamline. Then, for any point along the streamline passing through point A, the total head, H_A, is a constant; that is,

$$H_A = z_A + \frac{\gamma\, d_A}{\gamma} + \frac{v_A^{\,2}}{2g} \qquad \text{Eq. 7.3-1}$$

where z_A is the elevation of point A above the datum, v_A is the magnitude of the fluid velocity at point A, and d_A is the vertical distance of point A from the free surface. It is obvious that at any point on the vertical section passing through point A, $(z + d)$ is a constant. However, the velocity head varies, due to the nonuniformity of the velocity field at the section. In developing an empirical relation for describing the flow, this velocity variation is ignored, and the average velocity at this section is substituted into the formula. Then each point at the section has the same total head. It is also common practice to account for viscous effects by adding a head-loss term, h_f, to the Bernoulli equation, giving

$$\frac{V_1^{\,2}}{2g} + d_1 + z_1 = \frac{V_2^{\,2}}{2g} + d_2 + z_2 + h_f \qquad \text{Eq. 7.3-2}$$

Let us now consider a flow where the cross-sectional area, the floor slope, and the discharge are all constant. For this case, except for a short starting length, the free surface will be parallel to the channel floor. We will designate this flow state as *longitudinally uniform flow*. For such a flow,

$$d_2 = d_1 \qquad V_2 = V_1$$

and

$$h_f = z_1 - z_2 = SL \qquad \text{Eq. 7.3-3}$$

where S is the slope of the channel floor and L is the length of channel between sections 1 and 2. The head loss, h_f, can be taken to be a function of the same variables used in pipe flow. Then by dimensional analysis, h_f, can be expressed by the Darcy equation, Eq. 6.3-7. Since the flow is invariably turbulent and the flowing substance is water in a rather narrow temperature range, we may bypass the Darcy equation and go directly to one of the empirical formulas developed for open-channel flow. The one most often used in this country is the Chezy formula, named after the French hydraulics engineer who first proposed it in 1775; it expresses the bulk velocity, V, by the following formula

$$V = C\sqrt{R_H S} \qquad \text{Eq. 7.3-4}$$

where R_H is the hydraulic radius, S is the slope of the channel floor, and C is a coefficient, which depends on the roughness and the Reynolds number. Although there exist several different expressions for C, the one most often used in this country is the Manning formula, which is

$$C = \frac{A}{n} R_H{}^{1/6} \qquad \text{Eq. 7.3-5}$$

where n is the Manning roughness coefficient, which is tabulated in Table 7-1

TABLE 7-1 Manning Roughness Coefficient, n

Channel Surface Type	Condition	n
Glazed coating or enamel	Perfect order	0.010
Timber	Planed boards, carefully laid	0.010
	Planed boards, old or inferior	0.012
	Unplaned boards, carefully laid	0.012
	Unplaned boards, old or inferior	0.014
Metal	Smooth	0.010
	Riveted steel	0.015-0.020
	Corrugated metal	0.022-0.026
	Cast iron, new	0.015
	Cast iron, tuberculated	0.020
Masonry	Neat cement plaster	0.010
	Sand and cement plaster	0.012
	Finished concrete	0.012
	Unfinished concrete	0.014
	Concrete in bad condition	0.020
	Brick, good condition	0.013
	Brick, rough	0.020
Stonework	Smooth, dressed ashlar	0.013
	Rubble set in cement	0.017
	Fine, well-packed gravel	0.020
	Dry rubble	0.025
Earth	Regular, good condition	0.020
	Regular, fair condition	0.023
	Regular, some stones and weeds	0.028
	Regular, poor condition	0.035
	Winding, irregular, but clear	0.035
	Obstructed with debris and weeds	0.050-0.150

Source. From *Applied Hydraulics in Engineering*, 2nd ed., by H. M. Morris and J. M. Wiggert, Copyright © 1972, by the Ronald Press Company, New York.

and A is a constant, defined below. By substituting this expression into Eq. 7.3-4, we obtain

$$V = \frac{A}{n} R_H{}^{2/3} S^{1/2} \qquad \text{Eq. 7.3-6}$$

where the constant $A = 1.486$ when the variables are expressed in English units (V in ft/s and R_H in ft). However, $A = 1.00$ when the variables are expressed in SI units (V in m/s and R_H in m). S and n are dimensionless.

The design of channels for drainage and irrigation systems, and the like, is based on the equations developed for a uniform flow. Some typical problems encountered in the design of such channels are discussed in the problem section which follows.

Problems

7-1 A rectangular venturi flume with a horizontal floor is constructed in the natural bed of a stream for the purpose of measuring its flow rate. The widths of the upstream and throat sections of the flume are 1.20 m and 0.90 m, respectively. During a particular test, depths of 0.60 m and 0.55 m were measured at the upstream and throat sections, respectively. Determine the flow rate through the flume.

7-2 Consider a flow through a flume.

(*a*) Sketch typical curves for V versus x and h versus x for case (1) when the flow goes from subcritical to supercritical and case (2) when the flow remains subcritical over the entire length of the flume.

(b) Show that when the flow goes from subcritical to supercritical, $F = 1$ at the throat. Assume that the channel width, w, is a smooth function of x.

(c) Show that when the flow remains subcritical over the entire length of the flume, h has a relative minimum at the throat.

7-3 Suppose 3.4 m^3/s flows through the flume described in Problem 7-1. If it is observed that h decreases continuously through the flume, determine the water depth at the throat, if h at the upstream section is 1.6 m.

7-4 The approach and downstream channels to a sluice gate have widths of 9 m and 5m, respectively. If both channels have a rectangular cross section, the stream depth in the approach section is 4.5 m, and the flow rate under the gate is 54 m^3/s, determine the height of the stream just downstream from the gate.

7-5 Suppose that in Problem 7-4 the stream depth in the approach channel was fixed at 4.9 m, but the height of the sluice gate varied. What is the maximum flow rate that can pass under the gate?

7-6 Suppose that in the channel and sluice gate configuration, described in Problem 7-4, the depths in the approach and downstream channels are 3.0 m and 1.5 m, respectively. Determine: (*a*) the volume flow rate under the gate and; (*b*) whether a hydraulic jump occurs in the downstream channel. If your answer to this question is yes, then also determine the depth in the channel downstream from the hydraulic jump and the head loss which results from the jump.

7-7 A hydraulic jump occurs in a horizontal rectangular channel. The depth goes from 3 to 5 ft. Determine the discharge per foot of width and the head loss.

7-8 A vertical sharp-crested weir is 7.5 m high. The upstream water level is 4.5 m above the crest of the weir. Determine the y-coordinates (see Fig. 7-8) of the upper and lower nappes at $x = 3.0$ m.

7-9 If in Problem 7-8 the width of the weir is 9.0 m, determine the flow rate over the weir.

7-10 Water flows through a smooth, concrete box culvert, 3 ft high and 6 ft wide. If the water depth is uniform at 1 ft and the floor slope is 1%, determine the flow rate through the culvert.

7-11 Suppose that in Problem 7-10, Q was known to be 80 ft^3/s. Determine the depth of the flow if all other conditions remain unchanged.

7-12 A highway earth-drainage ditch in good condition has a cross section, as shown in Fig. 7-10. The longitudinal slope of the ditch is 2 m/100 m. If the depth is uniform at 0.3 m, determine the flow rate through the ditch.

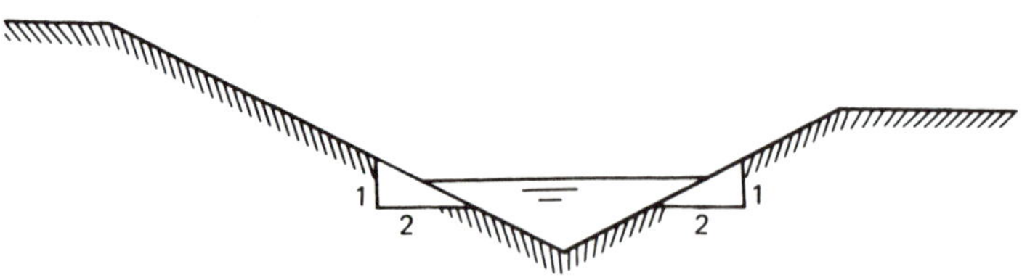

Fig. 7-10. Triangular drainage ditch.

7-13 A finished concrete channel having a rectangular cross section is subdivided into two sections, as shown in Fig. 7-11. The width of the channel is 1.2 m, and the floor slope and water depth of the first section are 5% and 7.3 cm, respectively. (*a*) Determine the Froude number in the first section; (*b*) determine the flow rate, Q, through the channel; and (*c*) assuming that the flow goes directly from a longitudinally uniform flow in the first section to a hydraulic jump to another longitudinally

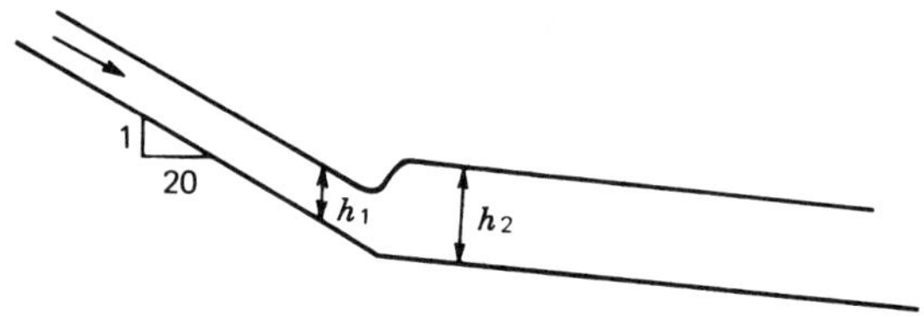

Fig. 7-11. Hydraulic jump resulting from change in floor slope.

uniform flow in the second section, determine the slope of the floor in the second section. Neglect elevation differences in the channel floor over the length of the hydraulic jump.

7-14 Consider a longitudinally uniform, open-channel flow through a culvert having a circular cross section (culvert is only partially filled, as shown in Fig. 7-12). (*a*) Determine expressions for the area that is covered by the liquid and the wetted

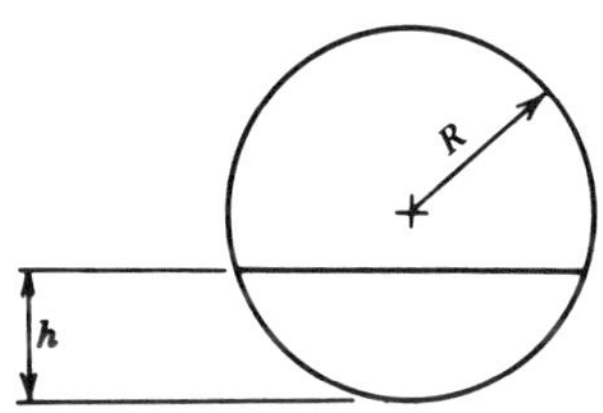

Fig. 7-12. Partially filled circular culvert.

perimeter as a function of the water depth, h. (*b*) Determine an equation for the value of h/R, where Q has a relative maximum. Consider the slope of the culvert to be fixed. (*c*) Plot Q/Q_0 as a function of h/R, for $0 \leq h/R \leq 2$, where Q_0 is the flow rate through the culvert when it is full. Assume that Q/Q_0 has a maximum at $h/R = 1.876$.

Chapter Eight

One-Dimensional, Non-Viscous, Compressible Gas Flow

8.1 INTRODUCTION

In this chapter we examine one-dimensional gas flows through ducts, nozzles, and diffusers. Rarefaction, compression, and shock waves are also briefly discussed. Nozzles are devices that speed up a flow, whereas diffusers slow it down. These devices are often found in turbines, compressors, propulsion systems, and other equipment. In general, the flow through nozzles, diffusers, and, in many cases, ducts of short length may be considered to be adiabatic and reversible (isentropic flow). The reversible behavior requires that the duct length be short, as it is in nozzles and diffusers, since then viscous effects are not significant. However, in such gas flows density and temperature variation are of sufficient magnitude and therefore must be considered in a proper analysis of the phenomenon. An important parameter that appears in the analysis is the Mach number, discussed in Chapter 5. Recall that the Mach number, M, was defined as the ratio of the magnitude of the fluid velocity, v, and the speed of sound in the fluid, c; that is,

$$M = \frac{v}{c} \qquad \text{Eq. 8.1-1}$$

Before analyzing gas flows through ducts, nozzles, and diffusers, we will briefly look at wave phenomena.

8.2 SOUND, RAREFACTION, COMPRESSION, AND SHOCK WAVES

A *sound wave* (one that is detectable by the human ear) is an oscillatory pressure disturbance of small amplitude in an ideal, isentropic, compressible fluid. Because of the elastic nature of a fluid, a pressure pulse will propagate through the fluid at a speed, c, as shown in Fig. 8-1. Although sound waves are normally not one-dimensional, they can be made so by placing a speaker inside a tube as shown. Using perturbation mathematics, one can readily model sound waves and demonstrate the propagation behavior of such pressure disturbances. Students interested in this treatment should see the work of Landau and Lifshitz, which is listed in the reference section at the end of the book. Although perturbation mathematics is beyond the scope of this text, we can obtain a relationship for the speed of sound in terms of other thermodynamic variables by examining a rarefaction wave that is created by the movement of a piston inside a long tube. In the following discussion, we have accepted the idea that a small pressure disturbance propogates at the speed of sound.

Suppose a piston inside a long tube is instantaneously brought from rest to speed, $V_P \ll c$, in the $(-x)$ direction, as shown in Fig. 8-2*a*. Experiments indicate that plots of p versus x, at times t_0, t_1, t_2, and so on, would appear as shown in Fig. 8-2*b*, 8-2*c*, and 8-2*d*. Notice that at each time $t_1, t_2, \ldots,$ the

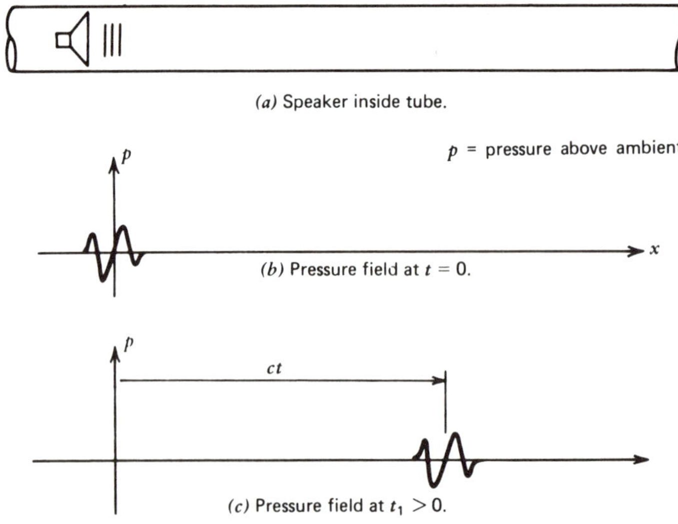

Fig. 8-1. Pressure field resulting from a sound-wave pulse from a speaker.

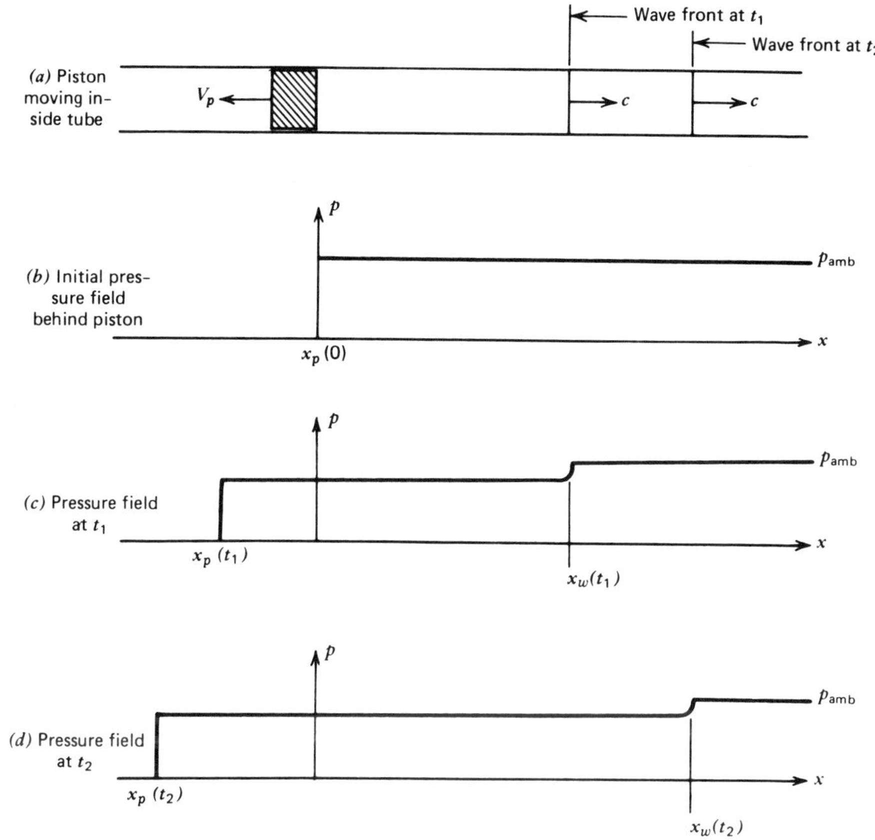

Fig. 8-2. Development of a rarefaction wave.

pressures change abruptly at x_w. This abrupt change in pressure is called a *wave front* or, more specifically, a *rarefaction wave*; the rarefaction designation indicates that the pressure behind the wave is less than the pressure in front of the wave. On the other hand, if the pressure behind the wave was greater than the pressure in front of the wave, it would be called a *compression wave*. Experiments indicate that when $V \ll c$, the wave front moves in the $(+x)$ direction at the speed of sound into the undisturbed region. The flow in the vicinity of the wave, as seen from a coordinate system attached to the wave front, is steady. Viewing the phenomenon from this vantage point simplifies the analysis. Select a control volume around the wave front, as shown in Fig. 8-3. Now

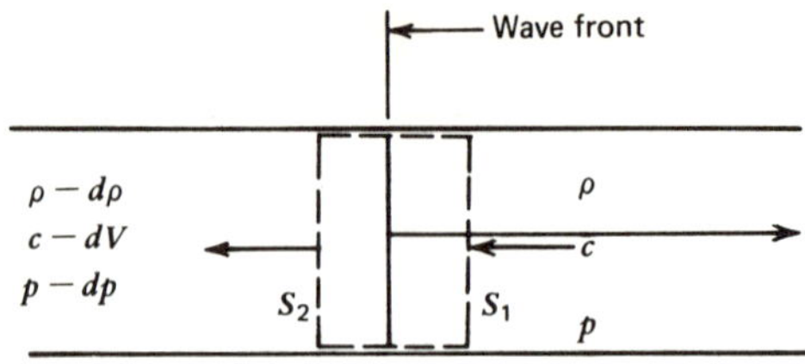

Fig. 8-3. Control volume around wave front.

apply equations of conservation of mass and Newton's second law to the control volume, Eqs. 4.3-5 and 4.7-5. Conservation of mass gives

$$-\rho cA + (\rho - d\rho)(c - dV)A = 0$$

which, after neglecting the higher order terms and simplifying, reduces to

$$c\,d\rho + \rho\,dV = 0 \qquad \text{Eq. 8.2-1}$$

Newton's second law gives

$$-pA + (p - dp)A = -\frac{\rho}{g_c}c(-cA) - \frac{(\rho - d\rho)}{g_c}(c - dV)(c - dV)A$$

which, after neglecting higher order terms and simplifying, reduces to

$$-dp = \frac{\rho c^2}{g_c} - \frac{(c - dV)}{g_c}[\rho c - \underbrace{(c\,d\rho + \rho\,dV)}_{\text{zero}}]$$

or

$$-dp = \frac{\rho c}{g_c}dV \qquad \text{Eq. 8.2-2}$$

Substituting Eq. 8.2-1 into Eq. 8.2-2, we obtain

$$dp = \frac{c^2}{g_c}d\rho$$

or

$$g_c\left(\frac{dp}{d\rho}\right) = c^2 \qquad \text{Eq. 8.2-3}$$

Because of the rapidity of the process and the very small changes that take place in pressure and temperature, the process may be considered to be isentropic. If in addition the process takes place in an ideal gas, then for an isentropic process, p/ρ^k = constant, and we may then write

$$c = \sqrt{g_c\left(\frac{dp}{d\rho}\right)_s} \qquad \text{Eq. 8.2-4}$$

where the subscript s indicates a path of constant entropy. Taking

$$\frac{p}{\rho^k} = K \qquad \text{or} \qquad p = K\rho^k$$

then

$$\frac{dp}{d\rho} = kK\rho^{k-1} = \frac{kK\rho^k}{\rho} = \frac{kp}{\rho}$$

But, from the ideal gas law,

$$\frac{p}{\rho} = RT$$

Substituting these expressions into Eq. 8.2-4, we obtain

$$c = \sqrt{g_c kRT} \qquad \text{Eq. 8.2-5}$$

If in the experiment just discussed, the piston was almost instantaneously brought up to a speed close to, but less than, the speed of sound, then the phenomenon can be modeled by a series of waves close together, as shown in Fig. 8-4. Only the furthest wave to the right (the one adjacent to the undisturbed region) would move at the speed of sound in the undisturbed region. The others would move at successively lower speeds and, thus, the wave region would tend to spread out as it moves further down the tube. This behavior results from the fact that as the piston picks up speed, it creates disturbances in fluids of lower and lower densities. Since the sound speed decreases as the density decreases, each successive wave moves down the tube at a lower speed than its neighbor to the right. This phenomenon can be mathematically modeled by the method of

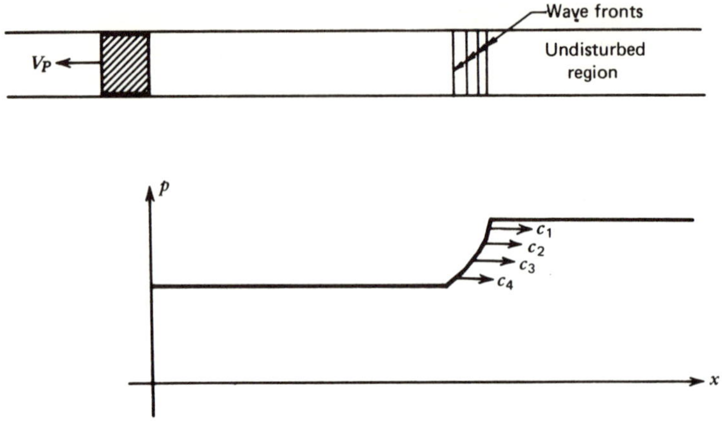

Fig. 8-4. Piston receding in a gas.

characteristics, a subject beyond the scope of this text. Students interested in this mathematical technique should see Courant and Friedrichs.[1]

Now suppose that the piston in Fig. 8-4 moved into the gas of interest at a speed close to the speed of sound (in the undisturbed region). Once again, at the beginning of the flow field development, a series of waves would be formed, as shown in Fig. 8-5. However, in this case, the waves would be *compression* waves, since the pressure behind the waves is greater than the pressure in the undisturbed region. Furthermore, as the piston accelerates, it creates disturbances in fluids of higher and higher densities, causing each successive wave to move faster than its preceding neighbor. In this way, the rear waves overtake the forward waves, thus developing a very thin region across which the fluid properties jump in value. This very thin region is called a *shock wave*, or shock front. The shock front in this case would be normal to the flow direction and therefore is designated as a normal shock. Although the shock front moves ahead of the piston at a speed greater than the speed of sound (in the undisturbed region), it will approach sound speed when $V_p \ll c$. The relationship between the shock speed and the piston speed is developed later in this chapter. Let us summarize the important points of our discussion regarding the movement of a piston inside a long tube.

1. When the piston recedes in a fluid, a rarefaction wave region is created. This wave region is much broader than that of a shock wave. The fluid properties from wave to

[1] R. Courant and K. O. Friedrichs, *Supersonic Flow and Shock Waves,* Interscience Publishers, New York, 1948.

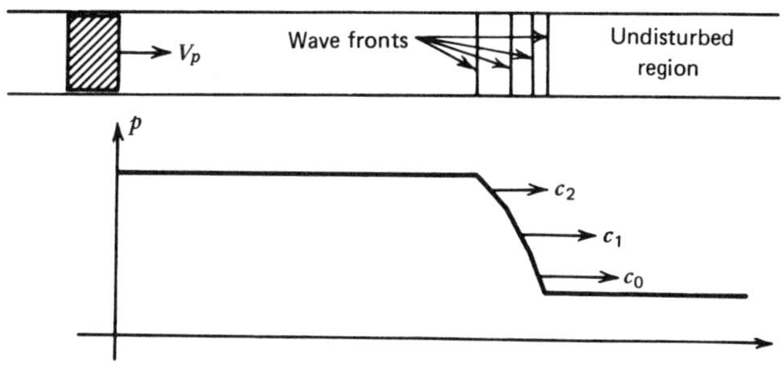

Fig. 8-5. Piston moving into a gas.

wave are mathematically continuous, but the derivatives with respect to distance are not.

2. When the piston moves into a fluid, a shock wave is created. The fluid properties jump in value across the shock. The shock front moves into the undisturbed region at a speed greater than the speed of sound, but approaches sound speed as $V_p \rightarrow 0$.

8.3 ONE-DIMENSIONAL FLOW IN DUCTS, NOZZLES, AND DIFFUSERS

The following assumptions are usually made when treating gas flows through short ducts, nozzles, and diffusers. (1) Properties are assumed to be uniform at each cross section. Property values are thus bulk or average values. (2) Viscous effects, heat transfer, and gravitational effects are negligible. (3) Air behaves as an ideal gas with constant specific heats over a wide range of conditions. (4) Flow is steady.

Now consider the variable cross-sectional area passage shown in Fig. 8-6 and construct a control volume as shown. In Sections 4.3 and 4.6, we applied

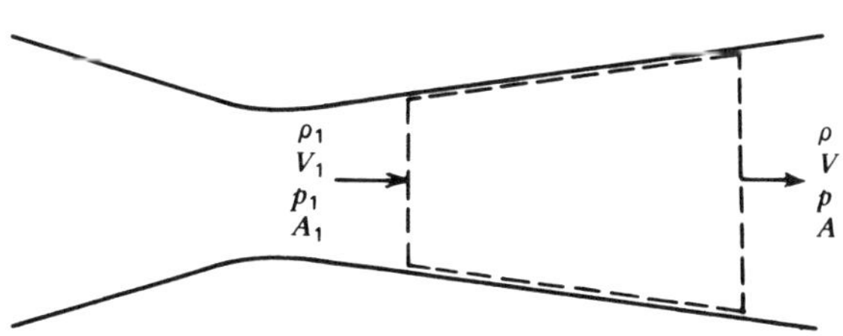

Fig. 8-6. Varying area passage.

the laws of mass conservation and the first law of thermodynamics to a one-dimensional channel flow and obtained Eqs. 4.4-4 and 4.6-23. These equations are valid for the problem under consideration. If, in these equations, we consider section 2 to be an arbitrary position within the passage, and section 1 as some fixed position, then these equations may be written as follows

$$\rho A V = \text{constant} \qquad \text{Eq. 8.3-1}$$

$$Jh + \frac{V^2}{2g_c} = \text{constant} \qquad \text{Eq. 8.3-2}$$

These equations may be written in differential form as follows

$$\rho A\, dV + \rho V\, dA + AV\, d\rho = 0$$

or

$$\frac{d\rho}{\rho} + \frac{dV}{V} + \frac{dA}{A} = 0 \qquad \text{Eq. 8.3-3}$$

and

$$J\, dh + \frac{V}{g_c}\, dV = 0 \qquad \text{Eq. 8.3-4}$$

To arrange Eqs. 8.3-3 and 8.3-4 in a desirable form, we use the thermodynamic equation, which combines the first and second laws of thermodynamics (see Appendix A.3, Eq. A.3-14); here it is written in terms of the intensive variable ρ and with the Joules constant inserted,

$$T\, ds + \frac{1}{J\rho}\, dp = dh \qquad \text{Eq. 8.3-5}$$

so that the variables have their commonly used dimensions. For an isentropic process, ds is zero and Eq. 8.3-5 reduces to

$$J\, dh = \frac{dp}{\rho} \qquad \text{Eq. 8.3-6}$$

Substituting the above equation into Eq. 8.3-4 and multiplying through by g_c/V^2, we obtain, after some rearranging,

$$\frac{dV}{V} = -\frac{g_c}{\rho V^2}\,dp \qquad \text{Eq. 8.3-7}$$

Then by substituting Eq. 8.3-7 into Eq. 8.3-3 and rearranging terms, we find that

$$\frac{dA}{A} = \frac{g_c}{\rho V^2}\,dp - \frac{d\rho}{\rho} = \frac{g_c}{\rho V^2}\left(1 - \frac{V^2}{g_c}\frac{d\rho}{dp}\right)dp \qquad \text{Eq. 8.3-8}$$

But by Eq. 8.2-3

$$\frac{1}{g_c}\frac{d\rho}{dp} = \frac{1}{c^2}$$

which, when substituted into Eq. 8.3-8, gives

$$\frac{dA}{A} = \frac{g_c}{\rho V^2}\left(1 - \frac{V^2}{c^2}\right)dp = \frac{g_c}{\rho V^2}(1 - M^2)\,dp$$

or

$$\frac{dA}{dp} = \frac{A g_c}{\rho V^2}(1 - M^2) \qquad \text{Eq. 8.3-9}$$

We may also combine Eqs. 8.3-7 and Eq. 8.3-9 to give

$$\frac{dV}{dA} = \frac{\dfrac{dV}{dp}}{\dfrac{dA}{dp}} = -\frac{V}{A(1 - M^2)} \qquad \text{Eq. 8.3-10}$$

From Eqs. 8.3-9 and 8.3-10 we may conclude

1. If $M < 1$: dV is negative if dA is positive, and dV is positive if dA is negative. Thus, if $M < 1$, the velocity decreases if the area increases (passage is a diffuser); and the velocity increases if the area decreases (passage is a nozzle).

2. If $M > 1$: dV is positive if dA is positive, and dV is negative if dA is negative. Thus, if $M > 1$, the velocity increases if the area also increases (passage is a nozzle); and the velocity decreases if the area decreases (passage is a diffuser).

These concepts are summarized in Fig. 8-7.

We now turn our attention to deriving the equations that are employed in developing the gas tables commonly used for solving problems involving isentropic flow. These equations give the temperature, pressure, and density in terms of the Mach number and the stagnation properties. Thus, *stagnation* property values, denoted by a zero subscript, serve as a reference state for isentropic compressible flows and are defined to be those properties of a fluid element that would be obtained if the flow was brought isentropically to rest. Since velocity depends on the frame of reference from which the fluid is being observed, stagnation property values also depend on the reference frame being used. On the other hand, the *static* property values of a fluid element, the values we have been using all along, are defined to be those property values that would be seen by an observer moving with the fluid element; therefore, they are independent of the reference frame being used.

Now let us express the constant on the right-hand side of Eq. 8.3-2 in terms

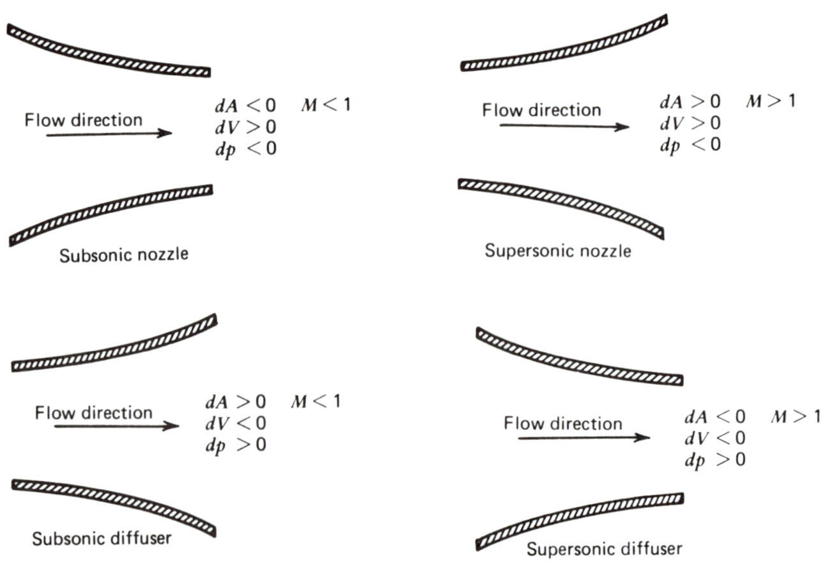

Fig. 8-7. Nozzles and diffusers.

of the stagnation enthalpy, h_0. Since by definition the stagnation velocity is zero, we have

$$Jh + \frac{V^2}{2g_c} = Jh_0 \qquad \text{Eq. 8.3-11}$$

or

$$V^2 = 2g_c J(h_0 - h)$$

which, by Eqs. 4.6-33 and 4.6-36, reduces to

$$V^2 = \frac{2k}{k-1} g_c R T_0 \left(1 - \frac{T}{T_0}\right) \qquad \text{Eq. 8.3-12}$$

The fluid velocity, V, can be expressed in terms of the pressure, p, by substituting the isentropic relation,

$$\frac{T}{T_0} = \left(\frac{p}{p_0}\right)^{(k-1)/k} \qquad \text{Eq. 8.3-13}$$

into Eq. 8.3-12, giving

$$V^2 = \frac{2k}{k-1} g_c R T_0 \left[1 - \left(\frac{p}{p_0}\right)^{(k-1)/k}\right] \qquad \text{Eq. 8.3-14}$$

Solving Eq. 8.3-12 for T/T_0, we obtain

$$\frac{T}{T_0} = 1 - \frac{k-1}{2}\left(\frac{V^2}{k g_c R T_0}\right)$$

If, in the above equation, we multiply and divide the last term by the temperature, T, and also substitute into it Eq. 8.2-5, we obtain

$$\frac{T}{T_0} = 1 - \frac{k-1}{2}\left(\frac{V^2}{c^2}\right)\frac{T}{T_0} = 1 - \frac{k-1}{2} M^2 \frac{T}{T_0}$$

or

$$\frac{T}{T_0} = \frac{1}{1 + \frac{k+1}{2} M^2} = \left(1 + \frac{k-1}{2} M^2\right)^{-1} \qquad \text{Eq. 8.3-15}$$

Substituting Eq. 8.3-13 into Eq. 8.3-15 and raising both sides to the $k/(k-1)$ power, we obtain

$$\frac{p}{p_0} = \left(1 + \frac{k-1}{2} M^2\right)^{-k/(k-1)} \qquad \text{Eq. 8.3-16}$$

We can obtain an expression for ρ/ρ_0 in terms of M by using the isentropic relationship

$$\frac{p}{p_0} = \left(\frac{\rho}{\rho_0}\right)^k \quad \text{or} \quad \frac{\rho}{\rho_0} = \left(\frac{p}{p_0}\right)^{1/k} \qquad \text{Eqs. 8.3-17}$$

Substituting the first of Eqs. 8.3-17 into 8.3-16, and then taking the kth root of the resulting equation, we find that

$$\frac{\rho}{\rho_0} = \left(1 + \frac{k-1}{2} M^2\right)^{-1/(k-1)} \qquad \text{Eq. 8.3-18}$$

It is also possible to express the ratio of two areas in the flow as functions of the Mach numbers at the two sections. Let A represent the area of some arbitrary section in the passage and A_1 the area of some reference section. Then, by Eq. 8.3-1,

$$\frac{A}{A_1} = \frac{\rho_1 V_1}{\rho V} = \frac{\rho_1 M_1 \sqrt{g_c k R T_1}}{\rho M \sqrt{g_c k R T}} \qquad \text{Eq. 8.3-19}$$

Noting that $\rho_1/\rho = (\rho_1/\rho_0)/(\rho/\rho_0)$ and that $T_1/T = (T_1/T_0)/(T/T_0)$, then the substitution of Eqs. 8.3-15 and 8.3-18 into Eq. 8.3-19 gives

$$\frac{A}{A_1} = \frac{M_1}{M} \left(\frac{1 + \dfrac{k-1}{2} M^2}{1 + \dfrac{k-1}{2} {M_1}^2}\right)^{(k+1)/2(k-1)} \qquad \text{Eq. 8.3-20}$$

The area of the section where stagnation conditions occur cannot be used as a reference area, since that area, by Eq. 8.3-1, must approach infinity. A convenient reference area is the area of the section where the Mach number is one. We will define this area as A^*; then letting $A_1 = A^*$ and $M_1 = 1$, Eq. 8.3-20 reduces to

$$\frac{A}{A^*} = \frac{1}{M}\left(\frac{1 + \dfrac{k-1}{2}M^2}{\dfrac{k+1}{2}}\right)^{(k+1)/2(k-1)} \qquad \text{Eq. 8.3-21}$$

The variations of A/A^*, p/p_0, T/T_0, and ρ/ρ_0 as functions of the Mach number are shown in Fig. 8-8. These values as functions of the Mach number are also

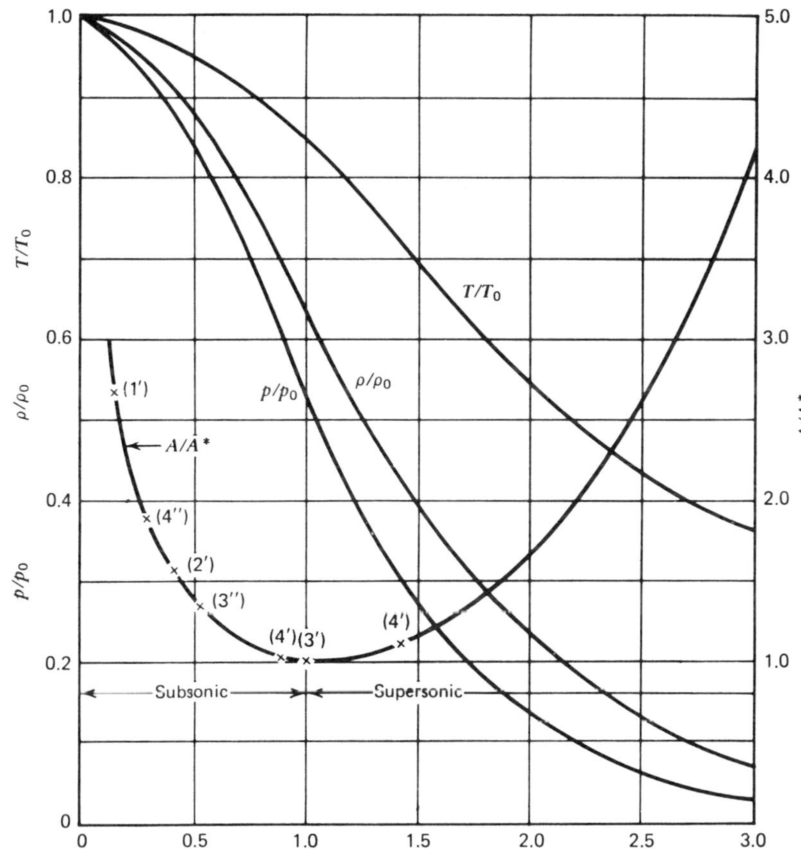

Fig. 8-8. Variations of the ratios T/T_0, p/p_0, ρ/ρ_0, and A/A^* as a function of the Mach number for a gas with $k = 1.40$.

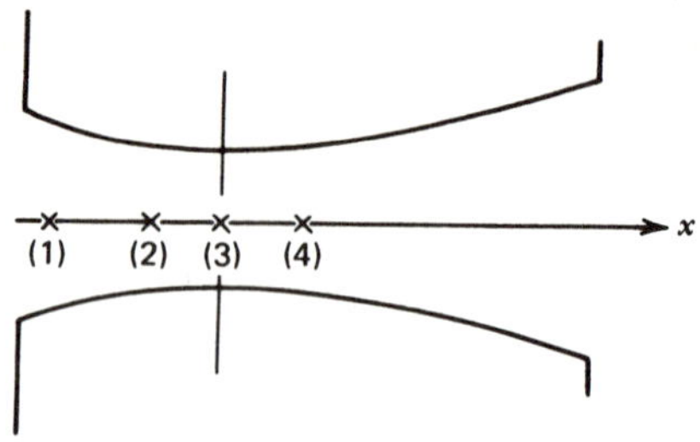

Fig. 8-9. A converging-diverging nozzle.

tabulated in the Appendix, Table A.6-8. From Fig. 8-8, we see that p/p_0, T/T_0, and ρ/ρ_0, all decrease as M increases. However, the characteristics of the A/A^* curve are significantly different. The following characteristics can be observed.

1. As M approaches either zero or infinity, A/A^* approaches infinity.

2. Except for $M = 1$, there exist two Mach numbers, one in the subsonic range and the other in the supersonic range, that give the same value for A/A^*.

3. A/A^* has a relative minimum at $M = 1$ and $(A/A^*)_{\text{rel min}}$ equals 1.

4. There are no values of A/A^* less than unity; that is,

$$\frac{A}{A^*} \geq 1$$

To further appreciate the significance of the A/A^* curve, let us consider a flow through a converging-diverging nozzle. We ask the following question: What points on the A/A^* curve correspond to the physical points in the nozzle (see Fig. 8-9). Consider point (1) in the nozzle, just downstream from the reservoir. Since M equals zero in the reservoir, its corresponding point on the A/A^* curve, say point (1′), must lie on the subsonic part of the curve, near $M = 0$, as shown in the Fig. 8-8. Now consider point (2), a point in the converging part of the nozzle, downstream from point (1). In the converging part of the nozzle, A/A^* decreases and thus point (2′) must lie to the right of point (1′) on the A/A^* curve, as shown in Fig. 8-8. At the throat,[2] point (3), A/A^* is a minimum.

[2] By definition the throat is the section where the cross-sectional area is a minimum.

If $(A_{th}/A^*) = 1$, then the minimum is a relative minimum and $M_{th} = 1.0$. Here the subscript *th* means the variable value at the throat. This is shown as point (3′) in Fig. 8-8. If $(A_{th}/A^*) > 1$, then (A_{th}/A^*) is an absolute minimum, but not a relative minimum, and $M_{th} < 1$. This point is shown as point (3″) in Fig. 8-8. The implication here is that M_{th} cannot be greater than 1. Now consider point (4), a point in the divergent part of the nozzle just downstream from the throat. Going from point (3) to point (4), A/A^* increases; that is, $(A/A^*)_{\text{point 4}} > (A/A^*)_{\text{point 3}}$. If $M_{th} = 1$ the above condition can be satisfied in a continuous fashion by either moving to the left or right of point (3′). The A/A^* values corresponding to this condition are shown as points (4′) in Fig. 8-8. This dual possibility is discussed in greater detail later in this section. However, if $M_{th} < 1$, then the point on the A/A^* curve corresponding to point (4), the point designated as (4″), must lie to the left of point (3″), since this is the only way we can satisfy, in a continuous fashion, the condition that $(A/A^*)_{\text{point 4}} > (A/A^*)_{\text{point 3}}$.

Now let us consider the following problem. Suppose either a converging or converging-diverging nozzle is attached to a very large reservoir, as shown in Fig. 8-10. Suppose we fix the pressure and temperature of the reservoir at p_0 and T_0, respectively, and also fix the nozzle cross-sectional area, $A(x)$, but vary the back pressure, p_b. The back pressure is defined as the pressure of the region into which the nozzle discharges. Suppose we select the throat section of the nozzle to be our observation section. Let the area at this section be A and the property values there be p, ρ, and V. As p_b is decreased from p_0, the Mach number at the throat and the mass flow rate through the nozzle increase from zero to values M and $\dot{m}$, respectively. We ask the following question: Does the mass flow rate, $\dot{m}$, passing through the nozzle have a relative maximum? Although in this problem we consider p_b to be the independent variable, we may take the Mach number at the throat to be implicitly related to p_b. The mass flow rate

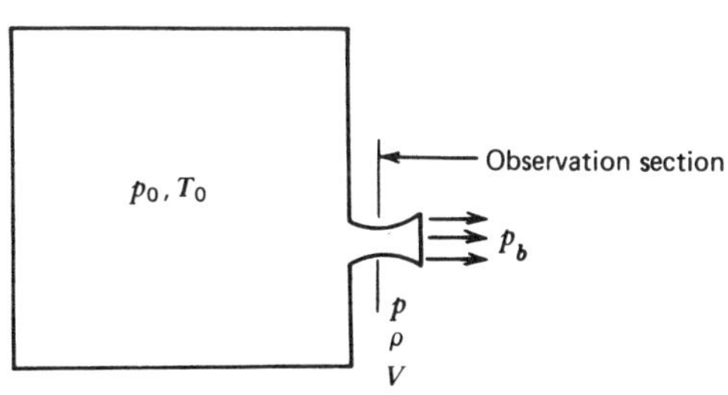

Fig. 8-10. A discharging nozzle.

through the nozzle can be determined from the properties at our observation section; that is,

$$\dot{m} = \rho AV = \frac{p}{RT} AcM = \frac{ApM}{RT} \sqrt{kg_c RT}$$

$$= A\sqrt{\frac{g_c k}{R}} \frac{Mp}{\sqrt{T}}$$

Substituting Eqs. 8.3-15 and 8.3-16 into the above equation, we obtain

$$\dot{m} = A\sqrt{\frac{g_c k}{R}} \frac{p_0}{\sqrt{T_0}} \frac{M}{\left(1 + \frac{k-1}{2} M^2\right)^{(k+1)/2(k-1)}} \qquad \text{Eq. 8.3-22}$$

If $\dot{m}$ has a relative extremum, then

$$\frac{d\dot{m}}{dM} = 0$$

Differentiating the right-hand side of Eq. 8.3-22, with respect to M, we obtain, after some algebraic manipulation, the following equation

$$\frac{d\dot{m}}{dM} = A\sqrt{\frac{g_c k}{R}} \frac{p_0}{\sqrt{T_0}} \left(1 + \frac{k-1}{2} M^2\right)^{-(k+1)/2(k-1)} \left[1 - \frac{(k+1)M^2}{2\left(1 + \frac{k-1}{2} M^2\right)}\right]$$

Thus $d\dot{m}/dM = 0$ when

$$1 - \frac{(k+1)M^2}{2\left(1 + \frac{k-1}{2} M^2\right)} = 0 \qquad \text{Eq. 8.3-23}$$

The above equation is satisfied when $M = 1$. We can see that the relative extremum is a relative maximum, by solving for

$$\frac{M}{\left(1 + \frac{k-1}{2} M^2\right)^{(k+1)/2(k-1)}}$$

from Eq. 8.3-21, and substituting the result into Eq. 8.3-22, giving

$$\dot{m} = A\sqrt{\frac{g_c k}{R}}\frac{p_0}{\sqrt{T_0}}\left(\frac{2}{k+1}\right)^{(k+1)/2(k-1)}\left(\frac{A^*}{A}\right) \qquad \text{Eq. 8.3-24}$$

From Fig. 8-8, we observe that (A/A^*) has a relative minimum at $M = 1$. Therefore, (A^*/A) has a relative maximum at $M = 1$. From Eq. 8.3-24, we can now conclude that the $\dot{m}$ extremum at $M = 1$ is a relative maximum.

The value of (p/p_0) at $M = 1$ is designated as the *critical pressure ratio.* From Eq. 8.3-16, we see that

$$\left(\frac{p}{p_0}\right)_c = \left(\frac{2}{k+1}\right)^{k/(k-1)} \qquad \text{Eq. 8.3-25}$$

For $k = 1.4$, $(p/p_0)_c = 0.528$. Since $M_{th} \leq 1$,

$$0.528 \leq (p/p_0)_{th} \leq 1$$

By plotting $\dot{m}$ versus $(p/p_0)_{th}$ for a particular nozzle, as shown in Fig. 8-11, we see that the extremum is not only a relative maximum, but also an absolute maximum.

In our initial description of the problem, we indicated that if the reservoir conditions (p_0 and T_0) and the nozzle cross-sectional area $A(x)$ are fixed, then the mass flow rate through the nozzle is determined by the back pressure. We now want to consider plots of p versus x for both the converging nozzle and the converging-diverging nozzle, for various values of back pressure, p_b. For the converging nozzle, we will assume that (dA/dx) is zero at the nozzle exit. For the converging-diverging nozzle, we will assume that $(dA/dx)_{th}$ is zero. A plot

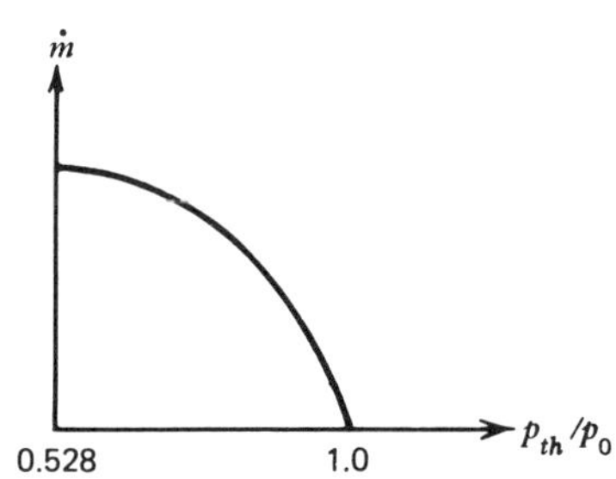

Fig. 8-11. A plot of m versus (p_{th}/p_0).

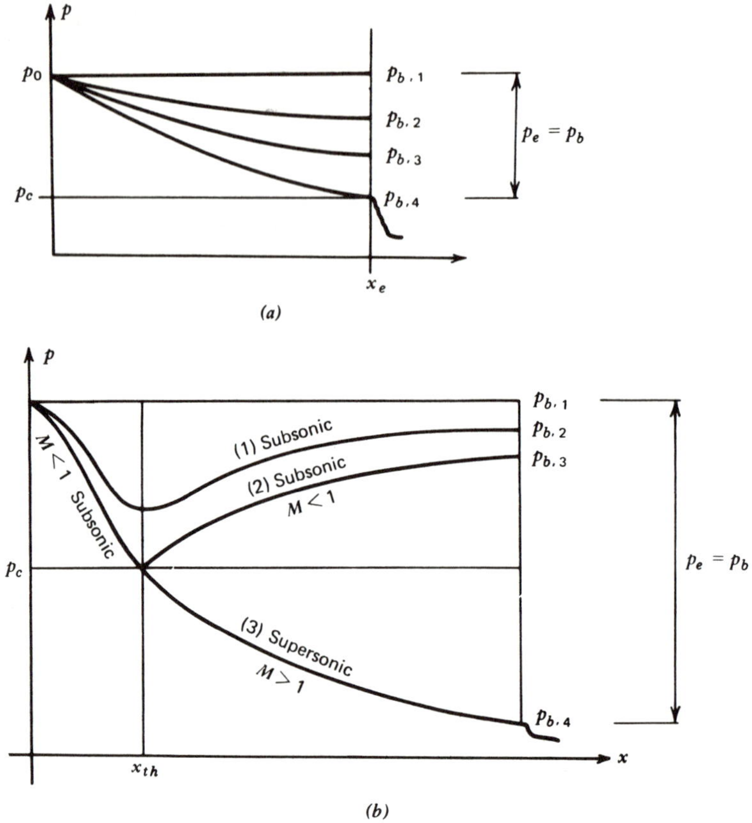

Fig. 8-12. Pressure versus *x*. (*a*) A converging nozzle. (*b*) A converging-diverging nozzle.

of p versus x for the converging nozzle is shown in Fig. 8-12*a*. Notice that when $p_b \geq p_c$, the nozzle exit pressure, p_e, equals p_b. When $p_b = p_c = 0.528p_0$, the nozzle is said to be *choked*, since any further drop in back pressure leaves both the mass flow rate and the exit pressure unaffected. A plot of p versus x for the converging-diverging nozzle configuration is shown in Fig. 8-12*b*. For this case, the nozzle becomes choked when the back pressure is somewhat higher than p_c (the pressure labeled $p_{b,3}$ in Fig. 8-12*b*). However, for this value of back pressure, the pressure at the throat, p_{th}, equals p_c. Any further drop in back pressure leaves both the mass flow rate and the pressure at the throat unaffected. It should also be noted that there exist two limiting curves that allow for isentropic flow to exist completely through the nozzle. These curves are labeled $p_{b,3}$ and $p_{b,4}$ in Fig. 8-12*b*. If the back pressure lies somewhere between $p_{b,3}$ and $p_{b,4}$, then a shock will be formed somewhere in the nozzle between x_{th} and x_e. This is discussed in greater detail later in this chapter. It is also appropriate at this time to give a word of caution with respect to the use of the isentropic tables. Recall that for every A_e/A^* there correspond two values for M_e. The subsonic value is to be used if $p_b \geq p_{b,3}$, and the supersonic value is to be used if $p_b \leq p_{b,4}$. Problems where $p_{b,4} < p_b < p_{b,3}$ are more complicated and are left for further discussion later in this chapter. Finally, it should be observed that for all pressure curves not passing through (x_{th}, p_c)

$$\left(\frac{dp}{dx}\right)_{th} = 0$$

We can see that this is a required condition by writing

$$\frac{dA}{dx} = \left(\frac{dA}{dp}\right)\left(\frac{dp}{dx}\right)$$

However, the geometry of the problem indicates that

$$\left(\frac{dA}{dx}\right)_{th} = 0 \qquad \text{Eq. 8.3-26}$$

For those curves that do not pass through (x_{th}, p_c), $M_{th} < 1$. Under these circumstances, we see by Eq. 8.3-9 that

$$\left(\frac{dA}{dp}\right)_{th} \neq 0$$

Therefore, $(dp/dx)_{th}$ must equal zero to satisfy Eq. 8.3-26. However, if the pressure curve passes through (x_{th}, p_c), then M_{th} equals 1 and by Eq. 8.3-9,

$$\left(\frac{dA}{dp}\right)_{th} = 0$$

Therefore, the condition that $(dA/dx)_{th}$ equals zero can still be satisfied without $(dp/dx)_{th}$ equals to zero. This is shown in Fig. 8-12*b*.

Example 8-1

Air at 70 F is stored in a large reservoir. If the stored air is allowed to escape through a converging nozzle, with the exit area equal to 1 in^2, into a region having a pressure equal to 14.5 psia, find the mass flow rate through the nozzle for the following conditions: (*a*) reservoir pressure equals 20 psia and (*b*) reservoir pressure equals 50 psia.

SOLUTION

First we determine the critical pressure ratio from the isentropic tables. From Table A.6-8, $p/p_0 = 0.528$ for $M = 1$. Thus, $p_c/p_0 = 0.528$. We know that if $p_b/p_0 > 0.528$, then the nozzle is not choked and the exit pressure, p_e, equals p_b.

For case (*a*)

$$\frac{p_b}{p_0} = \frac{14.5}{20.0} = 0.725 > 0.528$$

Thus, $p_e/p_0 = 0.725$. The Mach number at the exit, M_e, can also be determined from the isentropic tables. For $p_e/p_0 = 0.725$, we can determine, by interpolation, that $M_e = 0.694$ and that $T_e/T_0 = 0.9122$. Thus, the temperature at the exit, T_e, equals 483.2 R. We can then determine c_e; that is,

$$c_e = \sqrt{g_c kRT_e} = \sqrt{32.17(1.4)(53.35)(483.2)} = 1077.5 \text{ ft/s}$$

The mass flow rate can be determined by

$$\dot{m} = \rho_e A_e V_e = \frac{p_e}{RT_e} A_e M_e c_e = \frac{14.5(1)(0.694)(1077.5)}{53.35(483.2)}$$

$$= 0.421 \text{ lb}_\text{m}\text{/s}$$

For case (*b*)

$$\frac{p_b}{p_0} = \frac{14.5}{50.0} = 0.29 < 0.528$$

Therefore, the nozzle is choked, giving $M_e = 1$ and $p_e = 0.528(50) = 26.4$ psia. From the isentropic table, $T_e/T_0 = 0.8333$, giving $T_e = 441.4$ R. Thus

$$c_e = \sqrt{g_c kRT_e} = \sqrt{32.17(1.4)(53.35)(441.4)} = 1029.8 \text{ ft/s}$$

Then the mass flow rate is determined by

$$\dot{m} = \frac{p_e}{RT_e} A_e M_e c_e = \frac{26.4(1)(1)(1029.8)}{53.35(441.4)} = 1.15 \text{ lb}_m/\text{s}$$

Example 8-2

A converging-diverging nozzle, with an exit area equal to 1 ft^2, has been designed to operate isentropically at the following conditions: $M_e = 2.0$, $p_0 = 500$ psia, and $T_0 = 700$ R. (*a*) If the reservoir conditions are at the design values, find the maximum back pressure that will also result in the exit conditions being at the design values. (*b*) Determine the mass flow rate and the throat area, A_{th}, for the above specified conditions. (*c*) Determine the maximum back pressure that will still result in a choked nozzle (exit conditions will no longer be at the design values). (*d*) What is the mass flow rate for the conditions specified in part (*c*)?

SOLUTION

(*a*) In order for a flow through a converging-diverging nozzle to be completely isentropic (here we exclude conditions outside the nozzle), $p_b \leq p_{b,4}$, as shown in Fig. 8-12*b*. Thus, $p_{b,4}$ is the maximum back pressure that will result in the exit conditions being at the design values. From the isentropic table, for $M = 2.0$, $p/p_0 = 0.1278$. Since for this case $p_e = p_b$

$$p_e = 0.1278(500) = 63.9 \text{ psia}$$

(*b*) From Example 8-1, the mass flow rate is given by

$$\dot{m} = \frac{p_e}{RT_e} A_e M_e c_e$$

To obtain T_e and c_e observe that for $M = 2.0$, the isentropic table gives $T/T_0 = 0.5556$. Thus,

$$T_e = 0.5556(700) = 388.92 \text{ R}$$

and

$$c_e = \sqrt{32.17(1.4)(53.35)(388.9)} = 966.7 \text{ ft/s}$$

Then from the above formula for $\dot{m}$, we have

$$\dot{m} = \frac{63.9(144)(1)(2.0)(966.7)}{53.35(388.9)}$$

$$= 857.4 \text{ lb}_m/\text{s}$$

To determine A_{th} note that $M_e = 2.0$. From the isentropic table for $M = 2.0$, $A/A^* = 1.688$. Therefore, $A_e/A^* = A_e/A_{th} = 1.688$. Since $A_e = 1.0 \text{ ft}^2$, $A_{th} = 0.5924 \text{ ft}^2$.

(*c*) The maximum back pressure that will still result in a choked nozzle is $p_{b,3}$ in Fig. 8-12*b*. To determine this value we must determine A_e/A^* from the design conditions. This was determined in part (*b*); that is, for $M = 2.0$, $A/A^* = 1.688$. We now determine the subsonic Mach number, also from the isentropic tables, that corresponds to $A/A^* = 1.688$. From the tables, we see that the corresponding Mach number equals 0.372 (by interpolation). For the case $M_e = 0.372$, $p/p_0 = 0.9108$ and $T/T_0 = 0.9736$. Thus,

$$p_e = 0.9108(500) = 455.4 \text{ psia}$$

(*d*) Since the nozzle is choked for both parts (*a*) and (*c*), and T_0 and p_0 are the same in both parts, the mass flow rates are also the same. Thus, $\dot{m} = 857.4$ lb_m/s. It is left as a student exercise to show that the new exit conditions will give nearly the same answer as that given above (a small difference may result from round-off error).

Before we can proceed with an analysis of the flow for the back pressure conditions: $p_{b,4} < p_b < p_{b,3}$, as shown in Fig. 8-12*b*, we need to develop the Rankine-Hugoniot relations that are associated with normal shock waves. This is done in the next section.

8.4 SHOCK WAVES AND THE RANKINE–HUGONIOT RELATIONS

Let us again consider the problem of a piston moving inside a cylinder. Recall from our discussion in Section 8.2 that, when a piston moves into a gas, a shock wave is created that moves into the undisturbed region at a constant speed, V_s, which is greater than the speed of sound. Consider a coordinate

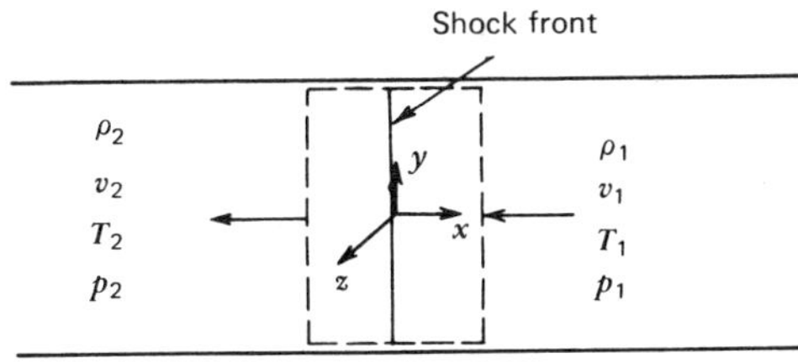

Fig. 8-13. The control volume around the shock front. The (x, y, z) coordinate system is attached to the shock front.

system (x, y, z) attached to the shock front, and the control volume shown as the dashed line in Fig. 8-13. Since V_s is constant, we may consider the flow to be steady. Let us now apply to the control volume the three fundamental laws of conservation of mass, Newton's second law, and the first law of thermodynamics. Application of Eqs. 4.3-3, 4.7-4, and 4.6-11 gives

$$-\rho_1 v_1 + \rho_2 v_2 = 0 \qquad \text{Eq. 8.4-1}$$

$$-p_1 A + p_2 A = \frac{\rho_1}{g_c} A v_1^{\,2} - \frac{\rho_2}{g_c} A v_2^{\,2} \qquad \text{Eq. 8.4-2}$$

and

$$-\rho_1 A v_1\left(J u_1 + \frac{v_1^{\,2}}{2g_c}\right) + \rho_2 A v_2\left(J u_2 + \frac{v_2^{\,2}}{2g_c}\right) = p_1 A v_1 - p_2 A v_2 \qquad \text{Eq. 8.4-3}$$

Since $\rho_1 v_1 = \rho_2 v_2$ and $h = u + p/J\rho$, the last equation reduces to

$$J h_1 + \frac{v_1^{\,2}}{2g_c} = J h_2 + \frac{v_2^{\,2}}{2g_c} \qquad \text{Eq. 8.4-4}$$

We now proceed to develop some of the working formulas used to construct the normal shock table, Table A.6-9. First, substitute the relations

$$h = c_p T \qquad \text{and} \qquad c_p = \frac{k}{k-1}\frac{R}{J}$$

into Eq. 8.4-4. Also introduce into Eq. 8.4-4 the stagnation temperatures on both sides of the shock,[3] $T_{0,1}$ and $T_{0,2}$, by the use of Eq. 8.3-12. This gives

$$\frac{kR}{k-1}T_1 + \frac{2kg_cRT_{0,1}}{2(k-1)g_c}\left(1 - \frac{T_{0,1}}{T_1}\right) = \frac{kR}{k-1}T_2 + \frac{2kg_cRT_{0,2}}{2(k-1)g_c}\left(1 - \frac{T_2}{T_{0,2}}\right)$$

By eliminating those terms which cancel out, the above equation reduces to

$$T_{0,1} = T_{0,2} \qquad \text{Eq. 8.4-5}$$

Thus, the stagnation temperatures on both sides of the shock are the same. Then by Eq. 8.3-15,

$$T_{0,1} = T_1\left(1 + \frac{k-1}{2}M_1{}^2\right)$$

and

$$T_{0,2} = T_2\left(1 + \frac{k-1}{2}M_2{}^2\right)$$

Substituting these expressions into Eq. 8.4-5 gives, after some rearranging,

$$\frac{T_2}{T_1} = \frac{1 + \dfrac{k-1}{2}M_1{}^2}{1 + \dfrac{k-1}{2}M_2{}^2} \qquad \text{Eq. 8.4-6}$$

We can also obtain an expression for p_2/p_1 in terms of the Mach number by first canceling out the cross-sectional area, A, from each term in Eq. 8.4-2 and rearranging the terms as shown here

$$p_2\left(1 + \frac{\rho_2}{p_2}\frac{v_2{}^2}{g_c}\right) = p_1\left(1 + \frac{\rho_1}{p_1}\frac{v_1{}^2}{g_c}\right)$$

Substituting k/kRT for ρ/p gives

$$p_2\left(1 + \frac{kv_2{}^2}{kg_cRT_2}\right) = p_1\left(1 + \frac{kv_1{}^2}{kg_cRT_1}\right)$$

3 We do not know a priori that the stagnation temperatures are the same on both sides of the shock.

Noting that $c^2 = kg_c RT$ and that $M = v/c$, the above equation reduces to

$$\frac{p_2}{p_1} = \frac{1 + kM_1^2}{1 + kM_2^2} \qquad \text{Eq. 8.4-7}$$

Although the stagnation temperatures are the same on both sides of the shock, this is not the case with the stagnation pressures. To see that this is so, substitute Eq. 8.3-16, with the proper subscripts, for both p_2 and p_1, giving

$$\frac{p_{0,2}}{\left(1 + \frac{k-1}{2} M_2^2\right)^{k/(k-1)}} \cdot \frac{\left(1 + \frac{k-1}{2} M_1^2\right)^{k/(k-1)}}{p_{0,1}} = \frac{1 + kM_1^2}{1 + kM_2^2}$$

which reduces to

$$\frac{p_{0,2}}{p_{0,1}} = \frac{(1 + kM_1^2)\left(1 + \frac{k-1}{2} M_2^2\right)^{k/(k-1)}}{(1 + kM_2^2)\left(1 + \frac{k-1}{2} M_1^2\right)^{k/(k-1)}} \qquad \text{Eq. 8.4-8}$$

One additional relation is needed before the set used to construct the shock tables is complete. This last relationship gives M_2 in terms of M_1, and it is obtained as follows. First note that

$$\rho v = \frac{p}{RT} Mc = \frac{pM}{RT} \sqrt{kg_c RT} = \frac{pM\sqrt{kg_c}}{\sqrt{RT}}$$

Substituting this relation into Eq. 8.4-1 and rearranging the terms, we obtain

$$\frac{p_1 M_1 \sqrt{kg_c}}{\sqrt{RT_1}} = \frac{p_2 M_2 \sqrt{kg_c}}{\sqrt{RT_2}}$$

or

$$\frac{p_2}{p_1} = \frac{M_1}{M_2} \sqrt{\frac{T_2}{T_1}}$$

Substituting Eq. 8.4-7 and Eq. 8.4-6 into the above equation gives

$$\frac{1 + kM_1^2}{1 + kM_2^2} = \frac{M_1}{M_2} \frac{\sqrt{1 + \frac{k-1}{2} M_1^2}}{\sqrt{1 + \frac{k-1}{2} M_2^2}} \qquad \text{Eq. 8.4-9}$$

or

$$\frac{M_2\sqrt{1 + \frac{k-1}{2} M_2^2}}{1 + kM_2^2} = \frac{M_1\sqrt{1 + \frac{k-1}{2} M_1^2}}{1 + kM_1^2}$$

We can solve for M_2 in terms of M_1 by first squaring both sides of the above equation and then letting the right-hand side equal L, giving

$$\left(\frac{k-1}{2} - k^2 L\right) M_2^4 + (1 - 2kL) M_2^2 - L = 0$$

Using the quadratic formula, we obtain

$$M_2^2 = \frac{-(1 - 2kL) \pm \sqrt{(1 - 2kL)^2 - 4\left(\frac{k-1}{2} - k^2 L\right)(-L)}}{2\left(\frac{k-1}{2} - k^2 L\right)}$$

With some algebraic manipulation we can show that the terms inside the radical sign can be expressed as a perfect square; that is,

$$(1 - 2kL)^2 + 4L\left(\frac{k-1}{2} - k^2 L\right) = \frac{(M_1^2 - 1)^2}{(1 + kM_1^2)^2}$$

Thus,

$$M_2^2 = \frac{(-1 + 2kL) \pm \frac{M_1^2 - 1}{1 + kM_1^2}}{(k-1) - 2k^2 L} \qquad \text{Eq. 8.4-10}$$

Substituting for L in the above equation and with some algebraic manipulation, we can show that (1) if the negative sign in Eq. 8.4-10 is used, then

$$M_2^2 = M_1^2$$

and (2) if the positive sign in Eq. 8.4-10 is used, then

$$M_2^2 = \frac{M_1^2 + \dfrac{2}{k-1}}{\dfrac{2kM_1^2}{k-1} - 1} \qquad \text{Eq. 8.4-11}$$

Since $M_2 > 0$, and for a shock front $M_2 \neq M_1$, the positive sign in Eq. 8.4-10 is to be used. A plot of M_2 versus M_1, for $k = 1.4$, is shown in Fig. 8-14. Notice that when $M_1 < 1$, $M_2 > 1$; and when $M_1 > 1$, $M_2 < 1$. However, the second law of thermodynamics rules out the possibility that $M_1 < 1$. Recall from thermodynamics that for an ideal gas, with constant specific heats undergoing a process from state 1 to state 2 (see Eq. A.3-31),

$$s_2 - s_1 = c_p \ln \frac{T_2}{T_1} - \frac{R}{J} \ln \frac{p_2}{p_1} \qquad \text{Eq. 8.4-12}$$

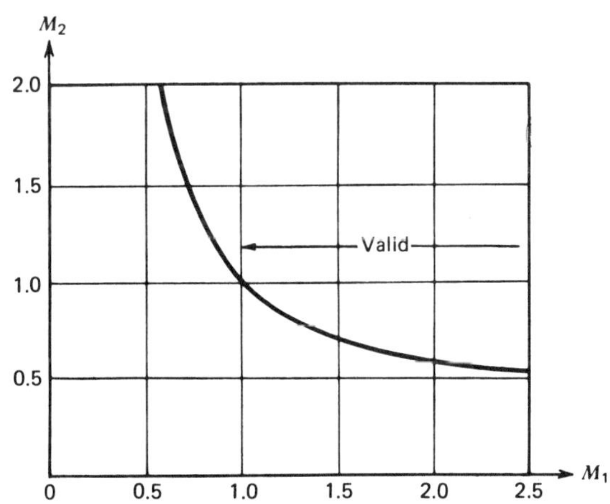

Fig. 8-14. Plot of M_2 versus M_1 for $k = 1.4$. (From *Gas Dynamics*, by J. E. A. John, Allyn and Bacon, Inc., 1969.)

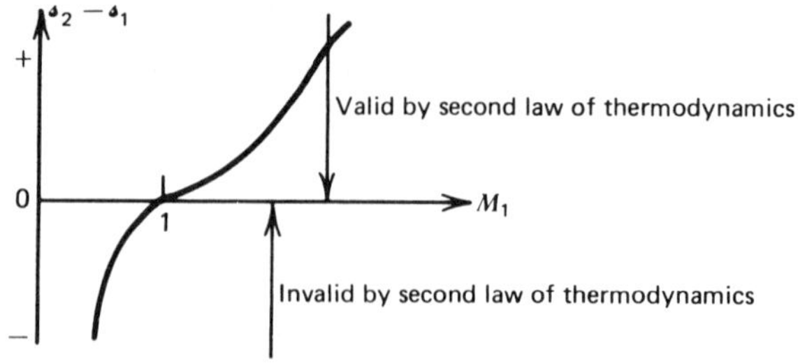

Fig. 8-15. Plot of $(s_2 - s_1)$ versus M_1. (From *Gas Dynamics*, by J. E. A. John, Allyn and Bacon, Inc., 1969.)

Substituting Eqs. 8.4-6, 8.4-7, and 8.4-11 into the above equation, $(s_2 - s_1)$ becomes only a function of M_1 and k. A plot of $(s_2 - s_1)$ versus M_1 would appear as shown in Fig. 8-15. A fluid element in state 1 (the undisturbed state) is considered to undergo an adiabatic, irreversible process to state 2 as the shock wave passes by. However, the second law of thermodynamics states that the change in entropy of an element undergoing such a process must be greater than zero; that is,

$$m\,ds > \frac{\delta Q}{T} = 0$$

Therefore, the second law rules out the lower half of the curve as a possible solution to the problem. A normal shock table is given in Appendix A.6-9. This table gives M_1 versus M_2, T_2/T_1, p_2/p_1, ρ_2/ρ_1, and $p_{0,2}/p_{0,1}$.

One final, useful relation for the normal shock will now be developed. We have already begun the development of this relation in Eq. 8.4-8, which is

$$\frac{p_{0,2}}{p_{0,1}} = \frac{(1 + kM_1^{\,2})\left(1 + \dfrac{k-1}{2}M_2^{\,2}\right)^{k/(k-1)}}{(1 + kM_2^{\,2})\left(1 + \dfrac{k-1}{2}M_1^{\,2}\right)^{k/(k-1)}} \qquad \text{Eq. 8.4-8}$$

We now substitute Eq. 8.4-9 into Eq. 8.4-8, giving

$$\frac{p_{0,2}}{p_{0,1}} = \frac{M_1\left(1 + \dfrac{k-1}{2}M_2^{\,2}\right)^{(k+1)/2(k-1)}}{M_2\left(1 + \dfrac{k-1}{2}M_1^{\,2}\right)^{(k+1)/2(k-1)}} \qquad \text{Eq. 8.4-13}$$

By applying Eq. 8.3-21 on both sides of the shock, we see that the A^* values are different on both sides of the shock. That is, if A is the cross-sectional area where the shock takes place, then by Eq. 8.3-21

$$\frac{\dfrac{A}{A_2^*}}{\dfrac{A}{A_1^*}} = \frac{A_1^*}{A_2^*} = \frac{\dfrac{1}{M_2}\left(\dfrac{2 + (k-1)M_2{}^2}{k+1}\right)^{(k+1)/2(k-1)}}{\dfrac{1}{M_1}\left(\dfrac{2 + (k-1)M_1{}^2}{k+1}\right)^{(k+1)/2(k-1)}}$$

or

$$\frac{A_1^*}{A_2^*} = \frac{M_1\left(1 + \dfrac{k-1}{2} M_2{}^2\right)^{(k+1)/2(k-1)}}{M_2\left(1 + \dfrac{k-1}{2} M_1{}^2\right)^{(k+1)/2(k-1)}}$$

Substituting the above equation into Eq. 8.4-13 gives

$$\frac{p_{0,2}}{p_{0,1}} = \frac{A_1^*}{A_2^*} \qquad \text{Eq. 8.4-14}$$

Let us look at an example problem illustrating the use of the shock table.

Example 8-3

A normal shock wave moves down a tube at 1695 ft/s. The undisturbed air has a temperature of 59 F and a pressure of 14.7 psia. Determine (*a*) the Mach number on both sides of the shock as seen from a coordinate system attached to the shock and (*b*) the temperature and pressure of the air after the shock passes by.

SOLUTION

As seen from a coordinate system attached to the shock, upstream conditions are $v_1 = 1695$ ft/s, $T_1 = 59$ F, and $p_1 = 1.47$ psia, as shown in Fig. 8-16. Then

$$c_1 = \sqrt{kg_c RT_1} = \sqrt{1.4(32.17)(53.35)(519)}$$
$$= 1116.7 \text{ ft/s}$$

and

$$M_1 = \frac{v_1}{c_1} = \frac{1695}{1116.7} = 1.518$$

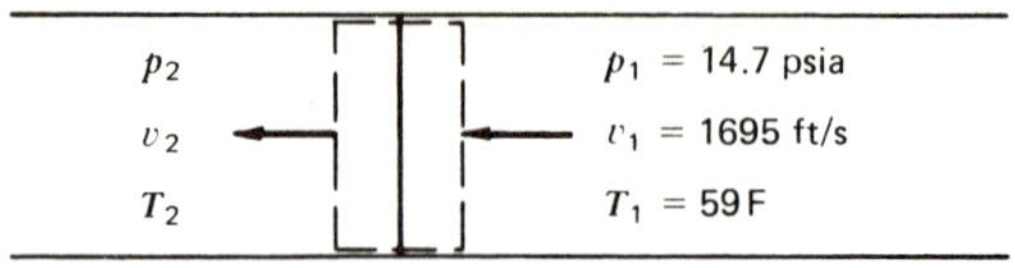

Fig. 8-16. Upstream and downstream conditions as seen from a coordinate system attached to a shock.

From the normal shock table, for $M_1 = 1.518$, we determine by interpolation that $M_2 = 0.6948$, $p_2/p_1 = 2.522$, and $T_2/T_1 = 1.333$. The answers to part (*a*) are: $M_1 = 1.518$ and $M_2 = 0.6948$. The answers to part (*b*) are: $T_2 = 1.333$ $(518.6) = 691.4$ R and $p_2 = 2.522(14.7) = 37.1$ psia.

We may now also determine the shock speed resulting from a piston moving at a uniform velocity into a gas-filled tube. This problem was discussed earlier in Section 8.2. Select (X, Y, Z) to be fixed in space, and let the velocity of the piston and shock front as seen from (X, Y, Z) be $V_p\mathbf{i}$ and $V_S\mathbf{i}$, respectively (see Fig. 8-17). Earlier in this section, the three fundamental laws—conservation of mass, Newton's second law, and the first law of thermodynamics—were applied

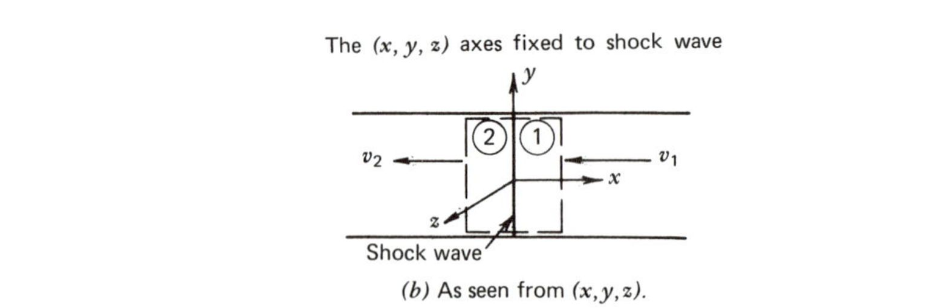

Fig. 8-17. A shock wave resulting from a piston moving into a gas.

to a control volume moving with the shock front, giving Eqs. 8.4-1, 8.4-2, and 8.4-4; that is,

$$\rho_1 v_1 = \rho_2 v_2$$

$$p_1 + \frac{\rho_1 v_1^2}{g_c} = p_2 + \frac{\rho_2 v_2^2}{g_c}$$

$$Jh_1 + \frac{v_1^2}{2g_c} = Jh_2 + \frac{v_2^2}{2g_c} \qquad \text{Eqs. 8.4-15}$$

The velocities v_1 and v_2 are those seen from a coordinate system, (x, y, z), attached to the shock front. Substituting $p = \rho RT$ into the second equation and $h = kRT/J(k-1)$ into the third equation in the above set, we obtain

$$\rho_1 RT_1 + \frac{\rho_1 v_1^2}{g_c} = \rho_2 RT_2 + \frac{\rho_2 v_2^2}{g_c} \qquad \text{Eq. 8.4-16}$$

and

$$\frac{Rk}{k-1} T_1 + \frac{v_1^2}{2g_c} = \frac{Rk}{k-1} T_2 + \frac{v_2^2}{2g_c} \qquad \text{Eq. 8.4-17}$$

Solving Eq. 8.4-16 for RT_2 gives

$$RT_2 = \frac{\rho_1}{\rho_2}\left(RT_1 + \frac{v_1^2}{g_c}\right) - \frac{v_2^2}{g_c}$$

Substituting this equation into Eq. 8.4-17 gives

$$\frac{Rk}{k-1} T_1 + \frac{v_1^2}{2g_c} = \frac{k}{k-1}\left[\frac{\rho_1}{\rho_2}\left(RT_1 + \frac{v_1^2}{g_c}\right) - \frac{v_2^2}{g_c}\right] + \frac{v_2^2}{2g_c}$$

$$= \frac{k}{k-1}\frac{\rho_1 RT_1}{\rho_2} + \frac{k}{k-1}\frac{\rho_1 v_1^2}{\rho_2 g_c} - \frac{k+1}{2(k-1)}\frac{v_2^2}{g_c}$$

Substituting for ρ_1/ρ_2 from the first of set Eqs. 8.4-15, we obtain

$$\frac{Rk}{k-1} T_1 + \frac{v_1^2}{2g_c} = \frac{kR}{k-1}\frac{v_2}{v_1} T_1 + \frac{k}{k-1}\frac{v_2 v_1}{g_c} - \frac{k+1}{2(k-1)}\frac{v_2^2}{g_c} \qquad \text{Eq. 8.4-18}$$

For the direction of the piston motion, shown in Fig. 8-17*a*, the fluid velocities relative to the (X, Y, Z) and (x, y, z) systems are respectively,

$$(\mathbf{v})_{XYZ} = V\mathbf{i}$$

and

$$(\mathbf{v})_{xyz} = -v\mathbf{i} \qquad \text{Eqs. 8.4-19}$$

In this way, v_1 and v_2 will be positive numbers. Since the velocity of (x, y, z) with respect to (X, Y, Z) is $V_S\mathbf{i}$, we know from dynamics that

$$(\mathbf{v})_{XYZ} = (\mathbf{v})_{xyz} + V_S\mathbf{i}$$

Substituting Eqs. 8.4-19 into the above equation, we obtain, after some rearrangement of the terms,

$$v = V_S - V$$

or

$$v_2 = V_S - V_2 \qquad \text{and} \qquad v_1 = V_S - V_1 \qquad \text{Eqs. 8.4-20}$$

But $V_1 = 0$, since the fluid in region 1 as seen from (X, Y, Z) is undisturbed. Since the fluid adjacent to the piston must move with the piston, $V_2 = V_p$. Then by Eqs. 8.4-20

$$v_1 = V_S \qquad \text{and} \qquad v_2 = V_S - V_p \qquad \text{Eqs. 8.4-21}$$

Substituting these values into Eq. 8.4-18 and eliminating those terms which cancel out, we obtain

$$\frac{2V_S}{g_c} - \frac{(k+1)V_p}{g_c} - \frac{2kRT_1}{V_S} = 0$$

or

$$V_S^2 - \frac{k+1}{2}V_pV_S - kRg_cT_1 = 0$$

Solving for V_S, by using the quadratic formula, we find that

$$V_S = \frac{1}{4}(k + 1)V_p + \frac{1}{2}\sqrt{\left(\frac{k + 1}{2}\right)^2 V_p^2 + 4kRg_c T_1} \qquad \text{Eq. 8.4-22}$$

The use of the negative sign in the quadratic formula would lead to a negative V_S, which is not possible in the given arrangement. We can now see that in Example 8-3, the shock front velocity of 1695 ft/s resulted from a piston moving at a velocity of 800 ft/s inside a tube. For then,

$$V_S = 0.25(2.4)(800) + 0.5\sqrt{0.25(2.4)^2(800)^2 + 4(1.4)(53.35)(32.17)(519)}$$
$$= 1695 \text{ ft/s}$$

We now return to our discussion of the converging-diverging nozzle.

8.5 SHOCK WAVES INSIDE CONVERGING-DIVERGING NOZZLES

Recall from Section 8.3 that when the back pressure, p_b, lies between $p_{b,3}$ and $p_{b,4}$, shown in Fig. 8-12, no completely isentropic solution exists that will satisfy the given conditions. This results in the development of a shock wave located between the throat and the exit, as shown in Fig. 8-18. The following example illustrates the method of solution for problems that involve a shock inside a converging-diverging nozzle.

Example 8-4

Air exhausts from a large reservoir through a convergent-divergent nozzle. The pressure and temperature in the tank are, respectively, 1000 psia and 1000 R. A normal shock wave is formed in the divergent section of the nozzle at a point where the static pressure immediately preceding the shock is 127.8 psia (see Fig. 8-19). Assuming that the flow before and after the shock is isentropic, find (*a*) the Mach number before and after the shock; (*b*) the static temperature before and after the shock; (*c*) the loss in stagnation pressure as a result of the shock process; and (*d*) the ratio A_1^*/A_2^*.

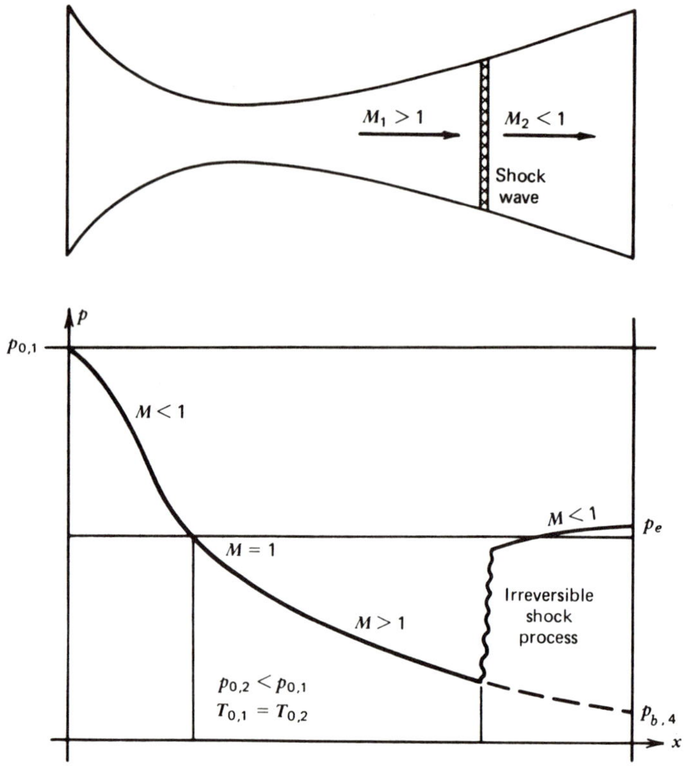

Fig. 8-18. A shock wave inside a converging-diverging nozzle.

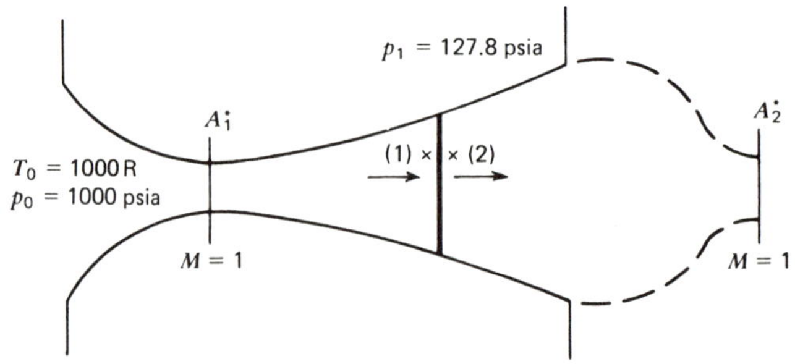

Fig. 8-19. Shock in diverging part of nozzle.

SOLUTION

Since the flow preceding the shock front is isentropic, we may use the isentropic tables to determine the unknown properties at section 1. First determine

$$\frac{p_1}{p_{0,1}} = \frac{127.8}{1000} = 0.1278$$

From the isentropic tables, we see that $M_1 = 2.0$ and $T_1/T_{0,1} = 0.5556$. Therefore, the Mach number and static temperature before the shock are 2.0 and 555.6 R, respectively. From the normal shock table, for the case $M_1 = 2.0$, $M_2 = 0.5774$, $T_2/T_1 = 1.688$, and $p_{0,2}/p_{0,1} = 0.7209$. Thus, the Mach number and static temperature after the shock are 0.5774 and 937.9 R, respectively. The stagnation pressure after the shock is 720.9 psia and, therefore, the loss in stagnation pressure resulting from the shock process is 279.1 psia. From Eq. 8.4-14, $A_1^*/A_2^* = p_{0,2}/p_{0,1} = 0.7209$.

Example 8-5

Suppose the converging-diverging nozzle described in Example 8-2 was run with the reservoir conditions at the design values, but with a back pressure equal to 375 psia. Determine whether a normal shock wave occurs inside the nozzle and, if so, the Mach number before and after the shock, the ratio of the area where the shock occurs and the exit area.

SOLUTION

In the paragraph preceding Example 8-1, it was pointed out that when $p_{b,4} < p_b < p_{b,3}$, as shown in Fig. 8-12*b*, a shock wave develops somewhere in the diverging part of the nozzle. From the solution to Example 8-2, we see that this occurs when 63.9 psia $< p_b <$ 455.4 psia. Since $p_b = 375$ psia, a shock wave does develop in the diverging part of the nozzle. We cannot obtain a direct solution to the remaining questions, since we do not know any of the following information: (1) the Mach number either just before or just after the shock, (2) the location where the shock occurs, or (3) the stagnation pressure of the fluid after the shock. The solution is obtained indirectly by assuming a value for the Mach number, M_1, just before the shock. Since we know that the supersonic, isentropic solution for the Mach number at the nozzle exit is 2.0, the assumed value for M_1 should lie between 1.0 and 2.0. After assuming a value for M_1, we can determine M_2 and $p_{0,2}/p_{0,1}$ from the shock tables. Then from Eq. 8.4-14 we can determine the ratio A_2^*/A_1^*. But $A_1^* = A_{th}$, which was determined in Example 8-2 to be 0.5924 ft^2. Thus, A_2^* can be determined. Knowing that $A_e = 1$ ft^2, we may then determine the ratio, A_e/A_2^*, from which M_e and

$p_e/p_{0,2}$ can be found by using the isentropic tables. Knowing $p_{0,1}$, $p_e/p_{0,2}$, and $p_{0,2}/p_{0,1}$, we can then determine p_e. That is,

$$p_e = \left(\frac{p_e}{p_{0,2}}\right)\left(\frac{p_{0,2}}{p_{0,1}}\right)(p_{0,1})$$

Repeating this process for different values of M_1, we can construct a plot of p_e versus M_1. The value of M_1 where $p_e = 375$ psia is the Mach number just upstream from the shock. Values for the terms described above are tabulated in Table 8-1. Figure 8-20 is a plot of p_e versus M_1, from which we determine that $M_1 = 1.68$ is the solution to our problem. To demonstrate that this is the correct solution, we may determine that $M_2 = 0.6458$, $p_{0,2}/p_{0,1} = 0.8640$, $A_2^* = 0.6856$ ft^2, $A_e/A_2^* = 1.4585$, $M_e = 0.446$, and $p_e = 376.8$ psia. The last value is sufficiently close to the given back pressure of 375 psia that we may take these values as the solution to our problem. From the isentropic table, for $M_1 = 1.68$, we can determine that $A_1/A_1^* = 1.319$; then the ratio of the cross-sectional area where shock occurs and A_e is given by

$$\frac{A_1}{A_e} = \frac{A_1}{A_1^*}\frac{A_1^*}{A_e} = 1.319\left(\frac{0.5924}{1.0}\right) = 0.7813$$

Two final comments regarding shock waves in nozzles. First, suppose in Example 8-5, $p_{b,4} < p_b < 287.6$ psia. We see from Table 8-1 that if a normal shock occurred at A_e, then $p_e = 287.6$ psia. Since $p_b < 287.6$, a normal shock at the exit cannot occur. In this case, a shock wave inclined to the flow direction develops. This type of shock wave, which is called an *oblique shock*, is weaker than a normal shock wave. As the back pressure is lowered further and further toward the supersonic isentropic solution $p_{b,4}$, the angle between the shock and the flow decreases, resulting in a weaker and weaker oblique shock.

TABLE 8-1 Graphical Solution for M_1

Assumed M_1	M_2	$\frac{p_{0,2}}{p_{0,1}}$	A_2^*	$\frac{A_e}{A_2^*}$	M_e	$\frac{p_e}{p_{0,2}}$	p_e
1.2	0.8422	0.9928	0.5967	1.6759	0.3754	0.9073	450.4
1.4	0.7397	0.9582	0.6182	1.6175	0.3918	0.8995	430.9
1.6	0.6684	0.8952	0.6618	1.5110	0.4263	0.8825	395.0
1.8	0.6165	0.8127	0.7289	1.3719	0.4840	0.8519	346.2
2.0	0.5774	0.7201	0.8217	1.2169	0.5771	0.7980	287.6

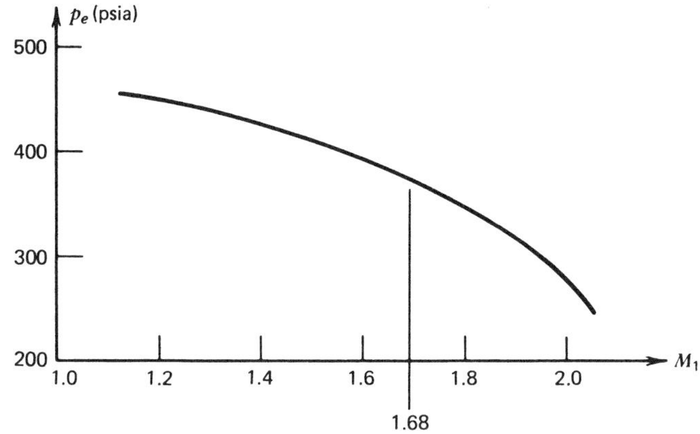

Fig. 8-20. Plot of p_e versus M_1.

8.6 A COMPUTER PROBLEM AS A CLASS PROJECT[4]

A small tank with its longitudinal axis in a vertical position is connected to an air supply, as shown in Fig. 8-21. The tank contains a converging nozzle, which is plugged when it is pressurized. Under this condition, the air supply provides a means for pressurizing the tank. We wish to predict the pressure and temperature time histories within the tank after the nozzle is unblocked. A mathematical model for solving the problem may be developed by applying the energy equation, given by Eq. 4.6-13, first to a control volume just inside the walls of the tank and across the nozzle exit, as indicated by the dashed line in Fig. 8-21, and then to a control volume enclosing the walls of the tank. In our model we will assume that the air properties inside the tank are uniform as far as the volume integration is concerned. In addition, we will also assume that the potential energy and the air velocity inside the tank are negligible. Under these circumstances Eq. 4.6-13, for the former control volume described above, reduces to

$$\frac{d}{dt}(m_a u_a) = -\dot{m}_e\left(u_e + \frac{p_e}{\rho_e J} + \frac{V_e^2}{2g_c J}\right) + \left(\frac{\delta Q}{\delta t}\right)_{w\to a} \qquad \text{Eq. 8.6-1}$$

[4] Much of this section first appeared in the article referenced below:
Bober, W., Heinzman, J. D. and Lapychak, D. M., "Non-steady Nozzle Flow: An Educational Experiment," *The International Journal of Mechanical Engineering Education,* Vol. 7, Aug. 1979, Copyright © 1979 by The Institute of Mechanical Engineers & UMIST. Much of the article is reprinted with the permission of the publisher.

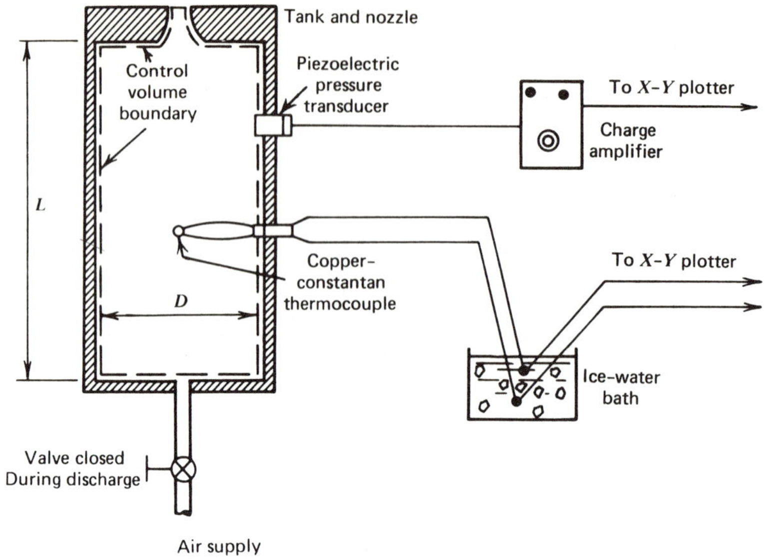

Fig. 8-21. Discharging tank with instrumentation for measuring pressure and temperature time histories.

where the subscript a implies air properties inside the tank and the subscript e implies air properties at the nozzle exit. The $(\delta Q/\delta t)_{\alpha \to w}$ value represents the rate of heat transfer from the air inside the tank to the tank walls. This term is significant and cannot be neglected. Utilizing the following thermodynamic relations

$$u_a = c_{v,a} T_a \qquad \text{Eq. 8.6-2}$$

$$h_e = u_e + \frac{p_e}{\rho_e J} = c_{p,a} T_e \qquad \text{Eqs. 8.6-3}$$

the energy equation reduces to

$$c_{v,a}\left(T_a \frac{dm_a}{dt} + m_a \frac{dT_a}{dt}\right) = -\dot{m}_e\left(c_{p,a} T_e + \frac{1}{J}\frac{V_e^2}{2g_c}\right) + \left(\frac{\delta Q}{\partial t}\right)_{w \to a} \qquad \text{Eq. 8.6-4}$$

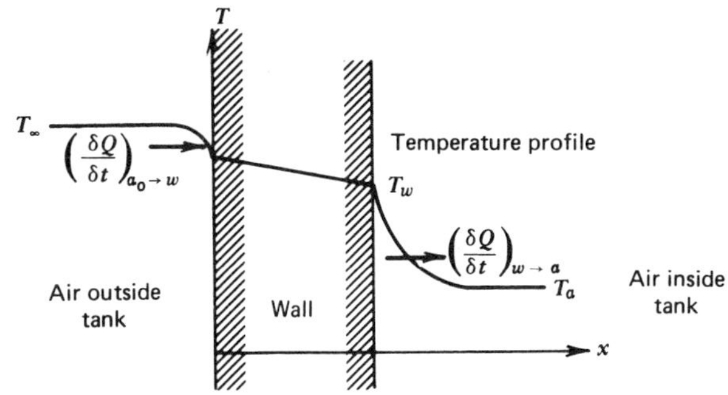

Fig. 8-22. Temperature profile.

For the control volume enclosing the walls of the tank, Eq. 4.6-13 reduces to the following equation when the temperature variation throughout the wall is neglected

$$m_w c_w \frac{dT_w}{dt} = -\left(\frac{\delta Q}{\delta t}\right)_{w\to a} - \left(\frac{\delta Q}{\delta t}\right)_{w\to a_o} \qquad \text{Eq. 8.6-5}$$

where the subscript w implies wall properties. The $(\delta Q/\delta t)_{w\to a}$ and $(\delta Q/\delta t)_{w\to a_o}$ values are, respectively, the heat transfer from the tank walls to the air inside the tank and the heat transfer from the tank walls to the air outside the tank. In dealing with problems involving heat transfer from solid walls to surrounding fluids (see Fig. 8-22), it is common practice to introduce the heat transfer coefficient, $\bar{h}$, which is defined by the following equation

$$\left(\frac{\delta Q}{\delta t}\right)_{w\to a} = \bar{h}A_s(T_w - T_a) \qquad \text{Eq. 8.6-6}$$

where A_s is the surface area of the wall. In the English system, the dimensions of $\bar{h}$ arc Btu/(s-ft^2-F). There are many empirical expressions for $\bar{h}$ under a variety of conditions. We will delay a more detailed discussion of $\bar{h}$ and return to the formulation of the problem.

Applying this concept to the two energy equations gives

$$\frac{dT_a}{dt} = -\frac{T_a}{m_a}\frac{dm_a}{dt} + \frac{A_s\bar{h}_i}{c_{v,a}m_a}(T_w - T_a) - \frac{\dot{m}_e}{m_a}\frac{c_{p,a}}{c_{v,a}}T_e - \frac{\dot{m}_e V_e^2}{2Jg_c c_{v,a} m_a}$$

$$= f_1(t; T_a, T_w, m_a) \qquad \text{Eq. 8.6-7}$$

$$\frac{dT_w}{dt} = \frac{A_s\bar{h}_o}{m_w c_w}(T_\infty - T_w) - \frac{A_s\bar{h}_i}{m_w c_w}(T_w - T_a) = f_2(t; T_a, T_w, m_a) \qquad \text{Eq. 8.6-8}$$

where $\bar{h}_i$ and $\bar{h}_o$ are, respectively, the inside and outside convective heat transfer coefficients, and T_∞ is the outside air temperature a short distance from the tank walls. Conservation of mass and the ideal gas law provide two additional equations; they are

$$\frac{dm_a}{dt} = -\dot{m}_e \qquad \text{Eq. 8.6-9}$$

$$p_a = \frac{m_a R_a T_a}{\mathcal{V}} \qquad \text{Eq. 8.6-10}$$

where R_a is the gas constant for air and $\mathcal{V}$ is the volume inside the tank.

The functional relation for $\dot{m}_e$ in terms of the other variables is obtained from one-dimensional, compressible flow. Two possible cases exist, depending on the ratio p_b/p_a, where p_b is the back pressure, the pressure in the region in which the nozzle empties. In this case, p_b is the room pressure. The two cases are as follows.

Case 1 $\quad \dfrac{p_b}{p_a} > 0.528$

then

$$p_e = p_b$$

$$M_e = \frac{2}{k-1}\left[1 - \left(\frac{p_e}{p_a}\right)^{(k-1)/k}\right]$$

$$T_e = \frac{T_a}{1 + \dfrac{k-1}{2} M_e^2}$$

$$c_e = \sqrt{k R_a g_c T_e}$$

$$V_e = M_e c_e$$

$$\dot{m}_e = \frac{p_e}{R_a T_e} A_e V_e \qquad \text{Eq. 8.6-11}$$

Case 2 $\dfrac{p_b}{p_a} \leq 0.528$

therefore, flow is choked. Then

$$p_e = 0.528\, p_a$$

$$M_e = 1.0$$

$$T_e = \frac{T_a}{1 + \dfrac{k-1}{2}}$$

$$V_e = c_e = \sqrt{kR_a g_c T_e}$$

$$\dot{m}_e = \frac{p_e}{R_a T_e} A_e V_e \qquad \text{Eq. 8.6-12}$$

From these relations we can see that $\dot{m}_e$ is implicitly a function of the variables T_a, T_w, m_a, and t. In substituting Eq. 8.6-9 into Eq. 8.6-7 we establish a system of three, first-order, coupled, ordinary differential equations of the form

$$\frac{dT_a}{dt} = f_1(t;\, T_a,\, T_w,\, m_a)$$

$$\frac{dT_w}{dt} = f_2(t;\, T_a,\, T_w,\, m_a)$$

$$\frac{dm_a}{dt} = f_3(t;\, T_a,\, T_w,\, m_a) \qquad \text{Eqs. 8.6-13}$$

This system may be numerically solved by the Runge–Kutta method, which is described later in this section. To complete the formulation of the problem, an estimate of the convective heat transfer coefficients, $\bar{h}_i$ and $\bar{h}_o$, still needs to be made. At this point, one may either make the crude approximation that $\bar{h}_i$ and $\bar{h}_o$ are constants, or delve deeper into the subject of heat transfer and obtain a more accurate expression for $\bar{h}_i$ and $\bar{h}_o$. For those students selecting the simple route, take $\bar{h}_i = 0.0025$ Btu/(s-ft^2-F) and $\bar{h}_o = 0.0005$ Btu/(s-ft^2-F) and skip directly to the paragraph that describes the Runge–Kutta method. A more accurate means for estimating $\bar{h}_i$ and $\bar{h}_o$ is now described.

Heat transfer in a fluid results from two basic phenomena—one being heat conduction and the other being fluid motion and energy storage. This combination is called *convection*. Convection is further subdivided into forced and natural. Forced convection occurs when the fluid motion is caused by some external mechanism such as a fan, a pump, and the like. Natural convection occurs when the fluid motion results from buoyancy effects caused by density changes in the fluid, which result from the heat transfer. Empirical relations for the convective heat transfer coefficient exist for both forced and natural convection. We may safely assume that $\bar{h}_o$ results from natural convection. The determination of which empirical relation to use for the evaluation of $\bar{h}_i$ is not quite so obvious. First, we should realize that the cross-sectional area of the tank is much greater than the cross-sectional area of the nozzle. Thus, the velocity of the airstream inside the tank will be very small, despite the fact that over part of the run the velocity of the airstream at the exit of the nozzle will be sonic. Thus, it is reasonable to assume that heat transfer will take place from the walls to the air inside the tank predominantly by natural convection. The empirical relation for $\bar{h}$ for natural convection[5] from vertical plates and cylinders is

$$\frac{\bar{h}L}{k_t} = \begin{cases} 0.68 + \dfrac{0.670(\mathrm{Gr}\cdot\mathrm{Pr})^{1/4}}{[1 + (0.492/\mathrm{Pr})^{9/16}]^{4/9}} & \text{for } (\mathrm{Gr}\cdot\mathrm{Pr}) \le 10^9 \\[2ex] \left[0.825 + \dfrac{0.387(\mathrm{Gr}\cdot\mathrm{Pr})^{1/6}}{[1 + (0.492/\mathrm{Pr})^{9/16}]^{8/27}}\right]^2 & \text{for } (\mathrm{Gr}\cdot\mathrm{Pr}) > 10^9 \end{cases} \qquad \text{Eq. 8.6-14}$$

where

$$\mathrm{Gr} = \frac{g\beta(T_w - T_a)L^3}{\nu^2} = \text{Grashof number} \qquad \text{Eq. 8.6-15}$$

Pr is the fluid Prandtl number; k_t is the fluid thermal conductivity; L is the plate or cylinder length; and β is the fluid coefficient of thermal expansion (which for air may be taken to be $1/T$). For air, k_t and Pr are tabulated as functions of temperature (see Table A.6-10). Properties k_t, Pr, ν, and β are all to be evaluated at the film temperature, T_f, which is defined to be

$$T_f = \tfrac{1}{2}(T_w + T_a) \qquad \text{Eq. 8.6-16}$$

[5] See B. V. Karlekar and R. M. Desmond, *Engineering Heat Transfer, West Publishing Co., New York, 1977.*

As a matter of convenience, account for the heat transfer from the horizontal surfaces by increasing the length of the vertical wall by a length equal to D.

To apply the Runge–Kutta formula to our problem, subdivide the time axis into N intervals. Then the values for the dependent variables at time step, t_{i+1}, in terms of the values of the variables at time step, t_i, are

$$T_{a,i+1} = T_{a,i} + \frac{\Delta t}{6}(1_1 + 21_2 + 21_3 + 1_4)$$

$$T_{w,i+1} = T_{w,i} + \frac{\Delta t}{6}(q_1 + 2q_2 + 2q_3 + q_4)$$

$$m_{a,i+1} = m_{a,i} + \frac{\Delta t}{6}(r_1 + 2r_2 + 2r_3 + r_4)$$

$$p_{a,i+1} = \frac{R_a}{V\!\!\!/} m_{a,i+1} T_{a,i+1} \qquad \text{Eq. 8.6-17}$$

where

$$1_1 = f_1(t_i;\ T_{a,i},\ T_{w,i},\ m_{a,i})$$

$$1_2 = f_1\left(t_i + \frac{\Delta t}{2};\ T_{a,i} + \frac{\Delta t}{2} 1_1,\ T_{w,i} + \frac{\Delta t}{2} q_1,\ m_{a,i} + \frac{\Delta t}{2} r_1\right)$$

$$1_3 = f_1\left(t_i + \frac{\Delta t}{2};\ T_{a,i} + \frac{\Delta t}{2} 1_2,\ T_{w,i} + \frac{\Delta t}{2} q_2,\ m_{a,i} + \frac{\Delta t}{2} r_2\right)$$

$$1_4 = f_1(t_i + \Delta t;\ T_{a,i} + \Delta t\, 1_3,\ T_{w,i} + \Delta t\, q_3,\ m_{a,i} + \Delta t\, r_3)$$

The $r_1, r_2, \ldots,$ values are analogous to $1_1, 1_2, \ldots$. The initial conditions, $T_a(0)$, $T_w(0)$, $m_a(0)$, and $p_a(0)$ must also be specified. In the computer program the variations of the thermal properties of air with respect to temperature should be taken into account.

Students interested in developing their computer skills, should write a computer program for solving this problem. Use the following data in the program: nozzle diameter at exit section = 0.05 in.; step size = 0.01 s; p_b = 29.49 in. of

Hg.; $T_\infty = 72$ F; $T_a(0) = 72$ F; $p_a(0) = 50.0$ psig; diameter of tank, $D = 0.229$ ft; tank length, $L = 0.854$ ft; and tank wall thickness, $b = 1/6$ in. The mass density, ρ_w, and the specific heat, c_w, of the tank wall material are 74.8 $\text{lb}_\text{m}/\text{ft}^3$ and 0.30 Btu/(lb_m − F), respectively. Since in the temperature range of interest, $-100\ \text{F} < T_a < 80\ \text{F}$, c_p for air varies insignificantly, we may take $c_{p,a} =$ 0.2402 Btu/(lb_m − F) and $c_{v,a} = 0.1716$ Btu/(lb_m − F). Use the following two subroutines: CHTC(T_a, T_w, h) and THERMP(T_f, ν, Pr, k_t). Subroutine CHTC is to calculate the convective heat-transfer coefficients, $\bar{h}_i$ and $\bar{h}_o$. Subroutine THERMP is to calculate the thermodynamic properties ν, Pr, and k_T at the temperature T_f by linear interpolation from table values. The thermodynamic table values are to be entered into the main program by a "read statement" and placed in common with the subroutine programs. Continue the program for 2000 time steps.

Finally, it should be mentioned that the above problem mathematically models an actual experimental situation carried out at the Rochester Institute of Technology. The tank and nozzle were fabricated by an RIT student. The instrumentation for the experiment included a piezoelectric transducer, a charge amplifier, a thermocouple made from wires 0.001 in. in diameter, and two (x, y) plotters. The mathematical solution agreed reasonably well with the experimental results that are shown in Figs. 8-23 and 8-24.

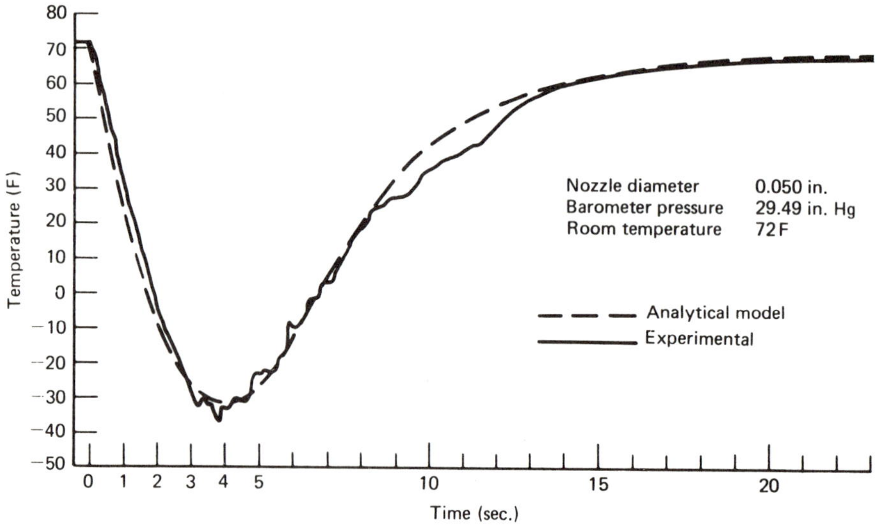

Fig. 8-23. Time history for tank air temperature.

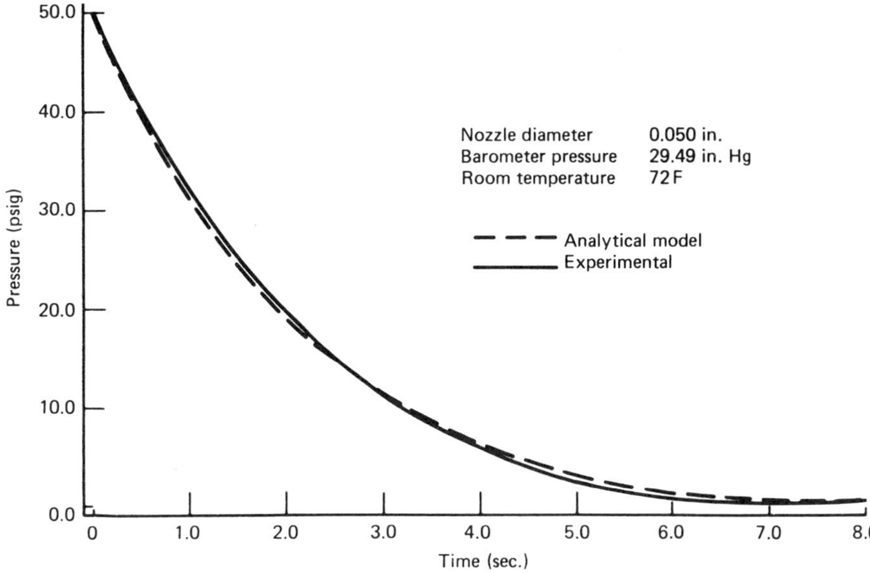

Fig. 8-24. Time history for tank air pressure.

Problems

8-1 Determine the speed of sound in air at (*a*) −20 F, (*b*) 20 C, and (*c*) 200 C.

8-2 Determine the speed of sound in hydrogen at (*a*) 0 C, (*b*) 50 C, and (*c*) 200 C.

8-3 The isothermal compressibility was defined by Eq. 1.4-27; that is

$$\alpha_T = -\frac{1}{\forall}\left(\frac{\partial \forall}{\partial p}\right)_T$$

Similarly, the compressibility at constant entropy is defined by

$$\alpha_S - -\frac{1}{\forall}\left(\frac{\partial \forall}{\partial p}\right)_S$$

For liquids subjected to small pressure disturbances, as in a sound wave, we may take $\rho = \rho(p)$. Under these circumstances, $\alpha_S = \alpha_T$.

(*a*) Show that:

$$\alpha_S = \frac{1}{\rho}\left(\frac{d\rho}{dp}\right)_S \quad \text{and} \quad c = \sqrt{\frac{g_c}{\rho \alpha_S}}$$

Hint. Take $\forall = m/\rho$ and m = constant.

(*b*) Determine the speed of sound in water at 60 F. For water at 1.0 atm and 60 F, $\alpha_T = 3.17 \times 10^{-6}$ in.^{2}/lb$_f$.

8-4 At a given point in a passage of varying area, the static temperature is 40 C and the stagnation temperature is 200 C. The fluid flowing in the passage is air. If the pressure is decreasing in the direction of flow, state whether the passage is convergent or divergent, whether the velocity is increasing or decreasing in the direction of flow, and give the value of the Mach number at the given point.

8-5 Air flows in a pipe of circular cross section under conditions of isentropic flow. At state *A* the static pressure and temperature are, respectively, 3.45×10^5 N/m^2 abs. and 330 K, and the velocity is 145 m/s. The pipe diameter at state *A* is 8.9 cm. At state *B* the velocity has increased to 210 m/s. At state *B* find the following: (*a*) the mass flow rate, (*b*) the pipe diameter, (*c*) the stagnation pressure, and (*d*) the static temperature.

8-6 Nitrogen flows through a converging nozzle from a large reservoir where the temperature is 10 C. The nozzle has an exit area of 0.30×10^{-3} m^2 and discharges into a room where the pressure is 1.0 atm. Find the mass flow rate through the nozzle if the reservoir absolute pressure is (*a*) 1.36 atm and (*b*) 3.40 atm.

8-7 Suppose that in Problem 8-6 the back pressure was reduced to a complete vacuum. Determine the mass flow rate through the nozzle for each case.

8-8 Air enters a convergent-divergent nozzle from a large tank in which the stagnation pressure and temperature are, respectively, 7.0×10^5 N/m^2 abs. and 415 K. If the back pressure is 1.013×10^5 N/m^2 abs. and the mass flow rate is 0.45 kg/s, find the Mach number and area of the exit section if the flow through the entire nozzle is isentropic and the exit pressure is the same as the back pressure.

8-9 Repeat Problem 8-8 for an exhaust pressure of 5.2×10^5 N/m^2 abs. and a mass flow rate of 1.35 kg/s.

8-10 At state 1 in a convergent-divergent nozzle with a mass flow rate of air of 2 lb$_m$/s, the cross-sectional area is 7 in.^{2}. The static pressure and temperature at state 1 are, respectively, 30 psia and 1200 F. Assuming the flow to be isentropic, find the area of the nozzle at a point where the Mach number is 1.30.

8-11 A converging-diverging nozzle connects two very large reservoirs containing air. The exit section of the nozzle is circular with a diameter of 0.075 m. If the nozzle is designed to operate *isentropically* at an exit Mach number of 2.00 when the absolute pressure and temperature of the discharging reservoir are at 5.5×10^6 N/m^2 and -6.5 C, respectively, (*a*) find the maximum pressure in the second reservoir to choke the nozzle, (*b*) determine the mass flow rate when the absolute pressure in the second reservoir is 3.5×10^6 N/m^2, (*c*) determine the mass flow rate when the absolute pressure in the second reservoir is 0.5×10^6 N/m^2.

8-12 A converging-diverging nozzle is used to discharge air from a large tank. The air temperature in the tank is 65°C. The exit and throat areas of the nozzle are 1.5 cm^2 and 1.0 cm^2, respectively. The nozzle back pressure is 1.013×10^5 N/m^2 abs. (*a*) Determine the minimum tank pressure to choke the nozzle. (*b*) Determine the mass flow rate if the tank pressure is 3.45×10^5 N/m^2 abs. (*c*) Determine the mass flow rate if the tank pressure is 6.90×10^5 N/m^2 abs.

8-13 A converging-diverging nozzle is to be used to develop a high-velocity discharge from a rocket engine. The nozzle is designed to operate at an altitude of 30,000 ft, with the chamber pressure and temperature equal to 350 psia and 3000 R, respectively. Under these conditions the flow through the nozzle is to be isentropic, with p_e equal to p_b. The discharging gas can be assumed to have a ratio of specific heats equal to 1.4 and a molecular weight of 20.0. The nozzle exit area is 2.0 ft^2. Determine the thrust, F_T, of the engine, if F_T is defined by

$$F_T = \frac{\rho_e}{g_c} A_e V_e^2$$

8-14 A normal shock wave moves down a tube at a speed of 490 m/s and into undisturbed air at 1.0 atm and 20 C. Determine the air speed, the static and stagnation pressures, and the temperatures behind the shock (as seen from a coordinate system attached to the shock wave).

8-15 Suppose a piston in a very long tube moves at a speed of 240 m/s into air whose undisturbed pressure and temperature are 1.013×10^5 N/m^2 abs. and 15 C, respectively. (*a*) What will be the speed of the normal shock wave moving down the tube? (*b*) What will the air density, temperature, and speed be in the vicinity of the piston.

8-16 Suppose that in Problem 8-8, the back pressure is raised until a normal shock wave forms in the divergent section at a point where the static pressure is 2.46×10^5 N/m^2 abs. Find the conditions immediately following the shock wave. Find the exit pressure and Mach number under these conditions.

8-17 Air speed and temperature measurements at two sections in the divergent part of a converging-diverging nozzle indicate the Mach numbers at sections *a* and *b* are 2.2 and 0.58, respectively (see Fig. 8-25). If the ratio of the cross-sectional areas, A_b/A_a, is 1.8, find p_2/p_1 and p_a/p_b.

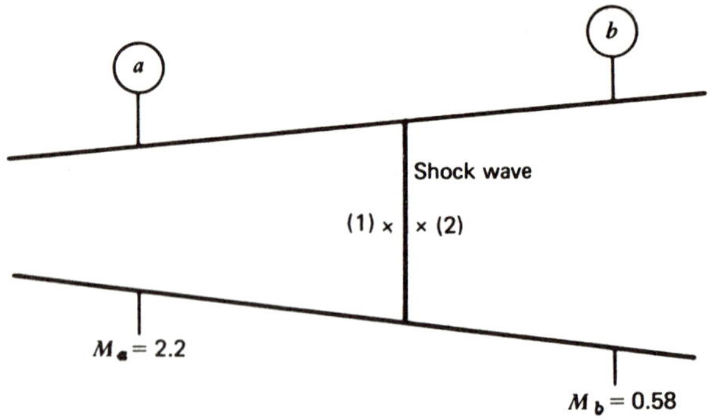

Fig. 8-25. A shock wave in the divergent section of a nozzle.

8-18 A tank containing air at a pressure and temperature of 100 psia and 600 R, respectively, discharges through a convergent-divergent nozzle. A normal shock wave is formed at a point where the static pressure is 27.24 psia. For a mass flow rate of 1.0 lb_m/s find the exit pressure and area if the exit Mach number is 0.50.

8-19 If, for the identical nozzle of Problem 8-8, the exhaust pressure is raised to a value of 5.86×10^5 N/m^2 abs., find the Mach number before and after the shock wave that is formed. (This is a trial-and-error solution.)

8-20 Suppose that in Problem 8-11 the absolute pressure in the second reservoir was 4.0×10^6 N/m^2. (*a*) Determine whether a normal shock wave occurs inside the nozzle. If your answer to part (*a*) is yes, determine (*b*) the Mach number before and after the shock and (*c*) the ratio of the area where the shock occurs and the exit area.

Chapter Nine

An Introduction to Subsonic Aerodynamics

9.1 INTRODUCTION

One of the most important problems in aerodynamics is to determine the lift and drag on an airplane wing. Recall from Chapter 5 that the component of the aerodynamic force which points in the direction of the free stream is defined as the drag, whereas the component normal to the free stream, in the upward direction, is defined as the lift. When the flight speed and altitude result in a free-stream Mach number less than 0.5, the lift on the wing or *airfoil* may be determined by assuming the flow field to be incompressible and nonviscous (ideal flow). The drag, on the other hand, cannot be determined by making this assumption. In fact, the ideal flow assumption in this instance would lead to the incorrect conclusion that the drag is zero (d'Alembert's paradox). Flows around a wing are usually treated as two-dimensional problems. This corresponds to a flow around a wing of infinite span length. In a two-dimensional flow we need only examine what happens at a cross section to understand what is happening in the general flow field. For example, to determine the flow field around an infinite circular cylinder, we need only examine the flow pattern around a circle. Finite length wings are treated by applying a correction term to the two-dimensional solution to account for the finite length. Concepts used in treating the two-dimensional lift problem are developed in Sections 9.1.1 to 9.8. A brief discussion of the finite wing problem is given in Section 9.9. The chapter ends with the treatment of the skin-friction drag problem on a flat plate wing by the von Kármán-Pohlhausen method.

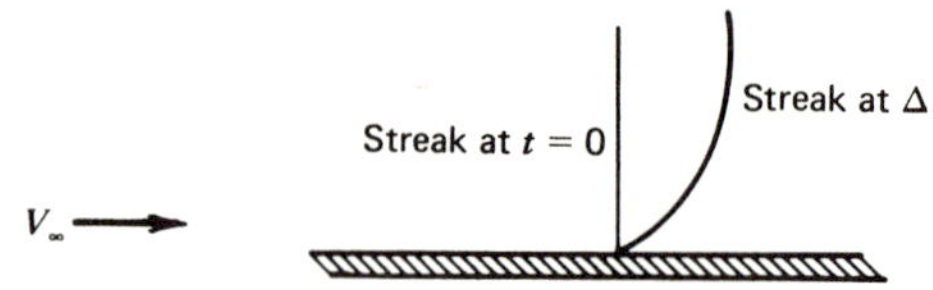

Fig. 9-1. Flat plate in a water tunnel.

9.1.1 THE GENERAL, TWO-DIMENSIONAL FLOW FIELD

To obtain some understanding of the behavior of the flow field around a body, such as a wing, we turn to experimental observations. Suppose a flat plate was installed in the test section of a water tunnel. Suppose at $t = 0$, there is no flow and, at this time, by the use of a dye injector, a vertical streak is placed in the fluid perpendicular to the plate as shown in Fig. 9-1. Then the flow is started. A short time later, at $t = \Delta t$, the streak is distorted and appears as the curve labeled "Streak at Δt" in Fig. 9-1. If a steady flow is established and the velocity profile measured, we would find that $v_1 = 0$ at $y = 0$ and $v_1 \to V_\infty$ at $y = \delta(x)$, where $\delta(x)$ is a short distance[1] above the plate (see Fig. 9-2). For an airflow over a 15 ft long flat plate, with a free-stream velocity of 200 ft/s, δ at the end of the plate[2] would be 2.3 in. The region, $0 \leq y < \delta(x)$, is called the *boundary layer.* Outside the boundary layer, viscous effects are *not* important. However, inside the boundary layer, viscous effects are very important. Furthermore, if the pressure is measured across the boundary layer (in the direction perpendicular to the plate), it would be found that there is no significant pressure variation. Now if an airfoil was placed in a flow at a small angle of attack, once again a boundary layer would be established around the airfoil, as shown in Fig. 9-3. On a typical wing, the flow in the boundary layer is usually laminar near the leading edge. However, the flow tends to become turbulent toward the rear half of the wing. Behind the airfoil is a low pressure, turbulent region, called the

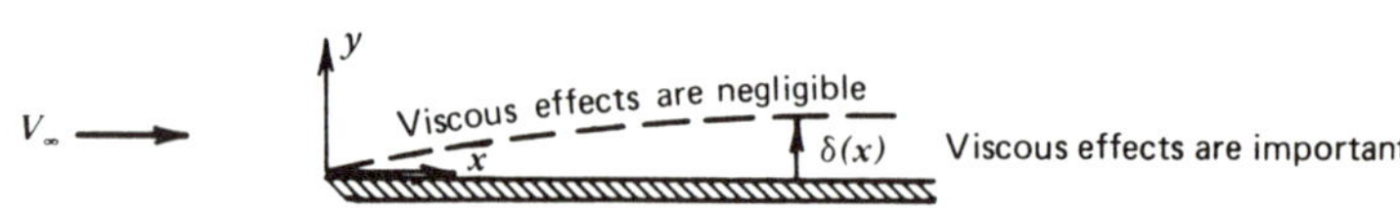

Fig. 9-2. Boundary layer on a flat plate.

[1] It is customary to define δ as the y value where $v_1 = 0.99V_\infty$.

[2] See H. Schlichting, *Boundary-Layer Theory*, 6th ed., McGraw-Hill, New York, 1968, p. 41.

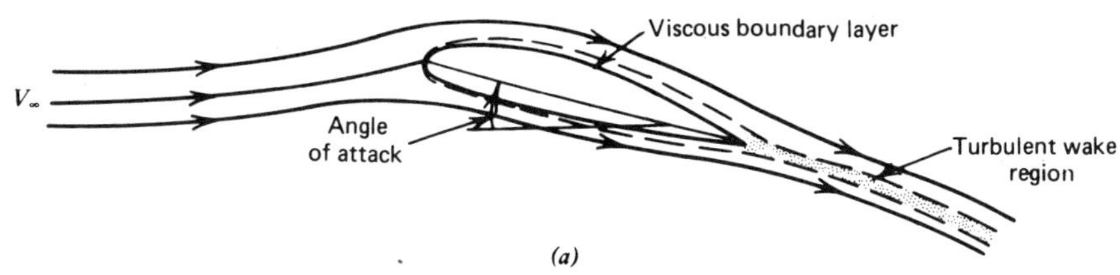

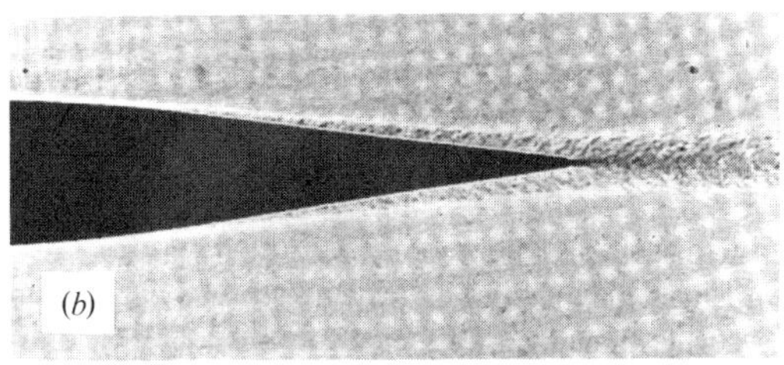

Fig. 9-3. The airfoil at a small angle of attack. (*a*) A sketch of a general flow field. (*b*) Shadow graph of the boundary layer on the rear half of the airfoil. Note that the boundary layer is laminar at the midsection, then turns turbulent toward the rear. Also note that the wake is turbulent. (From *Introduction to Fluid Mechanics* by J. A. Owczarek. Courtesy of the National Physical Laboratory, Teddington, Middlesex, England.

wake. If the airfoil is well designed, the wake is a thin sheet behind the airfoil. The thin sheet is unstable and breaks up. (This is discussed in greater detail in Section 9.9.) Again, the pressure across the boundary layer is constant. This implies that a good approximation for the lift on a wing can be obtained by considering the flow to be *nonviscous*. However, to obtain the drag on the wing, the boundary layer effects must be considered, since skin-friction drag, which is essentially a viscous effect, is a significant factor in the overall drag on a wing. In addition to the skin-friction drag, there is also a pressure drag. The pressure drag on streamlined bodies at small angles of attack is not usually significant, but it is significant at larger angles of attack. We break down the flow field in this manner because there does not exist a general, exact, closed-form solution for the general fluid-flow equations which include viscous effects.

Associated with a boundary layer is the phenomenon called boundary-layer separation. This occurs when a fluid element flowing through the boundary layer is slowed down, not only by viscous effects, but also by an adverse pressure gradient (the pressure increases as we move toward the trailing edge). This occurs on the upper surface of a typical airfoil, as shown in Fig. 9-4. When the boundary layer separates from the wing it causes the wake to become many folds thicker, as shown in Fig. 9-5. If we compare an airfoil under conditions

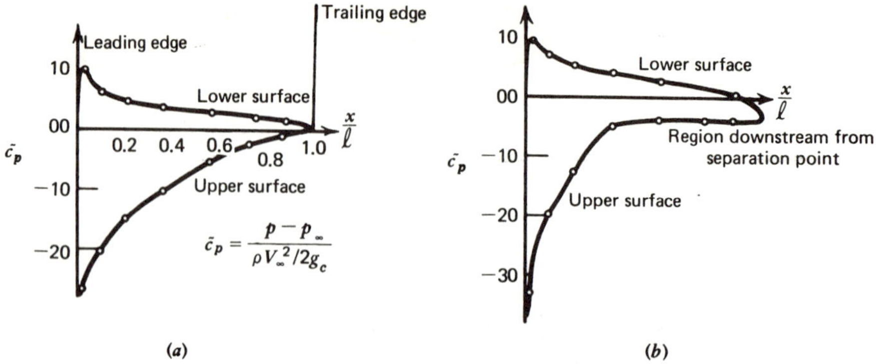

Fig. 9-4. Pressure distribution on the upper and lower surfaces of a typical airfoil. (*a*) Boundary layer has not been separated: $\alpha = 11.6°$. (*b*) Boundary layer has separated: $\alpha = 14.6°$. (From W. F. Durand ed., *Aerodynamic Theory*, Vol. IV, Copyright © 1963, by Dover Publications, Inc.)

just prior to separation and just after separation, we would find that for the latter case the pressure recovery in the region downstream of the separation point is not as great as that which would exist in the former case. Typical pressure-distribution curves for nonseparated and separated flows are shown in Figs. 9-4*a* and 9-4*b*. As a result of this phenomenon, we find that separation causes a significant increase in the pressure drag.

Normally, if we hold the free-stream velocity, V_∞, constant and increase the angle of attach, α, the lift, L, will also increase. However, when separation occurs, the lift begins to drop, as shown in Fig. 9-6. The angle where L starts to decrease as α increases is called the *stall angle*.

For blunt bodies boundary-layer separation is avoided only at very low Reynolds numbers. At moderate and large Reynolds numbers, but below the critical Reynolds number, the pressure drag becomes increasingly important as the Reynolds number of the flow increases. This is so because the pressure in the wake region is much lower than the pressure near the front stagnation

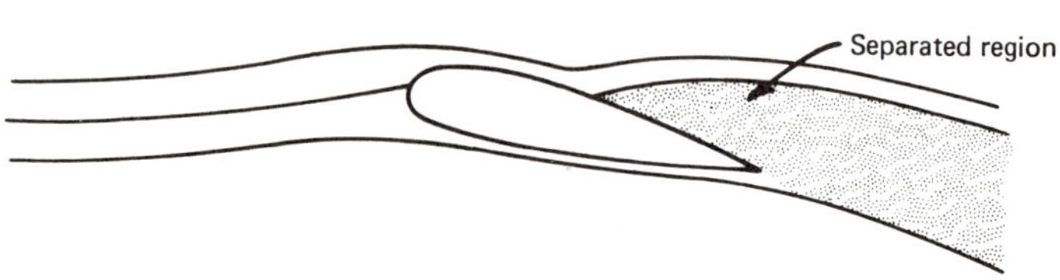

Fig. 9-5. An airfoil with the separated region well up on the airfoil.

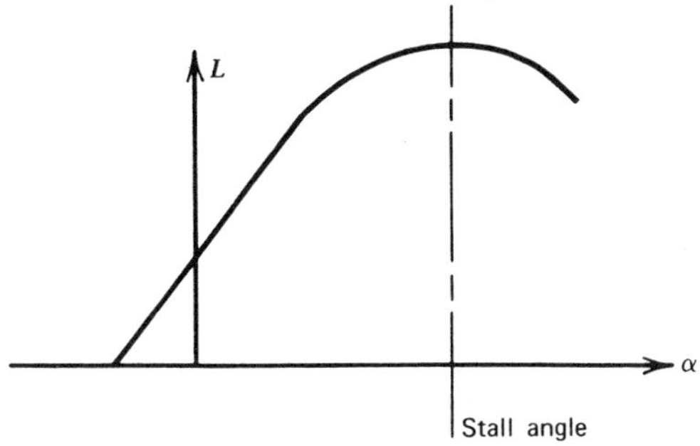

Fig. 9-6. Plot of lift versus angle of attack.

point. Consider the flow over a circular cylinder. Experiments and ideal flow theory show good agreement on the front half of the cylinder, but deviate significantly on the rear half, as shown in Fig. 9-7. Notice the sharp break in the experimental subcritical curve at about 70°. This point represents the position where separation begins. The subcritical label refers to the fact that the boundary layer on the front half is laminar. The supercritical label implies that the boundary layer is turbulent. The strong transverse mixing character of turbulence delays the separation point and also increases the pressure recovery in the wake region. In either case, the relatively low pressure region in the rear of the cylinder results in a significant pressure drag on the cylinder. The transition from a laminar boundary layer to a turbulent boundary layer accounts for the sudden drop in the overall drag coefficient, C_D, at a Reynolds number approximately equal to 4×10^5, as shown in Fig. 9-8.

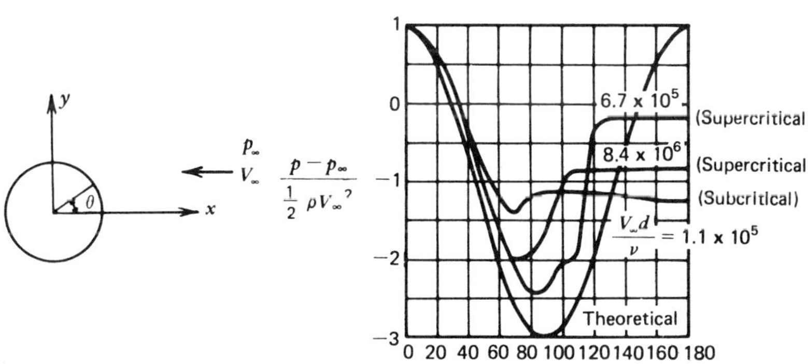

Fig. 9-7. Pressure distribution around a circular cylinder. (From *An Introduction to Fluid Dynamics,* by G. K. Batchelor, Cambridge University Press, New York, 1967.)

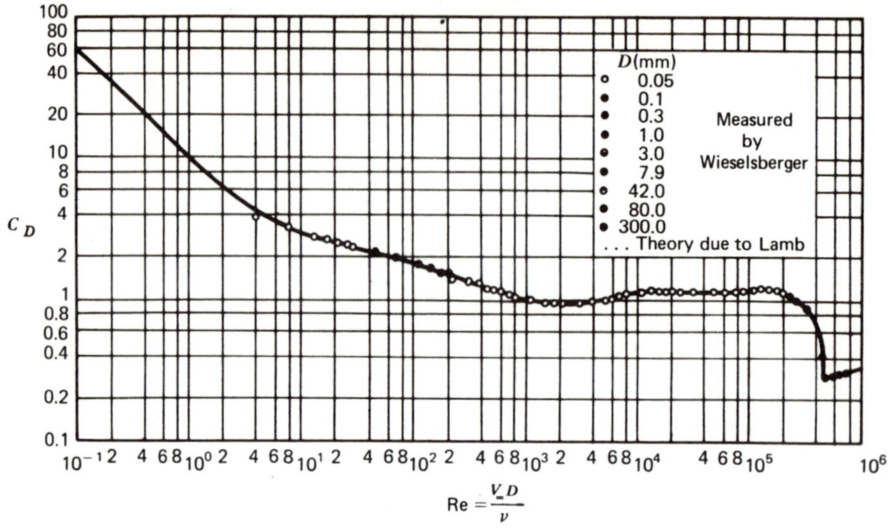

Fig. 9-8. Drag coefficient for circular cylinders as a function of the Reynolds number. (From *Boundary-Layer Theory* by H. Schlichting. Copyright © 1968 by McGraw-Hill. Used with permission of McGraw-Hill Book Company.)

9.2 THE BASIC EQUATIONS FOR THE FLOW FIELD OUTSIDE THE BOUNDARY LAYER

In the next few sections we direct our attention to the concepts that are utilized in developing a method for predicting the lift on an airfoil. As we mentioned earlier, this can be accomplished by considering the flow field to be ideal (inviscid and incompressible). Invariably, the additional assumption is made that the flow field is irrotational. That is,

$$\nabla \times \mathbf{v} = 0 \qquad \text{Eq. 9.2-1}$$

Justification for this assumption is given in Section 9.2.1. The irrotational assumption has the significance of simplifying the mathematical treatment of the problem.

The first governing equation describing the flow field is the continuity equation, Eq. 4.5-2; that is,

$$\frac{\partial \rho}{\partial t} + \nabla \cdot (\rho \mathbf{v}) = 0$$

Since the incompressible flow assumption gives good results for Mach numbers < 0.5, we may take ρ to be a constant for the problem under consideration. Thus, Eq. 4.5-2 reduces to

$$\nabla \cdot \mathbf{v} = 0 \qquad \text{Eq. 9.2-2}$$

Recall that the divergence of a vector $\mathbf{A}$ is given in Chapter 2, Eq. 2.2-12; that is,

$$\nabla \cdot \mathbf{A} = \frac{\partial A_1}{\partial x} + \frac{\partial A_2}{\partial y} + \frac{\partial A_3}{\partial z}$$

Thus Eq. 9.2-2 becomes

$$\frac{\partial v_1}{\partial x} + \frac{\partial v_2}{\partial y} + \frac{\partial v_3}{\partial z} = 0 \qquad \text{Eq. 9.2-3}$$

For a two-dimensional flow, where

$$\mathbf{v} = v_1(x, y)\mathbf{i} + v_2(x, y)\mathbf{j} \qquad \text{Eq. 9.2-4}$$

the above equation reduces to

$$\frac{\partial v_1}{\partial x} + \frac{\partial v_2}{\partial y} = 0 \qquad \text{Eq. 9.2-5}$$

The second basic equation is a statement that the flow field is irrotational; that is, $\nabla \times \mathbf{v} = 0$. For a two-dimensional flow field, where $\mathbf{v}$ is given by Eq. 9.2-4, we see from Eq. 2.2-19 that

$$\nabla \times \mathbf{v} = \mathbf{k}\left(\frac{\partial v_2}{\partial x} - \frac{\partial v_1}{\partial y}\right) = 0 \quad \text{or} \quad \frac{\partial v_2}{\partial x} - \frac{\partial v_1}{\partial y} = 0 \qquad \text{Eq. 9.2-6}$$

The third governing equation to be considered is Euler's equation, which was derived in Chapter 4, and it is given by Eq. 4.8-7. For airflows, gravitational effects are negligible; thus, Eq. 4.8-7 reduces to

$$\frac{\partial \mathbf{v}}{\partial t} + (\mathbf{v} \cdot \nabla)\mathbf{v} = -\frac{g_c}{\rho}\nabla p \qquad \text{Eq. 9.2-7}$$

We now apply the vector identity, given in Eq. 2.2-21,

$$(\mathbf{v} \cdot \nabla)\mathbf{v} = \tfrac{1}{2}\nabla(\mathbf{v} \cdot \mathbf{v}) - \mathbf{v} \times (\nabla \times \mathbf{v})$$

to the second term in Eq. 9.2-7. Also note that for *irrotational* flow, $\nabla \times \mathbf{v} = 0$ and, thus, the second term on the right-hand side of the above equation is zero. Then Eq. 9.2-7 becomes

$$\frac{\partial \mathbf{v}}{\partial t} + \tfrac{1}{2}\nabla(\mathbf{v} \cdot \mathbf{v}) = -\frac{g_c}{\rho}\nabla p \qquad \text{Eq. 9.2-8}$$

For steady flow,

$$\frac{\partial \mathbf{v}}{\partial t} = 0 \qquad \text{Eq. 9.2-9}$$

For ρ equal to a constant,

$$\frac{g_c}{\rho}\nabla p = \nabla\left(\frac{g_c p}{\rho}\right) \qquad \text{Eq. 9.2-10}$$

Substituting Eqs. 9.2-9 and 9.2-10 into Eq. 9.2-8, we obtain

$$\tfrac{1}{2}\nabla(\mathbf{v} \cdot \mathbf{v}) + \nabla\left(\frac{g_c p}{\rho}\right) = 0$$

or

$$\nabla\left(\tfrac{1}{2}(\mathbf{v} \cdot \mathbf{v}) + \frac{g_c p}{\rho}\right) = 0$$

Thus,

$$\tfrac{1}{2}(\mathbf{v} \cdot \mathbf{v}) + \frac{g_c p}{\rho} = f(t) \qquad \text{Eq. 9.2-11}$$

where $f(t)$ is a function to be determined by an initial condition. But for steady flow,

$$f(t) = \text{constant}$$

Dividing through by g_c, and noting that $\mathbf{v} \cdot \mathbf{v} = v^2$, Eq. 9.2-11 reduces to

$$\frac{v^2}{2g_c} + \frac{p}{\rho} = \text{constant}$$

The constant on the right-hand side can be determined if the undisturbed stream conditions are known. In particular, if the stream conditions far upstream from an airfoil are $p = p_\infty$ and $|\mathbf{v}| = V_\infty$, then

$$\frac{v^2}{2g_c} + \frac{p}{\rho} = \frac{V_\infty{}^2}{2g_c} + \frac{p_\infty}{\rho}$$

or

$$p = p_\infty + \frac{\rho}{2g_c}(V_\infty{}^2 - v^2) \qquad \text{Eq. 9.2-12}$$

Thus, knowing the velocity field, the pressure field can be determined.

9.2.1 CONCEPTS LEADING TO THE IRROTATIONAL FLOW ASSUMPTION OUTSIDE THE BOUNDARY-LAYER AND WAKE REGIONS

First, consider the circulation around an arbitrary surface element, S, in space, as shown in Fig. 9-9. Let c be the curve enclosing surface S. The surface S is taken to consist of the same fluid particles and, thus, moves with the flow. The shape of S may change with time. The circulation, Γ, is defined by

$$\Gamma = \oint_c \mathbf{v} \cdot d\mathbf{l} \qquad \text{Eq. 9.2-13}$$

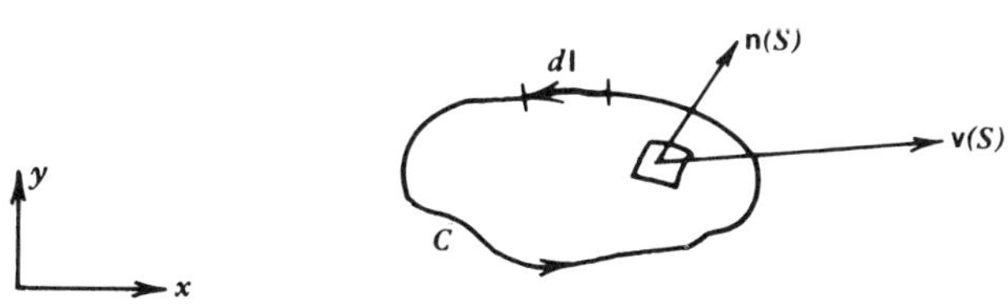

Fig. 9-9. Surface element in the flow field.

By Stokes Theorem,

$$\oint_c \mathbf{v} \cdot d\mathbf{l} = \oiint_S (\nabla \times \mathbf{v}) \cdot d\mathbf{S} \qquad \text{Eq. 9.2-14}$$

The $\nabla \times \mathbf{v}$ is defined as the vorticity vector, ω; that is,

$$\boldsymbol{\omega} \equiv \nabla \times \mathbf{v} \qquad \text{Eq. 9.2-15}$$

It can be shown that *viscosity generates vorticity*. The proof of this fact is beyond the scope of this book. Furthermore, it can also be shown that $d\Gamma/dt = 0$, if the absolute viscosity, μ, is zero. This last statement is Kelvin's theorem. Proof of Kelvin's theorem can be found in the book by Milne–Thomson (see reference section).

Now consider a small arbitrary surface element, S, far upstream from an airfoil at $t = t_0$, as shown in Fig. 9-10. At this time, the velocity, $\mathbf{v}$, of all elements on the surface can be taken to be $V_\infty \mathbf{i}$. Thus, $(\nabla \times \mathbf{v})$ at each point on the surface is zero. Then the circulation, Γ, around the bounding curve of S is zero by Eq. 9.2-14. Suppose at time t, surface element S is in the vicinity of the airfoil, but *has not passed* through the boundary layer. Then, according to Kelvin's theorem, $d\Gamma/dt = 0$. Thus,

$$\Gamma(t) = \Gamma(t_0) = 0 \qquad \text{Eq. 9.2-16}$$

But

$$\Gamma(t) = \oiint_S (\nabla \times \mathbf{v}) \cdot d\mathbf{S} = 0.$$

Since S is arbitrary, $\nabla \times \mathbf{v} = 0$. This is true for all surface elements that do not pass through the boundary layer, where vorticity is generated. This covers the

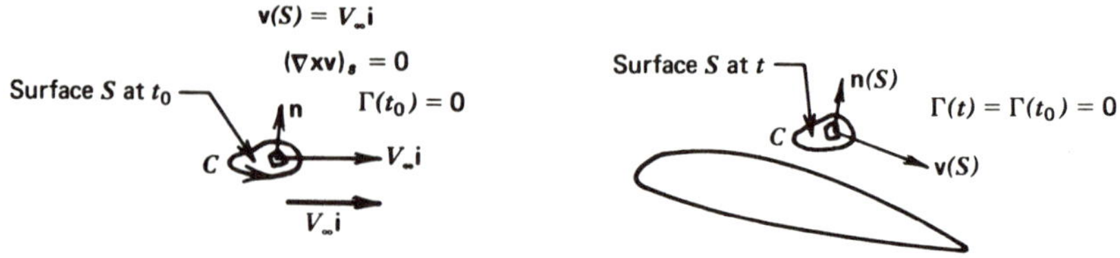

Fig. 9-10. A surface element in the vicinity of the airfoil at time t is located far upstream from the airfoil at time t_0.

entire flow field that is outside the boundary layer and the wake. This, then, is the justification for considering the flow field outside the boundary layer and the wake to be irrotational.

A word of caution is in order. It should be realized that if the circulation is taken around a curve enclosing the airfoil, then $\Gamma \neq 0$. This is so because the curve encloses fluid that is located in the boundary layer where $\nabla \times \mathbf{v}$ is not zero. Thus, if c_2 encloses the airfoil, then, in general,

$$\Gamma(c_2) = \oint_{c_2} \mathbf{v} \cdot d\mathbf{S} \neq 0$$

9.3 THE VELOCITY POTENTIAL AND THE STREAM FUNCTION FOR TWO-DIMENSIONAL FLOW

The velocity potential, Φ, or the stream function, Ψ, are often introduced into fluid-flow problems because they frequently reduce the difficulty in obtaining a solution to a particular problem.

From mathematics, we know that a necessary and sufficient condition for the existence of a scalar function, $f(x, y, z)$, which satisfies the relation

$$\nabla f = \mathbf{F}$$

is that

$$\nabla \times \mathbf{F} = 0$$

A proof of the necessary condition part of the theorem is given in Appendix A.1.10. Since, for the problem under discussion, the velocity flow field is irrotational, we know that $\mathbf{v}$ can be expressed as the gradient of a scalar, Φ; that is,

$$\mathbf{v} = \nabla \Phi \qquad \text{Eq. 9.3-1}$$

where Φ is called the *velocity potential.* In the next paragraph we see how a problem may be simplified by introducing velocity potential.

Introducing the velocity potential, Φ, given in Eq. 9.3-1, into Eq. 9.2-2, we obtain

$$\nabla \cdot \nabla \Phi = 0 \qquad \text{Eq. 9.3-2}$$

Note that $\nabla \cdot \nabla\Phi$ is designated as $\nabla^2\Phi$ and is called the Laplacian. Also recall that

$$\nabla\Phi = \mathbf{i}\frac{\partial\Phi}{\partial x} + \mathbf{j}\frac{\partial\Phi}{\partial y} + \mathbf{k}\frac{\partial\Phi}{\partial z} \qquad \text{Eq. 9.3-3}$$

Introducing Eq. 9.3-3 into Eq. 9.3-2, we obtain

$$\nabla \cdot \nabla\Phi = \frac{\partial}{\partial x}\left(\frac{\partial\Phi}{\partial x}\right) + \frac{\partial}{\partial y}\left(\frac{\partial\Phi}{\partial Y}\right) + \frac{\partial}{\partial z}\left(\frac{\partial\Phi}{\partial z}\right) = 0$$

or

$$\nabla^2\Phi = \frac{\partial^2\Phi}{\partial x^2} + \frac{\partial^2\Phi}{\partial y^2} + \frac{\partial^2\Phi}{\partial z^2} = 0 \qquad \text{Eq. 9.3-4}$$

For a two-dimensional field, where $\Phi = \Phi(x, y)$, Eq. 9.3-4 reduces to

$$\frac{\partial^2\Phi}{\partial x^2} + \frac{\partial^2\Phi}{\partial y^2} = 0 \qquad \text{Eq. 9.3-5}$$

It should be noted that the introduction of the velocity potential, Φ, results in a single complete equation, as indicated by Eq. 9.3-5. This single equation replaces the coupled partial differential equations, Eqs. 9.2-5 and 9.2-6, in the two unknowns, v_1 and v_2.

To determine the forces acting on an airfoil, one procedure is to solve Eq. 9.3-5 with the proper boundary conditions, obtaining a solution to the velocity field adjacent to the airfoil. Then Eq. 9.2-12 would give the pressure acting on the airfoil, from which the forces acting on the airfoil could be determined. Although the general airfoil problem is not an easy one, there exist two classes of airfoils for which solutions can be obtained by a mathematical technique called *conformal mapping*. The first type is classified as the Joukowski airfoils and the second is the Kármán–Treffitz airfoils. In this technique, the solution to the problem of an airflow around an airfoil is related through conformal mapping to the solution to the problem of an airflow around a circular cylinder. The solution to the latter problem can readily be obtained, either by the superposition of three simple flows or by the method of separation of variables. Since the latter method, with the addition of the Fourier series, can also be used to solve a wide class of problems, we have chosen to solve the problem of an ideal flow around a circular cylinder by the method of separation of variables. This is done in Section 9.5. In Section 9.4, we first introduce the reader to several simple flow configurations.

The existence of the stream function for a two-dimensional flow rests on the mathematical theorem which states that the order of the differentiation in obtaining the mixed derivatives of a continuous function, $f(x, y)$, does not affect the end result; that is,

$$\frac{\partial^2 f}{\partial x\, \partial y} = \frac{\partial^2 f}{\partial y\, \partial x}$$

If a function Ψ is defined such that

$$v_1 = \frac{\partial \Psi}{\partial y} \quad \text{and} \quad v_2 = -\frac{\partial \Psi}{\partial x} \qquad \text{Eqs. 9.3-6}$$

then it is required by the above theorem that

$$\frac{\partial v_1}{\partial x} = -\frac{\partial v_2}{\partial y}$$

or

$$\frac{\partial v_1}{\partial x} + \frac{\partial v_2}{\partial y} = 0$$

The above equation is satisfied if the flow is incompressible, which is the case for the problem under consideration. The Ψ value is called the stream function. It can be shown that the curves $\Psi = c_i$, where the c_i values are constants, are streamlines. Recall that a streamline is a curve whose tangent at every point on the curve points in the same direction as the fluid-velocity vector at that point. Observe that on the curves $\Psi = c_i$

$$d\Psi = \frac{\partial \Psi}{\partial x} dx + \frac{\partial \Psi}{\partial y} dy = 0 \qquad \text{Eq. 9.3-7}$$

Thus, the slope of the tangent line to the curve $\Psi = c_i$, denoted as $(dy/dx)_{\Psi=c_i}$, is given by

$$\left(\frac{dy}{dx}\right)_{\Psi=c_i} = -\frac{\dfrac{\partial\Psi}{\partial x}}{\dfrac{\partial\Psi}{\partial y}} \qquad \text{Eq. 9.3-8}$$

Substituting Eqs. 9.3-6 into 9.3-8, we obtain

$$\left(\frac{dy}{dx}\right)_{\Psi=c_i} = \frac{v_2}{v_1} \qquad \text{Eq. 9.3-9}$$

Now consider the slope of the tangent to a streamline. Since the tangent to a streamline points in the same direction as the velocity vector, as shown in Fig. 9-11,

$$\left(\frac{dy}{dx}\right)_{\text{streamline}} = \frac{v_2}{v_1} \qquad \text{Eq. 9.3-10}$$

Comparing Eqs. 9.3-9 and 9.3-10, we see that the curves of both families have the same slope. Therefore, both curve families coincide. We may thus conclude that the curves, $\Psi = c_i$, are also streamlines.

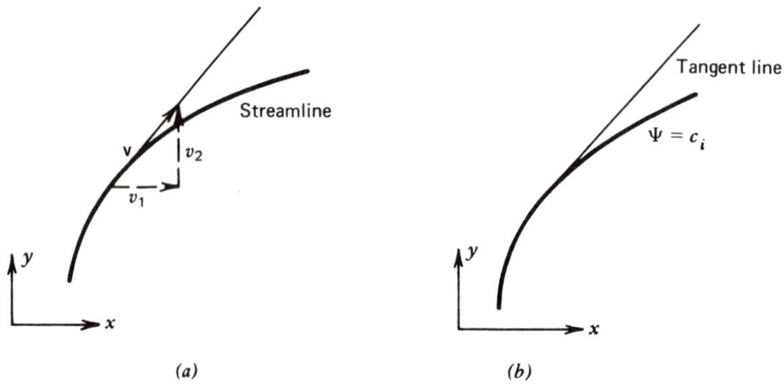

Fig. 9-11. Streamline and velocity vector.

Now let us consider the curves $\phi = k_i$, where the k_i values are constants. On these *equipotential* curves,

$$d\phi = \frac{\partial \phi}{\partial x} dx + \frac{\partial \phi}{\partial y} dy = 0$$

Thus, the slopes of the equipotential curves are given by

$$\left(\frac{dy}{dx}\right)_{\phi = k_i} = -\frac{\dfrac{\partial \phi}{\partial x}}{\dfrac{\partial \phi}{\partial y}} = -\frac{v_1}{v_2} \qquad \text{Eq. 9.3-11}$$

Comparing Eqs. 9.3-9 and 9.3-10, we see that the slopes of the two curve families are negative reciprocals, indicating that the two families are orthogonal to each other, as shown in Fig. 9-12; that is, the curves $\Psi = c_i$ are orthogonal to the curves $\phi = k_i$.

It should also be observed that a difference in the values of Ψ, between two streamline curves, gives the mass rate of flow per unit depth through a surface connecting the two streamlines. Note that if $\dot{m}_s$ equals the mass rate of flow through a surface, S, connecting curves Ψ_1 and Ψ_2 as shown in Fig. 9-13, then

$$\dot{m}_s = \oiint_S \rho \mathbf{v} \cdot d\mathbf{S} = \oiint_S \rho \mathbf{v} \cdot \mathbf{n}\, dl\, dz$$

$$= \rho(\Delta z) \int_P^Q \left(\frac{\partial \Psi}{\partial y} \mathbf{i} - \frac{\partial \Psi}{\partial x} \mathbf{j}\right) \cdot \mathbf{n}\, dl$$

$$- \rho(\Delta z) \int_P^Q \left(\frac{\partial \Psi}{\partial y} \mathbf{i} - \frac{\partial \Psi}{\partial x} \mathbf{j}\right) \cdot (\sin \theta \mathbf{i} \quad \cos \theta \mathbf{j})\, dl$$

$$= \rho(\Delta z) \int_P^Q \left(\frac{\partial \psi}{\partial y} \sin \theta + \frac{\partial \psi}{\partial x} \cos \theta\right) dl$$

But $\sin \theta\, dl = dy$ and $\cos \theta\, dl = dx$; thus

$$\dot{m}_s = \rho(\Delta z) \int_P^Q \left(\frac{\partial \Psi}{\partial x} dx + \frac{\partial \Psi}{\partial y} dy\right) = \rho(\Delta z) \int_{\Psi_2}^{\Psi_2} d\Psi$$

$$= \rho(\Delta z)(\Psi_2 - \Psi_1) \qquad \text{Eq. 9.3-12}$$

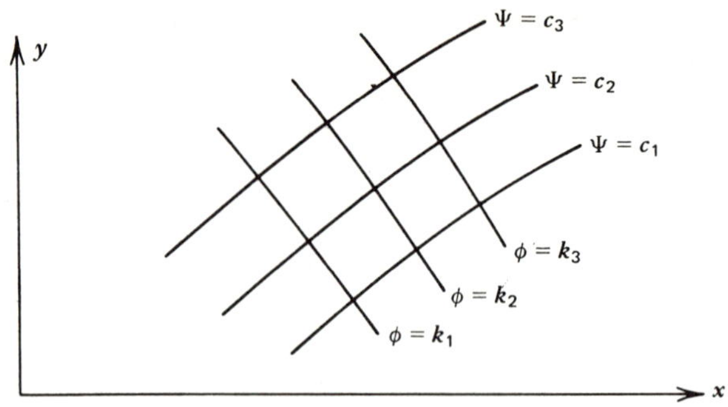

Fig. 9-12. Curves $\Psi = c_i$ are orthogonal to curves $\phi = k_i$.

It can readily be shown that the stream function Ψ satisfies Laplace's equation for an irrotational field. The equation, $\nabla \times \mathbf{v} = 0$, for a two-dimensional flow reduces to

$$\frac{\partial v_2}{\partial x} - \frac{\partial v_1}{\partial y} = 0 \qquad \text{Eq. 9.3-13}$$

Substituting Eqs. 9.3-6 into Eq. 9.3-13 gives

$$\frac{\partial^2 \Psi}{\partial x^2} + \frac{\partial^2 \Psi}{\partial y^2} = 0 \qquad \text{Eq. 9.3-14}$$

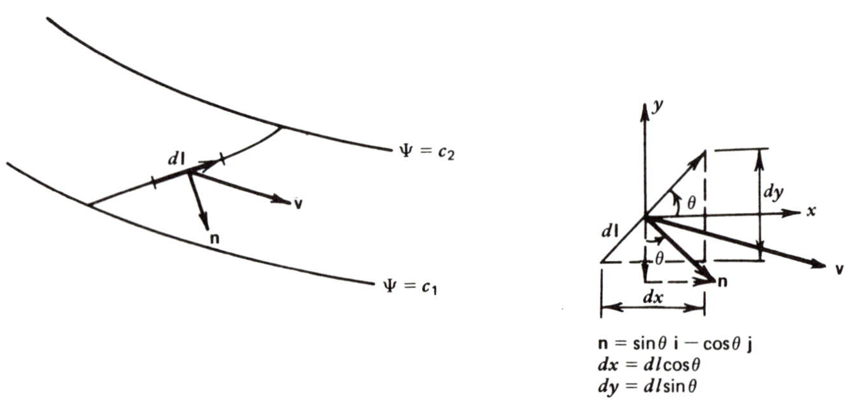

Fig. 9-13. Flow between two streamlines.

For problems that involve circular boundaries, it is useful to express the stream function and the velocity potential in polar coordinates. By suppressing the z-variable in Eqs. 2.2-25 and 2.2-26, we obtained the transformation relationship between the polar coordinate components of a vector and its Cartesian components. If (v_r, v_θ) are the polar coordinate components of the velocity vector, then

$$\mathbf{v} = v_r \mathbf{e}_r + v_\theta \mathbf{e}_\theta \qquad \text{Eq. 9.3-15}$$

and, by Eqs. 2.2-26,

$$\begin{aligned} v_r &= v_1 \cos\theta + v_2 \sin\theta \\ v_\theta &= v_1 \sin\theta + v_2 \cos\theta \end{aligned} \qquad \text{Eqs. 9.3-16}$$

Substituting Eqs. 9.3-6 into Eqs. 9.3-16, we obtain

$$\begin{aligned} v_r &= \frac{\partial \Psi}{\partial y} \cos\theta - \frac{\partial \Psi}{\partial x} \sin\theta \\ v_\theta &= -\frac{\partial \Psi}{\partial y} \sin\theta - \frac{\partial \Psi}{\partial x} \cos\theta \end{aligned} \qquad \text{Eqs. 9.3-17}$$

If we now write an expression for $\partial\Psi/\partial\theta$ and $\partial\Psi/\partial r$ in terms of $\partial\Psi/\partial x$ and $\partial\Psi/\partial y$, we can show, by the use of the chain rule, that

$$\begin{aligned} \frac{1}{r}\frac{\partial \Psi}{\partial \theta} &= -\frac{\partial \Psi}{\partial x} \sin\theta + \frac{\partial \Psi}{\partial y} \cos\theta \\ \frac{\partial \Psi}{\partial r} &= \frac{\partial \Psi}{\partial x} \cos\theta + \frac{\partial \Psi}{\partial y} \sin\theta \end{aligned} \qquad \text{Eqs. 9.3-18}$$

Comparing Eqs. 9.3-18 and 9.3-17, we see that

$$v_r = \frac{1}{r}\frac{\partial \Psi}{\partial \theta} \quad \text{and} \quad v_\theta = -\frac{\partial \Psi}{\partial r} \qquad \text{Eqs. 9.3-19}$$

or

$$\mathbf{v} = \frac{1}{r}\frac{\partial \Psi}{\partial \theta}\mathbf{e}_r - \frac{\partial \Psi}{\partial r}\mathbf{e}_\theta \qquad \text{Eq. 9.3-20}$$

The relationship between the r- and θ-components of the velocity vector and the derivatives of velocity potential, with respect to r and θ, can be obtained by substituting Eqs. 2.2-27 (with the z-component suppressed) and Eq. 9.3-15 into Eq. 9.3-1, giving

$$v_r = \frac{\partial \phi}{\partial r} \quad \text{and} \quad v_\theta = \frac{1}{r}\frac{\partial \phi}{\partial \theta} \qquad \text{Eqs. 9.3-21}$$

and

$$\mathbf{v} = \frac{\partial \phi}{\partial r}\mathbf{e}_r + \frac{1}{r}\frac{\partial \phi}{\partial \theta}\mathbf{e}_\theta \qquad \text{Eq. 9.3-22}$$

9.4 SOME SIMPLE PLANE FLOW CONFIGURATIONS

The superposition of some simple flow configurations may often lead to a streamline pattern that coincides, in a certain region, with that of a flow of interest. Since there is no flow across a streamline, any streamline can be considered to be a solid boundary. Also, recall that the curves, Ψ = constant, are also streamlines and, therefore, a solid boundary may also be represented by a curve, Ψ = constant. Finally, it should be noted that these simple flows, as is true with all ideal flow situations, must satisfy Eqs. 9.3-14 and 9.3-5 (Laplace's equations for both the stream function and the velocity potential).

9.4.1 UNIFORM FLOW

Recall that at the beginning of Chapter 4, we defined a uniform flow to be one where the fluid properties are the same everywhere. Thus, in a uniform flow, the fluid velocity is a constant vector everywhere in the flow field. This results in a family of streamlines that are parallel straight lines. It is obvious that the stream function

$$\psi = Ax + By$$

yields a streamline pattern that satisfies this requirement. That is, if we set $\psi = \psi_i$ = constant, then we obtain the equation of a streamline, which is expressed

$$y = -\frac{A}{B}x + \frac{\psi_i}{B}$$

The constants A and B are determined by the velocity magnitude and direction, or its components. Suppose

$$\mathbf{v} = V_1\mathbf{i} + V_2\mathbf{j}$$

where V_1 and V_2 are both constants, then by Eq. 9.3-6

$$v_1 = V_1 = \frac{\partial\psi}{\partial y} = B$$

and

$$v_2 = V_2 = -\frac{\partial\psi}{\partial x} = -A$$

Thus,

$$\psi = -V_2 x + V_1 y \qquad \text{Eq. 9.4-1}$$

The velocity potential, ϕ, can be obtained from Eq. 9.3-1; that is,

$$\frac{\partial\phi}{\partial x} = V_1 \qquad \text{and} \qquad \frac{\partial\phi}{\partial y} = V_2 \qquad \text{Eq. 9.4-2}$$

Integrating the first equation gives

$$\phi = V_1 x + f(y)$$

Differentiating this equation with respect to y and equating the result to the second equation in the above set (Eqs. 9.4-2), we obtain

$$\frac{\partial\phi}{\partial y} = f'(y) = V_2$$

Thus,

$$f(y) = V_2 y + C$$

and

$$\phi = V_1 x + V_2 y + C$$

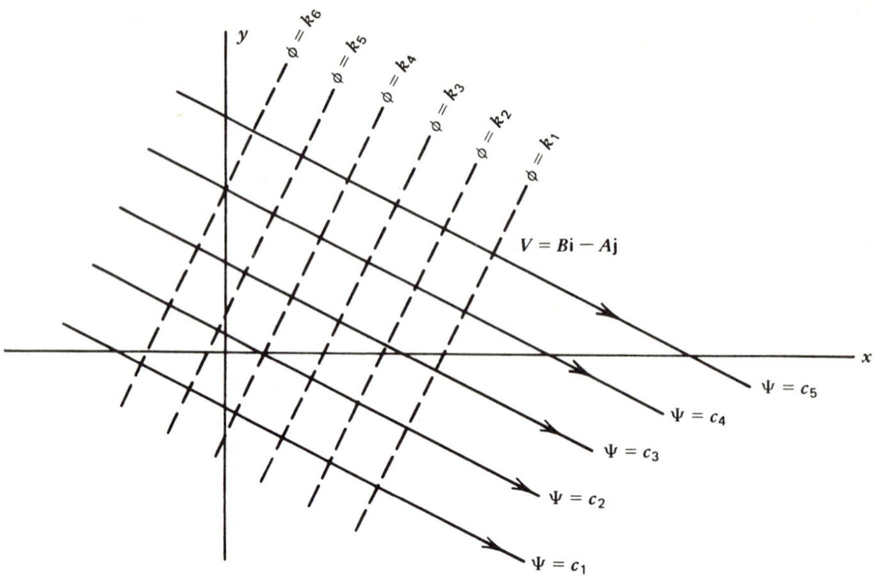

Fig. 9-14. A uniform flow.

It should be noted that in nearly all fluid-flow problems, the velocity potential can only be defined up to an arbitrary constant. This is so because arbitrary constants are usually determined from the velocity field, and the velocity field only defines the derivatives of ϕ. Since the addition or subtraction of an arbitrary constant to ϕ does not affect the flow-field velocity, the arbitrary constant may be taken to be zero. Thus,

$$\phi_1 = V_1 x + V_2 y \qquad \text{Eq. 9.4-3}$$

It should be obvious that ψ and ϕ, as defined by Eqs. 9.4-1 and 9.4-3, both satisfy Laplace's equation. It is left as a student exercise to show that the family of curves $\phi = k_i$ are orthogonal to the family $\psi = c_j$ (see Fig. 9.14).

9.4.2 TWO-DIMENSIONAL SOURCES AND SINKS

A source represents a streamline pattern that emanates radially outward from a single point, say the origin, as shown in Fig. 9-15. This streamline pattern can obviously be represented in polar coordinates by

$$\Psi = A\theta. \qquad \text{Eq. 9.4-4}$$

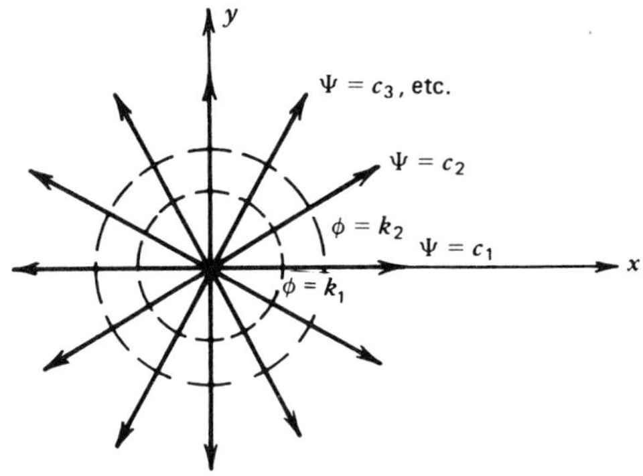

Fig. 9-15. A source flow.

The solution is obvious since Ψ = constant implies θ = constant, which is the equation of a line emanating from the origin in a radial direction. Then

$$\frac{\partial \Psi}{\partial r} = 0 \qquad \text{and} \qquad \frac{\partial \Psi}{\partial \theta} = A \qquad \text{Eqs. 9.4-5}$$

Thus, by Eq. 9.3-20,

$$\mathbf{v} = \frac{A}{r}\,\mathbf{e}_r \qquad \text{Eq. 9.4-6}$$

Since $\mathbf{v} = \nabla\phi$, we obtain by Eq. 9.3-22,

$$\frac{\partial \phi}{\partial r} = \frac{A}{r} \qquad \text{and} \qquad \frac{\partial \phi}{\partial \theta} = 0$$

Integration gives

$$\phi = A \ln r$$

Here, we have taken the constant of integration to be zero. The source strength,

Q, is defined to be the rate at which fluid crosses a cylinder of unit depth whose axis coincides with the z-axis. If the radius of the cylinder is R, then

$$Q = \oiint_S \mathbf{v} \cdot d\mathbf{S} = \int_0^{2\pi} \int_0^1 \frac{A}{R} \mathbf{e}_r \cdot \mathbf{e}_r R \, d\theta \, dz$$

or

$$Q = 2\pi A$$

We see that the source strength is independent of the radius of the cylinder that is considered in its determination. In terms of the source strength, ϕ and Ψ are given by

$$\phi = \frac{Q}{2\pi} \ln r$$

$$\Psi = \frac{Q}{2\pi} \theta \qquad \text{Eqs. 9.4-7}$$

We see that ϕ is not defined at the origin. Finally, when the flow direction is toward the center, we designate the center as a *sink*. One can readily show that ϕ and ψ, as given by Eqs. 9.4-7, both satisfy Laplace's equation.

9.4.3 VORTEX FLOW

A flow pattern that may be used to construct a flow pattern resembling those of tornadoes and whirlpools is designated as a line vortex, or simply a vortex. The streamline pattern of a vortex is represented by concentric circles, along which the magnitude of the velocity is a constant. However, velocity magnitudes on different streamlines are different. Recall that we earlier showed that the curves $\Psi = c_i$ are orthogonal to the curves $\phi = k_i$. Since the streamlines are concentric circles, as shown in Fig. 9-16, the curves $\phi = k_i$ are radial lines emanating from the origin. Thus, the streamlines of the source become the equipotential lines of the vortex, and vice versa. Therefore, in vortex flow,

$$\phi = A\theta$$

Then by Eq. 9.3-22, $\mathbf{v} = (A/r)\mathbf{e}_\theta$. If the direction of flow on the concentric circles is in the positive θ direction, then A is positive. We can determine the stream function for this flow by use of Eq. 9.3-20; then

$$\frac{\partial \Psi}{\partial r} = -\frac{A}{r} \qquad \text{and} \qquad \frac{\partial \psi}{\partial \theta} = 0$$

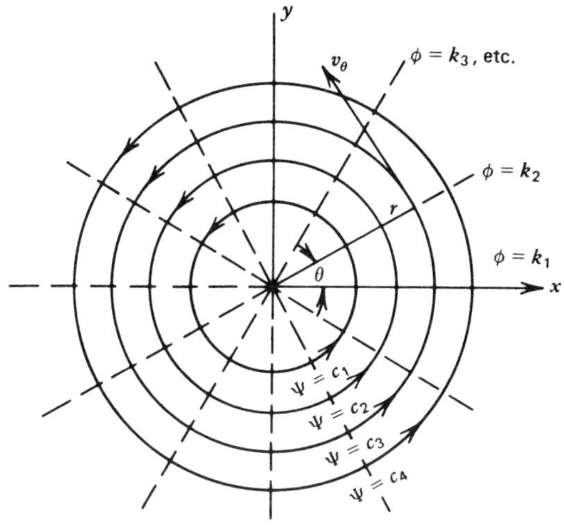

Fig. 9-16. The flow pattern of a vortex.

Integration gives

$$\psi = -A \ln r$$

The strength of a vortex, Γ, is established by the circulation around one of the concentric circles. Then

$$\Gamma = \oint_C \mathbf{v} \cdot d\mathbf{l} = \int_0^{2\pi} \frac{A}{R} \mathbf{e}_\theta \cdot \mathbf{e}_\theta \, R \, d\theta = 2\pi A \qquad \text{Eq. 9.4-8}$$

Again we see that the strength of a vortex is independent of the radius of the circle on which the circulation is evaluated. In terms of the vortex strength, the velocity potential and the stream function become

$$\phi = \frac{\Gamma}{2\pi} \theta$$

$$\Psi = -\frac{\Gamma}{2\pi} \ln r$$

$$\mathbf{v} = \frac{\Gamma}{2\pi} \frac{1}{r} \mathbf{e}_\theta \qquad \text{Eqs. 9.4-9}$$

One can readily show that ϕ and ψ, as given by Eqs. 9.4-9, both satisfy Laplace's equation. Notice that in the vortex the velocity is singular at the origin. This does not disturb us, since in nature, such as in tornadoes and whirlpools, the velocity field is not two-dimensional near the center. Under these circumstances, the vortex would only be used to model the flow field outside the core. When the core shrinks to a line, we have the flow pattern described by the vortex.

Finally, it should be observed that the flow field of a vortex is irrotational except at the center. Recall that if a flow is irrotational in a region, then $\nabla \times \mathbf{v} = 0$ everywhere in that region. The curl of a vector in polar coordinates can be obtained from Eq. 2.2-29 by suppressing the $\mathbf{e}_r$ and $\mathbf{e}_\theta$ components; thus, for a two-dimensional flow

$$\nabla \times \mathbf{v} = \left(\frac{\partial v_\theta}{\partial r} + \frac{v_\theta}{r} - \frac{1}{r}\frac{\partial v_r}{\partial \theta}\right)\mathbf{e}_z \qquad \text{Eq. 9.4-10}$$

By substituting the last of the three equations in the set labeled Eqs. 9.4-9 into Eq. 9.4-10, we see that for the vortex $\nabla \times \mathbf{v} = 0$ everywhere, except at the center where it is not defined.

The streamline pattern of a tornado that is outside the central core region appears as the solid lines in Fig. 9-17a. This pattern can be approximated by the superposition of a vortex and a sink. The stream function and velocity

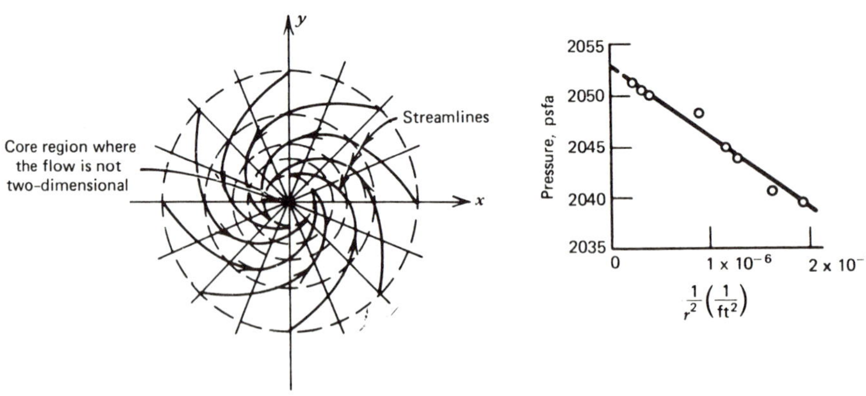

Fig. 9-17. The flow field of a tornado. (*a*) Streamline pattern. (*b*) Plot of p versus $(1/r^2)$. (Extracted from W. Lewis and P. J. Perkins, "Recorded Pressure Distribution in the Outer Portion of a Tornado Vortex," *Monthly Weather Review*, Vol. 81, 1953, p. 379. Also from *Essentials of Engineering Fluid Mechanics*, 3rd ed., by R. M. Olson: (Intext Educational Publishers) Fig. 6-17 and Fig. 6-18 (p. 202), Copyright © 1973 by Harper & Row. Reprinted by permission of the publisher.)

potential for this flow field can be obtained by the use of Eqs. 9.4-7 and 9.4-9; these are

$$\Psi = -\frac{\Gamma}{2\pi}\ln r - \frac{Q}{2\pi}\theta$$

$$\phi = \frac{\Gamma}{2\pi}\theta - \frac{Q}{2\pi}\ln r \qquad \text{Eqs. 9.4-11}$$

The velocity field can be obtained from either Eq. 9.3-20 or 9.3-22, giving

$$\mathbf{v} = -\frac{Q}{2\pi r}\mathbf{e}_r + \frac{\Gamma}{2\pi r}\mathbf{e}_\theta \qquad \text{Eq. 9.4-12}$$

From Eq. 9.4-12, we see that as $r \to \infty$, $|\mathbf{v}| \to 0$; or $V_\infty = 0$. We can also determine that

$$v^2 = \frac{Q^2}{4\pi^2 r^2} + \frac{\Gamma^2}{4\pi^2 r^2} \qquad \text{Eq. 9.4-13}$$

Thus, by Eq. 9.2-12,

$$p(r) = p_\infty - \frac{\rho}{8\pi^2 g_c}(Q^2 + \Gamma^2)\frac{1}{r^2} \qquad \text{Eq. 9.4-14}$$

Thus we see that the pressure field is linear with respect to $(1/r^2)$. A plot of actual pressure measurements in the outer portion of a typical tornado is shown in Fig. 9-17*b*. We see that the pressure is indeed linear with respect to $(1/r^2)$.

There exists one additional flow pattern that may be used in constructing a two-dimensional streamline pattern of interest. This pattern is that of the doublet, which consists of a source-sink pair of equal strength. In the problem section, the student is guided through the determination of the velocity potential and stream function for this pattern.

Finally, it should be noted that the flow field around a circular cylinder can be represented by the superposition of a uniform flow, a doublet, and a vortex. Since this problem is solved by the method of *separation of variables* in the next section, the validity of the above statement is left as a student exercise (see Problem 9-8).

9.5 IDEAL TWO-DIMENSIONAL FLOW AROUND A CIRCULAR CYLINDER

In this section, we solve the problem of an ideal, two-dimensional flow around a circular cylinder by the method of separation of variables. Consider a cylinder of radius, a, in a flow having a uniform velocity, $-V_\infty \mathbf{i}$, far upstream from the cylinder, as shown in Fig. 9-18. Let $\mathbf{v} = \nabla\Phi$, which by Eq. 9.3-22 becomes

$$\mathbf{v} = \frac{\partial \Phi}{\partial r}\mathbf{e}_r + \frac{1}{r}\frac{\partial \Phi}{\partial \phi}\mathbf{e}_\theta \qquad \text{Eq. 9.5-1}$$

The governing equation for the velocity potential is given by Eq. 9.3-5, which in polar coordinates becomes (by Eq. 2.2-30, with the last term on the right-hand side suppressed),

$$\nabla^2\Phi = \frac{\partial^2 \Phi}{\partial r^2} + \frac{1}{r}\frac{\partial \Phi}{\partial r} + \frac{1}{r^2}\frac{\partial^2 \Phi}{\partial \theta^2} = 0 \qquad \text{Eq. 9.5-2}$$

The first boundary condition to be employed is a statement that the velocity vector on the surface of the cylinder must have no normal component; that is,

$$\mathbf{v}(B)\cdot\mathbf{e}_r = 0$$

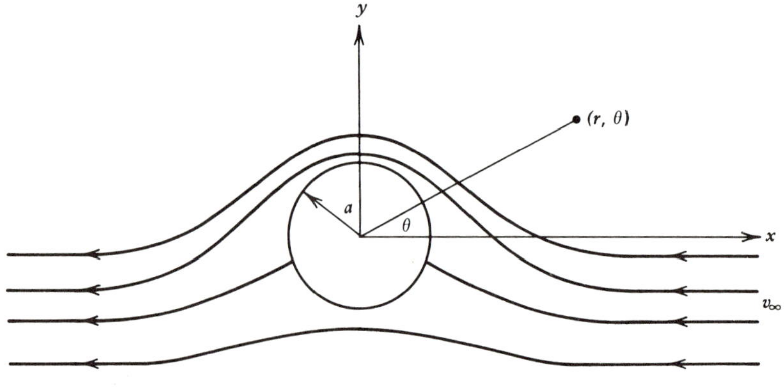

Fig. 9-18. Flow field around a circular cyclinder.

In terms of Φ, the boundary condition becomes

$$\left(\frac{\partial \Phi}{\partial r}\mathbf{e}_r + \frac{1}{r}\frac{\partial \Phi}{\partial \theta}\mathbf{e}_\theta\right)_{r=a} \cdot \mathbf{e}_r = 0$$

or

$$\frac{\partial \Phi}{\partial r}(a, \theta) = 0 \qquad \text{Eq. 9.5-3}$$

The second boundary condition is a statement that the velocity of the fluid that is far away from the cylinder is undisturbed; that is,

$$\mathbf{v}(\infty, \theta) = -V_\infty \mathbf{i} \qquad \text{Eq. 9.5-4}$$

In terms of Φ, Eq. 9.5-4 becomes

$$\nabla\Phi(\infty, \theta) = \left(\frac{\partial \Phi}{\partial r}\mathbf{e}_r + \frac{1}{r}\frac{\partial \Phi}{\partial \theta}\mathbf{e}_\theta\right)_{(\infty, \theta)} = -V_\infty(\cos\theta\,\mathbf{e}_r - \sin\theta\,\mathbf{e}_\theta) \qquad \text{Eq. 9.5-5}$$

Equating the $\mathbf{e}_r$-components on both sides of the equation and also the $\mathbf{e}_\theta$-components, we obtain

$$\frac{\partial \Phi}{\partial r}(\infty, \theta) = -V_\infty \cos\theta \qquad \text{Eq. 9.5-6}$$

and

$$\lim_{r\to\infty}\left(\frac{1}{r}\frac{\partial \Phi}{\partial \theta}(r, \theta)\right) = V_\infty \sin\theta \qquad \text{Eq. 9.5-7}$$

The problem is to solve Eq. 9.5-2, subject to boundary conditions given by Eqs. 9.5-3, 9.5-6, and 9.5-7. The solution can be obtained by the method of separation of variables. In this method, we examine whether the function ϕ of separable form (as indicated below) can satisfy both the partial differential equation and the boundary conditions. Assume that

$$\Phi = f(r)g(\theta)$$

then

$$\frac{\partial \Phi}{\partial r} = f'g \qquad \text{and} \qquad \frac{\partial \Phi}{\partial \theta} = fg'$$

$$\frac{\partial^2 \Phi}{\partial r^2} = f''g \qquad \text{and} \qquad \frac{\partial^2 \Phi}{\partial \theta^2} = fg'' \qquad \text{Eqs. 9.5-8}$$

where $f' = df/dr$, $f'' = d^2f/dr^2$, $g' = dg/d\theta$, and $g'' = d^2g/d\theta^2$. Substituting Eqs. 9.5-8 into Eq. 9.5-2, we obtain

$$f''g + \frac{1}{r} f'g + \frac{1}{r^2} fg'' = 0$$

Dividing through by fg and multiplying by r^2, we find that

$$\frac{r^2 f''}{f} + r\frac{f'}{f} = -\frac{g''}{g} \qquad \text{Eq. 9.5-9}$$

Note that the left-hand side of Eq. 9.5-9 is a pure function of r and that the right-hand side is a pure function of θ. The only way a pure function of r can equal a pure function of θ is for both to be equal to a constant, or

$$\frac{r^2 f''}{f} + r\frac{f'}{f} = -\frac{g''}{g} = \pm\lambda^2 \qquad \text{Rq. 9.5-10}$$

where $\lambda =$ constant. The use of the positive sign leads to $g(\theta) = g(\cos\lambda\theta, \sin\lambda\theta)$, whereas the use of the negative sign leads to $g(\theta) = g(\cosh\lambda\theta, \sinh\lambda\theta)$. The former is periodic; the latter is not. The boundary conditions (Eqs. 9.5-6 and 9.5-7) require that the solution be periodic. Thus, the positive sign is to be used. Equation 9.5-10 yields two equations. They are

$$g'' + \lambda^2 g = 0 \qquad \text{Eq. 9.5-11}$$

$$r^2 f'' + rf' - \lambda^2 f = 0 \qquad \text{Eq. 9.5-12}$$

The indicial equation for Eq. 9.5-11 is

$$m^2 + \lambda^2 = 0 \qquad \text{or} \qquad m = \pm\lambda i$$

This can be readily obtained by assuming that $g = Ce^{m\theta}$. The solution for $g(\theta)$ is

$$g = c_1 \cos\lambda\theta + c_2 \sin\lambda\theta \qquad \text{Eq. 9.5-13}$$

with

$$g'(\theta) = \lambda(-c_1 \sin\lambda\theta + c_2 \cos\lambda\theta) \qquad \text{Eq. 9.5-14}$$

Equation 9.5-12 is a linear, *variable coefficient*, differential equation. This form can be identified as Euler's differential equation, and it is readily solved by a transformation that will turn it into a linear, constant coefficient equation. The transformation

$$r = e^z \qquad \text{or} \qquad z = \ln r \qquad \text{Eqs. 9.5-15}$$

will transform Eq. 9.5-12 into a linear, constant coefficient, differential equation; that is,

$$\frac{df}{dr} = \frac{df}{dz}\frac{dz}{dr} = \frac{1}{r}\frac{df}{dz}$$

$$\frac{d^2f}{dr^2} = \frac{d}{dr}\left(\frac{1}{r}\frac{df}{dz}\right) = -\frac{1}{r^2}\frac{df}{dz} + \frac{1}{r}\frac{d}{dr}\left(\frac{df}{dz}\right)$$

$$= \frac{1}{r}\left[\frac{d}{dz}\left(\frac{df}{dz}\right)\right]\left(\frac{dz}{dr}\right) - \frac{1}{r^2}\frac{df}{dz}$$

$$= \frac{1}{r^2}\left(\frac{d^2f}{dz^2} - \frac{df}{dz}\right) \qquad \text{Eqs. 9.5-16}$$

Substituting Eqs. 9.5-16 into Eq. 9.5-12, we obtain

$$\frac{r^2}{r^2}\left(\frac{d^2f}{dz^2} - \frac{df}{dz}\right) + \frac{r}{r}\frac{df}{dz} - \lambda^2 f = 0$$

or

$$\frac{d^2f}{dz^2} - \lambda^2 f = 0 \qquad \text{Eq. 9.5-17}$$

Equation 9.5-17 is a linear, constant coefficient, differential equation. The corresponding indicial equation is

$$m^2 - \lambda^2 = 0 \qquad \text{or} \qquad m = \pm\lambda$$

Therefore,

$$f(z) = c_3 e^{\lambda z} + c_4 e^{-\lambda z} \qquad \text{Eq. 9.5-18}$$

But,

$$e^{\lambda z} = r^{\lambda} \qquad \text{and} \qquad e^{-\lambda z} = r^{-\lambda} = \frac{1}{r^{\lambda}} \qquad \text{Eqs. 9.5-19}$$

Substituting Eqs. 9.5-19 into Eq. 9.5-18, we find that

$$f(r) = c_3 r^\lambda + \frac{c_4}{r^\lambda} \qquad \text{Eq. 9.5-20}$$

and

$$f'(r) = \lambda\left(c_3 r^{\lambda-1} - \frac{c_4}{r^{\lambda+1}}\right) \qquad \text{Eq. 9.5-21}$$

In the general case, $\lambda = 0$ must also be included. For $\lambda = 0$, Eqs. 9.5-11 and 9.5-12 reduce to

$$g'' = 0 \qquad \text{and} \qquad rf'' + f' = 0 \qquad \text{Eqs. 9.5-22}$$

The solutions to these equations are

$$g = c_5\theta + c_6 \qquad \text{and} \qquad f = c_7 \ln r + c_8 \qquad \text{Eqs. 9.5-23}$$

The product of f and g as given by Eqs. 9.5-20 and 9.5-13 represents one solution for Φ, and the product of f and g as given by Eqs. 9.5-23 represents a second solution for Φ. The general solution is the sum of the two solutions; that is,

$$\Phi(r, \theta) = \left(c_3 r^\lambda + \frac{c_4}{r^\lambda}\right)(c_1 \cos \lambda\theta + c_2 \sin \lambda\theta) + (c_7 \ln r + c_8)(c_5\theta + c_6) \qquad \text{Eq. 9.5-24}$$

The constants c_1 through c_8 and λ are evaluated so that the boundary conditions are satisfied. To accomplish this, the derivatives of Φ are needed. They are

$$\frac{\partial\Phi}{\partial r}(r, \theta) = \lambda\left(c_3 r^{\lambda-1} - \frac{c_4}{r^{\lambda+1}}\right)(c_1 \cos \lambda\theta + c_2 \sin \lambda\theta) + \frac{c_7}{r}(c_5\theta + c_6) \qquad \text{Eq. 9.5-25}$$

and

$$\frac{1}{r}\frac{\partial\Phi}{\partial\theta}(r, \theta) = \frac{\lambda}{r}\left(c_3 r^\lambda + \frac{c_4}{r^\lambda}\right)(-c_1 \sin \lambda\theta + c_2 \cos \lambda\theta) + \frac{c_5}{r}(c_7 \ln r + c_8) \qquad \text{Eq. 9.5-26}$$

The boundary condition given by Eq. 9.5-6 requires that

$$\lambda(c_1 \cos \lambda\theta + c_2 \sin \lambda\theta)\left(c_3 r^{\lambda-1} - \frac{c_4}{r^{\lambda+1}}\right)_{r=\infty} + c_7(c_5\theta + c_6)\left(\frac{1}{r}\right)_{r=\infty}$$
$$= -V_\infty \cos\theta \qquad \text{Eq. 9.5-27}$$

It can be seen that if $\lambda = 1$, $c_2 = 0$, and $c_1 c_3 = -V_\infty$, Eq. 9.5-27 will be satisfied. Substituting these values into Eq. 9.5-25, we obtain

$$\frac{\partial \Phi}{\partial r} = \left(-V_\infty - \frac{c_1 c_4}{r^2}\right)\cos\theta + \frac{c_7}{r}(c_5\theta + c_6) \qquad \text{Eq. 9.5-28}$$

The boundary condition given by Eq. 9.5-6 requires that

$$\left(-V_\infty - \frac{c_1 c_4}{a^2}\right)\cos\theta + \frac{c_7}{a}(c_5\theta + c_6) = 0 \qquad \text{Eq. 9.5-29}$$

Equation 9.5-29 will be satisfied for all θ, if

$$c_1 c_4 = -V_\infty a^2 \qquad \text{and} \qquad c_7 = 0 \qquad \text{Eqs. 9.5-30}$$

Substitution of Eqs. 9.5-30 into Eq. 9.5-24 gives

$$\Phi(r, \theta) = -\left(V_\infty r + \frac{V_\infty a^2}{r}\right)\cos\theta + c_5 c_8 \theta + c_6 c_8 \qquad \text{Eq. 9.5-31}$$

Since Φ, through its boundary conditions, is determined by its derivatives, it can only be determined up to an arbitrary constant. Thus we may take $c_6 c_8$ to be zero. We will also take $c_5 c_8 = A$; then

$$\Phi(r, \theta) = -V_\infty\left(r + \frac{a^2}{r}\right)\cos\theta + A\theta \qquad \text{Eq. 9.5-32}$$

It can readily be seen that Eq. 9.5-32 also satisfies the boundary condition given by Eq. 9.5-7; that is, take the derivative of Eq. 9.5-32 with respect to θ and divide by r, giving

$$\frac{1}{r}\frac{\partial \Phi}{\partial \theta} = V_\infty\left(1 + \frac{a^2}{r^2}\right)\sin\theta + \frac{A}{r}$$

Then

$$\lim_{r\to\infty} \frac{1}{r}\frac{\partial\Phi}{\partial\theta} = V_\infty \sin\theta$$

which agrees with Eq. 9.5-7. It also should be noticed that the velocity potential given by Eq. 9.5-32 is multivalued. This results from the presence of the $A\theta$ term. It is this term that destroys the velocity field symmetry between the upper and lower surfaces of the cylinder. A nonzero A is required if the flow around a cylinder is to be related to that around an airfoil that has a lift. The velocity field for the flow around the cylinder can now be determined from ϕ, since $\mathbf{v} = \nabla\phi$. Substituting the derivatives of Φ into Eq. 9.5-1, we obtain

$$\mathbf{v} = -V_\infty\left(1 - \frac{a^2}{r^2}\right)\cos\theta\,\mathbf{e}_r + \left[\left(1 + \frac{a^2}{r^2}\right)V_\infty \sin\theta + \frac{A}{r}\right]\mathbf{e}_\theta \qquad \text{Eq. 9.5-33}$$

It is left as a student exercise to show that the arbitrary constant A is related to a circulation, Γ, around the cylinder (see Problem 9-12) by the following equation

$$A = \frac{\Gamma}{2\pi}$$

From the general velocity field, we can now determine: (1) $\mathbf{v}(a, \theta)$, (2) $p(a, \theta)$, and (3) the lift, L, and drag, D, acting on the cylinder.[3] It is left as a student exercise (see Problem 9-13) to show that

$$\frac{L}{\ell_S} = \frac{\rho V_\infty \Gamma}{g_c}$$

and Eqs. 9.5-34

$$D = 0$$

where ℓ_S is the span length of the cylinder.

Let us now attempt to bring some physical significance to the circulation term. Recall that the circulation was defined by the following equation

$$\Gamma = \oint_c \mathbf{v}\cdot d\mathbf{l} = \int_0^{2\pi} \mathbf{v}\cdot\mathbf{e}_\theta a\, d\theta$$

[3] See Section 9.1 for the definition of lift and drag.

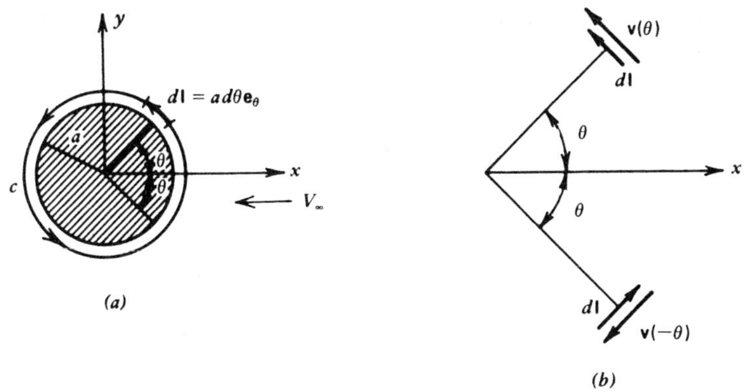

Fig. 9-19. The circulation around a cylinder. (*a*) Curve for evaluating circulation. (*b*) Sketch of **v** and $d\mathbf{l}$ at symmetrical points about the *x*-axis.

where c is the curve around the cylinder. Also note that the integration is counterclockwise (see Fig. 9-19). We can see that if the velocity field is symmetrical about the x-axis then the integration of $\mathbf{v} \cdot d\mathbf{l}$ around the cylinder would be zero, since

$$(\mathbf{v} \cdot d\mathbf{l})_{\theta} = -(\mathbf{v} \cdot d\mathbf{l})_{-\theta}$$

Therefore, the circulation can be thought of as a measure of the nonsymmetry of the velocity field around the cylinder. Of course, the nonsymmetry of the velocity field results in a nonsymmetrical pressure field, which leads to a net lift force on the cylinder. Then it is also not surprising that the lift can be related to the circulation, Γ. Although a nonzero circulation destroys the symmetry of the velocity field with respect to the x-axis, it does not do so with respect to the y-axis. Thus, the net drag force on the cylinder is zero.

9.6 THE STREAM FUNCTION SOLUTION FOR THE FLOW AROUND A CIRCULAR CYLINDER

Once the velocity potential for the flow around a cylinder is determined, the stream function, Ψ, can readily be obtained. Recall by Eq. 9.3-20 that

$$\mathbf{v} = \frac{1}{r}\frac{\partial \Psi}{\partial \theta}\mathbf{e}_r - \frac{\partial \Psi}{\partial r}\mathbf{e}_\theta \qquad \text{Eq. 9.6-1}$$

Equating the $\mathbf{e}_r$ and $\mathbf{e}_\theta$ components in Eq. 9.6-1 to the corresponding $\mathbf{e}_r$ and $\mathbf{e}_\theta$ components in Eq. 9.4-33, we obtain

$$\frac{1}{r}\frac{\partial \Psi}{\partial \theta} = -V_\infty\left(1 - \frac{a^2}{r^2}\right)\cos\theta$$

and Eqs. 9.6-2

$$-\frac{\partial \Psi}{\partial r} = \left[V_\infty\left(1 + \frac{a^2}{r^2}\right)\sin\theta + \frac{A}{r}\right]$$

Integrating the first equation in the set of Eqs. 9.6-2, we obtain

$$\Psi = -V_\infty\left(r - \frac{a^2}{r}\right)\sin\theta + f(r) \qquad \text{Eq. 9.6-3}$$

and

$$\frac{\partial \Psi}{\partial r} = -V_\infty\left(1 + \frac{a^2}{r^2}\right)\sin\theta + f'(r) \qquad \text{Eq. 9.6-4}$$

Equating Eq. 9.6-4 to Eq. 9.6-2, we see that

$$f'(r) = -\frac{A}{r} \qquad \text{and} \qquad f(r) = -A\ln r + B$$

Then,

$$\Psi = -V_\infty\left(r - \frac{a^2}{r}\right)\sin\theta - A\ln r + B \qquad \text{Eq. 9.6-5}$$

Again Ψ can only be determined up to an arbitrary constant; therefore, we may take $\Psi(a, \theta) = 0$, since the boundary $r = a$ is a streamline. Applying this to Eq. 9.6-5, we obtain

$$\Psi(a, \theta) = -A\ln a + B = 0$$

or

$$B = A\ln a$$

Equation 9.6-5 now becomes

$$\Psi = -V_\infty\left(r - \frac{a^2}{r}\right)\sin\theta - A\ln\left(\frac{r}{a}\right) \qquad \text{Eq. 9.6-6}$$

Observe that the velocity potential, Φ, was obtained by the method of separation of variables. The solution for Ψ, the stream function, was then obtained from the solution of Φ. Of course, the problem could have been done in reverse. That is, Ψ could have been obtained by the method of separation of variables and Φ obtained from Ψ.

We have now shown, by Eqs. 9.5-32 and 9.6-6, that for the flow problem, as shown in Fig. 9-16,

$$\Phi(r, \theta) = -V_\infty\left(r + \frac{a^2}{r}\right)\cos\theta + A\theta$$

and Eqs. 9.6-7

$$\Psi(r, \theta) = -V_\infty\left(r - \frac{a^2}{r}\right)\sin\theta - A\ln\left(\frac{r}{a}\right)$$

9.7 THE PRESSURE FIELD ON A FLAT PLATE INCLINED TO THE DIRECTION OF THE FREE STREAM

By the use of complex variables and conformal mapping, the velocity field around certain types of airfoils can be related to the velocity field around a circular cylinder. It is this reason that makes the potential flow solution of the flow around a circular cylinder an important problem. This becomes more apparent when one realizes that the potential flow solution around a circular cylinder does not correctly model the actual flow situation (see Fig. 9.7). This is so because the potential flow solution does not account for the wake region behind the cylinder. In the Reynolds number range of interest, the wake behind the cylinder is quite extensive. However, this is not a major problem with a streamlined airfoil, since in this case the wake region is quite thin. There are two classes of airfoils that may be solved by conformal mapping. These are (1) Joukowski airfoils and (2) Kármán–Treffitz airfoils. A flat-plate airfoil can be considered a special case of a Joukowski airfoil. Thus, by conformal mapping, a cylinder can be mapped into a flat plate, represented by the slit shown in Fig. 9-20*a*. Since the use of conformal mapping is beyond the scope of this book, we will give the solution to the velocity field around the plate without proof.

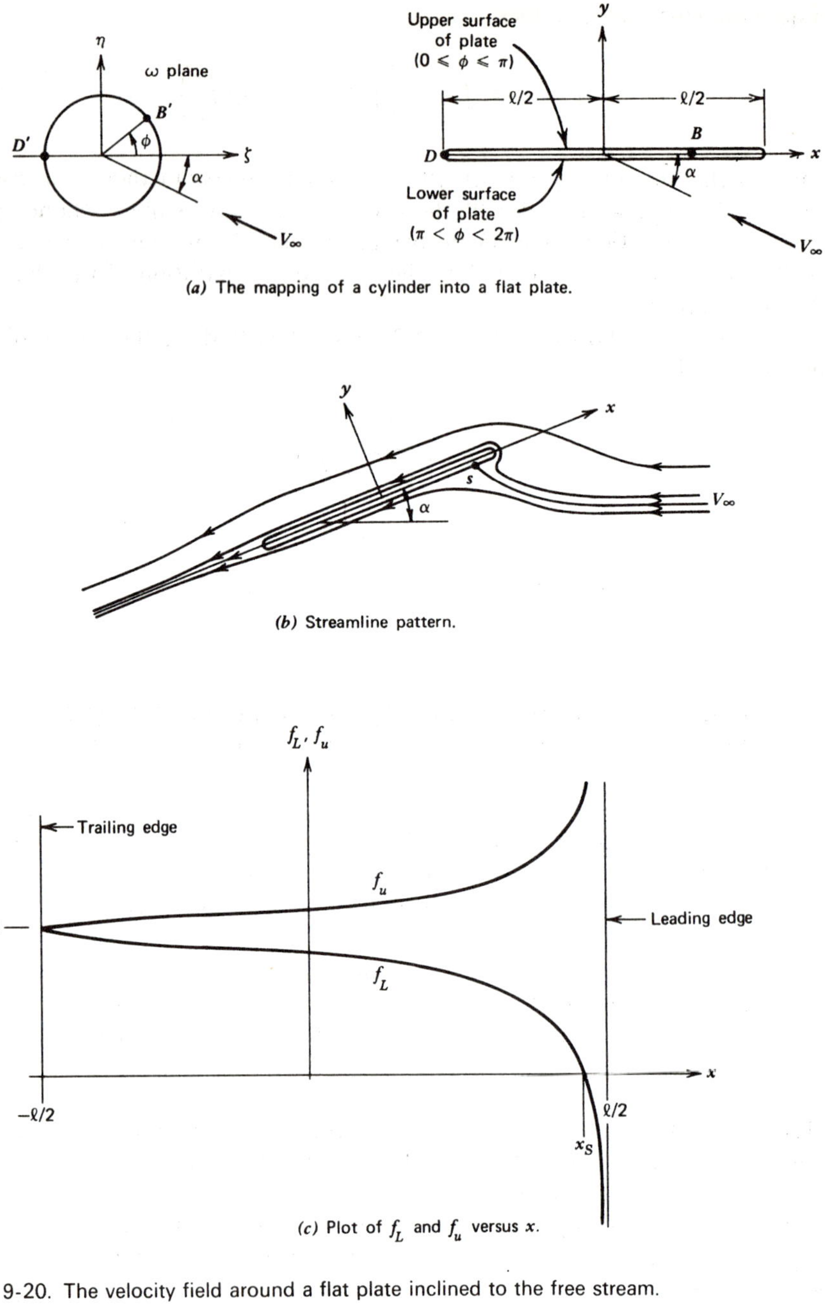

(a) The mapping of a cylinder into a flat plate.

(b) Streamline pattern.

(c) Plot of f_L and f_u versus x.

Fig. 9-20. The velocity field around a flat plate inclined to the free stream.

Students interested in this development should see the work of Batchelor, which is listed in the reference section.

By the use of conformal mapping, the Kutta condition (described in this section) and Eqs. 9.6-7, the following formula for the velocity field on a flat plate can be derived; that is,

$$(\mathbf{v})_{\text{on plate}} = \begin{cases} -V_\infty\left(\cos\alpha - \sin\alpha\sqrt{\dfrac{\ell + 2x}{\ell - 2x}}\right)\mathbf{i}, & \text{for lower surface} \\ -V_\infty\left(\cos\alpha + \sin\alpha\sqrt{\dfrac{\ell + 2x}{\ell - 2x}}\right)\mathbf{i}, & \text{for upper surface} \end{cases} \qquad \text{Eq. 9.7-1}$$

The streamline pattern of the flow around the plate is shown in Fig. 9-20*b*. Note that there exists a forward stagnation point on the lower surface at point *S*. To gain a knowledge of the velocity distribution on the upper and lower surfaces, let

$$f_L = \cos\alpha - \sin\alpha\sqrt{\frac{\ell + 2x}{\ell - 2x}}$$

and Eq. 9.7-2

$$f_u = \cos\alpha + \sin\alpha\sqrt{\frac{\ell + 2x}{\ell - 2x}}$$

Plots of f_L and f_u as functions of x are shown in Fig. 9-20*c*. The position of the stagnation point on the lower surface, x_s, can be established by determining the x value that satisfies $f_L = 0$. One can readily show that

$$x_s = \frac{\ell}{2}(\cos^2\alpha - \sin^2\alpha) \qquad \text{Eq. 9.7-3}$$

Also observe that $f_L(x)$ becomes negative for $x_s < x \leq \ell/2$. Thus, the direction of the velocity on the lower surface, between x_s and the leading edge, is in the positive x direction. We observe that, on a theoretical basis, the velocity magnitude of the fluid particle negotiating the turn from the lower surface to the upper surface around the sharp leading edge goes to infinity. Of course, in a real situation the leading edge would have some degree of roundness, thus avoiding this singularity problem.

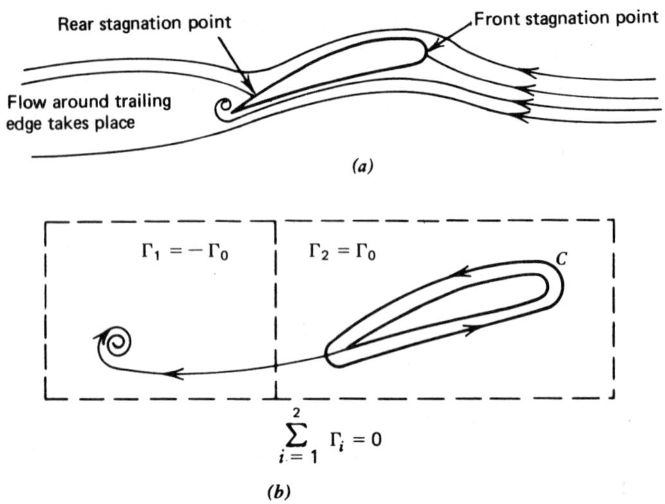

Fig. 9-21. Circulation development around an airfoil. (*a*) Early stages of an airfoil rapidly accelerated from rest. (*b*) Circulation develops around an airfoil to counteract circulation of the shed vortex.

A qualitative explanation of the Kutta condition, which is used in deriving Eq. 9.7-1, can be expressed in this way: When an airfoil with a rounded leading edge and a sharp trailing edge is accelerated rapidly from rest to a steady velocity, the fluid motion in its early development is irrotational everywhere. This is so because vorticity is generated at a finite rate by viscosity. Therefore, the initial circulation around an airfoil is zero. Also, in the early stages, the rear stagnation point is not located at the sharp trailing edge. This results in a flow taking place on the lower surface, around the trailing edge and toward the rear stagnation point, as shown in Fig. 9-21*a*. However, the fluid here decelerates sharply as it moves toward the rear stagnation point, causing the flow to separate and a clockwise vortex to form. Photographs indicate that the starting vortex is shed from the airfoil, as shown in Fig. 9-22. However, by Kelvin's theorem, the circulation around the airfoil must develop to counteract the circulation of the shed vortex, as shown in Fig. 9-21*b*. This circulation affects the velocity field such that the fluid velocity on the upper and lower surfaces has the same finite value at the trailing edge. This condition is known as the Kutta condition.[4] Although ideal flow theory requires that the starting vortex remain intact,

[4] When we relate the flow problem around a cylinder to that around an airfoil, it is the Kutta condition that enables us to evaluate the arbitrary constant *A* appearing in Eqs. 9.6-7.

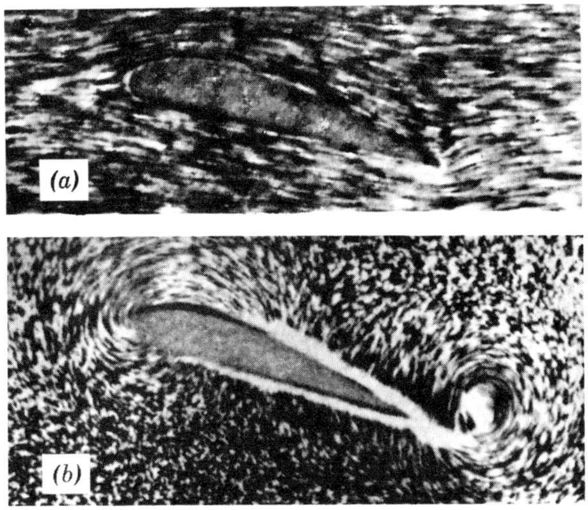

Fig. 9-22. Photographs showing the flow around an airfoil brought suddenly into motion. (*a*) Shows the streamlines at the moment the airfoil is brought impulsively into motion—camera moving with the airfoil. (*b*) Shows the shedding of the starting vortex—camera at rest, relative to the undisturbed fluid. (From *Applied Hydro- and Aeromechanics* by L. Prandtl and O. G. Tietjens. Courtesy of United Engineering Trustees, Inc.)

viscous effects cause it to dissipate. In spite of this, the circulation around the airfoil remains unchanged as long as the airfoil continues to move at constant speed. Any speed change results in additional vortex shedding.

Once the fluid velocity on the plate is known, the pressure on the plate can be determined from Eq. 9.2-12, that is,

$$p = p_\infty + \frac{\rho}{2g_c}(V_\infty{}^2 - v^2) \qquad \text{Eq. 9.2-12}$$

Substituting Eq. 9.7-1 into 9.2-12, we obtain an expression for the pressure on the lower surface of the plate, p_L,

$$p_L = p_\infty + \frac{\rho V_\infty{}^2}{2g_c}\left[1 - \cos^2\alpha - \sin^2\alpha\left(\frac{\ell + 2x}{\ell - 2x}\right) + 2\sin\alpha\cos\alpha\sqrt{\frac{\ell + 2x}{\ell - 2x}}\right]$$

Eq. 9.7-4

or, in nondimensional form,

$$\frac{p_L - p_\infty}{\dfrac{\rho V_\infty{}^2}{2g_c}} = \left[1 - \cos^2\alpha - \sin^2\alpha\left(\frac{1 + 2\dfrac{x}{\ell}}{1 - 2\dfrac{x}{\ell}}\right) + 2\sin\alpha\cos\alpha\sqrt{\frac{1 + 2\dfrac{x}{\ell}}{1 - 2\dfrac{x}{\ell}}}\,\right]$$

Eq. 9.7-5

In a similar manner, the pressure on the upper surface of the plate, p_u, is given by

$$p_u = p_\infty + \frac{\rho V_\infty{}^2}{2g_c}\left[1 - \cos^2\alpha - \sin^2\alpha\left(\frac{\ell + 2x}{\ell - 2x}\right) - 2\sin\alpha\cos\alpha\sqrt{\frac{\ell + 2x}{\ell - 2x}}\right]$$

Eq. 9.7-6

or, in nondimensional form,

$$\frac{p_u - p_\infty}{\dfrac{\rho V_\infty{}^2}{2g_c}} = \left[1 - \cos^2\alpha - \sin^2\alpha\left(\frac{1 + 2\dfrac{x}{\ell}}{1 - 2\dfrac{x}{\ell}}\right) - 2\sin\alpha\cos\alpha\sqrt{\frac{1 + 2\dfrac{x}{\ell}}{1 - 2\dfrac{x}{\ell}}}\,\right]$$

Eq. 9.7-7

Subtracting Eq. 9.7-6 from Eq. 9.7-4, we obtain the difference between the pressure on the lower and upper surfaces at x, which we will designate as Δp; that is,

$$\Delta p(x) = p_L(x) - p_u(x) = \frac{\rho V_\infty{}^2}{2g_c}\left(4\sin\alpha\cos\alpha\sqrt{\frac{\ell + 2x}{\ell - 2x}}\right) \qquad \text{Eq. 9.7-8}$$

or, in nondimensional form,

$$\frac{\Delta p(x)}{\dfrac{\rho V_\infty{}^2}{2g_c}} = 4\sin\alpha\cos\alpha\sqrt{\frac{1 + 2\dfrac{x}{\ell}}{1 - 2\dfrac{x}{\ell}}} \qquad \text{Eq. 9.7-9}$$

Aerodynamicists often designate the nondimensional pressure term found on the left-hand side of Eqs. 9.7-5 and 9.7-7 as the local pressure coefficient $\tilde{c}_p$; that is,

$$\tilde{c}_p = \frac{p - p_\infty}{\dfrac{\rho V_\infty^2}{2g_c}} \qquad \text{Eq. 9.7-10}$$

Then

$$\frac{p_u - p_\infty}{\dfrac{\rho V_\infty^2}{2g_c}} = \tilde{c}_{p,u} \quad \text{and} \quad \frac{p_L - p_\infty}{\dfrac{\rho V_\infty^2}{2g}} = \tilde{c}_{p,L}$$

The following example illustrates typical $\tilde{c}_p$ values for a flat-plate airfoil.

TABLE 9-1 Nondimensional Pressure on a Flat Plate Airfoil making an Angle of Attack of 10° with the Free Stream

$\frac{x}{l}$	$\tilde{c}_{p,L}$	$\tilde{c}_{p,u}$	$\frac{\Delta p}{\rho V_\infty^2 / 2g_c}$
0.499	−20.665	−42.285	21.620
0.49	0.448	−6.359	6.807
0.47	0.934	−1.936	2.870
0.45	0.948	−2.034	2.982
0.4	0.785	−1.267	2.052
0.3	0.593	−0.775	1.368
0.2	0.482	−0.568	1.045
0.1	0.404	−0.434	0.838
0.0	0.341	−0.343	0.684
−0.1	0.289	−0.269	0.558
−0.2	0.241	−0.207	0.448
−0.3	0.194	−0.148	0.342
−0.4	0.141	−0.087	0.228
−0.5	0.030	+0.030	0.000

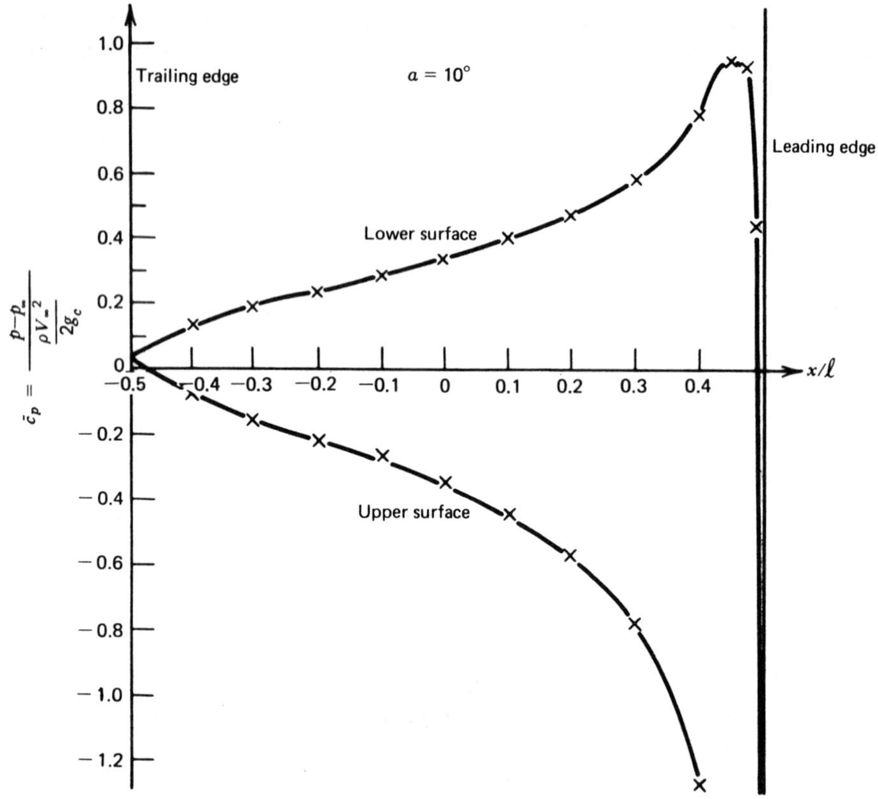

Fig. 9-23. Nondimensionalized pressure curve on a flat plate.

Example 9-1

Consider a flat-plate airfoil moving at 200 mi/h relative to the still air, as shown in Fig. 9-20*b*. If the plate makes an angle of attack of 10° with the free stream, tabulate and plot $\tilde{c}_{p,L}$, $\tilde{c}_{p,u}$, and $\Delta p/(\rho V_\infty^{\,2}/2g_c)$, for

$$\frac{x}{\ell} = 0.499,\ 0.49, 0.47, 0.45, 0.4, 0.3, \ldots, 0.0,\ -0.1, \ldots,\ -0.5$$

Assume that the free stream is at standard atmospheric conditions.

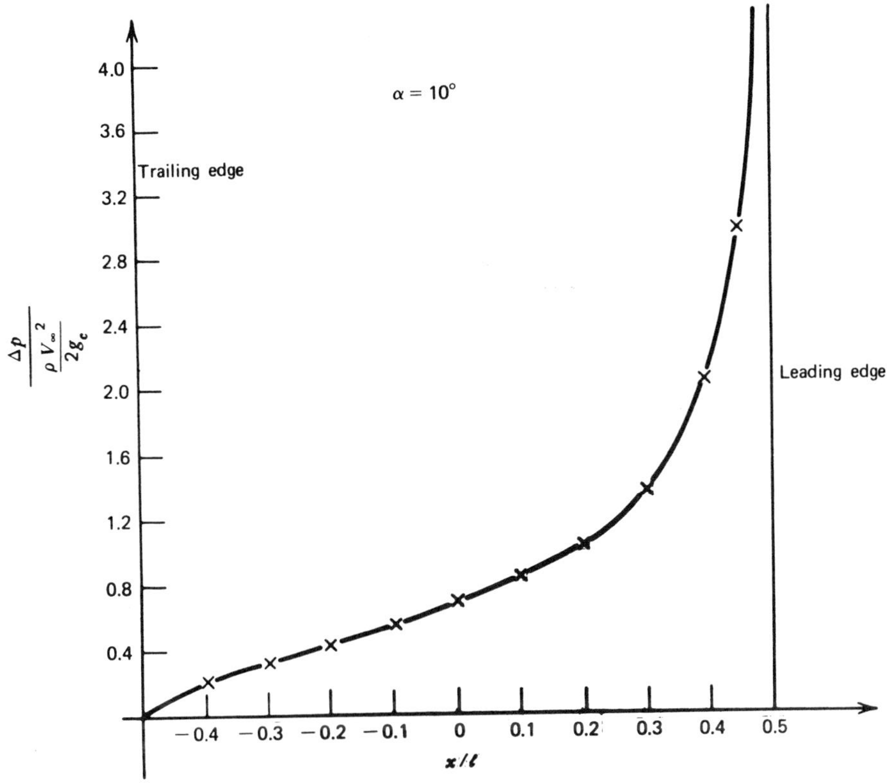

Fig. 9-24. Non dimensionalized pressure difference on a flat plate.

SOLUTION

The condition that the free stream is at standard atmospheric conditions implies that $\rho = 0.07650\ \text{lb}_m/\text{ft}^3$ and $p_\infty = 2116.2\ \text{lb}_f/\text{ft}^2$. Utilizing Eqs. 9.7-5, 9.7-7, and 9.7-9, Table 9.1 can be constructed.

Plots of nondimensional pressure versus x/ℓ are shown in Figs. 9-23 and 9-24. These plots give the pressure that acts on a flat plate at an angle of attack of 10°.

Although our discussion of the velocity and pressure fields around airfoils has been limited, for the most part, to the flat-plate airfoil, theoretical methods for determining the pressure distribution around airfoils of general shape have been developed. A comparison of theory and experiment for the NACA (National Advisory Committee for Aeronautics) 4412 airfoil is shown in Fig. 9-25.

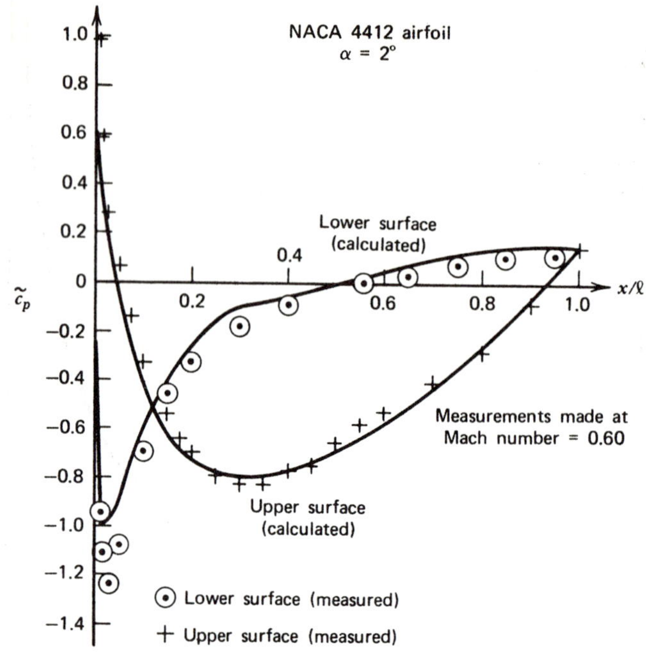

Fig. 9-25. Local pressure coefficient versus x/l for NACA 4412 airfoil. Comparison between theory and experiment. (From *Elementary Applied Aerodynamics* by P. E. Hemke, Copyright © 1946 by Prentice-Hall, Inc. Reprinted by permission of the publisher.)

9.8 FORCES ON AN AIRFOIL

The ultimate objective in our analysis of the flow field around an airfoil is the determination of the lift and drag forces acting on the airfoil. The lift is defined to be the component of the resultant aerodynamic force that points in the direction perpendicular to the free-stream direction, as shown in Fig. 9-26. The drag, on the other hand, is the component of the resultant force that points in the direction of the free stream.

Let us first consider the flat-plate airfoil. The aerodynamic force can be decomposed into either the lift and drag components or into components that point in the directions along and perpendicular to the plate, as shown in Fig. 9-27. That is,

$$\mathbf{F} = F_1\mathbf{i} + F_2\mathbf{j} = -D\mathbf{i}' + L\mathbf{j}' \qquad \text{Eq. 9.8-1}$$

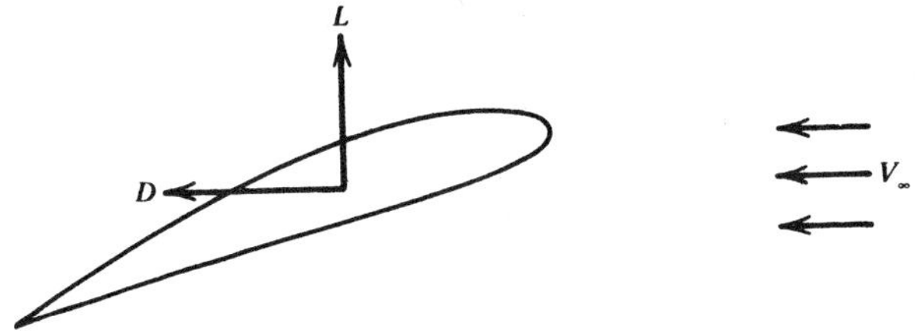

Fig. 9-26. Lift and drag on an airfoil.

To obtain L and D in terms of F_1 and F_2, let us express $(\mathbf{i}, \mathbf{j})$ in terms of $(\mathbf{i}', \mathbf{j}')$. That is,

$$\mathbf{i} = \cos\alpha\,\mathbf{i}' + \sin\alpha\,\mathbf{j}'$$
$$\mathbf{j} = -\sin\alpha\,\mathbf{i}' + \cos\alpha\,\mathbf{j}' \qquad \text{Eqs. 9.8-2}$$

Substituting the above relations into Eq. 9.8-1, we obtain

$$F_1(\cos\alpha\,\mathbf{i}' + \sin\alpha\,\mathbf{j}') + F_2(-\sin\alpha\,\mathbf{i}' + \cos\alpha\,\mathbf{j}') = D\mathbf{i}' + L\mathbf{j}' \qquad \text{Eq. 9.8-3}$$

By equating $\mathbf{i}'$-components on both sides of the above equation, we find that

$$-D = F_1\cos\alpha - F_2\sin\alpha \qquad \text{Eq. 9.8-4}$$

By equating $\mathbf{j}'$-components on both sides of Eq. 9.8-3, we find that

$$L = F_1\sin\alpha + F_2\cos\alpha \qquad \text{Eq. 9.8-5}$$

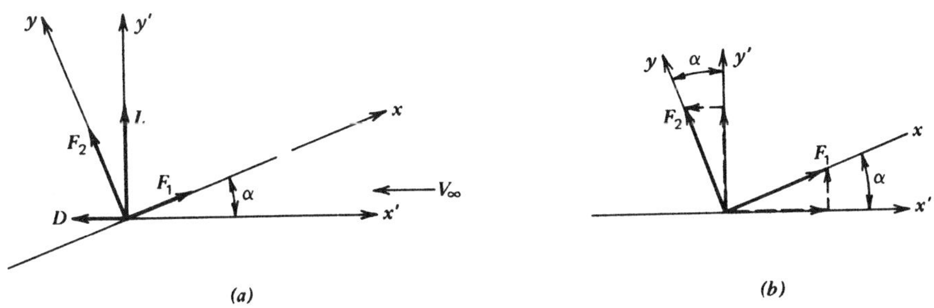

Fig. 9-27. Forces on a flat-plate airfoil. (a) Forces L, D, F_1, and F_2. (b) (x', y') components of $\mathbf{F}_1$ and $\mathbf{F}_2$.

Having already determined the pressure field on the flat-plate airfoil, we can readily determine the F_2 force component. However, the F_1 force component is not so easily obtained. This results from the fact that the flat-plate airfoil has a sharp leading edge at the nose, which in our mathematical model results in the pressure field being singular at $x = \ell/2$ ($p \to \infty$ as $x \to \ell/2$). However, this singularity does not cause any great difficulty in evaluating F_2. We thus have to rely on some additional information to complete the formulation of the problem. By the use of complex variable theory, it can be shown that for a general airfoil with a round nose and a sharp trailing edge, the drag, D, equals zero. If we allow our flat-plate airfoil to have some rounding at the nose, we can then substitute this information into Eq. 9.8-4, giving

$$F_1 = \frac{F_2 \sin \alpha}{\cos \alpha}$$

Substituting this equation into Eq. 9.8-5 gives the lift in terms of F_2; that is,

$$L = \frac{F_2 \sin^2 \alpha}{\cos \alpha} + F_2 \cos \alpha = \frac{F_2}{\cos \alpha} \qquad \text{Eq. 9.8-6}$$

Let us now evaluate F_2 utilizing the information obtained in Section 9.7. The integral shown below gives F_2; that is,

$$F_2 = \int_{-l/2}^{\ell/2} \int_0^{\ell_s} \Delta p \, dx \, dz \qquad \text{Eq. 9.8-7}$$

where ℓ is the chord length and ℓ_s is the span length of the airfoil. Now, Δp is given by Eq. 9.7-8; that is,

$$\Delta p = p_L(x) - p_u(x) = \frac{2\rho V_\infty{}^2}{g_c} \sin \alpha \cos \alpha \sqrt{\frac{\ell + 2x}{\ell - 2x}} \qquad \text{Eq. 9.7-8}$$

Substituting the above equation into Eq. 9.8-6 and integrating with respect to z, we obtain

$$F_2 = \frac{2\ell_s \rho V_\infty{}^2}{g_c} \sin \alpha \cos \alpha \int_{-\ell/2}^{\ell/2} \sqrt{\frac{\ell + 2x}{\ell - 2x}} \, dx \qquad \text{Eq. 9.8-8}$$

But

$$\sqrt{\frac{\ell + 2x}{\ell - 2x}} = \frac{\sqrt{1 + \frac{2x}{\ell}}}{\sqrt{1 - \frac{2x}{\ell}}} \times \frac{\sqrt{1 + \frac{2x}{\ell}}}{\sqrt{1 + \frac{2x}{\ell}}} = \frac{1 + \frac{2x}{\ell}}{\sqrt{1 - 4\left(\frac{x}{\ell}\right)^2}}$$

Now, let $\xi = 2x/\ell$; then $\xi^2 = 4x^2/\ell^2$ and $d\xi = (2/\ell)dx$. Substituting these expressions into the above integral and appropriately changing the limits of integration, we obtain

$$\int_{-\ell/2}^{\ell/2} \sqrt{\frac{\ell + 2x}{\ell - 2x}}\, dx = \frac{\ell}{2} \int_{-1}^{1} \frac{1 + \xi}{\sqrt{1 - \xi^2}}\, d\xi$$

$$= \frac{\ell}{2} \left(\int_{-1}^{1} \frac{d\xi}{\sqrt{1 - \xi^2}} + \int_{-1}^{1} \frac{\xi\, d\xi}{\sqrt{1 - \xi^2}} \right)$$

or

$$\int_{-\ell/2}^{\ell/2} \sqrt{\frac{\ell + 2x}{\ell - 2x}}\, dx = \frac{\ell}{2} \left[\sin^{-1} \xi - \sqrt{1 - \xi^2} \right]_{-1}^{1} = \frac{\pi \ell}{2} \qquad \text{Eq. 9.8-9}$$

By substituting the above equation into Eq.9.8-8, we find that

$$F_2 = \frac{\rho V_\infty^2}{g_c} \pi \ell \ell_s \sin \alpha \cos \alpha \qquad \text{Eq. 9.8-10}$$

We can now determine the lift on the flat-plate airfoil from Eq. 9.8-6; that is,

$$L = \frac{\rho V_\infty^2}{g_c} \pi \ell \ell_s \sin \alpha \qquad \text{Eq. 9.8-11}$$

Aerodynamicists usually express the lift in terms of the circulation, Γ. Let us evaluate the circulation for the flat-plate airfoil. Recall that the circulation, Γ, was defined by Eq. 9.2-13 (also Eq. 9.4-8); that is,

$$\Gamma = \oint_c \mathbf{v} \cdot d\mathbf{l} \qquad \text{Eq. 9.2-13}$$

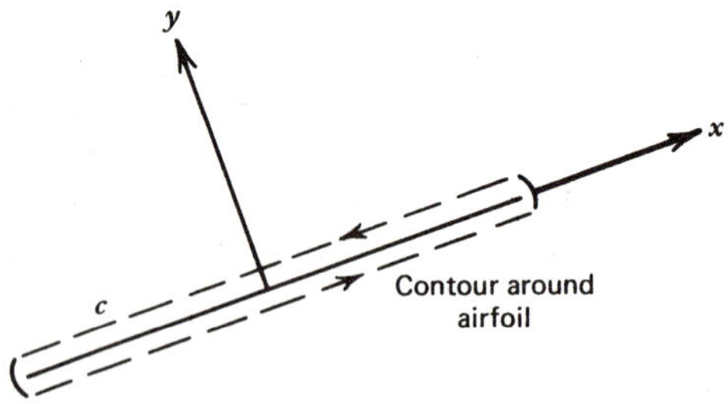

Fig. 9-28. Contour around airfoil.

Taking c to be the contour, shown in Fig. 9-28, the circulation becomes

$$\Gamma = \int_{-\ell/2}^{\ell/2} \mathbf{v}_L \cdot \mathbf{i}\, dx + \int_{\ell/2}^{-\ell/2} \mathbf{v}_u \cdot \mathbf{i}\, dx = \int_{-\ell/2}^{\ell/2} \mathbf{v}_L \cdot \mathbf{i}\, dx - \int_{-\ell/2}^{\ell/2} \mathbf{v}_u \cdot \mathbf{i}\, dx$$

Substituting the expression for $\mathbf{v}_L$ and $\mathbf{v}_u$, as given by Eq. 9.7-1, yields

$$\Gamma = V_\infty \int_{\ell/2}^{\ell/2} \left[\left(-\cos\alpha + \sin\alpha \sqrt{\frac{\ell + 2x}{\ell - 2x}} \right) - \left(-\cos\alpha - \sin\alpha \sqrt{\frac{\ell + 2x}{\ell - 2x}} \right) \right] dx$$

$$= 2V_\infty \sin\alpha \int_{-\ell/2}^{\ell/2} \sqrt{\frac{\ell + 2x}{\ell - 2x}}\, dx$$

However, we previously determined that

$$\int_{-\ell/2}^{\ell/2} \sqrt{\frac{\ell + 2x}{\ell - 2x}}\, dx = \frac{\pi\ell}{2} \qquad \text{Eq. 9.8-9}$$

Therefore,

$$\Gamma = V_\infty \pi\ell \sin\alpha \qquad \text{Eq. 9.8-12}$$

By substituting the above equation into Eq. 9.8-11, we can express the lift in terms of the circulation. That is,

$$\frac{L}{\ell_s} = \frac{\rho V_\infty \Gamma}{g_c} \qquad \text{(for flat-plate airfoil)} \qquad \text{Eq. 9.8-13}$$

Let us now compare the above expression with the expression for the lift on a cylinder. By Eqs. 9.5-34 we see that

$$\frac{L}{\ell_s} = \frac{\rho V_\infty \Gamma}{g_c} \qquad \text{(for circular cylinder)} \qquad \text{Eq. 9.5-34}$$

Since the cylinder and the flat-plate airfoil are two totally different geometrically shaped bodies, one might suspect that Eq. 9.8-13 is a general expression; and so it is. It can be shown by the use of complex variables that the lift and drag on any airfoil with a rounded nose and a sharp trailing edge are given by

$$\frac{L}{\ell_s} = \frac{\rho V_\infty \Gamma}{g_c} \qquad \text{(in general)}$$

and Eqs. 9.8-14

$$D = 0 \qquad \text{(in general)}$$

Aerodynamicists often express the lift on an airfoil in terms of the non-dimensional quantity called the lift coefficient, C_L, defined by the equation

$$L = \frac{\rho V_\infty{}^2}{2g_c} AC_L$$

where A is the plan area of the wing. If ℓ_s is the span length and ℓ is the chord length, then

$$C_L = \frac{2g_c L}{\rho V_\infty{}^2 \ell_s \ell}$$

Then by Eq. 9.8-11, we obtain for the *flat-plate airfoil*

$$C_L = 2\pi \sin \alpha$$

Similarly, the drag coefficient, C_D, is defined by the equation

$$D = \frac{\rho V_\infty{}^2}{2g_c} AC_D$$

Then

$$C_D = \frac{2g_c D}{\rho V_\infty{}^2 \ell_s \ell}$$

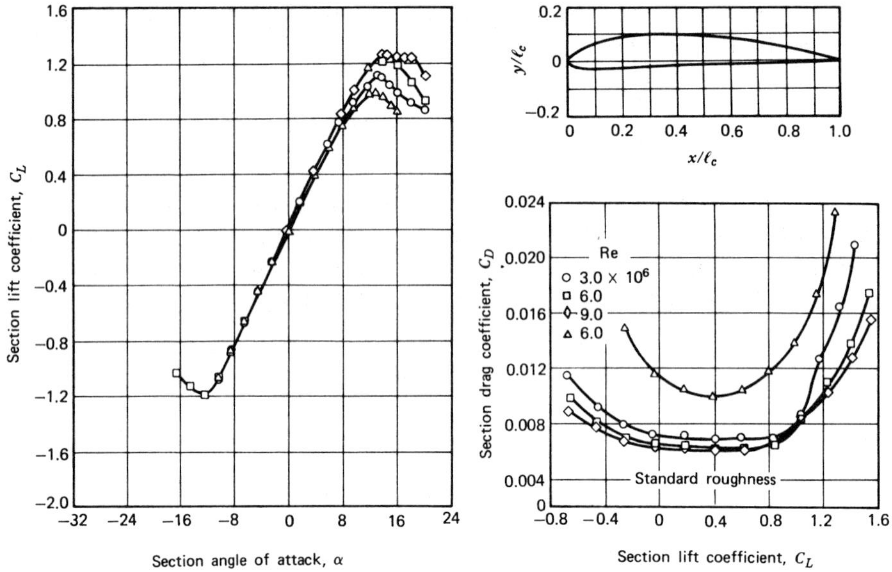

Fig. 9-29. Plots of C_L versus α and C_D versus C_L for the NACA 4412 airfoil. (From *Theory of Wing Sections, Including a Summary of Airfoil Data* by I. H. Abbott and A. E. Von Doenhoff, Dover Publications, Inc., 1959.)

Plots of C_L versus α and C_D versus C_L for the NACA 4412 airfoil at several different Reynolds numbers are shown in Fig. 9-29. Observe that at small angles of attack C_L is *not* a function of the Reynolds number. The Reynolds number only affects C_L when boundary-layer separation becomes an important factor. That is, the Reynolds number determines where and when boundary-layer separation will occur. As a result, the stall angle depends on the Reynolds number. This can be seen in Fig. 9-29. On the other hand, C_D is a function of the Reynolds number for all angles of attack.

9.9 AIRFOILS OF FINITE SPAN

In wings and airfoils of finite span, end effects require that some modification be made to the two-dimensional solution of the corresponding wing. If the ratio of the span length to chord length, denoted as the aspect ratio, is large, then the 3-D effects are small. Consider the plan view of the wing shown in Fig. 9-30. For an airfoil experiencing a lifting force, the pressure on the lower surface is greater than the pressure on the upper surface. Near the ends of the

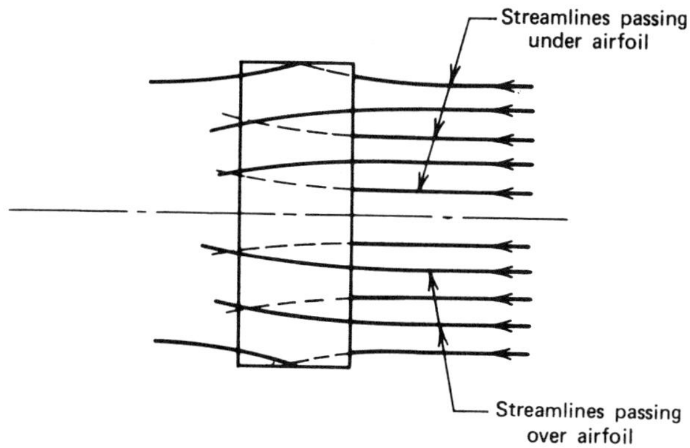

Fig. 9-30. Streamline pattern over and under the airfoil.

span of the airfoil, this pressure difference causes a flow to take place from the lower surface to upper surface around the ends. This flow situation is denoted by the dashed line, which becomes a solid line, in Fig. 9-30. Thus, the fluid velocity adjacent to the surface develops a small laterally outward component on the lower surface and an inward component on the upper surface. This spanwise momentum is maintained as the fluid moves downstream from the airfoil. This phenomenon can be modeled by considering the wing to be a *bound vortex sheet*[5] (a vortex sheet attached to the airfoil), and the wake region to be a trailing, free vortex sheet (a vortex sheet free to move with the free stream).

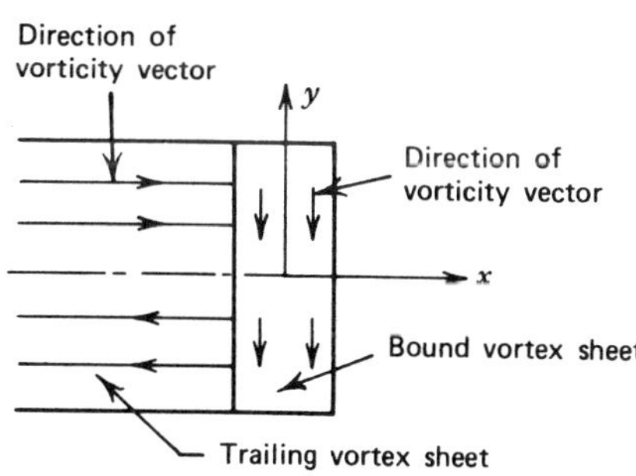

Fig. 9-31. Vortex sheet.

[5] See Problem 9-18.

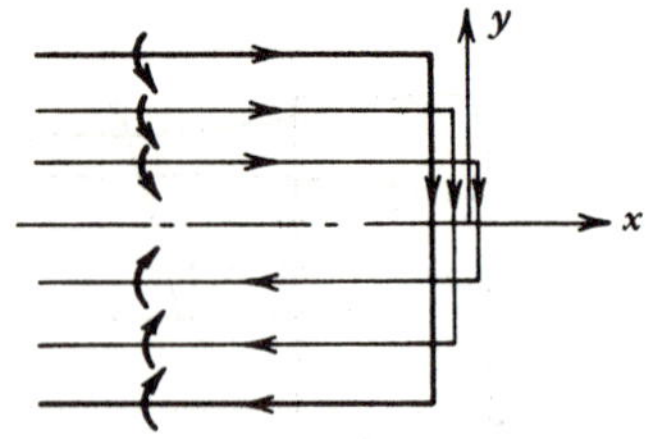

Fig. 9-32. Superposition of a series of horseshoe vortices.

These concepts were first introduced by Lanchester and Prandtl. In its simplest form, the theory is denoted as *lifting line* theory. In the general case, the vorticity vector of the bound vortex sheet points in the y direction and has a strength that is a maximum at the center and that decreases to zero at the ends. The trailing vortex sheet is taken to be straight, with the vorticity vector pointing in a direction parallel to the direction of flight.[6] This is shown in Fig. 9-31. This system can be approximated by the superposition of a number of simple *horseshoe* vortices, as shown in Fig. 9-32. It has been observed that the trailing vortex sheet is unstable and rolls up into two trailing line vortices, as shown in Fig. 9-33. In all of the vortex systems, the trailing vortices end with the vortex shed at the start of the airfoil motion. However, for all practical purposes, the trailing vortices may be considered to extend an infinite distance downstream.

The effect of the trailing vortex system on the fluid adjacent to the airfoil is to add a velocity component that is normal to both the direction of flight and to the direction of the span—the effect of which is to superimpose onto the two-dimensional solution a small lift and drag component. This additional drag component is called the *induced drag.* The mathematical details involved in the theory are presented in the work of both Glauert and Batchelor; both references are listed in the end-of-book reference section. The net result is that an elliptic planform for a wing (or two semiellipses with different minor axes) gives a minimum induced drag for a given lift. That is, if ℓ is the chord length as a function of y, then

$$\ell = \ell_0\left(1 - \frac{y^2}{\ell_s^2}\right)^{1/2}$$

[6] In a real flight situation, the trailing vortices move with the stream and become inclined to the flight direction.

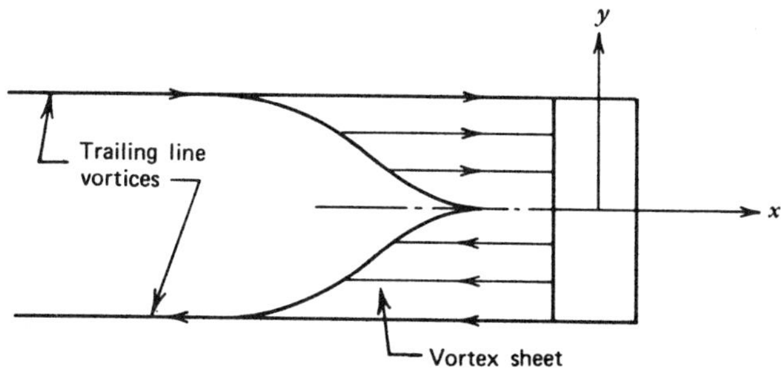

Fig. 9-33. The roll up of the trailing vortex sheet.

where ℓ_0 is the chord length at the centerline of the wing and ℓ_s is the span length. Then the actual lift, L_{act}, is given by

$$L_{act} = \left(1 + \frac{\beta l_0}{8\ell_s}\right)^{-1} L_{2\text{-dim}}$$

where $\beta = dC_L/d\alpha$.

9.10 SKIN-FRICTION DRAG

In the previous sections our attention was primarily directed toward developing a method for predicting the *lift* on an airplane wing. In this section our efforts are directed toward drag prediction; in particular, the prediction of the viscous or skin-friction drag on a wing. It was mentioned earlier that for a well-designed wing, at not too large an angle of attack, the dominant type of drag force is skin-friction drag (as distinguished from pressure drag). Recall that in our previous discussion of the flow field around a wing, it was mentioned that viscous effects are important only in a thin layer adjacent to the wing's surface. This thin layer was designated as the *boundary layer* (see Fig. 9-3). Outside the boundary layer, viscous effects are negligible and the irrotational, incompressible[7] flow development, discussed in the previous sections, becomes valid. In addition, Bernoulli's equation also becomes valid outside the boundary layer. Assuming that the irrotational flow problem can be solved, the pressure and velocity fields at the

[7] Of course, we are limiting our discussion to flows where the Mach number is less than 0.5.

boundary-layer limits become known, say $P(x)$ and $U(x)$, respectively. Then by Bernoulli's equation

$$\frac{P}{\rho} + \frac{U^2}{2g_c} = \frac{p_\infty}{\rho} + \frac{U_\infty{}^2}{2g_c}$$

where p_∞ and U_∞ are, respectively, the pressure and the velocity far from the wing. For example, for a flat plate at a nonzero angle of attack, expressions for P and U can be taken to be those given by Eqs. 9.7-5, 9.7-7, and 9.7-1, respectively. In those equations, x was the distance along the plate. It was also mentioned that experiments indicate that there is no significant pressure variation across the boundary layer. Therefore, if (x, y) is an arbitrary point within the boundary layer, then

$$p(x, y) = P(x)$$

This provides some simplification in treating the boundary-layer region. However, even then, boundary-layer theory for a general wing shape is too complicated a subject for an introductory textbook. However, we can obtain an understanding of the boundary-layer concept by treating the flat plate at a zero angle of attack. For then we reduce the complexity, since

$$P = p_\infty \qquad \text{and} \qquad U = U_\infty$$

Approximate methods, as well as more exact methods, have been developed for treating the boundary-layer region. The approximate method developed by Von Kármán and Pohlhausen[8] is at an appropriate mathematical level to be covered in this book. By applying Newton's second law to the control volume, shown in Fig. 9-34, we can obtain an expression for the shear stress on the flat plate. Newton's second law is given by Eq. 4.7-4; that is,

$$\underset{\text{C.V.}}{\iiint} \frac{\partial}{\partial t}\left(\frac{\rho}{g_c}\mathbf{v}\right) dS + \underset{\text{C.S.}}{\oint\!\!\oint} \frac{\rho}{g_c}\mathbf{v}(\mathbf{v} \cdot d\mathbf{S}) = \underset{\text{C.S.}}{\oint\!\!\oint} \boldsymbol{\tau}\, dS + \underset{\text{C.V.}}{\iiint} \mathbf{F}\, d\mathcal{V}$$

[8] See H. Schlichting, *Boundary-Layer Theory*, 6th ed., McGraw-Hill, New York, 1968.

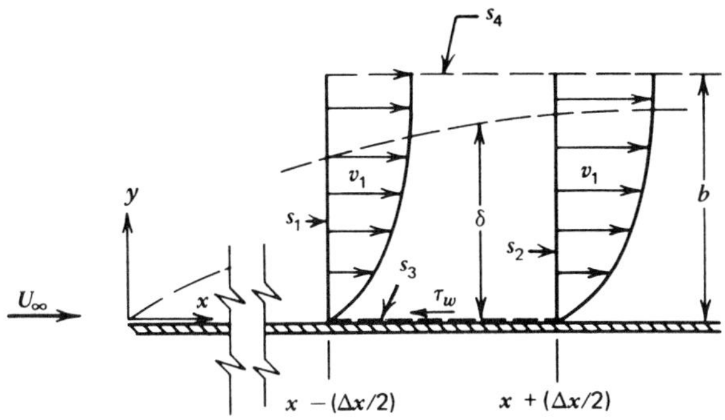

Fig. 9-34. An infinitesimal section of the boundary layer. Inside the boundary layer $\mathbf{v} = v_1(x, y)\mathbf{i} + v_2(x, y)\mathbf{j}$. (From the law of mass conservation we see that $v_2(x, y) \neq 0$. However, we should expect that $v_2 \ll U_\infty$ everywhere.) Take $\boldsymbol{\tau}(S_3) = -\tau_w(x)\mathbf{i}$.

For steady incompressible flow, the **i**-component of the above equation per unit depth is

$$-\frac{\rho}{g_c}\int_0^b \left[v_1\left(x-\frac{\Delta x}{2}, y\right)\right]^2 dy + \frac{\rho}{g_c} U_\infty v_2(x, b)\,\Delta x$$

$$+\frac{\rho}{g_c}\int_0^b \left[v_1\left(x+\frac{\Delta x}{2}, y\right)\right]^2 dy$$

$$= -\tau_w(x)\,\Delta x + \int_0^b p\left(x-\frac{\Delta x}{2}, y\right) dy$$

$$-\int_0^b p\left(x+\frac{\Delta x}{2}, y\right) dy \qquad \text{Eq. 9.10-1}$$

For the flat plate at zero angle of attack,

$$p\left(x-\frac{\Delta x}{2}, y\right) - p\left(x+\frac{\Delta x}{2}, y\right) = p_\infty \qquad \text{and} \qquad v_1(b, y) = U_\infty$$

Therefore, the last two terms on the right-hand side of Eq. 9.10-1 cancel. We can also apply the law of mass conservation to the control volume shown in Fig. 9-34. The law of mass conservation is given by

$$\iiint\limits_{\text{C.V.}} \frac{\partial \rho}{\partial t}\, d\mathcal{V} + \oiint\limits_{\text{C.S.}} \rho \mathbf{v} \cdot d\mathbf{S} = 0$$

For steady incompressible flow, the above equation per unit depth reduces to

$$-\rho \int_0^b v_1\left(x - \frac{\Delta x}{2}, y\right) dy + \rho \int_0^b v_1\left(x + \frac{\Delta x}{2}, y\right) dy + \rho v_2(x, b)\, \Delta x = 0$$

or

$$\rho v_2(x, b)\, \Delta x = -\rho \int_0^b \left[v_1\left(x + \frac{\Delta x}{2}, y\right) - v_1\left(x - \frac{\Delta x}{2}, y\right)\right] dy \qquad \text{Eq. 9.10-2}$$

Substituting the above equation into Eq. 9.10-1 and solving for $\tau_w(x)$, we obtain

$$\tau_w(x) = -\frac{\rho}{g_c} \int_0^b \left\{ \frac{\left[v_1\left(x + \frac{\Delta x}{2}, y\right)\right]^2 - \left[v_1\left(x - \frac{\Delta x}{2}, y\right)\right]^2}{\Delta x} \right\} dy$$

$$+ \rho \frac{U_\infty}{g_c} \int_0^b \left\{ \frac{v_1\left(x + \frac{\Delta x}{2}, y\right) - v_1\left(x - \frac{\Delta x}{2}\right)}{\Delta x} \right\} dy$$

Taking $\lim_{\Delta x \to 0}$ of both sides of the above equation, we find that

$$\tau_w(x) = -\frac{\rho}{g_c}\left[\int_0^b \frac{\partial (v_1{}^2)}{\partial x}\, dy + \int_0^b U_\infty \frac{\partial v_1}{\partial x}\, dy\right]$$

$$= -\frac{\rho}{g_c}\frac{d}{dx}\int_0^b (v_1{}^2 - U_\infty v_1)\, dy = \frac{\rho}{g_c}\frac{d}{dx}\int_0^b v_1(U_\infty - v_1)\, dy \qquad \text{Eq. 9.10-3}$$

In the above equation, $v_1 = v_1(x, y)$. When $y \geq \delta$, $v_1(x, y) = U_\infty$. Therefore, the upper limit of the integral on the right-hand side of Eq. 9.10-3 can be replaced by δ. Multiplying both sides of Eq. 9.10-3 by g_c/ρ, it becomes

$$\frac{g_c \tau_w(x)}{\rho} = \frac{d}{dx} \int_0^\delta v_1(x, y)[U_\infty - v_1(x, y)]\, dy$$

or

$$\frac{g_c \tau_w(x)}{\rho U_\infty^2} = \frac{d}{dx} \int_0^\delta \frac{v_1(x, y)}{U_\infty} \left(1 - \frac{v_1(x, y)}{U_\infty}\right) dy \qquad \text{Eq. 9.10-4}$$

When the flow inside the boundary is laminar, the velocity distribution can be approximated by a three-term cubic polynomial in y/δ. We use three terms because there are three boundary conditions specified for the problem. These boundary conditions determine the arbitrary constants in the cubic. That is, if

$$\frac{v_1}{U_\infty} = a + b\left(\frac{y}{\delta}\right) + c\left(\frac{y}{\delta}\right)^3 \qquad \text{Eq. 9.10-5}$$

then the constants a, b, and c are determined by the following conditions

$$\text{At } y = 0, \qquad \frac{v_1}{U_\infty} = 0$$

$$\text{At } y = \delta, \qquad v_1 = U_\infty \quad \text{and} \quad \frac{dv_1}{dy} = 0 \qquad \text{Eqs. 9.10-6}$$

Therefore,

$$a = 0, \qquad b = \tfrac{3}{2}, \qquad \text{and} \qquad c = -\tfrac{1}{2}$$

and

$$\frac{v_1}{U_\infty} = \frac{3}{2}\frac{y}{\delta} - \frac{1}{2}\left(\frac{y}{\delta}\right)^3 \qquad \text{Eq. 9.10-7}$$

Then,

$$\frac{g_c \tau_w}{\rho U_\infty^{\ 2}} = \frac{d}{dx} \int_0^\delta \left[\frac{3}{2}\frac{y}{\delta} - \frac{1}{2}\left(\frac{y}{8}\right)^3\right]\left[1 - \frac{3}{2}\frac{y}{\delta} + \frac{1}{2}\left(\frac{y}{\delta}\right)^3\right] dy$$

$$= \frac{d}{dx} \int_0^\delta \left[-\frac{1}{4}\left(\frac{y}{\delta}\right)^6 + \frac{3}{2}\left(\frac{y}{\delta}\right)^4 - \frac{1}{2}\left(\frac{y}{\delta}\right)^3 - \frac{9}{4}\left(\frac{y}{\delta}\right)^2 + \frac{3}{2}\left(\frac{y}{\delta}\right)\right] dy$$

To simplify the integration, let

$$\eta = \frac{y}{\delta}; \qquad \text{then } d\eta = \frac{1}{\delta}\, dy$$

when $y = 0, \eta = 0$, and when $y = \delta, \eta = 1$. Then,

$$\frac{g_c \tau_w}{\rho U_\infty^{\ 2}} = \frac{d}{dx}\,\delta \int_0^1 \left(-\frac{\eta^6}{4} + \frac{3}{2}\eta^4 - \frac{1}{2}\eta^3 - \frac{9}{4}\eta^2 + \frac{3}{2}\eta\right) d\eta$$

or

$$\frac{g_c \tau_w}{pU_\infty^{\ 2}} = \frac{39}{280}\frac{d\delta}{dx} \qquad \text{Eq. 9.10-8}$$

But

$$\tau_w = \frac{\mu}{g_c}\frac{\partial v_1}{\partial y}(0, x) = \frac{3}{2}\frac{\mu}{g_c} U_\infty \left(\frac{1}{\delta} - \frac{y^2}{\delta^3}\right)_{y=0} = \frac{3}{2}\frac{\mu U_\infty}{g_c \delta} \qquad \text{Eq. 9.10-9}$$

Substituting Eq. 9.10-9 into Eq. 9.10-8, we obtain

$$\frac{3}{2}\frac{\mu}{\rho U_\infty \delta} = \frac{39}{280}\frac{d\delta}{dx}$$

or

$$\delta \frac{d\delta}{dx} = \frac{140}{13}\frac{\mu}{\rho U_\infty} = \frac{140}{13}\frac{\nu}{U_\infty} \qquad \text{Eq. 9.10-10}$$

Then,

$$\int \delta \, d\delta = \frac{140}{13} \frac{\nu}{U_\infty} \int dx$$

or

$$\frac{\delta^2}{2} = \frac{140}{13} \frac{\nu}{U_\infty} x + c$$

When $x = 0$, $\delta = 0$. Therefore, $c = 0$, and

$$\delta = \sqrt{\frac{280}{13}} \sqrt{\frac{\nu x}{U_\infty}} = 4.64 \sqrt{\frac{\nu x}{U_\infty}} \qquad \text{Eq. 9.10-11}$$

Then, by Eq. 9.10-9,

$$\tau_w = \frac{3}{2} \frac{\nu U_\infty \rho}{g_c(4.64)} \sqrt{\frac{U_\infty}{\nu x}} = \frac{0.323 U_\infty \rho}{g_c} \sqrt{\frac{U_\infty \nu}{x}} \qquad \text{Eq. 9.10-12}$$

The local skin-friction coefficient, c_f, is defined by

$$c_f = \frac{\tau_w}{\dfrac{\rho U_\infty^2}{2 g_c}} \qquad \text{Eq. 9.10-13}$$

By substituting the above expression for τ_w into Eq. 9.10-12, we find that

$$c_f = 0.646 \sqrt{\frac{\nu}{U_\infty x}} \qquad \text{Eq. 9.10-14}$$

If the flow in the boundary layer remains laminar over the entire plate, then the drag, D, for a plate of width b is given by

$$D = b \int_0^L \tau_w(x) \, dx = \frac{0.323 b U_\infty \rho \sqrt{U_\infty \nu}}{g_c} \int_0^L \frac{dx}{x}$$

$$= \frac{0.646 \rho U_\infty^2 b L}{g_c} \sqrt{\frac{\nu}{U_\infty L}} \qquad \text{Eq. 9.10-15}$$

The average skin-friction coefficient is defined by

$$C_f = \frac{D}{A\dfrac{\rho U_\infty^{\,2}}{2g_c}} = \frac{D}{\dfrac{bL\rho U_\infty^{\,2}}{2g_c}} \qquad \text{Eq. 9.10-16}$$

Substituting Eq. 9.10-15 into the above equation, we obtain

$$C_f = 1.292\sqrt{\frac{\nu}{U_\infty L}} = 2c_f(L) \qquad \text{Eq. 9.10-17}$$

In boundary-layer flow it is convenient to introduce a local Reynolds number, Re_x, defined by the following equation

$$\mathrm{Re}_x = \frac{U_\infty x}{\nu} \qquad \text{Eq. 9.10-18}$$

We see that c_f and C_f can then be expressed in terms of Re_x and Re_L, respectively; that is,

$$c_f = \frac{0.646}{\sqrt{\mathrm{Re}_x}} \quad \text{and} \quad C_f = \frac{1.292}{\sqrt{\mathrm{Re}_L}} \qquad \text{Eqs. 9.10-19}$$

A more rigorous analysis of this problem was carried out by Blasius.[9] He obtained slightly different values for the constants in the above relationships; these are

$$c_f = \frac{0.664}{\sqrt{\mathrm{Re}_x}} \quad \text{and} \quad C_f = \frac{1.328}{\sqrt{\mathrm{Re}_L}} \qquad \text{Eqs. 9.10-20}$$

The corresponding changes in the constants in the δ, τ_w, and D relationships can be expressed

$$\delta = 4.9\sqrt{\frac{\nu x}{U_\infty}} \qquad \text{Eq. 9.10-21}$$

$$\tau_w = \frac{0.332\rho U_\infty}{g_c}\sqrt{\frac{U_\infty \nu}{x}} \qquad \text{Eq. 9.10-22}$$

[9] See Schlichting, p. 128.

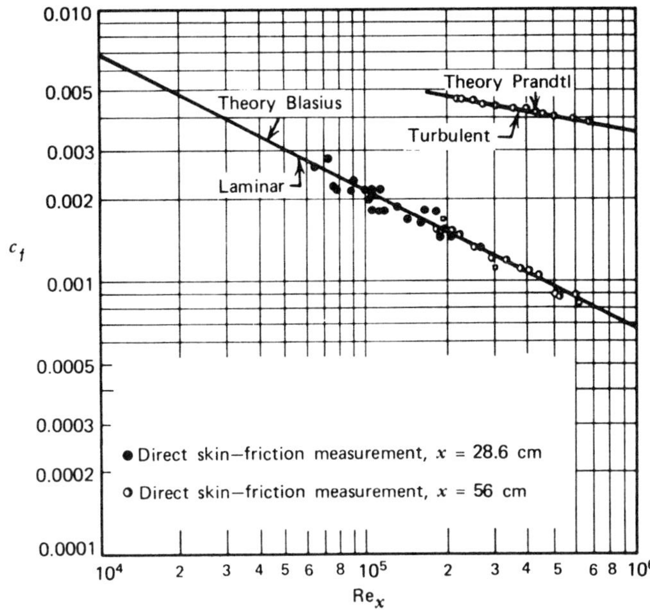

Fig. 9-35. Skin-friction coefficient. A comparison between theory and experiment. (From Direct Measurements of Skin Friction by S. Dhawan, NACA Rep. 1121 (1953). Courtesy of NACA.

and

$$D = \frac{0.664 \rho U_\infty{}^2 bL}{g_c} \sqrt{\frac{\nu}{U_\infty L}} \qquad \text{Eq. 9.10-23}$$

A comparison between theory and experiment is shown in Fig. 9-35. It has been found that the transition to a turbulent boundary layer occurs in a range of Re_x between 3×10^5 to 5×10^5, depending on the level and frequency of disturbances present in the flow.[10] However, if the level of disturbances is kept very low, the transition point can be delayed to a point where $(\mathrm{Re}_x)_{\text{transition}} \approx 2.8 \times 10^6$.

[10] See Schlichting, p. 458.

9.11 THE TURBULENT BOUNDARY LAYER OVER A FLAT PLATE

It has been found that the Von Kármán approximate method, described in Section 9.10, gives good results even when the boundary layer is turbulent. However, in this case, the velocity, v_1, is to be interpreted as a time-averaged value. As in the case of the laminar boundary layer, the method requires a knowledge of the velocity profile. The following velocity profile agrees reasonably well with experimental observations, except for positions very close to the wall; that is,

$$\frac{v_1}{U_\infty} \approx \left(\frac{y}{\delta}\right)^{1/7} \qquad \text{Eq. 9.11-1}$$

We can see that this profile is not appropriate at the wall by observing that the above profile would lead us to the incorrect conclusion that the shear stress at the wall is infinite. That is,

$$\tau_w = \frac{\mu}{g_c}\frac{\partial v_1}{\partial y}(x, 0) = \frac{\mu}{7g_c} \cdot \frac{1}{\delta^{1/7}}\left(\frac{1}{y^{6/7}}\right)_{y=0} = \infty$$

In order to obtain a solution to the problem, it is necessary to add an additional empirical relationship. Blasius found that

$$\tau_w = 0.0225\,\frac{\rho}{g_c}\,U_\infty{}^2\left(\frac{\nu}{U_\infty \delta}\right)^{1/4} \qquad \text{Eq. 9.11-2}$$

Substituting Eq. 9.11-1 and 9.11-2 into Eq. 9.10-4, we obtain

$$0.0225\left(\frac{\nu}{U_\infty \delta}\right)^{1/4} = \frac{d}{dx}\int_0^\delta \left(\frac{y}{\delta}\right)^{1/7}\left[1 - \left(\frac{y}{\delta}\right)^{1/7}\right] dy$$

Letting $\eta = y/\delta$, the above equation becomes

$$0.0225\left(\frac{\nu}{U_\infty \delta}\right)^{1/4} = \frac{d\delta}{dx}\int_0^1 (\eta^{1/7} - \eta^{2/7})\, d\eta$$

$$= \frac{7}{72}\frac{d\delta}{dx}$$

or

$$\int \delta^{1/4}\, d\delta = 0.232\left(\frac{\nu}{U_\infty}\right)^{1/4} \int dx + C$$

Thus,

$$\tfrac{4}{5}\delta^{5/4} = 0.232\left(\frac{\nu}{U_\infty}\right)^{1/4} x + C$$

Evaluation of the constant of integration, C, requires a known initial condition. Boundary layers always start out in the laminar region, then turn turbulent past the transition point. As was mentioned earlier, the exact location of the transition point depends a great deal on the intensity and frequency of disturbances in the flow field. Therefore, the location of the transition point cannot be firmly specified. Prandtl avoided this problem by assuming that the turbulent boundary layer started at the leading edge. This condition gives $\delta = 0$ at $x = 0$. Thus, $C = 0$ and

$$\delta = 0.371\left(\frac{\nu}{U_\infty}\right)^{1/5} x^{4/5} = 0.371\left(\frac{\nu}{U_\infty x}\right)^{1/5} x \qquad \text{Eq. 9.11-3}$$

Substituting this value for δ into Eq. 9.11-2, we find that

$$\tau_w(x) = 0.0288\,\frac{\rho U_\infty{}^2}{g_c}\left(\frac{\nu}{U_\infty x}\right)^{1/5}$$

Then, by Eq. 9.10-13,

$$c_f = 0.0576\left(\frac{\nu}{U_\infty x}\right)^{1/5}$$

The drag, D, on a plate of width b is then determined from the following equation

$$D = b\int_0^L \tau_w(x)\, dx = 0.0288\,\frac{\rho U_\infty{}^2 b}{g_c}\left(\frac{\nu}{U_\infty}\right)^{1/5}\int_0^L \frac{dx}{x^{1/5}}$$

$$= 0.036\,\frac{\rho U_\infty{}^2 bL}{g_c}\left(\frac{\nu}{U_\infty L}\right)^{1/5}$$

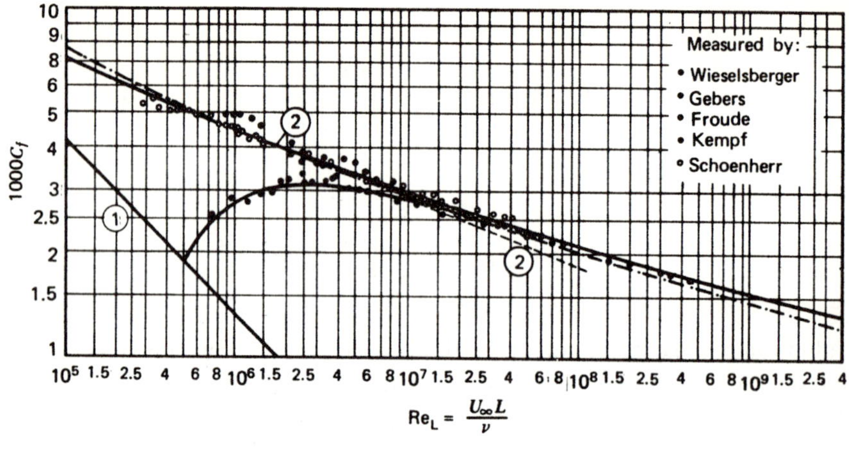

Fig. 9-36. Average skin-friction coefficient for smooth flat plate at zero incidence; comparison between theory and measurement. Curve 1 is from Eq. 9.10-20 and curve 2 is from Eq. 9.11-4. (From *Boundary-Layer Theory* by H. Schlichting, Copyright © 1968 by McGraw-Hill. Used with permission of McGraw-Hill Book Company.)

Then, by Eq. 9.10-16,

$$C_f = 0.072\left(\frac{\nu}{U_\infty L}\right)^{1/5}$$

However, a better agreement with experimental results is obtained (see Fig. 9-36) if the 0.072 in the above formula is replaced with 0.074; that is,

$$C_f = 0.074\left(\frac{\nu}{U_\infty L}\right)^{1/5}, \qquad \text{for } 5 \times 10^5 < \mathrm{Re}_L < 10^7 \qquad \text{Eq. 9.11-4}$$

Then,

$$D = 0.037\,\frac{\rho U_\infty{}^2 bL}{g_c}\left(\frac{\nu}{U_\infty L}\right)^{1/5} \qquad \text{Eq. 9.11-5}$$

The corresponding change in the local drag coefficient gives

$$c_f = 0.0592\left(\frac{\nu}{U_\infty x}\right)^{1/5} \qquad \text{Eq. 9.11-6}$$

and

$$\tau_w = 0.0296 \frac{\rho U_\infty^2}{g_c} \left(\frac{\nu}{U_\infty x} \right)^{1/5} \qquad \text{Eq. 9.11-7}$$

Since, in reality, the flow in the boundary layer is turbulent only on the downstream side of the transition point, accuracy[11] can be improved by subtracting from the drag determined by Eq. 9.11-5—the difference in drag between a turbulent boundary layer and a laminar boundary layer from the leading edge to the point of transition, x_T. Therefore,

$$D = \frac{\rho U_\infty^2 b}{g_c} \left\{ 0.037 \left[L \left(\frac{\nu}{U_\infty L} \right)^{1/5} - x_T \left(\frac{\nu}{U_\infty x_T} \right)^{1/5} \right] + 0.664 x_T \left(\frac{\nu}{U_\infty x_T} \right)^{1/2} \right\}$$

Eq. 9.11-8

Problems

9-1 An airfoil model having a chord length of 10.0 cm is tested in a wind tunnel, with the free-stream pressure and temperature equal to 1.00×10^5 N/m^2 and 40 C, respectively. The free-stream Reynolds number is 10^5 and the angle of attack is 11.6°. Fig. 9-4a shows the pressure distribution obtained in the test. (*a*) Determine the pressure on the upper and lower surfaces of the airfoil at $x/L = 0.2$ and $x/L = 0.8$. (*b*) Is there an adverse pressure gradient on either the upper or lower surface for $0.2 \le x/L \le 0.8$? Explain.

9-2 The boundary-layer thickness, δ, and the shear stress, τ, for a flow over a flat plate at zero angle of attack are determined from boundary-layer theory. These quantities are given by the following equations

$$\delta = 4.9 \sqrt{\frac{\nu x}{V_\infty}} \qquad \text{and} \qquad \tau = 0.332 \frac{\rho V_\infty^2}{g_c} \sqrt{\frac{\nu}{V_\infty x}}$$

(*a*) Determine the boundary-layer thickness and shear stress at a location of 8 cm from the leading edge for a flat-plate model airfoil at a zero angle of attack when the airstream velocity, temperature, and pressure in the test section are 75 m/s, 40 C, and 1.00×10^5 N/m^2, respectively. (*b*) Integrate the shear stress $\tau(x)$ and determine the skin-friction drag per meter of wing span, D/b, if the chord length of the model airfoil is 8 cm.

[11] See Schlichting, p. 600.

9-3 The critical Reynolds number for the flow over a flat plate is approximately 3.0×10^5. In this case, the characteristic length appearing in the Reynolds number is the chord length. However, if the plate is smooth and the intensity of the disturbance in the stream is kept low (conditions compatible with many wind-tunnel situations), transition to turbulence can be delayed until $\text{Re} \approx 3 \times 10^6$. (*a*) Using $(\text{Re})_c$ for the determination of the transition point, determine the maximum length of plate for a completely laminar boundary layer for the airstream conditions described in Problem 9-2. (*b*) Repeat part (*a*), but use the delayed-transition Reynolds number mentioned above.

9-4 Air flows over a 1.0-in. diameter circular cylinder. The free-stream temperature and pressure are 70 F and 14.7 psia, respectively. (*a*) Determine the pressure on the surface of the cylinder at $\theta = 0°$, $90°$, and $180°$ for the three Reynolds numbers indicated in Fig. 9-7. (*b*) Determine the drag on the cylinder per unit depth at these three Reynolds numbers.

9-5 (*a*) Determine the velocity potential and stream function for the flow field described in Example 4-1. (*b*) Show that the equipotential curve and the streamline curve passing through the point (x, y) are orthogonal to each other. (*c*) Determine the mass flow per unit depth between the wall $y = 0$ and the streamline passing through $(3, 4)$.

9-6 (*a*) What is the velocity potential and stream function for a uniform flow in the x direction of the velocity V_∞? (*b*) Determine the x- and y-components of the velocity vector for a field consisting of the superposition of the uniform flow in the x direction of a velocity V_∞ and a source of strength Q. (*c*) Show that the velocity potential and stream function obtained for both the source and the line vortex satisfy Laplace's equation.

9-7 (*a*) Determine the velocity potential for the superposition of a source-sink pair of equal strength. Take the distance between the source and sink to be $2b$, as shown in Fig. 9-37. (*b*) Show that the velocity potential for a doublet is given by

$$\phi = -\frac{K}{\pi}\frac{\cos\theta}{r}$$

Hint. A doublet is obtained by taking the limit of $(\phi_{\text{source}} + \phi_{\text{sink}})$ as $b \to 0$ and $Q \to \infty$, constrained by $bQ = K$. Utilize the fact that $\ln(1 + u) = u - \frac{1}{2}u^2 + \frac{1}{3}u^3, \ldots$, and $\ln(1 - u) = -u - \frac{1}{2}u^2 - \frac{1}{3}u^3, \ldots$.
(*c*) Determine the stream function for the doublet. (*d*) Sketch the streamline pattern for the doublet.

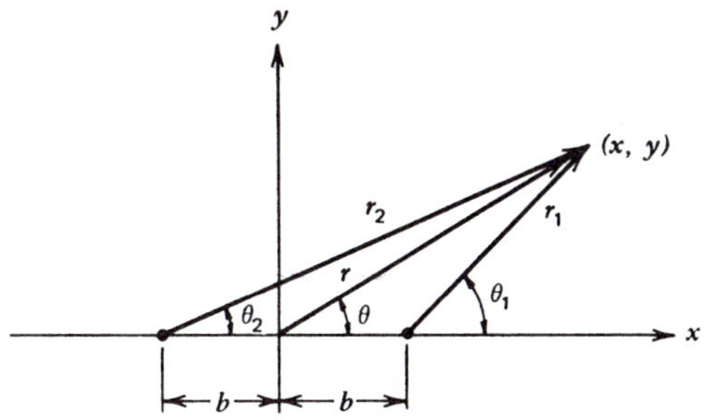

Fig. 9-37. A source-sink pair.

9-8 (a) Determine the stream function resulting from the superposition of a doublet and a uniform flow in the $(-x)$ direction of velocity V_∞. (b) Show that the shape of the curve obtained by setting $\Psi = 0$ is that of a circle. (c) Determine the strength of the doublet that will give a circle of radius a for the curve found in part (b). (d) Substitute the strength of the doublet obtained in part (c) into your answer for part (a). (e) Consider a vortex where the stream function, Ψ, is zero on a circle of radius a. To obtain the stream function for this case, we need to add a constant of integration to the equation obtained earlier. The equation of the stream function is given by Eq. 9.4-9; it is modified to reflect this integration constant

$$\Psi = -\frac{\Gamma}{2\pi}\ln\frac{r}{a}$$

Now superimpose this equation and the equation obtained in part (d) and compare with Eq. 9.6-6. What conclusion can be drawn regarding the superposition of a uniform flow, a doublet, and a vortex?

9-9 Show that the velocity potential, ϕ, for a three-dimensional point source having a strength Q and located at the origin is $\phi = Q/(4\pi r)$.

9-10 Simultaneous pressure measurements at two points in a tornado field were recorded as follows:

$$p = 2051.2 \text{ psfa}, \qquad \text{at } r = 1902 \text{ ft}$$

$$p = 2039.3 \text{ psfa}, \qquad \text{at } r = 715 \text{ ft}$$

where r is the distance from the center of the core. If the air density is $0.0697 \text{ lb}_m/\text{ft}^3$, determine the magnitude of air velocity at $r = 715$ ft and $r = 1902$ ft.

9-11 Consider an airflow around a cylinder with no circulation. (*a*) Using Eqs. 9.2-12 and 9.5-33, determine $(p - p_\infty)/q$, on the surface of the cylinder for $\theta = 0°$, 90°, 180°, and 270° ($q = \rho V_\infty^2/(2g_c)$). Does your answer agree with the theoretical curve shown in Fig. 9-7? (*b*) What is the difference in pressure, in pounds per square inch, between the theoretical value and the measured value at $\theta = 180°$ for a Reynolds number flow of 1.1×10^5? Take the free-stream pressure and temperature to be 14.7 psia and 80 F, respectively. Assume that the cylinder has a diameter of 1.0 ft.

9-12 Air flows over a cylinder, which has a diameter of 0.3 m. The free-stream conditions are 1.0132×10^5 N/m^2 abs., 300 K, and 60 m/s. (*a*) Determine the circulation, Γ, around the cylinder in terms of the variables present in Eq. 9.5-33. (*b*) If $\Gamma = 40.5$ m^2/s, determine the angular position of the stagnation points. (*c*) Using ideal flow theory, determine (1) the pressure at the stagnation points, and (2) the magnitude of the velocity and the pressure on the cylinder at θ equals $\pi/2$ and $3\pi/2$.

9-13 The lift on a body is defined to be the component of the resultant aerodynamic force in the direction perpendicular to the free-stream velocity, whereas the drag is the component in the same direction as the free stream. Using the theoretical pressure distribution for a flow around a cylinder, (*a*) show that the lift and drag on a cylinder with no circulation are both zero, and (*b*) determine the lift and drag on the cylinder described in Problem 9-12.

9-14 A flat-plate wing, as shown in Fig. 9-20*b*, flies at 200 mi/h at an angle of attack of 10°. Take the free stream to be at standard atmospheric conditions. From the results of potential flow theory, determine: (*a*) the magnitude of the velocity of the airstream on the upper and lower surfaces of the wing at x equals $-(l/2)$, 0, and $l/2$; (*b*) the position on the lower surface where the pressure has a relative maximum (see Fig. 9-23); (*c*) the pressure on the lower surface at the position found in (*b*); (*d*) the pressure difference $(p_L - p_u)$ at the position found in (*b*).

9-15 A flat-plate wing flies in air at standard atmospheric pressure, at 300 K, and at an angle of attack of 10°. The flight Reynolds number is 9.0×10^6. Determine: (*a*) the lift per unit span length if the chord length is 1.5 m; (*b*) the required span length to support a total weight of 90,000 N.

9-16 The chord length of the airfoil whose characteristics are plotted in Fig. 9-29 is 1.5 m. The airfoil flies in air at standard atmospheric pressure, at 300 K, and at an angle of attack of 10°. The flight Reynolds number is 9.0×10^6. Determine: (*a*) the span length required to support a total weight of 90,000 N; (*b*) the thrust required by the engines to overcome the drag created by the airfoil. Assume the thrust is directed in the opposite direction to the free stream.

9-17 A second approach for constructing the flow field associated with a line vortex in a two-dimensional field is as follows. Consider a flow field where the streamline pattern is that of concentric circles in the (x, y) plane and where the vorticity vector, $\boldsymbol{\omega}$, is a nonzero constant vector in the z direction inside a core of radius b and zero outside the core; that is,

$$\boldsymbol{\omega} = \begin{cases} \omega_0 \mathbf{k}, & \text{for } 0 \leq r \leq b \\ \mathbf{0}, & \text{for } r > b \end{cases}$$

(*a*) Using Stoke's theorem show that

$$\mathbf{v} = \begin{cases} \dfrac{\omega b^2}{2r}\, \mathbf{e}_\theta, & \text{for } r \geq b \\[2ex] \dfrac{\omega r}{2}\, \mathbf{e}_\theta, & \text{for } r \leq b \end{cases}$$

(*b*) The velocity field of the line vortex, as given by the third equation in the set of Eqs. 9.4-9, is obtained by letting $\omega \to \infty$, $b \to 0$, in such a manner that ωb^2 is a constant. Relate ωb^2 to the strength of the vortex, Γ, described in Eqs. 9.4-9.

9-18 Consider point Q_u on the upper surface of an airfoil and Q_L, directly below Q_u, on the lower surface of the airfoil. As discussed at the end of Section 9.7, the finite length of the wing results in a difference in the velocity components in the spanwise direction at these two points. This effect can be modeled by considering the wing and wake region to be a vortex sheet. The vortex sheet is developed in a similar manner as the line vortex described in Problem 9-17. Consider a core region where the velocity vector has only x- and y-components. Take the vorticity vector, $\boldsymbol{\omega}$, in the infinitesimal region of the core, shown in Fig. 9-38, to be a constant vector. Using the form of the divergence theorem given by the second equation in the set of Eqs. 2.2-17, show that

$$\boldsymbol{\omega}\varepsilon = \mathbf{k} \times (\mathbf{v}_u - \mathbf{v}_L)$$

A vortex sheet is obtained by letting $|\boldsymbol{\omega}| \to \infty$ and $\varepsilon \to 0$, in such a manner that $\boldsymbol{\omega}\varepsilon$ is a constant vector.

9-19 Water at 10 C flows over a submerged flat plate at a zero angle of attack. The free-stream velocity is 0.2 m/s. (*a*) Determine the velocity of the fluid 9.3 m from the leading edge and 9.5 cm from the plate surface. (*b*) Find the wall shear stress 0.3 m from the leading edge. (*c*) Find the drag on that part of the plate surface that extends from the leading edge to 0.3 m downstream from the leading edge. Assume that the

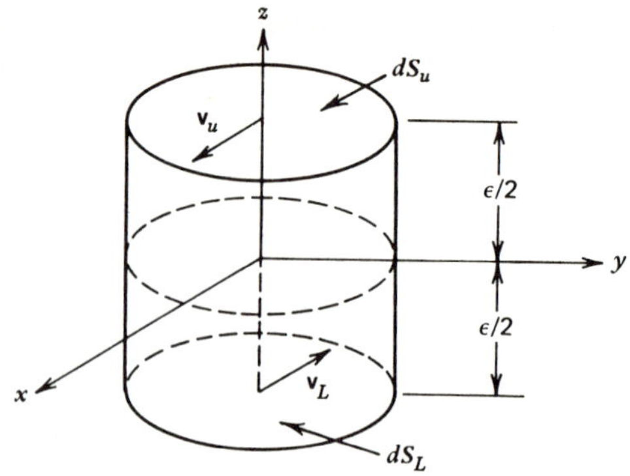

Fig. 9-38. The construction of a vortex sheet.

plate has both surfaces exposed to the flow. Take the plate width to be 1 m. (*d*) Determine the position from the leading edge where the boundary-layer transition occurs. Assume transition occurs at the lower Re_x limit.

9-20 Air, having a free-stream pressure and temperature equal to 1.013×10^5 N/m^2 and 300 K, respectively, flows over a flat-plate model airfoil at a zero angle of attack. If the chord length of the model is 7.5 cm, determine the maximum free-stream velocity that will ensure a laminar boundary layer over the entire length of the airfoil.

9-21 For the airfoil model described in Problem 9-20, (*a*) determine the shear stress and the local skin-friction coefficient at the midchord position. Assume that the free-stream velocity is 25 m/s. (*b*) Determine the drag and the average skin-friction coefficient on the airfoil model. Assume that the free-stream velocity is 25 m/s and the span length is 0.5 m.

9-22 It is desired to obtain a first-guess estimate of the drag on an airplane wing by assuming that it is a flat-plate airfoil. The airplane is to fly at an altitude of 20,000 ft at a speed of 200 mi/h. Assume that the atmosphere corresponds to the U.S. Standard Atmosphere. The wing has a chord length of 5 ft and a span length of 30 ft. Estimate the drag on the wing and the average drag coefficient by using: (*a*) Eqs. 9.11-5 and 9.11-4; (*b*) Eqs. 9.11-8 and 9.10-16.

9-23 Determine the drag per unit span length for the flat-plate wing and flight conditions described in Problem 9-15, except take the angle of attack to be 0°.

9-24 The boundary layer on a thin airfoil may be approximated as a flat plate with a pressure gradient. In this case, the velocity just outside the boundary layer can be taken as $U(x)$. (*a*) Using the von Kármán and Pohlhausen method, show that the boundary-layer equation for a thin airfoil can be expressed by

$$\frac{g_c \tau_w}{\rho} = -\frac{d}{dx}\int_0^b v_1{}^2\,dy + U\frac{d}{dx}\int_0^b v_1\,dy + bU\frac{dU}{dx}$$

(*b*) Noting that

(1) $bU\dfrac{dU}{dx} = \displaystyle\int_0^b U\frac{dU}{dx}\,dy$

(2) $U\dfrac{d}{dx}\displaystyle\int_0^b v_1\,dy = \int_0^b U\frac{\partial v_1}{\partial x}\,dy = \int_0^b \frac{\partial\,(Uv_1)}{\partial x}\,dy - \int_0^b v_1\frac{dU}{dx}\,dy$

$$= \frac{d}{dx}\int_0^b Uv_1\,dy - \frac{dU}{dx}\int_0^b v_1\,dy$$

(3) $v_1/U = 1, \qquad \text{for } \delta \le y \le b$

show that the equation given in part (*a*) can be rewritten as follows

$$\frac{g_c \tau_w}{\rho} = \frac{d}{dx}\left[U^2\int_0^\delta \frac{v_1}{U}\left(1 - \frac{v_1}{U}\right)dy\right] + U\frac{dU}{dx}\int_0^\delta\left(1 - \frac{v_1}{U}\right)dy$$

9-25 An approximate solution to the boundary-layer equation can be obtained by assuming a velocity distribution of the form

$$\frac{v_1}{U} = A\left(\frac{y}{\delta}\right) + B\left(\frac{y}{\delta}\right)^2 + C\left(\frac{y}{\delta}\right)^3 + D\left(\frac{y}{\delta}\right)^4$$

where A, B, C, and D are, at most, a function of x. The variables A, B, C, and D are related to each other by the boundary conditions, which are

(1) $v_1(x, 0) = v_2(x, 0) = 0$

(2) $\nu\dfrac{\partial^2 v_1}{\partial y^2}(0, x) = \dfrac{g_c}{\rho}\dfrac{dP}{dx} = -U\dfrac{dU}{dx}$

(3) $v_1(x, \delta) = U(x)$

(4) $\dfrac{\partial v_1}{\partial y}(x, \delta) = 0$

(5) $\dfrac{\partial^2 v_1}{\partial y^2}(x, \delta) = 0$

Boundary condition (1) results from the fact that the velocity is zero at the wall. Condition (2) is arrived at from the boundary-layer field equations.[12] Condition (3) implies that, at $y = \delta$, the velocity equals the value obtained from the incompressible solution. Condition (4) states that, at $y = \delta$, the shear stress goes to zero. Condition (5) states that, at $y = \delta$, Euler's equations are valid.[13] The boundary conditions provide a means for relating the variables A, B, C, and D to δ and dU/dx.

(*a*) Show that boundary condition (2) gives

$$B = -\frac{\delta^2}{2\nu}\frac{dU}{dx}$$

(*b*) Let $\Lambda = (\delta^2/\nu)(dU/dx)$. Show that the variables A, B, C, and D are related to Λ by the following set of equations.

$$A = 2 + \frac{\Lambda}{6} \qquad B = -\frac{\Lambda}{2} \qquad C = -2 + \frac{\Lambda}{2} \qquad D = 1 - \frac{\Lambda}{6}$$

(*c*) Let $\eta = y/\delta$. Show that v_1/U can be expressed as follows

$$\frac{v_1}{U} = (2\eta - 2\eta^3 + \eta^4) + \frac{\Lambda}{6}(\eta - 3\eta^2 + 3\eta^3 - \eta^4) = F(\eta) + \Lambda G(\eta)$$

(*d*) Show that $\tau_w = \mu U(2 + \Lambda/6)/\delta g_c$

(*e*) Show that the boundary-layer equation developed in Problem 9-24 reduces to

$$\frac{\nu U}{\delta}\left(2 + \frac{\Lambda}{6}\right) = \frac{d}{dx}\left\{\delta U^2 \int_0^1 [F(\eta) + \Lambda G(\eta)][1 - F(\eta) - \Lambda G(\eta)]\, d\eta\right\}$$
$$+ U\frac{dU}{dx}\delta \int_0^1 [1 - F(\eta) - \Lambda G(\eta)]\, d\eta$$

[12] The boundary-layer field equations (whose development is beyond the scope of this book) are developed in Schlichting. They are

$$v_1\frac{\partial v_1}{\partial x} + v_2\frac{\partial v_1}{\partial y} = -\frac{g_c}{\rho}\frac{\partial p}{\partial x} + \nu\frac{\partial^2 v_1}{\partial y^2} \qquad \text{Eq. (1)}$$

$$\frac{\partial p}{\partial y} = 0 \qquad \text{Eq. (2)}$$

We see that if $\partial^2 v_1/\partial y^2 = 0$, then Eq. (1) reduces to the x-component of Eq. 9.2-7 for the case of steady flow.

[13] See footnote 12.

9-26 (*a*) Using the boundary-layer equation developed in Problem 9-25, and the following definitions

$$I_1 = \int_0^1 [F(\eta) - F^2(\eta)]\, d\eta \qquad I_2 = \int_0^1 [2F(\eta)G(\eta) + G(\eta)]\, d\eta$$

$$I_3 = \int_0^1 G^2(\eta)\, d\eta \qquad I_4 = \int_0^1 F(\eta)\, d\eta$$

$$I_5 = \int_0^1 G(\eta)\, d\eta$$

where $F(\eta) = 2\eta - 2\eta^3 + \eta^4$ and $G(\eta) = \frac{1}{6}(\eta - 3\eta^2 + 3\eta^3 - \eta^4)$, show that the differential equation for $\delta(x)$ can be expressed as follows

$$\frac{d\delta}{dx} = \frac{\nu}{\delta U(I_1 + 3I_2\Lambda - 5I_3\Lambda^2)} \times \left\{ 2I_3\Lambda^3 + (I_5 - 2I_2)\Lambda^2 + (I_4 - 2I_1 - \tfrac{5}{6})\Lambda + 2 - \frac{\delta^4 U}{\nu^2}\frac{d^2U}{dx^2}(I_2 - 2I_3\Lambda) \right\}$$

where

$$\Lambda = \frac{\delta^2}{\nu}\frac{dU}{dx}$$

(*b*) Show that the values for $I_1, I_2, \ldots, I_5$, as defined in part (*a*), are as follows

$$I_1 = \frac{37}{315} \qquad I_2 = -\frac{1}{945} \qquad I_3 = \frac{1}{9072}$$

$$I_4 = \frac{7}{10} \qquad I_5 = \frac{1}{120}$$

(*c*) A solution to the differential equation given in part (*a*) can be obtained by numerical methods. The integration is started at the stagnation point, where the differential equation is singular (U is zero at the stagnation point). This singularity can be eliminated by requiring the following equation to be valid at the stagnation point

$$2I_3\Lambda^3 + (I_5 - 2I_2)\Lambda^2 + (I_4 - 2I_1 - \tfrac{5}{6})\Lambda + 2 = 0$$

Write a computer program that uses Newton's method[14] for solving the roots of an equation of the form $F(\Lambda) = 0$. Use this program to show that $\Lambda_0 = 7.052$, where Λ_0 is the value of Λ at the stagnation point. From the definition of Λ and the value Λ_0, the value of δ at the stagnation point can be determined, if dU/dx is known.

(*d*) Using L'Hopital's rule show that

$$\left(\frac{d\delta}{dx}\right)_0 = \frac{\delta_0\left(\dfrac{d^2U}{dx^2}\right)_0 [4I_3\Lambda_0{}^2 + (2I_5 - 3I_2)\Lambda_0 + (I_4 - 2I_1 - \frac{5}{6})]}{\left(\dfrac{dU}{dx}\right)_0 [-17I_3\Lambda_0{}^2 + (3I_2 - 4I_5 + 8I_2)\Lambda_0 - 2I_4 + 5I_1 - \frac{5}{3}]}$$

where the subscript zero indicates the value of the variable at the stagnation point.

Computer Problems

9-27 Consider the flat-plate airfoil described in Section 9.7 (see Fig. 9-20*b*). Our interest is in developing a numerical method for determining $\delta(x)$ and $\tau_w(x)$ on the lower surface of the airfoil. As a matter of convenience, let us select the stagnation point to be the origin at the coordinate system for our boundary-layer analysis (see Fig. 9-39). (*a*) Let (x^*, y^*) be the coordinates used in developing the velocity field on the flat-plate airfoil. Under these circumstances, x^* replaces x in Eq. 9.7-1. Show that the velocity field on the lower surface of the airfoil, expressed in terms of the boundary-layer variables, becomes

$$U(x) = V_\infty\left[\cos\alpha - \sin\alpha\sqrt{\frac{\ell + 2(x_s^* - x)}{\ell - 2(x_s^* - x)}}\right]$$

where

$$x_s^* = \frac{\ell}{2}(\cos^2\alpha - \sin^2\alpha)$$

[14] Newton's method gives successive approximations for a root of an equation of the form $F(x) = 0$ by the following algorithm

$$x_{n+1} = x_n - \frac{F(x_n)}{F'(x_n)} \qquad n = 1, 2, 3, \ldots.$$

The successive approximations continue until $|x_{n+1} - x_n|$ is sufficiently small to give the root to the desired accuracy. The method may require a starting value, x_1, which is not too far from the correct value.

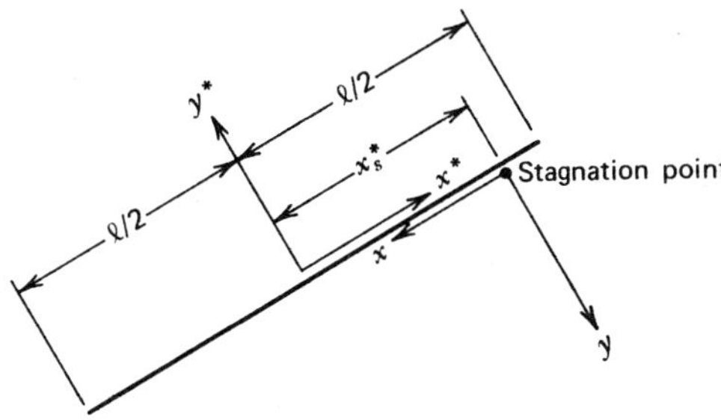

Fig. 9-39. The coordinate system for a boundary-layer analysis.

(*b*) Show that on the lower surface of the airfoil, dU/dx and d^2U/dx^2 can be expressed by the following relations

$$\frac{dU}{dx} = \frac{2\ell V_\infty \sin\alpha}{[\ell - 2(x_s^* - x)]\sqrt{\ell^2 - 4(x_s^* - x)^2}}$$

and

$$\frac{d^2U}{dx^2} = -4\ell V_\infty \sin\alpha \left\{ \frac{1}{[\ell - 2(x_s^* - x)]^2\sqrt{\ell^2 - 4(x_s^* - x)^2}} + \frac{2(x_s^* - x)}{[\ell - 2(x_s^* - x)][\ell^2 - 4(x_s^* - x)^2]^{3/2}} \right\}$$

(*c*) Consider a flat-plate model airfoil that is tested in a wind tunnel, under the following conditions

$$V_\infty = 150 \text{ ft/s} \qquad \ell = 3 \text{ in.} \qquad \alpha = 10°$$

$$p_\infty = 14.5 \text{ psia} \qquad T_\infty = 95 \text{ F}$$

Using the values of $I_1, I_2, \ldots, I_5$, Λ_0 and $(d\delta/dx)_0$, given in Problem 9-26(*b*), (*c*), and (*d*), and the Runge–Kutta method for solving a first-order differential equation, determine δ and τ_w at

$$\frac{x}{\frac{\ell}{2} + x_s^*} = 0, 0.1, 0.2, \ldots, 0.98$$

Note: For the flat plate airfoil at a non-zero angle of attack, dU/dx is singular at the trailing edge. Thus, the evaluation of δ and τ_w at the trailing edge ($x = l/2 + x_s^*$) requires special treatment.

(d) Using numerical integration by Simpson's rule and the values of τ_w obtained in part (c), determine the skin-friction drag on the lower surface of the flat-plate airfoil from

$$0 \le x \le 0.98\left(\frac{\ell}{2} + x_s^*\right)$$

9-28 The Joukowski transformation transforms a circle into a cross-sectional shape that may be used as an airfoil, provided the constants in the transformation are within the proper limits. Consider the transformation

$$z = \omega + \frac{b^2}{\omega} \qquad \text{Eq. (1)}$$

where z and ω are complex variables, and b is real; that is,

$$z = x + iy \qquad \text{and} \qquad \omega = \zeta + i\eta \qquad \text{Eq. (2)}$$

The polar representations of z and ω are

$$z = re^{i\theta} \qquad \text{and} \qquad \omega = \mathrm{R}e^{i\phi} \qquad \text{Eq. (3)}$$

The circle C', shown in Fig. 9-40a, is represented in the ω plane as follows

$$\omega(C') = (\zeta_0 + i\eta_0) + ae^{i\phi} \qquad \text{Eq. (4)}$$

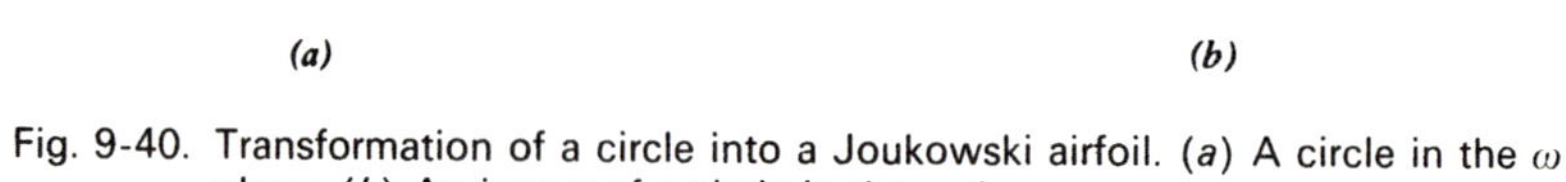

(a) (b)

Fig. 9-40. Transformation of a circle into a Joukowski airfoil. (a) A circle in the ω plane. (b) An image of a circle in the z plane.

Substituting Eq. (4) into Eq. (1), we obtain the image of C' in the z plane (see Fig. 9-40b); that is,

$$z(C) = (\zeta_0 + i\eta_0) + ae^{i\phi} + \frac{b^2}{(\zeta_0 + i\eta_0) + ae^{i\phi}}$$

$$= (\zeta_0 + a\cos\phi) + i(\eta_0 + a\sin\phi)$$

$$+ \frac{b^2[(\zeta_0 + a\cos\phi) - i(\eta_0 + a\sin\phi)]}{(\zeta_0 + a\cos\phi)^2 + (\eta_0 + a\sin\phi)^2}$$

Equating the real and imaginary parts on both sides of the above equation gives

$$x(C) = (\zeta_0 + a\cos\phi) + \frac{b^2(\zeta_0 + a\cos\phi)}{a^2 + \rho_0{}^2 + 2a(\zeta_0\cos\phi + \eta_0\sin\phi)}$$

and

$$y(C) = (\zeta_0 + a\sin\phi) - \frac{b^2(\eta_0 + a\sin\phi)}{a^2 + \rho_0{}^2 + 2a(\zeta_0\cos\phi + \eta_0\sin\phi)} \qquad \text{Eqs. (5)}$$

where $\rho_0{}^2 = \zeta_0{}^2 + \eta_0{}^2$. If the displacement η_0 is small, then the image curves will look like airfoils. (a) Show that the length of the airfoil, l, is given by

$$\ell = \frac{(2b + \zeta_0)^2}{(b + \zeta_0)}$$

(b) Using the expression given in part (a), show that

$$\left(\frac{\zeta_0}{b}\right) = -2 + \frac{\ell}{2b} \pm \frac{\ell}{2b}\sqrt{1 - \frac{4b}{\ell}}$$

Thus, $4b/\ell \leq 1$. Airfoil shapes occur when $1 - 4b/\ell \ll 1$. (c) Show that if $\ell = 8$ units (inches, centimeters, feet, etc.), then

$$\zeta_0 = -2b + 4\left(1 + \sqrt{1 - \frac{b}{2}}\right)$$

Thus, for $\ell = 8$ units, $b \leq 2$ units. Note that the plus sign was used in the expression for ζ_0/b. (d) Use a computer plot subroutine to obtain the airfoil shapes for the following set of values

$$\ell = 8 \text{ in.} \qquad b = 1.98,\ 1.9928,\ 1.995$$

$$\eta_0 = 0.0,\ 0.1,\ 0.2, \text{ and } 0.3$$

Start $x(C)$, $y(C)$ calculations at $\phi = -\pi + \beta$, where

$$\beta = \tan^{-1} \frac{\eta_0}{b + \zeta_0}$$

Observe that $a^2 = (\zeta_0 + b)^2 + \eta_0{}^2$

Observation of the resulting plots should indicate the following:

1. The airfoil thickness increases as b decreases.

2. The airfoil camber[15] increases as η_0 increases.

[15] The maximum distance between the chord and the mean line of the airfoil is defined as the camber (see Fig. 9-39*b*).

Appendix A.1

Elements of Vector Calculus

A.1.1 INTRODUCTION

This appendix has been written for the convenience of those students who desire or require a more thorough presentation of the concepts of vector calculus than is given by the summary in Section 2.2 of Chapter 2. We begin with the geometric interpretation of the partial derivatives, Eqs. 2.2-3 and 2.2-4, and develop the concepts in the same order as they are listed in Section 2.2.

A.1.2 THE DIRECTIONAL DERIVATIVE AND THE GRADIENT

Consider the scalar field $T(x, y)$. Geometrically, $T(x, y)$ can be represented as a surface. The partial derivatives

$$\frac{\partial T}{\partial x}(x_0, y_0) \qquad \text{and} \qquad \frac{\partial T}{\partial y}(x_0, y_0)$$

are defined by

$$\frac{\partial T}{\partial x}(x_0, y_0) = \lim_{x \to x_0} \frac{T(x, y_0) - T(x_0, y_0)}{x - x_0}$$

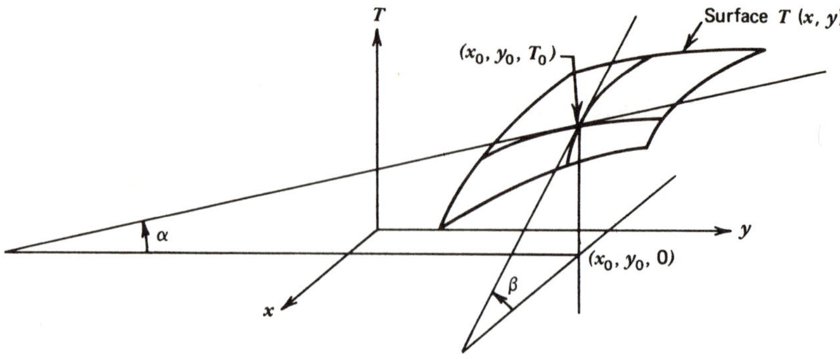

Fig. A.1-1. Geometric representation of the partial derivative.

and

$$\frac{\partial T}{\partial y}(x_0, y_0) = \lim_{y \to y_0} \frac{T(x_0, y) - T(x_0, y_0)}{y - y_0} \qquad \text{Eqs. A.1-1}$$

A geometric interpretation for these derivatives may be made. These derivatives may be considered to represent the slope of the line tangent to two special curves at (x_0, y_0, T_0) where $T_0 = T(x_0, y_0)$. One of these curves is made by the intersection of the surface $T(x, y)$ and the plane $x = x_0$. The other is made by the intersection of the surface $T(x, y)$ and the plane $y = y_0$ (see Fig. A.1-1).

One may also make the interpretation that these derivatives respectively give the rate of change of T with respect to distance in the x direction and in the y direction. However, a more general description of the rate of change of T with respect to distance is desirable. This is provided by the *directional derivative.* Let S be the distance from $P(x_0, y_0)$ to $Q(x, y)$ and θ be angle between the line PQ and the x-axis (see Fig. A.1-2). Define the directional derivative $\partial f/\partial S$ at (x_0, y_0) by

$$\frac{\partial f}{\partial S}(x_0, y_0) = \lim_{S \to 0} \frac{T(x, y) - T(x_0, y_0)}{S} \qquad \text{Eq. A.1-2}$$

It should be understood that θ is to be held constant in the limiting process. Observe that the value of the directional derivative not only depends on the function $T(x, y)$, but also on the direction.

One can readily relate the directional derivative at (x_0, y_0) to the partial derivatives $\partial T/\partial x$ and $\partial T/\partial y$ at (x_0, y_0). This can be accomplished by considering

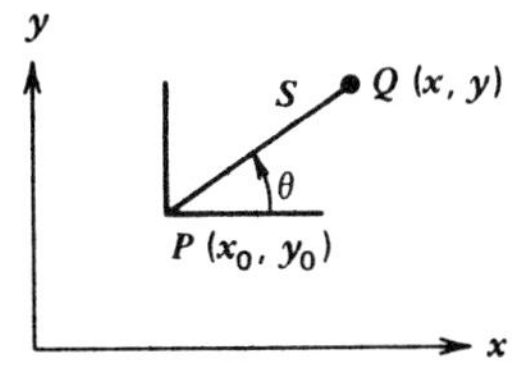

Fig. A.1-2. Geometry for the directional derivative.

a change of independent variable from (x, y) to (S, θ). The two sets of variables are related by

$$x = x_0 + S\cos\theta \rightarrow \frac{\partial x}{\partial S} = \cos\theta \qquad \text{Eqs. A.1-3}$$

$$y = y_0 + S\sin\theta \rightarrow \frac{\partial y}{\partial S} = \sin\theta$$

Use of the chain rule gives

$$\frac{\partial T}{\partial S} = \frac{\partial T}{\partial x}\frac{\partial x}{\partial S} + \frac{\partial T}{\partial y}\frac{\partial y}{\partial S} = \cos\theta\frac{\partial T}{\partial x} + \sin\theta\frac{\partial T}{\partial y} \qquad \text{Eq. A.1-4}$$

Thus, knowing $\partial T/\partial x$ and $\partial T/\partial y$ at (x_0, y_0) and specifying θ, $\partial T/\delta S$ at (x_0, y_0) can be determined.

In the geometric representation, the directional derivative represents the slope of the line tangent to the curve at (x_0, y_0, T_0) made by the intersection of the surface $T(x, y)$ and a vertical plane making an angle θ with the plane $y = y_0$.

The directional derivative can also be obtained by the use of the vector dot product. If a unit vector in the PQ direction is dotted with the vector quantity called the *gradient of T*, the directional derivative is obtained. The gradient is extensively used in physics and engineering. The gradient is defined by

$$\text{Gradient } T \equiv \nabla T \equiv \frac{\partial T}{\partial x}\mathbf{i} + \frac{\partial T}{\partial y}\mathbf{j} \qquad \text{Eq. A.1-5}$$

The unit vector, $\mathbf{e}_S$, in the direction from P to Q, is given by

$$\mathbf{e}_S = \cos\theta\,\mathbf{i} + \sin\theta\,\mathbf{j} \qquad \text{Eq. A.1-6}$$

The dot product of $\mathbf{e}_S$ and ∇T is

$$\mathbf{e}_S \cdot \nabla T = (\cos\theta\,\mathbf{i} + \sin\theta\ \mathbf{j}) \cdot \left(\frac{\partial T}{\partial x}\mathbf{i} + \frac{\partial T}{\partial y}\mathbf{j}\right)$$

$$= \cos\theta\,\frac{\partial T}{\partial x} + \sin\theta\,\frac{\partial T}{\partial y} \qquad \text{Eq. A.1-7}$$

Comparing Eqs. A.1-7 and A.1-4, it can be seen that

$$\frac{\partial T}{\partial S} = \mathbf{e}_S \cdot \nabla T \qquad \text{Eq. A.1-8}$$

The significance of ∇T can be discerned if we consider point P to be fixed while allowing θ to vary. Applying the definition of the dot product to Eq. A.1-8, we obtain

$$\frac{\partial T}{\partial S} = |\mathbf{e}_S||\nabla T|\cos\beta \qquad \text{Eq. A.1-9}$$

where β equals the angle between ∇T and $\mathbf{e}_S$ (see Fig. A.1-3*a*). Although $\nabla T(x_0, y_0)$ is independent of θ, $\partial T/\partial S$ at (x_0, y_0) varies with θ (see Fig. A.1-3*b*). Since

$$-1 \le \cos\beta \le 1$$

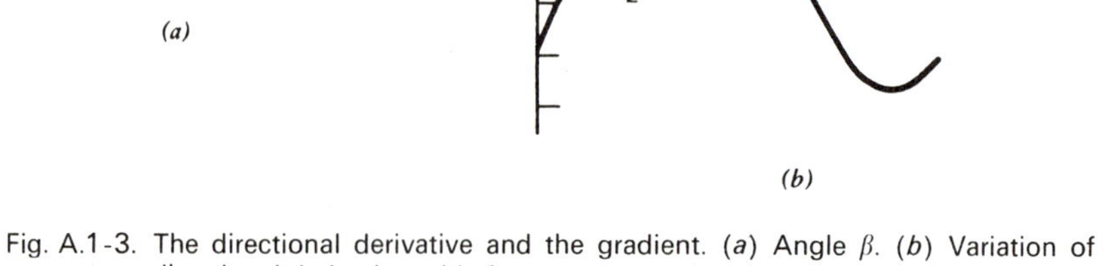

Fig. A.1-3. The directional derivative and the gradient. (*a*) Angle β. (*b*) Variation of directional derivative with θ.

$\partial T/\partial S$ at (x_0, y_0) will have a maximum value when $\cos \beta = 1$, *or when* $\beta = 0$. In other words, when PQ lines up with $\nabla T(x_0, y_0)$, $\partial T/\partial S$ at (x_0, y_0) has a maximum value; that is,

$$\left[\frac{\partial T}{\partial S}(x_0, y_0)\right]_{\max} = |\nabla T(x_0, y_0)| \qquad \text{Eq. A.1-10}$$

If we reverse our thinking, we can say that $\nabla T(x_0, y_0)$ is a vector whose magnitude equals the maximum rate of change of T with distance at (x_0, y_0) and points in the direction of the maximum directional derivative. Frequently in nature we find that some physical quantity will flow either in the direction of the gradient or in the opposite direction. For example, if a temperature field exists in an isotropic material, heat will flow in the opposite direction to ∇T; or if a raindrop falls on a smooth impervious hill, it will flow in the opposite direction to ∇Z, where $Z(x, y)$ is the elevation of the hill.

These concepts may readily be generalized for functions of three independent variables. For example, if $w = w(x, y, z)$,

$$\frac{\partial w}{\partial S} = \cos \alpha \frac{\partial w}{\partial x} + \cos \beta \frac{\partial w}{\partial y} + \cos \gamma \frac{\partial w}{\partial z} \qquad \text{Eq. A.1-11}$$

Where $(\cos \alpha, \cos \beta, \cos \gamma)$ are the direction cosines of the line PQ. The gradient of w would then be defined by

$$\nabla w = \frac{\partial w}{\partial x}\mathbf{i} + \frac{\partial w}{\partial y}\mathbf{j} + \frac{\partial w}{\partial z}\mathbf{k} \qquad \text{Eq. A.1-12}$$

A.1.3 LINE AND SURFACE INTEGRALS INVOLVING VECTOR FIELDS

Line and surface integrals involving vector fields are extensively used in engineering and physics. In fact, in mechanics, work is defined in terms of a line integral. Surface integrals appear in the mathematical representation of the fundamental laws of fluid mechanics. These integrals are defined below for the continuous vector field $\mathbf{F}(x, y, z)$.

Let C be any curve in space. Subdivide curve C into N segments, the end-points being designated as $Q_1, Q_2, Q_3, \ldots, Q_{N+1}$ (see Fig. A.1-4). Let $\Delta \mathbf{r}_j$ be the displacement vector connecting the endpoints of the jth segment. Also let Q'_j be somewhere on $\Delta\mathbf{r}_j$ and $\mathbf{F}_j = \mathbf{F}(Q'_j)$. Define I_L by

$$I_L = \sum_{j=1}^{N} \mathbf{F}_j \cdot \Delta\mathbf{r}_j$$

The line integral, designated by the term on the left-hand side of Eq. A.1-13, is defined to be $\lim_{\max|\Delta\mathbf{r}_j|\to 0} (I_L)$; that is,

$$\int_{\substack{C \\ (Q_1 \to Q_{N+1})}} \mathbf{F} \cdot d\mathbf{r} = \lim_{\max|\Delta\mathbf{r}_j|\to 0} \sum_{j=1}^{N} \mathbf{F}_j \cdot \Delta\mathbf{r}_j \qquad \text{Eq. A.1-13}$$

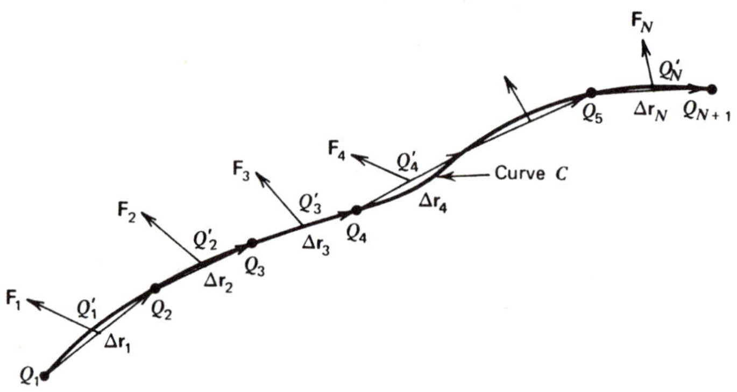

Fig. A.1-4. Subdivision of curve C.

The sign of the line integral depends on the direction of integration. Use of the line integral will be made later in this chapter.

Let S be any surface in space. Subdivide S into N surface elements. This could be accomplished by intersecting the surface with planes parallel to any two of the coordinate planes. Let ΔS_j be the jth surface element and let Q_j lie on ΔS_j (see Fig. A.1-5). Let $\mathbf{F}_j = \mathbf{F}(Q_j)$ and $\mathbf{n}_j$ be the unit normal vector to the surface S at Q_j. Define

$$I_S = \sum_{j=1}^{N} \mathbf{F}_j \cdot \mathbf{n}_j \, \Delta S_j$$

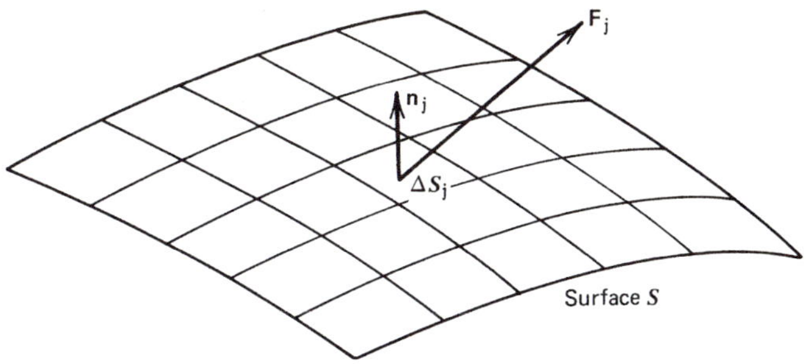

Fig. A.1-5. Subdivision of surface S.

The surface integral, designated by the term on the left-hand side of Eq. A.1-14, is defined to be $\lim_{\max \Delta S_j \to 0} (I_S)$; that is,

$$\oint\!\!\oint_S \mathbf{F} \cdot \mathbf{n}\, dS = \lim_{\max \Delta S_j \to 0} \sum_{j=1}^{N} \mathbf{F}_j \cdot \mathbf{n}_j\, \Delta S_j \qquad \text{Eq. A.1-14}$$

When the surface of integration encloses a region R, the unit normal vector appearing in the integral is to be taken as the outward normal. Frequently $\mathbf{n}\, dS$ is written as the single vector $d\mathbf{S}$.

A.1.4 THE DIVERGENCE OF A VECTOR

Let surface S enclose some infinitesimal region, R, which encloses the point $P(x, y, z)$, as shown in Fig. A.1-6. Recall from Section 2.2 that if $\mathbf{A}$ is a flux vector, then $\oint\!\!\oint_S \mathbf{A} \cdot d\mathbf{S}$ is equal to the net rate of flow of some physical quantity (such as mass, heat, energy, etc.) out of the enclosed region. Thus, by evaluating $\oint\!\!\oint_S \mathbf{A} \cdot d\mathbf{S}$ on a surface enclosing an infinitesimal volume around a point P, dividing through by ΔV, and taking the $\lim_{\Delta V \to 0}$, we can determine whether there is a net accumulation of some physical quantity per unit volume at point P. This process is essentially the integral definition of the *divergence* of $\mathbf{A}$, written div A or $\nabla \cdot \mathbf{A}$. That is,

$$\text{div}\, \mathbf{A} = \nabla \cdot \mathbf{A} = \lim_{\Delta V \to 0} \frac{1}{\Delta V} \oint\!\!\oint_S \mathbf{A} \cdot d\mathbf{S} \qquad \text{Eq. A.1-15}$$

where S encloses ΔV.

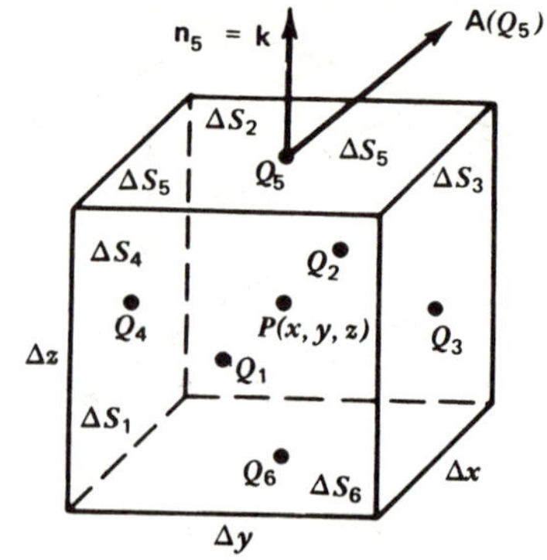

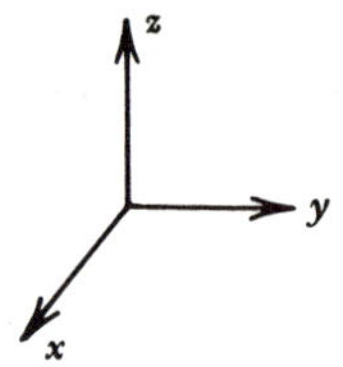

Fig. A.1-6. Infinitesimal volume in the continuum. Note that $\mathbf{n}(Q_1) = \mathbf{i}$, $\mathbf{n}(Q_2) = -\mathbf{i}$, $\mathbf{n}(Q_3) = \mathbf{j}$, $\mathbf{n}(Q_4) = -\mathbf{j}$, $\mathbf{n}(Q_5) = \mathbf{k}$, $\mathbf{n}(Q_6) = -\mathbf{k}$; and that $\Delta S_1 = \Delta S_2 = \Delta y\, \Delta z$, $\Delta S_3 = \Delta S_4 = \Delta x\, \Delta z$, $\Delta S_5 = \Delta S_6 = \Delta x\, \Delta y$.

To evaluate $\nabla \cdot \mathbf{A}$ at point $P(x, y, z)$ in Cartesian coordinates, take ΔV to be an infinitesimal cube enclosing point P (see Fig. A.1-6). Then

$$\oint\!\!\oint_S \mathbf{A} \cdot d\mathbf{S} = \sum_{i=1}^{6} \mathbf{A}(Q_i) \cdot \Delta \mathbf{S}_i \qquad \text{Eq. A.1-16}$$

where Q_i are the points at the center of surfaces ΔS_i. By evaluating $\mathbf{A}$ at Q_i on surface ΔS_i, an approximate average value of $\mathbf{A}$ on that surface is obtained

$$\text{div}\, \mathbf{A}(P) = \lim_{\Delta x \to 0} \lim_{\Delta y \to 0} \lim_{\Delta z \to 0} \frac{1}{\Delta x\, \Delta y\, \Delta z} \sum_{i=1}^{6} \mathbf{A}(Q_i) \cdot \mathbf{n}(Q_i)\, \Delta S_i \qquad \text{Eq. A.1-17}$$

Now

$$\mathbf{A}(Q_1) \cdot \mathbf{n}(Q_1)\, \Delta S_1 = [A_1(Q_1)\mathbf{i} + A_2(Q_1)\mathbf{j} + A_3(Q_1)\mathbf{k}] \cdot [\mathbf{i}\, \Delta y\, \Delta z]$$

$$= A_1\left(x + \frac{\Delta x}{2}, y, z\right) \Delta y\, \Delta z$$

Similarly

$$\mathbf{A}(Q_2)\cdot\mathbf{n}(Q_2)\,\Delta S_2 = -A_1\left(x - \frac{\Delta x}{2}, y, z\right)\Delta y\,\Delta z$$

The negative sign arises due to the fact that $\mathbf{n}(Q_2) = -\mathbf{i}$. Similarly

$$\mathbf{A}(Q_3)\cdot\mathbf{n}(Q_3)\,\Delta S_3 = A_2\left(x, y + \frac{\Delta y}{2}, z\right)\Delta x\,\Delta z$$

$$\mathbf{A}(Q_4)\cdot\mathbf{n}(Q_4)\,\Delta S_4 = -A_2\left(x, y - \frac{\Delta y}{2}, z\right)\Delta x\,\Delta z$$

$$\mathbf{A}(Q_5)\cdot\mathbf{n}(Q_5)\,\Delta S_5 = A_3\left(x, y, z + \frac{\Delta z}{2}\right)\Delta x\,\Delta y$$

$$\mathbf{A}(Q_6)\cdot\mathbf{n}(Q_6)\,\Delta S_6 = -A_3\left(x, y, z - \frac{\Delta z}{2}\right)\Delta x\,\Delta y$$

Thus,

$$\begin{aligned}\operatorname{div}\mathbf{A} &= \lim_{\Delta z\to 0}\lim_{\Delta y\to 0}\lim_{\Delta x\to 0}\frac{1}{\Delta x\,\Delta y\,\Delta z}\\ &\quad\times\left\{\left[A_1\left(x + \frac{\Delta x}{2}, y, z\right) - A_1\left(x - \frac{\Delta x}{2}, y, z\right)\right]\Delta y\,\Delta z\right.\\ &\quad+\left[A_2\left(x, y + \frac{\Delta y}{2}, z\right) - A_2\left(x, y - \frac{\Delta y}{2}, z\right)\right]\Delta x\,\Delta z\\ &\quad\left.+\left[A_3\left(x, y, z + \frac{\Delta z}{2}\right) - A_3\left(x, y, z - \frac{\Delta z}{2}\right)\right]\Delta x\,\Delta y\right\}\\ &= \lim_{\Delta z\to 0}\lim_{\Delta y\to 0}\lim_{\Delta x\to 0}\left\{\frac{1}{\Delta x}\left[A_1\left(x + \frac{\Delta x}{2}, y, z\right) - A_1\left(x - \frac{\Delta x}{2}, y, z\right)\right]\right.\\ &\quad+\frac{1}{\Delta y}\left[A_2\left(x, y + \frac{\Delta y}{2}, z\right) - A_2\left(x, y - \frac{\Delta y}{2}, z\right)\right]\\ &\quad\left.+\frac{1}{\Delta z}\left[A_3\left(x, y, z + \frac{\Delta z}{2}\right) - A_3\left(x, y, z - \frac{\Delta z}{2}\right)\right]\right\}\end{aligned}$$ Eq. A.1-18

$$\operatorname{div}\mathbf{A} = \frac{\partial A_1}{\partial x} + \frac{\partial A_2}{\partial y} + \frac{\partial A_3}{\partial z} = \boldsymbol{\nabla}\cdot\mathbf{A}$$ Eq. A.1-19

where all derivatives are evaluated at the point (x, y, z).

The divergence of a vector appears in many areas of engineering. For example, a basic equation in fluid mechanics is the continuity equation

$$-\frac{\partial \rho}{\partial t} = \nabla \cdot \rho \mathbf{v}$$

where ρ is the mass density of the fluid and $\mathbf{v}$ is the velocity. The equation can be given the following interpretation. The right-hand side represents the net rate at which mass flows out of the region surrounding the point (x, y, z) per unit volume. The left-hand side represents the rate of decrease of mass per unit volume.

A.1.5 THE DIVERGENCE THEOREM

The Divergence Theorem relates the volume integration of the divergence of a vector, $\mathbf{A}$, to the surface integration of the dot product of that vector, $\mathbf{A}$, with the normal to the surface. That is, if $\mathcal{V}$ is the volume bounded by the closed surface S and $\mathbf{A}$ is a vector field, then

$$\oiiint_{\mathcal{V}} (\nabla \cdot \mathbf{A})\, d\mathcal{V} = \oiint_{S} \mathbf{A} \cdot d\mathbf{S} = \oiint_{S} \mathbf{A} \cdot \mathbf{n}\, dS \qquad \text{Eq. A.1-20}$$

This theorem is extensively used in this book in the development of the fundamental field equation of fluid mechanics. A proof based on the integral definition of the divergence theorem follows.

Construct a grid dividing $\mathcal{V}$ into cubes (see Fig. A.1-7). Let P_k be a point somewhere within $\Delta\mathcal{V}_k$ and P_{k+1} be a point somewhere within $\Delta\mathcal{V}_{k+1}$. From the definition of volume integration

$$\oiiint_{\mathcal{V}} (\nabla \cdot \mathbf{A})\, d\mathcal{V} = \lim_{\max \Delta\mathcal{V}_k \to 0} \sum_k (\nabla \cdot \mathbf{A})_{P_k} \Delta\mathcal{V}_k$$

Recall the definition of $\mathbf{A}$; that is,

$$\nabla \cdot \mathbf{A} = \lim_{\Delta\mathcal{V} \to 0} \frac{1}{\Delta\mathcal{V}} \oiint_{S} \mathbf{A} \cdot d\mathbf{S}$$

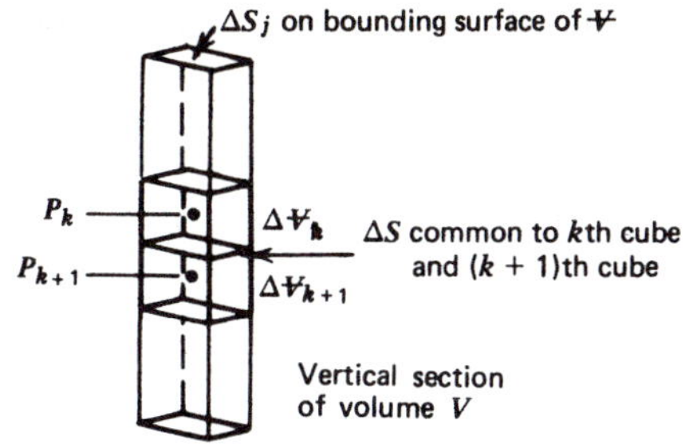

Fig. A.1-7. Geometry for the divergence theorem.

Then,

$$(\nabla \cdot \mathbf{A})_{P_k} \approx \frac{1}{\Delta \mathcal{V}_k} \oiint_{S \text{ of } \Delta \mathcal{V}_k} \mathbf{A} \cdot d\mathbf{S}$$

and

$$\sum_k (\nabla \cdot \mathbf{A})_{P_k} \Delta \mathcal{V}_k \approx \sum_k \oiint_{S \text{ of } \Delta \mathcal{V}_k} \mathbf{A} \cdot d\mathbf{S} \qquad \text{A.1-21}$$

Now, the

$$\sum_k \oiint_{S \text{ of } \Delta \mathcal{V}_k} \mathbf{A} \cdot d\mathbf{S}$$

results in the addition of the integration of $\mathbf{A} \cdot d\mathbf{S}$ over the bottom surface of the kth cube and $\mathbf{A} \cdot d\mathbf{S}$ over the top surface of the $(k + 1)$th cube. These two surfaces are the *same surface*; but $\mathbf{n}$ on the bottom surface of the kth cube is in the *opposite direction* to $\mathbf{n}$ on the top surface of the $(k + 1)$th cube. Therefore, the addition of these two integrations cancel each other out. This is true for all interior surfaces. Therefore,

$$\sum_k \oiint_{S \text{ of } \Delta \mathcal{V}_k} \mathbf{A} \cdot d\mathbf{S} \approx \sum_j \oiint_{\Delta S_j} \mathbf{A} \cdot d\mathbf{S} \qquad \text{Eq. A.1-22}$$

where ΔS_j lies on the exterior surface. By including the proper limits in the development, we would obtain

$$\sum_j \oiint_{\Delta S_j} \mathbf{A} \cdot d\mathbf{S} = \oiint_S \mathbf{A} \cdot d\mathbf{S} \qquad \text{Eq. A.1-23}$$

Thus,

$$\oiiint_{V} (\nabla \cdot \mathbf{A})\, dV = \oiint_S \mathbf{A} \cdot d\mathbf{S} \qquad \text{Eq. A.1-24}$$

A.1.6 OTHER USEFUL RELATIONS INVOLVING THE DIVERGENCE

A form that often appears as a result of the development of the field equations in various disciplines in engineering is the Laplacian of a scalar quantity, say ϕ. The Laplacian of $\phi(x, y, z)$, written $\nabla^2\phi$, is defined by

$$\nabla^2\phi = \nabla \cdot \nabla\phi \qquad \text{Eq. A.1-25}$$

In Cartesian coordinates,

$$\nabla^2\phi = \nabla \cdot \left(\mathbf{i}\frac{\partial\phi}{\partial x} + j\frac{\partial\phi}{\partial y} + \mathbf{k}\frac{\partial\phi}{\partial z}\right)$$

$$= \frac{\partial}{\partial x}\left(\frac{\partial\phi}{\partial x}\right) + \frac{\partial}{\partial y}\left(\frac{\partial\phi}{\partial y}\right) + \frac{\partial}{\partial z}\left(\frac{\partial\phi}{\partial z}\right)$$

$$\nabla^2\phi = \frac{\partial^2\phi}{\partial x^2} + \frac{\partial^2\phi}{\partial y^2} + \frac{\partial^2\phi}{\partial y^2} \qquad \text{Eq. A.1-26}$$

A vector identity that we will find very useful is

$$\nabla \cdot \phi\mathbf{A} = \phi(\nabla \cdot \mathbf{A}) + \mathbf{A} \cdot \nabla\phi \qquad \text{Eq. A.1-27}$$

This relation can readily be proved by carrying out, in Cartesian coordinates, the operation on the left-hand side and rearranging the terms to establish the right-hand side; that is,

$$\phi\mathbf{A} = \phi A_1\mathbf{i} + \phi A_2\mathbf{j} + \phi A_3\mathbf{k}$$

and

$$\nabla \cdot \phi\mathbf{A} = \frac{\partial}{\partial x}(\phi A_1) + \frac{\partial}{\partial y}(\phi A_2) + \frac{\partial}{\partial z}(\phi A_3)$$

Expanding, we obtain

$$\begin{aligned}\nabla \cdot \phi\mathbf{A} &= \phi\frac{\partial A_1}{\partial x} + A_1\frac{\partial \phi}{\partial x} + \phi\frac{\partial A_2}{\partial y} + A_2\frac{\partial \phi}{\partial y} + \phi\frac{\partial A_3}{\partial z} + A_3\frac{\partial \phi}{\partial z} \\ &= \phi\left(\frac{\partial A_1}{\partial x} + \frac{\partial A_2}{\partial y} + \frac{\partial A_3}{\partial z}\right) + A_1\frac{\partial \phi}{\partial x} + A_2\frac{\partial \phi}{\partial y} + A_3\frac{\partial \phi}{\partial z} \\ &= \phi(\nabla \cdot \mathbf{A}) + \mathbf{A} \cdot \nabla\phi\end{aligned}$$

Another useful relationship is one between a volume integration and a surface integration. It can be developed by use of the divergence theorem. If we let $\mathbf{A} = \phi\mathbf{e}$, where ϕ is a scalar function and $\mathbf{e}$ is any fixed vector quantity, and substitute for $\mathbf{A}$ in Eq. A.1-20, we obtain

$$\iiint_{V} (\nabla \cdot \phi\mathbf{e})\, dV = \oint\!\!\oint_{S} \phi\mathbf{e} \cdot d\mathbf{S} \qquad \text{Eq. A.1-28}$$

From Eq. A.1-27,

$$\nabla \cdot \phi\mathbf{e} = \phi\nabla \cdot \mathbf{e} + \mathbf{e} \cdot \nabla\phi \qquad \text{Eq. A.1-29}$$

The divergence of a constant vector is zero; therefore,

$$\iiint_{V} \mathbf{e} \cdot \nabla\phi\, dV = \oint\!\!\oint_{S} \mathbf{e}\phi \cdot \mathbf{n}\, dS = \oint\!\!\oint_{S} \mathbf{e} \cdot \mathbf{n}\phi\, dS \qquad \text{Eq. A.1-30}$$

$$\mathbf{e} \cdot \iiint_{V} \nabla\phi\, dV = \mathbf{e} \cdot \oint\!\!\oint_{S} \mathbf{n}\phi\, dS \qquad \text{Eq. A.1-31}$$

Then,

$$B_3(P) = \lim_{\Delta y \to 0} \lim_{\Delta x \to 0} \frac{1}{\Delta x \, \Delta y}$$

$$\times \left[F_1\left(x + \alpha_1 \frac{\Delta x}{2}, y - \frac{\Delta y}{2}, z\right) - F_1\left(x + \alpha_3 \frac{\Delta x}{2}, y + \frac{\Delta y}{2}, z\right) \right] \Delta x$$

$$+ \lim_{\Delta x \to 0} \lim_{\Delta y \to 0} \frac{1}{\Delta x \, \Delta y}$$

$$\times \left[F_2\left(x + \frac{\Delta x}{2}, y + \alpha_2 \frac{\Delta y}{2}, z\right) - F_2\left(x - \frac{\Delta x}{2}, y + \alpha_4 \frac{\Delta y}{2}, z\right) \right] \Delta y$$

$$= - \lim_{\Delta y \to 0} \frac{F_1\left(x, y + \frac{\Delta y}{2}, z\right) - F_1\left(x, y - \frac{\Delta y}{2}, z\right)}{\Delta y}$$

$$+ \lim_{\Delta x \to 0} \frac{F_2\left(x + \frac{\Delta x}{2}, y, z\right) - F_2\left(x - \frac{\Delta x}{2}, y, z\right)}{\Delta x}$$

or

$$B_3(P) = B_3(x, y, z) = \frac{\partial F_2}{\partial x}(x, y, z) - \frac{\partial F_1}{\partial y}(x, y, z)$$

or

$$B_3 = \frac{\partial F_2}{\partial x} - \frac{\partial F_1}{\partial y}$$

Of course in the above expression it is understood that B_3, $\partial F_2/\partial x$, and $\partial F_1/\partial y$ are all evaluated at (x, y, z). Similarly,

$$B_1 = \frac{\partial F_3}{\partial y} - \frac{\partial F_2}{\partial z}$$

$$B_2 = \frac{\partial F_1}{\partial z} - \frac{\partial F_3}{\partial x} \qquad \text{Eqs. A.1-36}$$

Therefore, $\nabla \times \mathbf{F}$, written as a vector is

$$\nabla \times \mathbf{A} = \left(\frac{\partial F_3}{\partial y} - \frac{\partial F_2}{\partial z}\right)\mathbf{i} + \left(\frac{\partial F_1}{\partial z} - \frac{\partial F_3}{\partial x}\right)\mathbf{j} + \left(\frac{\partial F_2}{\partial x} - \frac{\partial F_1}{\partial y}\right)\mathbf{k} \quad \text{Eq. A.1-37}$$

A.1.8 SOME VECTOR IDENTITIES INVOLVING THE CURL

Here we discuss two vector identities that involve the curl. Both are frequently used in fluid mechanics. The first is a statement that the curl of the gradient of a scalar, $f(x, y, z)$, is identically zero; that is,

$$\nabla \times \nabla f \equiv 0 \quad \text{Eq. A.1-38}$$

By carrying out the vector operations, the identity can readily be proven. That is, let

$$\mathbf{F} = \nabla f = \frac{\partial f}{\partial x}\mathbf{i} + \frac{\partial f}{\partial y}\mathbf{j} + \frac{\partial f}{\partial z}\mathbf{k}$$

then

$$F_1 = \frac{\partial f}{\partial x}, \qquad F_2 = \frac{\partial f}{\partial y}, \qquad \text{and} \qquad F_3 = \frac{\partial f}{\partial z} \quad \text{Eqs. A.1-39}$$

Substitute Eqs. A.1-39 into A.1-37, then

$$\nabla \times \nabla f = \left(\frac{\partial^2 f}{\partial y\, \partial z} - \frac{\partial^2 f}{\partial z\, \partial y}\right)\mathbf{i} + \left(\frac{\partial^2 f}{\partial z\, \partial x} - \frac{\partial^2 f}{\partial x\, \partial z}\right)\mathbf{j} + \left(\frac{\partial^2 f}{\partial x\, \partial y} - \frac{\partial^2 f}{\partial y\, \partial x}\right)\mathbf{k}$$

Eq. A.1-40

Since reversing the order of differentiation of a mixed derivative term does not affect the end result, each component in Eq. A.1-40 is identically zero and the identity is proven.

The second vector identity involves the expression $(\mathbf{v} \cdot \nabla)\mathbf{A}$. The term inside the parentheses is to be considered as an operator acting on each component of the vector, $\mathbf{A}$. That is, the x-component of the vector $(\mathbf{v} \cdot \nabla)\mathbf{A}$ is given by the

$$x\text{-component of } [(\mathbf{v} \cdot \nabla)\mathbf{A}] = v_1 \frac{\partial A_1}{\partial x} + v_2 \frac{\partial A_1}{\partial y} + v_3 \frac{\partial A_1}{\partial z} \quad \text{Eq. A.1-41}$$

The other components of $[(\mathbf{v} \cdot \nabla)\mathbf{A}]$ are expressed in a similar fashion. When $\mathbf{A} = \mathbf{v}$, the identity becomes of special interest in fluid mechanics. Under these circumstances, the vector identity becomes

$$(\mathbf{v} \cdot \nabla)\mathbf{v} = \tfrac{1}{2}\nabla(v^2) - \mathbf{v} \times (\nabla \times \mathbf{v}) \quad \text{Eq. A.1-42}$$

Applying the relation given in Eq. A.1-41 and the corresponding relations for the other components to the left-hand side of Eq. A.1-42, we obtain

$$\begin{aligned}(\mathbf{v} \cdot \nabla)\mathbf{v} = {} & \left(v_1 \frac{\partial v_1}{\partial x} + v_2 \frac{\partial v_1}{\partial y} + v_3 \frac{\partial v_1}{\partial z}\right)\mathbf{i} \\ & + \left(v_1 \frac{\partial v_2}{\partial x} + v_2 \frac{\partial v_2}{\partial y} + v_3 \frac{\partial v_2}{\partial z}\right)\mathbf{j} \\ & + \left(v_1 \frac{\partial v_3}{\partial x} + v_2 \frac{\partial v_3}{\partial y} + v_3 \frac{\partial v_3}{\partial z}\right)\mathbf{k} \end{aligned} \quad \text{Eq. A.1-43}$$

It can readily be shown that the right-hand side of Eq. A.1-42 can be put into the same form as the expanded form of the left-hand side, as indicated by Eq. A.1-43. This is done as follows

$$\begin{aligned}\frac{1}{2}\nabla(v^2) &= \frac{1}{2}\nabla(v_1^2 + v_2^2 + v_3^2) \\ &= \frac{1}{2}\left[\left(2v_1 \frac{\partial v_1}{\partial x} + 2v_2 \frac{\partial v_2}{\partial x} + 2v_3 \frac{\partial v_3}{\partial x}\right)\mathbf{i}\right. \\ &\quad + \left(2v_1 \frac{\partial v_1}{\partial y} + 2v_2 \frac{\partial v_2}{\partial y} + 2v_3 \frac{\partial v_3}{\partial y}\right)\mathbf{j} \\ &\quad \left. + \left(2v_1 \frac{\partial v_1}{\partial z} + 2v_2 \frac{\partial v_2}{\partial z} + 2v_3 \frac{\partial v_3}{\partial z}\right)\mathbf{k}\right] \end{aligned} \quad \text{Eq. A.1-44}$$

We now carry out the operations indicated in the second term on the right-hand side of Eq. A.1-42,

$$\nabla \times \mathbf{v} = \left(\frac{\partial v_3}{\partial y} - \frac{\partial v_2}{\partial z}\right)\mathbf{i} + \left(\frac{\partial v_1}{\partial z} - \frac{\partial v_3}{\partial x}\right)\mathbf{j} + \left(\frac{\partial v_2}{\partial x} - \frac{\partial v_1}{\partial y}\right)\mathbf{k} \qquad \text{Eq. A.1-45}$$

and

$$\begin{aligned} \mathbf{v} \times (\nabla \times \mathbf{v}) = &\left[v_2\left(\frac{\partial v_2}{\partial x} - \frac{\partial v_1}{\partial y}\right) - v_3\left(\frac{\partial v_1}{\partial z} - \frac{\partial v_3}{\partial x}\right)\right]\mathbf{i} \\ &+ \left[v_3\left(\frac{\partial v_3}{\partial y} - \frac{\partial v_2}{\partial z}\right) - v_1\left(\frac{\partial v_2}{\partial x} - \frac{\partial v_1}{\partial y}\right)\right]\mathbf{j} \\ &+ \left[v_1\left(\frac{\partial v_1}{\partial z} - \frac{\partial v_3}{\partial x}\right) - v_2\left(\frac{\partial v_3}{\partial y} - \frac{\partial v_2}{\partial z}\right)\right]\mathbf{k} \end{aligned} \qquad \text{Eq. A.1-46}$$

Combining Eqs. A.1-44 and A.1-46 as indicated on the right-hand side of Eq. A.1-42 and canceling out the like terms, but with different signs, we obtain

$$\begin{aligned} \tfrac{1}{2}\nabla(v^2) - \mathbf{v} \times (\nabla \times \mathbf{v}) = &\left(v_1\frac{\partial v_1}{\partial x} + v_2\frac{\partial v_1}{\partial y} + v_3\frac{\partial v_1}{\partial z}\right)\mathbf{i} \\ &+ \left(v_1\frac{\partial v_2}{\partial x} + v_2\frac{\partial v_2}{\partial y} + v_3\frac{\partial v_2}{\partial z}\right)\mathbf{j} \\ &+ \left(v_1\frac{\partial v_3}{\partial x} + v_2\frac{\partial v_3}{\partial y} + v_3\frac{\partial v_3}{\partial z}\right)\mathbf{k} \end{aligned} \qquad \text{Eq. A.1-47}$$

which is the same as the left-hand side of Eq. A.1-42, as indicated by A.1-43. Therefore, the identity is proven.

A.1.9 THE STOKES THEOREM

The Stokes theorem relates a surface integral to a line integral. Let S be a simple open surface with bounding curve C, as shown in Fig. A.1-9. Let the curl $\mathbf{F}$ be continuous everywhere on surface S. Then,

$$\oiint_S (\nabla \times \mathbf{F}) \cdot \mathbf{n}\, dS = \oint_C \mathbf{F} \cdot d\mathbf{r} \qquad \text{Eq. A.1-48}$$

A proof based on the integral definition of the curl follows.

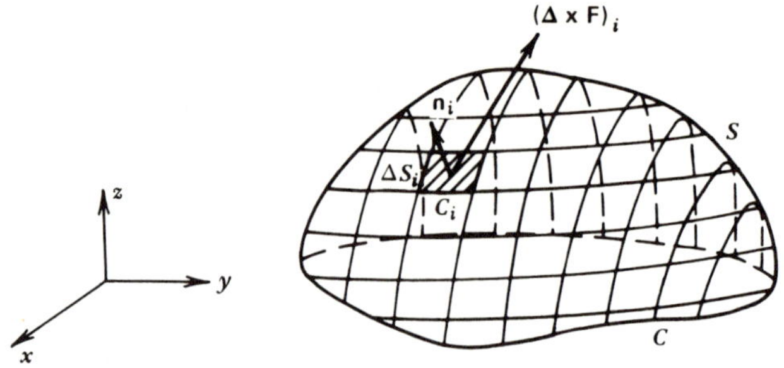

Fig. A.1-9. A simple open surface, S, with a bounding curve C.

Construct a subdivision of the surface by intersecting the surface with planes parallel to the (x, z) plane and planes parallel to the (x, y) plane. For a sufficiently large number of subdivision elements and by the definition of a surface integral,

$$\iint_S (\nabla \times \mathbf{F}) \cdot \mathbf{n}\, dS \approx \sum_i (\nabla \times \mathbf{F})_i \cdot \mathbf{n}_i\, \Delta S_i \qquad \text{Eq. A.1-49}$$

On the other hand, by the definition of the curl,

$$(\nabla \times \mathbf{F})_i \cdot \mathbf{n}_i \approx \frac{1}{\Delta S_i} \oint_{C_i} \mathbf{F} \cdot d\mathbf{r} \qquad \text{Eq. A.1-50}$$

and

$$(\nabla \times \mathbf{F})_i \cdot \mathbf{n}_i\, \Delta S_i \approx \oint_{C_i} \mathbf{F} \cdot d\mathbf{r} \qquad \text{Eq. A.1-51}$$

Thus,

$$\sum_i (\nabla \times \mathbf{F})_i \cdot \mathbf{n}_i\, \Delta S_i \approx \sum_i \oint_{C_i} \mathbf{F} \cdot d\mathbf{r} \qquad \text{Eq. A.1-52}$$

Consider the line integrals around two adjacent surface elements, as shown in Fig. A.1-10, say surface element j and $(j + 1)$. The line integration around C_j will result in the line integration of $(\mathbf{F} \cdot d\mathbf{r})$ from point (a) to point (d), and the

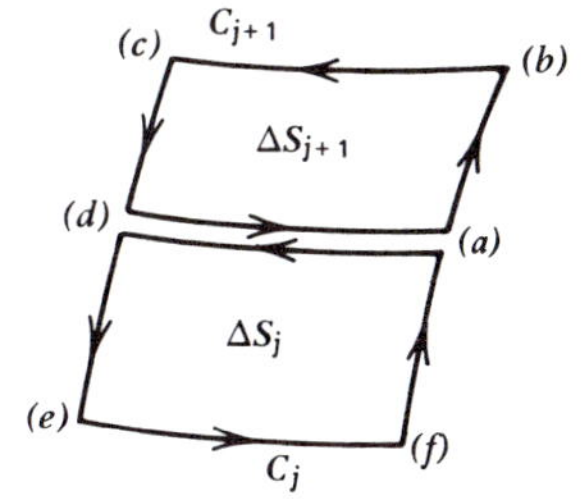

Fig. A.1-10. Adjacent surface elements on surface S.

line integration around C_{j+1} will result in the same line integration from point (d) to point (a). Since the sign of a line integral depends on the direction of integration, the sum of the two line integrations cancel out giving zero. This is true for all interior curves. Therefore,

$$\sum_i \oint_{C_i} \mathbf{F} \cdot d\mathbf{r} = \oint_C \mathbf{F} \cdot d\mathbf{r} \qquad \text{Eq. A.1-53}$$

where C is the bounding curve of S. Substituting Eqs. A.1-52 and A.1-53 into Eq. A.1-49, and using the proper limits, we obtain

$$\oiint_S (\nabla \times \mathbf{F}) \cdot \mathbf{n}\, dS = \oint_C \mathbf{F} \cdot d\mathbf{r}$$

A.1.10 MATHEMATICAL SIGNIFICANCE OF AN IRROTATIONAL FIELD

A vector field, $\mathbf{F}(x, y, z)$, is said to be irrotational in a region R, if everywhere in R, $\nabla \times \mathbf{F} = 0$. In the study of fluid mechanics we often come across situations where a large part of the velocity field can be considered to be irrotational. Problems involving the vector $\mathbf{F}(x, y, z)$ can often be simplified if $\mathbf{F}$ can be replaced by the gradient of a scalar, $f(x, y, z)$. A necessary condition that there exists a function $f(x, y, z)$ such that

$$\nabla f = \mathbf{F} \qquad \text{Eq. A.1-54}$$

is that $\nabla \times \mathbf{F} = 0$. The proof of this statement rests on the vector identity given in Eq. A.1-38; that is,

$$\nabla \times \nabla f \equiv 0 \qquad \text{Eq. A.1-55}$$

Take the curl of Eq. A.1-54; that is,

$$\nabla \times \nabla f = \nabla \times \mathbf{F} \qquad \text{Eq. A.1-56}$$

In order that Eq. A.1-56 be consistent with the vector identity given in Eq. A.1-38, it is required that

$$\nabla \times \mathbf{F} = 0 \qquad \text{Eq. A.1-57}$$

It can also be proven that $\nabla \times \mathbf{F} = 0$ is sufficient for the existence of a function $f(x, y, z)$, which satisfies Eq. A.1-54. Students interested in this proof are referred to almost any standard calculus textbook.

Appendix A.2

Projection of an Area onto a Plane

This section is concerned with the relationship between an arbitrarily oriented triangular area, NLM, shown in Fig. A.2-1, and its projection onto one of the coordinate planes. First, we prove the third equation in the set Eqs. 3.2-3; that is,

$$\Delta \mathbf{S} \cdot \mathbf{k} = \text{projection of } \Delta S \text{ onto the } (x, y) \text{ plane} = \Delta S_3$$

From this we can surmise the validity of the remaining two equations in the set. Other shaped elements can always be broken down into triangular elements.

The projection of ΔS onto the (x, y) plane is the triangle OLM whose area we designate as ΔS_3. The normal to ΔS is $\mathbf{n}$ and the normal to ΔS_3 is $\mathbf{k}$. Thus, we want to show that

$$\Delta S_3 = \Delta S \mathbf{n} \cdot \mathbf{k} = \Delta \mathbf{S} \cdot \mathbf{k} \qquad \text{Eq. A.2-1}$$

Consider the geometry shown in Fig. A.2-1. Take the plane perpendicular to the line $\overline{LM}$, which contains the origin. Since $\overline{LM}$ is a horizontal line, this plane will be a vertical plane and will thus contain the z-axis as shown. Since the plane NOP is perpendicular to the line $\overline{LM}$, every line in that plane is perpendicular to the line $\overline{LM}$. Thus, both lines $\overline{NP}$ and $\overline{OP}$ will be perpendicular to line $\overline{LM}$. Then the areas ΔS and ΔS_3 are given by

$$\Delta S = \tfrac{1}{2}\overline{LM} \cdot \overline{NP} \qquad \text{Eq. A.2-2}$$

$$\Delta S_3 = \tfrac{1}{2}\overline{LM} \cdot \overline{OP} \qquad \text{Eq. A.2-3}$$

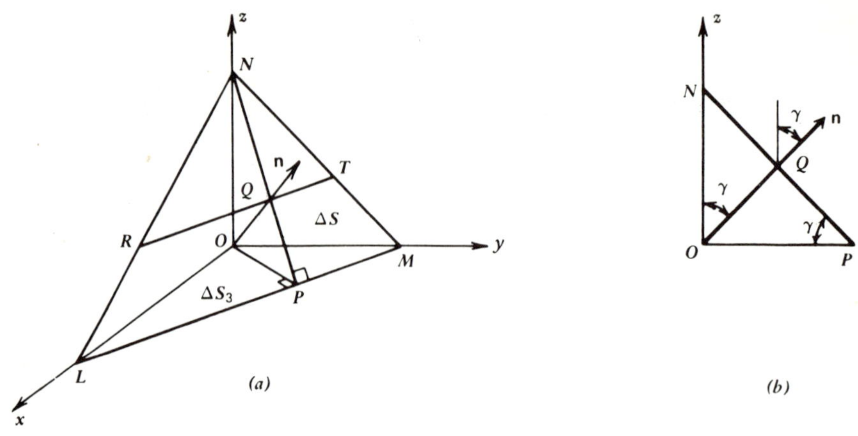

Fig. A.2-1. Geometry of a prismatic element. (*a*) Prismatic element. (*b*) Triangle *NOP*.

Now consider the triangle *NOP*. Take the line $\overline{OQ}$ to be perpendicular to line $\overline{NP}$. The normal to the plane *NOP* is in the direction of $\overline{OQ}$. To demonstrate that this is so, we must show that $\overline{OQ}$ is perpendicular to two nonparallel lines in the plane *NLM*. By construction, $\overline{OQ}$ is perpendicular to line $\overline{NP}$, which lies in the plane *NLM*. Now construct line $\overline{RT}$ parallel to line $\overline{LM}$ through point *Q*. Plane *NOP* is perpendicular to line $\overline{RT}$, since the plane was taken perpendicular to line $\overline{LM}$. Since $\overline{OQ}$ lies in plane *NOP*, it is perpendicular to line $\overline{RT}$. Therefore, $\overline{OQ}$ is perpendicular to two nonparallel lines in plane *NLM* and, thus, the normal is in the direction of $\overline{OQ}$. If the direction cosines of $\overline{OQ}$ are ($\cos\alpha$, $\cos\beta$, $\cos\gamma$), then

$$\mathbf{n} = \cos\alpha\,\mathbf{i} + \cos\beta\,\mathbf{j} + \cos\gamma\,\mathbf{k} \qquad \text{Eq. A.2-4}$$

where α, β, and γ are the angles that $\overline{OQ}$ makes with the *x*, *y*, and *z*-axis, respectively. By observing that $\measuredangle ONQ$ is complementary to both $\measuredangle NOQ$ and $\measuredangle OPN$, we can conclude that $\measuredangle OPN = \gamma$. Then,

$$\overline{OP} = \overline{NP} \cdot \cos\gamma \qquad \text{Eq. A.2-5}$$

Substituting Eq. A.2-5 into Eq. A.2-3, we obtain

$$\Delta S_3 = \tfrac{1}{2}\overline{LM} \cdot \overline{NP} \cdot \cos\gamma \qquad \text{Eq. A.2-6}$$

But,

$$\Delta \mathbf{S} \cdot \mathbf{k} = \Delta S(\cos\alpha\, \mathbf{i} + \cos\beta\, \mathbf{j} + \cos\gamma\, \mathbf{k}) \cdot \mathbf{k} \qquad \text{Eq. A.2-7}$$

$$= \tfrac{1}{2}\overline{LM} \cdot \overline{NP} \cdot \cos\gamma$$

By comparing Eqs. A.2-6 and A.2-7, we see that

$$\Delta S_3 = \Delta \mathbf{S} \cdot \mathbf{k}$$

In a similar fashion, the first two equations in the following set may be proved.

$$\Delta S_1 = \Delta \mathbf{S} \cdot \mathbf{i}$$

$$\Delta S_2 = \Delta \mathbf{S} \cdot \mathbf{j}$$

$$\Delta S_3 = \Delta \mathbf{S} \cdot \mathbf{k} \qquad \text{Eqs. A.2-8}$$

where

ΔS_1 = projection of ΔS on the (y, z) plane

ΔS_2 = projection of ΔS on the (x, z) plane

Appendix A.3

Elements of Basic Thermodynamics and Some Additional Relationships for the Ideal Gas

This brief section introduces the first and second laws of thermodynamics and demonstrates the applications of these two principles to the thermodynamics system. Utilizing the basic thermodynamic relationships derived from the first and second laws of thermodynamics, this section closes with the statement of a number of useful relationships for the ideal gas.

Thermodynamics, as a science, is based on an internally self-consistent set of definitions and a group of axioms, postulates, or laws. It is presumed that the student is already familiar with such thermodynamic terms as: state, property, process, path, cycle, and the like. Furthermore, it is assumed that the extensive parameters of mass, internal energy, volume, and mole number are clearly understood, as well as the intensive properties of temperature and pressure. Although "heat" and "work" are defined in Section 4.6.1, the definitions are repeated here to emphasize the unique nature of these two quantities.

Heat is energy in transit across the boundary of a system, without the attendant transfer of mass, due to a potential difference[1] in temperature between the system and its surroundings.

Work is energy in transit across the boundary of a system, without the attendant transfer of mass, due to a potential difference[1] between the system and surroundings other than temperature.

[1] The potential differences referred to in the definitions of heat and work are represented by differences in intensive parameters (such as temperature, pressure, magnetic field strength, electrical field potential, etc.) between the system and its surroundings. In the common case where the driving force is a pressure difference, we are led to the familiar expression for reversible work, $W_{rev} = \int p d\forall$.

Work and heat are seen to be transients that appear at the boundaries of a system (or between interacting systems) and only appear during the process through which systems change from one state to another. Heat and work are never stored in a system, as is internal energy (or gross potential and/or kinetic energies). Heat and work are path functions (inexact differentials) in contrast to internal energy, which is a point function or a state variable (exact differential). The change in internal energy for a system undergoing a process beginning at one specified point or state and ending at a second point or state is independent of the process by which the change of state occurs. Work and heat, on the other hand, are direct functions of the nature of the process or path.

The first law of thermodynamics is merely a statement of the principle of conservation of energy in terms of work, heat, and the various systems energies. For the closed or no-mass-flow system fixed in the space of the particular coordinate system, the first law is written in differential form as

$$\delta Q - \delta W = dU \qquad \text{Eq. A.3-1}$$

where

δQ = Amount of heat added to the system for the differential process.

δW = Amount of work done by the system on the surroundings for the differential process.

dU = The differential change in the system's internal energy.

Here we have used the usual thermodynamic sign conventions for heat and work in contrast to the continuum mechanics sign conventions used in Section 4.6. The δ notation on δQ and δW denotes the fact that δQ and δW are inexact differentials in contrast to dU, which is an exact differential.

When integrated over a finite process from states identified as (1) and (2), Eq. A.3-1 becomes

$$_1Q_2 - {_1W_2} = U_2 - U_1 \qquad \text{Eq. A.3-2}$$

It follows directly from Eqs. A.3-1 and A.3-2 that for the cyclic process

$$\oint(\delta Q - \delta W) = \oint dU = 0 \qquad \text{Eq. A.3-3}$$

and

$$\oint \delta Q = \oint \delta W \qquad \text{Eq. A.3-4}$$

For the simple closed system for which work is due solely to volume change and where the processes in question are reversible, the reader will recall the relationship

$$\delta W_{\text{rev}} = p(\text{\textit{V}})\, d\text{\textit{V}} \qquad \text{Eq. A.3-5}$$

where p is the system pressure (a function of the volume $\mathcal{V}$ in each equilibrium state along the path). For the integrated process from state (1) to state (2),

$$_1W_2 = \int_1^2 \delta W = \int_{\mathcal{V}_1}^{\mathcal{V}_2} p(\mathcal{V})\, d\mathcal{V} \qquad \text{Eq. A.3-6}$$

It is obvious that work is "path dependent," since Eq. A.3-6 cannot be evaluated without explicit knowledge of the relationship $p = p(\mathcal{V})$ for the path.

The second law of thermodynamics, as formulated by Carnot, Clausius, Planck, and others, denies the existence of a perpetual motion machine of the second kind. Simply stated, it is impossible to construct a cyclic heat engine that will continuously and completely convert heat from a reservoir to an equivalent amount of work (e.g., a heat engine cannot have a thermal efficiency of 100 %). The second law of thermodynamics leads to the concept of entropy, S, a property defined by the relationship

$$dS = \frac{\delta Q_{\text{rev}}}{T} \qquad \text{Eq. A.3-7}$$

where

S = Entropy (an extensive property).

Q_{rev} = Reversible heat transferred.

T = The system temperature at which the heat is transferred.

Equation A.3-7 may also be written as an inequality; that is,

$$dS \geq \frac{\delta Q}{T} \qquad \text{Eq. A.3-8}$$

where the equality holds for reversible processes and the inequality for irreversible (all real) processes. Since entropy is a property, it follows that

$$\oint dS = 0$$

such that

$$\oint \frac{\delta Q}{T} \leq 0 \qquad \text{Eq. A.3-9}$$

a relationship known as the Clausius inequality. For an isolated system ($Q = W = 0$, no interactions with the surroundings), it is readily seen that for all real processes the system entropy will increase. In fact, when all internal constraints on an isolated system are removed, the system will proceed to its ultimate equilibrium state—a state corresponding to maximum entropy. Equation A.3-7, when combined with Eqs. A.3-1 and A.3-5, leads to the following

$$\delta Q_{\text{rev}} - \delta W_{\text{rev}} = dU = T dS - p\,dV$$

or simply

$$dU = T\,dS - p\,d\mathcal{V} \qquad \text{Eq. A.3-10}$$

Since Eq. A.3-10 is merely a relationship among several thermodynamic properties on the equilibrium thermodynamic surface, it is valid for any process (real or reversible, flow or nonflow) whose end states lie on the thermodynamic equilibrium surface.

It is convenient to introduce the composite property known as enthalpy that is denoted by the symbol H and that is defined by

$$H = U + p\mathcal{V} \qquad \text{Eq. A.3-11}$$

Then

$$dH = dU + p\,d\mathcal{V} + \mathcal{V}\,dp$$

It is readily seen that Eq. A.3-10 leads to a second relationship between entropy and the other thermodynamic properties, one which involves enthalpy rather than internal energy. Specifically, by substituting for dU in Eq. 3-10, we obtain

$$dH = T\,dS + \mathcal{V}\,dp \qquad \text{Eq. A.3-12}$$

The reader recognizes, of course, that the relationships expressed by Eqs. A.3-10 and A.3-12 can also be written in terms of the corresponding specific intensive property variables; that is,

$$du = T\,d\mathscr{s} - p\,d\tilde{v} \qquad \text{Eq. A.3-13}$$

and

$$dh = T\,d\mathscr{s} + \tilde{v}\,dp \qquad \text{Eq. A.3-14}$$

The ideal gas is often used as a model in elementary studies in fluid mechanics and thermodynamics. It is therefore worthwhile to consider the application of several of the preceding thermodynamic concepts to the specific system known as the ideal gas. An ideal gas is one which behaves according to the ideal gas equation of state and whose internal energy per unit mass is a linear function of temperature. The ideal gas law can be written in one of several forms; these are

$$p\forall = NR^*T \qquad \text{Eq. A.3-15}a$$

$$p\forall = mRT \qquad \text{Eq. A.3-15}b$$

$$p\tilde{v} = RT \qquad \text{Eq. A.3-15}c$$

where N is the number of moles, R^* is the universal gas constant, and m is its mass. The reader recognizes that dimensional homogeneity is required in each form of Eq. A.3-15, as in all physical equations. The internal energy per unit mass can be expressed by an equation of the following form

$$u = c_1 T + b \qquad \text{Eq. A.3-16}$$

where c_1 and b are constants yet to be defined.

Two terms widely used in thermodynamics—the specific heat at constant volume, c_v, and the specific heat at constant pressure, c_p—are defined, respectively, as

$$c_v = \left(\frac{\partial u}{\partial T}\right)_v \qquad \text{Eq. A.3-17}$$

and

$$c_p = \left(\frac{\partial h}{\partial T}\right)_p \qquad \text{Eq. A.3-18}$$

In general, if the internal energy and enthalpy are functions of temperature and volume or temperature and pressure, then the specific heats are functions of the same variables. In the case of the ideal gas, Eq. A.3-16 indicates that internal energy is a function only of temperature. Thus,

$$c_v = \frac{du}{dT}$$

or

$$du = c_v\, dT \qquad \text{Eq. A.3-19}$$

And upon integration with c_v = constant,

$$u = c_v T + b \qquad \text{Eq. A.3-20}$$

where b is a constant of integration. Taking $u = 0$ when $T = 0$ gives $b = 0$ and

$$u = c_v T \qquad \text{Eq. A.3-21}$$

For the ideal gas, c_v is taken as a simple constant (equal to the c_1 of Eq. A.3-16) rather than as a function of temperature. We now recall that

$$h = u + p\tilde{v}$$

and that

$$p\tilde{v} = RT$$

Thus,

$$h = u + p\tilde{v} = c_v T + RT$$

or

$$h = (c_v + R)T \qquad \text{Eq. A.3-22}$$

suggesting that enthalpy, like internal energy, is a function only of temperature for the ideal gas. Applying Eq. A.3-18 to Eq. A.3-22, we obtain for the ideal gas

$$h = c_p T \qquad \text{Eq. A.3-23}$$

with c_p, like c_v, a simple constant. A comparison of Eqs. A.3-22 and A.3-23 reveals an important relationship for the ideal gas

$$c_v + R = c_p \qquad \text{or} \qquad c_p - c_v = R \qquad \text{Eqs. A.3-24}$$

Once again, dimensional homogeneity is required. Specifically for air when treated as an ideal gas,

$$c_v = 0.1713 \frac{\text{Btu}}{\text{lb}_\text{m}\text{-R}} = 717.4 \frac{\text{J}}{\text{kg-K}}$$

$$c_p = 0.240 \frac{\text{Btu}}{\text{lb}_\text{m}\text{-R}} = 1004 \frac{\text{J}}{\text{kg-K}}$$

$$R = 53.34 \frac{\text{lb}_\text{f}\text{-ft}}{\text{lb}_\text{m}\text{-R}} = 286.9 \frac{\text{J}}{\text{kg-K}}$$

For the dimensions of c_p, c_v and R as given above, Eqs. A.3-24 must be modified to include Joules constant, J. That is

$$c_p - c_v = \frac{R}{J}$$

where

$$J = 778.16 \frac{\text{ft-lb}_\text{f}}{\text{Btu}} = 1 \frac{\text{m} - \text{N}}{J}$$

Returning to Eq. A.3-13 and considering an ideal gas undergoing an isentropic process (a reversible and adiabatic process for which, by definition, $d\sigma = 0$), we have

$$du = -p\, d\tilde{v}$$

By applying the ideal gas relationships, we find that

$$c_v \, dT = -p \, d\tilde{v} = -\frac{RT}{\tilde{v}} \, d\tilde{v}$$

or

$$c_v \frac{dT}{T} = -R \frac{d\tilde{v}}{\tilde{v}}$$

Integration leads to the expression

$$\frac{c_v}{R} \ln\left(\frac{T_2}{T_1}\right) = -\ln\left(\frac{\tilde{v}_2}{\tilde{v}_1}\right) = \ln\left(\frac{\tilde{v}_1}{\tilde{v}_2}\right)$$

or, as it is more commonly written,

$$\frac{T_2}{T_1} = \left(\frac{\tilde{v}_1}{\tilde{v}_2}\right)^{R/c_v}$$

Introducing the ratio of specific heats, defined by

$$k = \frac{c_p}{c_v} \qquad \text{Eq. A.3-25}$$

and using the relationship of Eq. A.3-24, we obtain the final form

$$\frac{T_2}{T_1} = \left(\frac{\tilde{v}_1}{\tilde{v}_2}\right)^{k-1} \qquad \text{Eq. A.3-26}$$

Substituting for the ratio (T_2/T_1) in the above expression leads to an additional relationship for the perfect gas

$$\frac{T_2}{T_1} = \frac{p_2 \tilde{v}_2}{p_1 \tilde{v}_1} = \left(\frac{v_1}{\tilde{v}_2}\right)^{k-1}$$

or

$$\frac{p_2}{p_1} = \left(\frac{\tilde{v}_1}{\tilde{v}_2}\right)^{k} \qquad \text{A.3-27}a$$

or

$$p\tilde{v}^k = \text{constant} \qquad \text{A.3-27}b$$

Equation A.3-27*b* is a specific form of the general polytropic relationship

$$p\tilde{v}^n = \text{constant} \qquad \text{A.3-28}$$

with the exponent n assuming the following values for different types of processes or path relationships

$n = 0$ isobaric, $p = c$

$n = 1$ isothermal, $T = c$

$n = k$ isentropic, $s = c$

$n = \infty$ isometric, $\tilde{v} = c$

On the p-$\tilde{v}$ plane, for the ideal gas, these path relationships take the forms shown in Fig. A.3-1. By combining Eqs. A.3-26 and A.3-27*a*, or alternatively beginning with Eq. A.3-14 and repeating the analysis leading to Eq. A.3-26, we obtain a third relationship of the form

$$\frac{T_2}{T_1} = \left(\frac{p_2}{p_1}\right)^{k-1/k} \qquad \text{Eq. A.3-29}$$

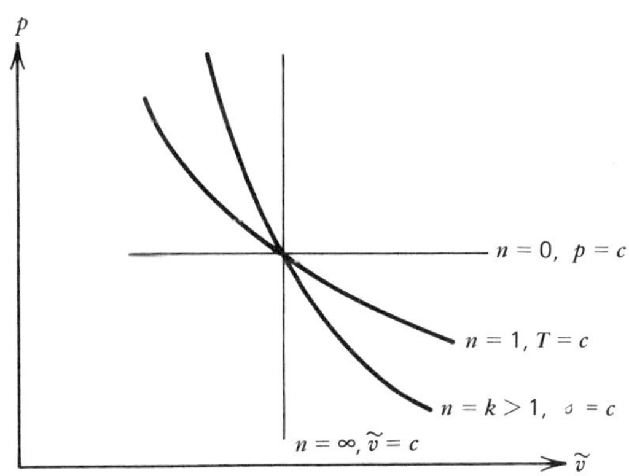

Fig. A.3-1. The p-$\tilde{v}$ diagram for several types of processes.

Finally, we develop two relationships that are used to determine the change in entropy for ideal gas processes. Equation A.3-13 may be rewritten as follows

$$ds = \frac{1}{T}\, du + \frac{p}{T}\, d\tilde{v}$$

Substitution of the various ideal gas relationships into the above equation leads to

$$ds = c_v \frac{dT}{T} + \frac{p}{T}\, d\tilde{v} = c_v \frac{dT}{T} + R\frac{d\tilde{v}}{\tilde{v}}$$

which upon integration leads to

$$s_2 - s_1 = c_v \ln\left(\frac{T_2}{T_1}\right) + R\ln\left(\frac{\tilde{v}_2}{\tilde{v}_1}\right) \qquad \text{Eq. A.3-30}$$

A similar manipulation of Eq. A.3-14 leads to

$$s_2 - s_1 = c_p \ln\left(\frac{T_2}{T_1}\right) - R\ln\left(\frac{p_2}{p_1}\right) \qquad \text{Eq. A.3-31}$$

The reader will note that substitution of the ratio

$$\frac{p_2\tilde{v}_2}{p_1\tilde{v}_1} = \frac{T_2}{T_1}$$

into either Eq. A.3-30 or A.3-31 will lead to a third relationship for which $(s_2 - s_1)$ is expressed as a function of p_2, p_1, $\tilde{v}_2$, and $\tilde{v}_1$.

Appendix A.4

Elements of Matrices

A.4.1 INTRODUCTION

In this section a brief introduction to matrices is given. Only those topics that are relevant to an understanding of Chapter 5 are discussed.

A *matrix* is an ordered set of numbers or functions, designated as an array, that obeys a prescribed set of laws of equality, addition, subtraction, and multiplication. These laws have been established so that matrices become a convenient tool for dealing with a system of linear equations (both algebraic and ordinary differential).

A matrix consisting of m rows and n columns is called an $m \times n$ matrix and is designated by

$$A = [a_{ij}] = \begin{bmatrix} a_{11} & a_{12} & \cdots & a_{1n} \\ a_{21} & a_{22} & \cdots & a_{2n} \\ \vdots & \vdots & \cdots & \vdots \\ a_{m1} & a_{m2} & \cdots & a_{mn} \end{bmatrix}$$

The element a_{ij} is the element in the ith row and jth column. A matrix that has the same number of rows as columns is called a *square* matrix.

A.4.2 SOME ELEMENTARY LAWS OF MATRICES

(a) Equality of two matrices.

Two matrices, A and B, are equal if and only if they contain the same number of rows and columns and $a_{ij} = b_{ij}$.

(*b*) Addition and subtraction of matrices.

The sum or difference of two matrices, A and B, where A and B have the same number of rows and columns, is a matrix whose elements are the sum or difference of the respective elements of A and B; that is, if $C = A + B$, then $c_{ij} = a_{ij} + b_{ij}$. Addition and subtraction of two matrices are *not* defined if the two matrices have a different number of rows or columns.

(*c*) Multiplication of a scalar and a matrix.

The product of a matrix A with a scalar k is a matrix whose elements are a product of k with the corresponding element of matrix A; that is, if $B = kA$, then

$$b_{ij} = ka_{ij}$$

(*d*) Matrix multiplication.

The multiplication of two matrices, A and B, resulting in the matrix C, is defined only when the number of columns of A equals the number of rows of B. Matrix C will have the same number of rows as matrix A and the same number of columns as matrix B. If

$$C = AB$$

then the element of C in the ith row and jth column, c_{ij}, is given by

$$c_{ij} = \sum_k a_{ik} b_{kj}$$

This is illustrated in the following example. If

$$A = \begin{bmatrix} 2 & 3 \\ 1 & -1 \\ 0 & 4 \end{bmatrix} \quad \text{and} \quad B = \begin{bmatrix} 5 & -2 & 4 & 7 \\ -6 & 1 & -3 & 0 \end{bmatrix}$$

then

$$C = \begin{bmatrix} [2(5) + 3(-6)] & [2(-2) + 3(1)] & [2(4) + 3(-3)] & [2(7) + 3(0)] \\ [1(5) - 1(-6)] & [1(-2) - 1(1)] & [1(4) - 1(-3)] & [1(7) - 1(0)] \\ [0(5) + 4(-6)] & [0(-2) + 4(1)] & [0(4) + 4(-3)] & [0(7) + 4(0)] \end{bmatrix}$$

$$= \begin{bmatrix} -8 & -1 & -1 & 14 \\ 11 & -3 & 7 & 7 \\ -24 & 4 & -12 & 0 \end{bmatrix}$$

A system of linear algebraic equations can readily be represented by the multiplication of two matrices. Observe that $AX = C$ where

$$A = \begin{bmatrix} a_{11} & a_{12} & a_{13} \\ a_{21} & a_{22} & a_{23} \\ a_{31} & a_{32} & a_{33} \end{bmatrix} \qquad X = \begin{bmatrix} x_1 \\ x_2 \\ x_3 \end{bmatrix} \qquad C = \begin{bmatrix} c_1 \\ c_2 \\ c_3 \end{bmatrix}$$

becomes

$$\begin{bmatrix} a_{11}x_1 + a_{12}x_2 + a_{13}x_3 \\ a_{21}x_1 + a_{22}x_2 + a_{23}x_3 \\ a_{31}x_1 + a_{32}x_2 + a_{33}x_3 \end{bmatrix} = \begin{bmatrix} c_1 \\ c_2 \\ c_3 \end{bmatrix}$$

Since two matrices are equal if and only if their elements are equal, we have

$$a_{11}x_1 + a_{12}x_2 + a_{13}x_3 = c_1$$

$$a_{21}x_1 + a_{22}x_2 + a_{23}x_3 = c_2$$

$$a_{31}x_1 + a_{32}x_2 + a_{33}x_3 = c_3$$

Matrix A is called the coefficient matrix. It should be noted that, in general, $AB \neq BA$.

(*e*) Rank of a matrix.

A submatrix of A is a matrix obtained by deleting certain rows or columns from matrix A. The *order* of the largest *square* submatrix of A whose determinant is not zero is the *rank* of matrix A. If the number of columns in a coefficient matrix is larger than the number of rows, then the order of the largest square submatrix equals the number of rows. Therefore, if we examine Eqs. 5.3-8, which result in the coefficient matrix A, given by the first of the two equations in Eqs. 5.3-9, we see that the number of rows in matrix A corresponds to the number of basic dimensions involved in the phenomenon of interest. Deleting the last two columns of A gives the largest order square submatrix, which is denoted as $\tilde{A}$, and whose determinant, $|\tilde{A}|$, is given by Eq. 5.3-10. Since $|\tilde{A}| \neq 0$, the rank of A is three.

Appendix A.5

The Mathematical Concept Behind Boundary-Layer Theory

In our discussion of the lift/drag problems of a thin wing at a small angle of attack (see Sections 9.1.1 and 9.10) it was mentioned that the lift problem is solved by considering the fluid to be inviscid, whereas the drag problem is solved by considering the fluid to be viscous. In the drag problem the pressure and velocity at the outer limits of the boundary layer are taken to be equal to the pressure and the velocity of the inviscid flow problem evaluated on the surface of the wing. In this section, we wish to illustrate the validity of this technique by examining the first approximation of the method known as *matched asymptotic expansions*. The method was developed in the 1950s by K. O. Friedricks, S. Kaplun, P. A. Langerstrom, and others. The whole subject is beautifully covered in Van Dyke's book.[1]

To illustrate the method, we will consider a simple problem, one where an exact solution can readily be obtained. The problem was first presented by Friedricks[2] in 1942. Consider the following differential equation

$$\varepsilon \frac{d^2 f}{dx^2} + \frac{df}{dx} = a \qquad \text{Eq. A.5-1}$$

with the following boundary conditions

$$f(0) = 0 \quad \text{and} \quad f(1) = 1 \qquad \text{Eqs. A.5-2}$$

[1] M. Van Dyke, *Perturbation Methods in Fluid Mechanics,* The Parabolic Press, Stanford, California, 1975.
[2] K. O. Friedricks, "Theory of Viscous Fluids," *Fluid Dynamics*, Chapter 4, Brown University, 1942.

Take ε = constant $\ll 1$. To obtain an approximate solution, one might first attempt to neglect the first term in the equation, since $\varepsilon \ll 1$. However, if this is done, one of the boundary conditions could not be satisfied. This results from the fact that the resulting equation is only of first order and, thus, only one boundary condition could be specified. It is this condition that classifies this problem as a *singular perturbation problem.* The method of matched asymptotic expansions provides a means for obtaining an approximate solution to the problem.

The method of matched asymptotic expansions consists of obtaining two approximate solutions—an inner solution that is valid near the inner boundary ($x = 0$) and an outer solution that is valid over the remaining portion of the x domain. The method will work if there is a small region where both solutions are valid. This region is designated as the overlap region. To obtain the outer solution, we neglect the term containing ε. Thus, the outer solution satisfies

$$\frac{df_0}{dx} = a \qquad \text{Eq. A.5-3}$$

and the boundary condition

$$f_0(1) = 1 \qquad \text{Eq. A.5-4}$$

The solution is

$$f_0 = (1 - a) + ax \qquad \text{Eq. A.5-5}$$

To obtain the inner solution, we make a change in the independent variable to amplify the inner region. To accomplish this let

$$X = \frac{x}{\varepsilon} \qquad \text{Eq. A.5-6}$$

The differential equation must now be expressed in terms of X. We now use the chain rule to obtain an equation involving the derivatives of f with respect to X; that is,

$$\frac{df}{dx} = \frac{df}{dX}\frac{dX}{dx} = \frac{1}{\varepsilon}\frac{df}{dX}$$

and

$$\frac{d^2f}{dx^2} = \frac{d}{dx}\left(\frac{1}{\varepsilon}\frac{df}{dX}\right) = \frac{1}{\varepsilon}\frac{d}{dX}\left(\frac{df}{dX}\right)\frac{dX}{dx} = \frac{1}{\varepsilon^2}\frac{d^2f}{dX^2}$$

Substituting the above equations into Eq. A.5-1, we obtain

$$\frac{\varepsilon}{\varepsilon^2}\frac{d^2f}{dX^2} + \frac{1}{\varepsilon}\frac{df}{dX} = a$$

or

$$\frac{d^2f}{dX^2} + \frac{df}{dX} = \varepsilon a \qquad \text{Eq. A.5-7}$$

We can obtain an approximate solution by neglecting the term containing ε. Thus, the inner approximate solution, which we designate as f_i, will satisfy

$$\frac{d^2f_i}{dX^2} + \frac{df_i}{dX} = 0 \qquad \text{Eq. A.5-8}$$

and the boundary condition

$$f_i(0) = 0 \qquad \text{Eq. A.5-9}$$

A second boundary condition will be specified by a matching process. The indicial equation associated with the above differential equation can be obtained by assuming that $f_i = ce^{mX}$. When this is done, the following indicial equation is obtained

$$(m^2 + m) = m(m + 1) = 0$$

Thus, $m = 0$ and $m = -1$, giving

$$f_i = c_1 + c_2 e^{-X}$$

Applying the boundary condition expressed by Eq. A.5-9, we obtain

$$f_i(0) = c_1 + c_2(1) = 0$$

or

$$c_2 = -c_1$$

Thus,

$$f_i(X) = c_1(1 - e^{-X}) \qquad \text{Eq. A.5-10}$$

The arbitrary constant c_1 is obtained by matching the inner and outer solutions in a special way. That is, we require that

$$\lim_{x \to 0} f_0 = \lim_{X \to \infty} f_i \qquad \text{Eq. A.5-11}$$

This matching procedure serves as the second boundary condition. To see that this method works, let us apply Eq. A.5-11 to our inner and outer solutions. From Eq. A.5-5, we see that

$$\lim_{x \to 0} f_0 = (1 - a)$$

From Eq. A.5-10, we see that

$$\lim_{X \to \infty} f_i = c_1$$

Then, by Eq. A.5-11,

$$c_1 = 1 - a$$

Thus, an approximate solution to Eq. A.5-1, with boundary conditions given by Eqs. A.5-2, is as follows

$$f(x) = \begin{cases} (1 - a)(1 - e^{-x/\varepsilon}); & \text{for the inner region} \\ (1 - a) + ax; & \text{for the outer region} \end{cases} \qquad \text{Eq. A.5-12}$$

To obtain some idea of the accuracy of our approximate solution, let us now determine the exact solution. Since Eq. A.5-1 is nonhomogeneous, the solution is obtained by taking

$$f = f_c + f_p$$

where f_c satisfies

$$\varepsilon \frac{d^2 f_c}{dx^2} + \frac{df_c}{dx} = 0$$

The indicial equation associated with the above differential equation is

$$\varepsilon m^2 + m = 0 \qquad \text{or} \qquad m(\varepsilon m + 1) = 0$$

Therefore,

$$m = 0 \qquad \text{and} \qquad m = -\frac{1}{\varepsilon}$$

Then,

$$f_c = c_1 + c_2 e^{-x/\varepsilon}$$

A particular solution can be obtained by the method of undetermined coefficients. Assume

$$f_p = Ax + B$$

Substituting the above form into Eq. A.5-1, we obtain

$$A = a$$

The B value is arbitrary and may be taken to be zero. Therefore,

$$f = c_1 + c_2 e^{-x/\varepsilon} + ax$$

Applying the boundary conditions given by Eqs. A.5-2, we find that

$$f(0) = 0 = c_1 + c_2$$

and

$$f(1) = 1 = c_1(1 - e^{-1/\varepsilon}) + a$$

Thus,

$$c_2 = -c_1$$

and

$$c_1 = \frac{1 - a}{1 - e^{-1/\varepsilon}}$$

Therefore, the exact solution is given by the following equation

$$f(x) = \frac{(1 - a)(1 - e^{-x/\varepsilon})}{(1 - e^{-1/\varepsilon})} + ax \qquad \text{Eq. A.5-13}$$

Let us now make a numerical comparison between the exact and approximate solutions. To do this, let us take

$$\varepsilon = 0.01 \qquad \text{and} \qquad a = 0.5$$

The Table A.5-1 gives the numerical comparisons.

Let us now indicate the similarity of solving the lift/drag problems of a thin airfoil at a small angle of attack and the simple mathematical problem described by Eqs. A.5-1 and A.5-2.

For simplicity sake, let us consider the flat-plate airfoil shown in Fig. 9-20*b*. To make the analogy, we must write the governing equations for a viscous fluid, which are the Navier–Stokes equations.[3] For steady flow, with gravitational effects neglected, these become

$$\nu\left(\frac{\partial^2 v_1}{\partial x^2} + \frac{\partial^2 v_1}{\partial y^2}\right) - v_1 \frac{\partial v_1}{\partial x} - v_2 \frac{\partial v_1}{\partial y} = \frac{1}{\rho}\frac{\partial p}{\partial x}$$

$$\nu\left(\frac{\partial^2 v_2}{\partial x^2} + \frac{\partial^2 v_2}{\partial y^2}\right) - v_1 \frac{\partial v_2}{\partial x} - v_2 \frac{\partial v_2}{\partial y} = \frac{1}{\rho}\frac{\partial p}{\partial y}$$

The boundary conditions for a flow around an airfoil, with the free stream in the $(-x)$ direction, would be

On the boundary of the airfoil, $\mathbf{v} = 0$

At an infinite distance from the airfoil, $\mathbf{v} = -V_\infty \mathbf{i}$

In this problem, ν takes the place of ε in the previous mathematical example. Therefore, for the outer solution, we neglect the terms containing ν, and this leads to the governing equations for an inviscid fluid. Since we have dropped the highest order derivatives from the governing equations, we must also drop one of the boundary conditions. The condition that is dropped is the zero-

[3] The derivation of the Navier–Stokes equations is beyond the scope of this textbook. The derivation can be found in Schlichting's *Boundary-Layer Theory*.

TABLE A.5-1 Comparison Between the Exact and Approximate Solutions

x	x/ε	f_i	f_0	Exact f
0.000	0.00	0.00000	0.50000	0.00000
0.001	0.10	0.04758	0.50050	0.04808
0.002	0.20	0.09063	0.50100	0.09163
0.003	0.30	0.12959	0.50150	0.13109
0.004	0.40	0.16483	0.50200	0.16683
0.005	0.50	0.19673	0.50250	0.19923
0.006	0.60	0.22559	0.50300	0.22859
0.007	0.70	0.25170	0.50350	0.25520
0.008	0.80	0.27533	0.50400	0.27933
0.009	0.90	0.29671	0.50450	0.30121
0.010	1.00	0.31606	0.50500	0.32106
0.020	2.00	0.43233	0.51000	0.44233
0.030	3.00	0.47510	0.51500	0.49010
0.040	4.00	0.49084	0.52000	0.51084
0.050	5.00	0.49663	0.52500	0.52163
0.060	6.00	0.49876	0.53000	0.52876
0.070	7.00	0.49954	0.53500	0.53454
0.080	8.00	0.49983	0.54000	0.53983
0.090	9.00	0.49993	0.54500	0.54493
0.100	10.00	0.49997	0.55000	0.54997
0.200	20.00	0.50000	0.60000	0.60000
0.300	30.00	0.50000	0.65000	0.65000
0.400	40.00	0.50000	0.70000	0.70000
0.500	50.00	0.50000	0.75000	0.75000
0.600	60.00	0.50000	0.80000	0.80000
0.700	70.00	0.50000	0.85000	0.85000
0.800	80.00	0.50000	0.90000	0.90000
0.900	90.00	0.50000	0.95000	0.95000
1.000	100.00	0.50000	1.00000	1.00000

tangential-velocity component on the airfoil. Thus, the boundary conditions for the outer solution become

On the airfoil, $\mathbf{v} \cdot \mathbf{n} = 0$

At an infinite distance from the airfoil, $\mathbf{v} = -V_\infty \mathbf{i}$

where $\mathbf{n}$ = unit normal vector to the airfoil.

To obtain the inner solution, the boundary-layer equations are used, which include a term containing ν. The boundary-layer equations are obtained by neglecting a number of terms that appear in the Navier–Stokes equations. The boundary-layer equations are given in the footnote to Problem 9-25. The amplification of the inner region is accomplished by defining the variable $\eta = y/\delta$. The matching of the inner and outer solutions occurs when we take

$$\lim_{\eta \to \infty} v_{1,i}(x, \eta) = \lim_{y \to 0} v_{1,0}(x, y)$$

where $v_{1,i}$ is the inner solution and $v_{1,0}$ is the outer solution for v_1. Although it may not at first be obvious, this is the exact concept described in Problems 9-24 and 9-25. In fact, this concept mathematically justifies the decomposing of the fluid flow field into the viscous and inviscid regions, as described in Sections 9.1.1 and 9.10.

Appendix A.6

Physical Properties of Fluids

TABLE A.6-1 Physical Properties of Water in English Units

Temperature (F)	Density ρ (lb_m/ft^3)	Kinematic Viscosity $\nu \times 10^5$ (ft^2/s)	Surface Tension $\sigma \times 10^2$ (lb_f/ft)	Vapor-Pressure Head p_v/γ (ft)	Bulk Modulus of Elasticity $K \times 10^{-3}$ (lb_f/in^2)
32	62.42	1.931	0.518	0.20	293
40	62.43	1.664	0.514	0.28	294
50	62.41	1.410	0.509	0.41	305
60	62.37	1.217	0.504	0.59	311
70	62.30	1.059	0.500	0.84	320
80	62.22	0.930	0.492	1.17	322
90	62.11	0.826	0.486	1.61	323
100	62.00	0.739	0.480	2.19	327
110	61.86	0.667	0.473	2.95	331
120	61.71	0.609	0.465	3.91	333
130	61.55	0.558	0.460	5.13	334
140	61.38	0.514	0.454	6.67	330
150	61.20	0.476	0.447	8.58	328
160	61.00	0.442	0.441	10.95	326
170	60.80	0.413	0.433	13.83	322
180	60.58	0.385	0.426	17.33	313
190	60.36	0.362	0.419	21.55	313
200	60.12	0.341	0.412	26.59	308
212	59.83	0.319	0.404	33.90	300

Source. From Hydraulic Models by J. C. Stevens, chairman, *ASCE Man. Eng. Pract.* 25, 1942. Courtesy of ASCE.

TABLE A.6-2 Physical Properties of Water in SI Units

Temperature (C)	Density ρ (kg/m³)	Kinematic Viscosity $\nu \times 10^6$ (m²/s)	Surface Tension $\sigma \times 10^2$ (N/m)	Vapor-Pressure Head p_v/γ (m)	Bulk Modulus of Elasticity $K \times 10^{-7}$ (N/m²)
0	999.9	1.792	7.62	0.06	204
5	1000.0	1.519	7.54	0.09	206
10	999.7	1.308	7.48	0.12	211
15	999.1	1.141	7.41	0.17	214
20	998.2	1.007	7.36	0.25	220
25	997.1	0.897	7.26	0.33	222
30	995.7	0.804	7.18	0.44	223
35	994.1	0.727	7.10	0.58	224
40	992.2	0.661	7.01	0.76	227
45	990.2	0.605	6.92	0.98	229
50	988.1	0.556	6.82	1.26	230
55	985.7	0.513	6.74	1.61	231
60	983.2	0.477	6.68	2.03	228
65	980.6	0.444	6.58	2.56	226
70	977.8	0.415	6.50	3.20	225
75	974.9	0.390	6.40	3.96	223
80	971.8	0.367	6.30	4.86	221
85	968.6	0.347	6.20	5.93	217
90	965.3	0.328	6.12	7.18	216
95	961.9	0.311	6.02	8.62	211
100	958.4	0.296	5.94	10.33	207

Source. From *Fluid Mechanics* by V. Streeter and E. B. Wylie. Copyright © 1975 by McGraw-Hill. Used with permission of McGraw-Hill Book Company.

TABLE A.6-3 Properties of Gases at Low Pressures and 80F

Gas	Chemical Formula	Molecular Weight	Gas Constant R		Specific Heat, Btu/lb_m-R		Specific-Heat Ratio k
			m-N/kg-K	ft-lb_f/lb_m-R	c_p	c_v	
Air	—	29.0	287.1	53.3	0.240	0.171	1.40
Carbon monoxide	CO	28.0	297	55.2	0.249	0.178	1.40
Helium	He	4.00	2077	386	1.25	0.753	1.66
Hydrogen	H_2	2.02	4121	766	3.43	2.44	1.40
Nitrogen	N_2	28.0	297	55.2	0.248	0.177	1.40
Oxygen	O_2	32.0	260	48.3	0.219	0.157	1.40
Water vapor	H_2O	18.0	462	85.8	0.445	0.335	1.33

Source. From *Fluid Mechanics* by V. Streeter and E. B. Wylie. Copyright © 1975 by McGraw-Hill. Used with permission of McGraw-Hill Book Company.

TABLE A.6-4*a* Properties of the U.S. Standard Atmosphere (SI Units)

Geometric Altitude (m)	Temperature (K)	p/p_0	ρ/ρ_0
0	288.2	1.000[a]	1.000[b]
500	284.9	0.9421	0.9529
1,000	281.7	0.8870	0.9075
1,500	278.4	0.8345	0.8638
2,000	275.2	0.7846	0.8217
2,500	271.9	0.7372	0.7812
3,000	268.7	0.6920	0.7423
3,500	265.4	0.6492	0.7048
4,000	262.2	0.6085	0.6689
4,500	258.9	0.5700	0.6343
5,000	255.7	0.5334	0.6012
6,000	249.2	0.4660	0.5389
7,000	242.7	0.4057	0.4817
8,000	236.2	0.3519	0.4292
9,000	229.7	0.3040	0.3813
10,000	223.3	0.2615	0.3376
11,000	216.8	0.2240	0.2978
12,000	216.7	0.1915	0.2546
13,000	216.7	0.1636	0.2176
14,000	216.7	0.1399	0.1860
15,000	216.7	0.1195	0.1590
16,000	216.7	0.1022	0.1359
17,000	216.7	0.08734	0.1162
18,000	216.7	0.07466	0.09930
19,000	216.7	0.06383	0.08489
20,000	216.7	0.05457	0.07258
22,000	218.6	0.03995	0.05266
24,000	220.6	0.02933	0.03832
26,000	222.5	0.02160	0.02797
28,000	224.5	0.01595	0.02047
30,000	226.5	0.01181	0.01503
40,000	250.4	0.2834×10^{-2}	0.3262×10^{-2}
50,000	270.7	0.7874×10^{-3}	0.8383×10^{-3}
60,000	255.8	0.2217×10^{-3}	0.2497×10^{-3}
70,000	219.7	0.5448×10^{-4}	0.7146×10^{-4}
80,000	180.7	0.1023×10^{-4}	0.1632×10^{-4}
90,000	180.7	0.1622×10^{-5}	0.2588×10^{-5}

Source. From *Introduction to Fluid Mechanics* by R. W. Fox and A. T. McDonald.

[a] $p_0 = 1.01325 \times 10^5$ N/m^2 absolute

[b] $\rho_0 = 1.2250$ kg/m^3

TABLE A.6-4*b* Properties of the U.S. Standard Atmosphere (English Units)

Geometric Altitude (ft)	Temperature (F)	p/p_0	ρ/ρ_0
0	59.0	1.000[a]	1.000[b]
500	57.2	0.982	0.985
1,000	55.4	0.964	0.971
2,000	51.9	0.929	0.943
3,000	48.3	0.896	0.915
4,000	44.7	0.864	0.888
5,000	41.2	0.832	0.862
6,000	37.6	0.801	0.836
7,000	34.0	0.772	0.811
8,000	30.5	0.743	0.786
9,000	26.9	0.715	0.762
10,000	23.4	0.688	0.739
12,000	16.2	0.636	0.693
14,000	9.1	0.588	0.650
16,000	2.0	0.542	0.609
18,000	−5.1	0.499	0.570
20,000	−12.3	0.460	0.533
25,000	−30.0	0.371	0.449
30,000	−47.8	0.297	0.375
35,000	−65.6	0.236	0.311
40,000	−69.7	0.186	0.247
45,000	−69.7	0.149	0.195
50,000	−69.7	0.115	0.153
60,000	−69.7	0.713×10^{-1}	0.949×10^{-1}
70,000	−67.4	0.443×10^{-1}	0.585×10^{-1}
80,000	−62.0	0.276×10^{-1}	0.357×10^{-1}
90,000	−56.5	0.174×10^{-1}	0.224×10^{-1}
100,000	−51.1	0.110×10^{-1}	0.140×10^{-1}
200,000	−2.7	0.287×10^{-2}	0.220×10^{-3}

Source. From *Introduction to Fluid Mechanics* by R. W. Fox and A. T. McDonald. Copyright © 1978 by John Wiley & Sons. Reprinted by permission of the publisher.

[a] p_0 = 14.696 psia = 2116.2 psia.

[b] ρ_0 = 0.002378 slug/ft^3 = 0.07651 lb$_m$/ft^3.

TABLE A.6-5 Property Values of Engine Oil and Mercury

English Units			SI Units		
T (F)	ρ (lb_m/ft^3)	ν (ft^2/s)	T (C)	ρ (kg/m^3)	ν (m^2/s)
	Engine oil (unused)			Engine oil (unused)	
32	56.13	0.0461	0	899.12	0.00428
68	55.45	0.0097	20	888.23	0.00090
104	54.69	0.0026	40	876.05	0.00024
140	53.94	0.903×10^{-3}	60	864.04	0.839×10^{-4}
176	53.19	0.404×10^{-3}	100	852.02	0.375×10^{-4}
212	52.44	0.219×10^{-3}	100	840.01	0.203×10^{-4}
248	51.75	0.133×10^{-3}	120	828.96	0.124×10^{-4}
284	51.00	0.086×10^{-3}	140	816.94	0.080×10^{-4}
320	50.31	0.060×10^{-3}	160	805.89	0.056×10^{-4}
	Mercury (Hg)			Mercury (Hg)	
32	850.78	0.133×10^{-5}	0	13,628.22	0.124×10^{-6}
68	847.71	0.123×10^{-5}	20	13,579.04	0.114×10^{-6}
122	843.14	0.112×10^{-5}	50	13,505.84	0.104×10^{-6}
212	835.57	0.0999×10^{-5}	100	13,384.58	0.0928×10^{-6}
302	828.06	0.0918×10^{-5}	150	13,264.28	0.0853×10^{-6}
			200	13,144.94	0.0802×10^{-6}
			250	13,025.60	0.0765×10^{-6}
			315.5	12,847	0.0673×10^{-6}

Source. From *Analysis of Heat and Mass Transfer* by E. R. G. Eckert and R. M. Drake, Jr. Copyright © 1972 by McGraw-Hill. Used with permission of McGraw-Hill Book Company.

TABLE A.6-6 Surface Tension of Liquids in Contact with Air, Water, or Their Own Vapors

Substance	Surface Tension (lb_f/ft)	Surface Tension (N/m)
Benzene-air	0.0020	0.029
Carbon tetrachloride-air	0.0018	0.027
Water-air	0.0050	0.073
Mercury-air	0.030	0.435
Benzene-water	0.0024	0.035
Carbon tetrachloride-water	0.0031	0.045
Mercury-water	0.026	0.375
Benzene vapor	0.0020	0.029
Carbon tetrachloride vapor	0.0018	0.027
Water vapor	0.0050	0.073

Source. From *Essentials of Engineering Fluid Mechanics*, 3rd ed., by Reuben M. Olson: (Intext Educational Publishers), Table A-6 (p. 600), Copyright © 1961, 1963, 1973 by Harper & Row. Reprinted by permission of the publisher.

TABLE A.6-7 Vapor Pressure and Some Liquids

Temperature		Water		Mercury	Kerosene	Methyl Alcohol
(F)	(C)	($lb_f/in.^2$)	(N/m^2)	($lb_f/in.^2$)	($lb_f/in.^2$)	($lb_f/in.^2$)
32	0	0.088	0.0294	0.0000036	—	—
41	5	0.126	0.0421	—	0.33	0.74
50	10	0.178	0.0592	0.0000095	0.37	1.03
59	15	0.247	0.0822	—	—	1.40
68	20	0.339	0.113	0.000023	0.51	1.83
77	25	0.459	0.153	—	0.60	2.38
86	30	0.615	0.209	0.000054	0.70	3.1
95	35	0.815	0.271	—	0.80	3.9
104	40	1.070	0.355	0.000117	0.89	—
113	45	1.390	0.462	—	0.96	—
122	50	1.789	0.594	—	1.03	—
131	55	2.28	0.722	—	1.09	—
140	60	2.89	0.959	—	—	—
149	65	3.63	1.205	—	—	—
158	70	4.52	1.501	—	—	—

Source. From *Essentials of Engineering Fluid Mechanics*, 3rd ed., by Reuben M. Olson: (Intext Educational Publishers), Table A.7 (p. 600), Copyright © 1961, 1963, 1973 by Harper & Row. Reprinted by permission of the publisher.

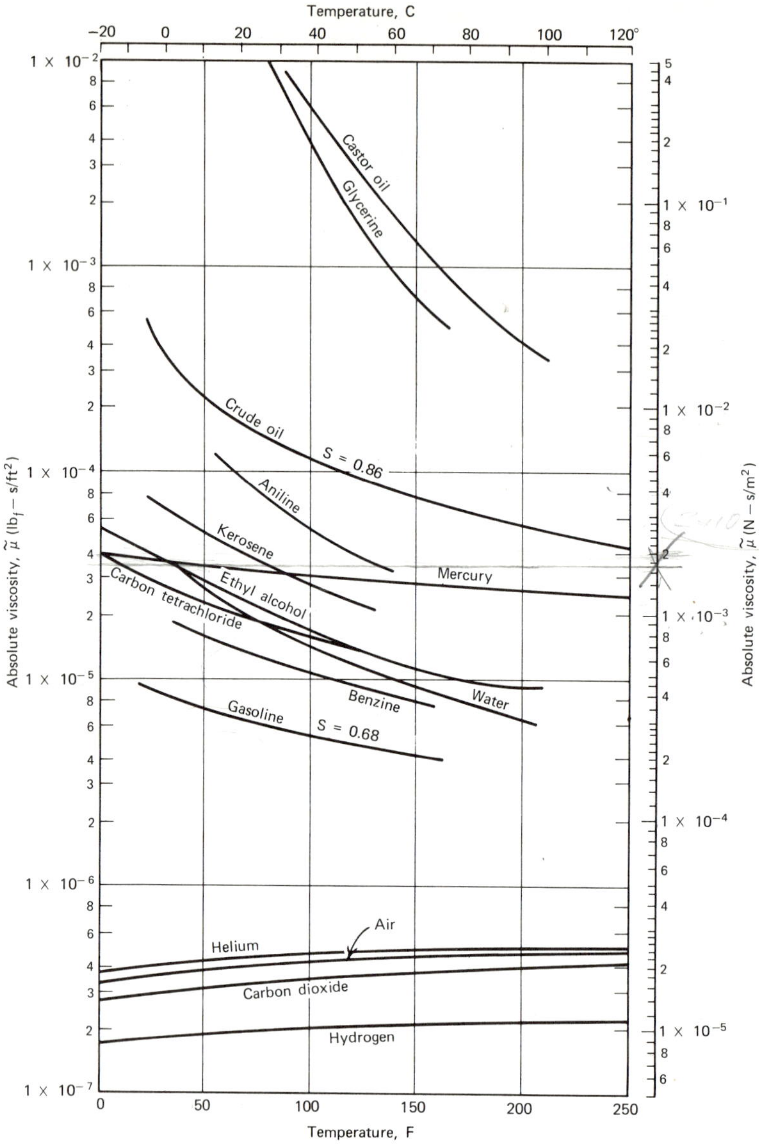

Fig. A.6-1. Absolute viscosities of certain gases and liquids. (From *Fluid Mechanics* by V. L. Streeter and E. B. Wylie. Copyright © 1975 by McGraw-Hill. Used with permission of McGraw-Hill Book Company.)

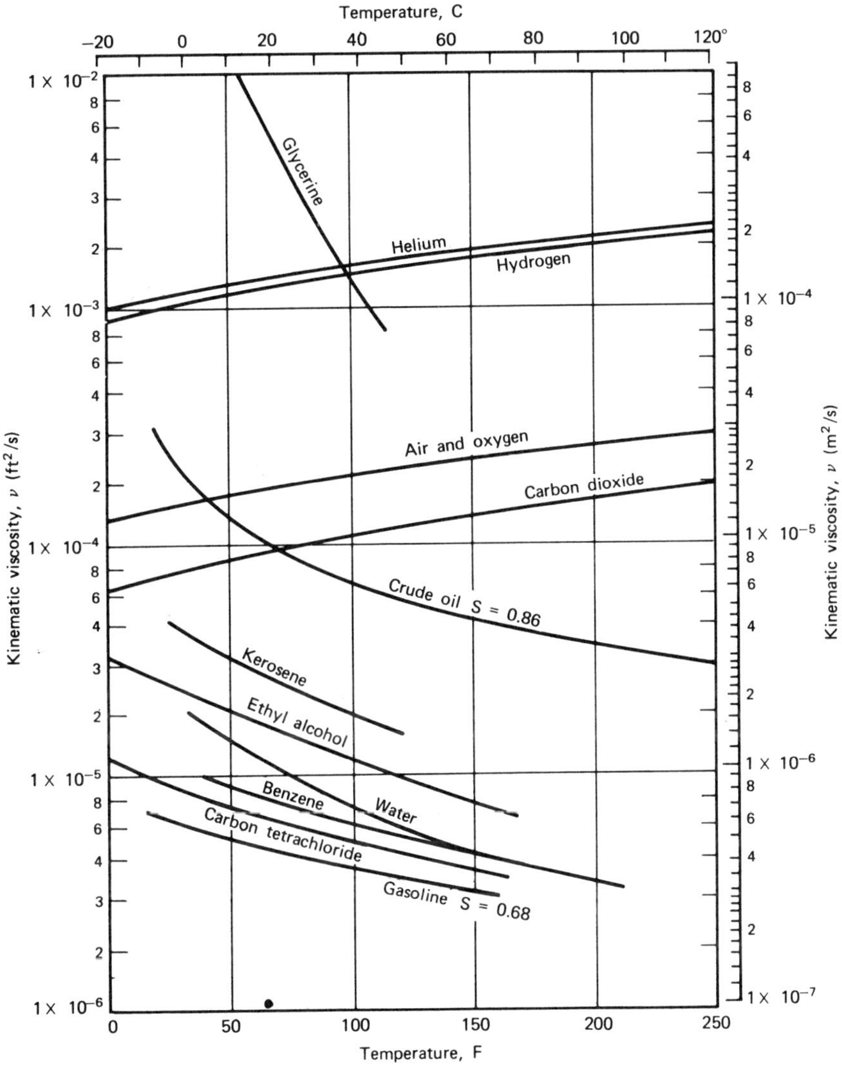

Fig. A.6-2. Kinematic viscosities of certain gases and liquids. The gases are at standard pressure. (From *Fluid Mechanics* by V. L. Streeter and E. B. Wylie. Copyright © 1975 by McGraw-Hill. Used with permission of McGraw-Hill Book Company.)

TABLE A.6-8 One-Dimensional Isentropic Relations (for a Perfect Gas with Constant Specific Heat, $k = 1.4$)

M	A/A^*	p/p_0	ρ/ρ_0	T/T_0	M	A/A^*	p/p_0	ρ/ρ_0	T/T_0
0.00	—	1.000	1.000	1.000	1.02	1.00	0.516	0.623	0.828
0.01	57.87	0.9999	0.9999	0.9999	1.04	1.00	0.504	0.613	0.822
0.02	28.94	0.9997	0.9999	0.9999	1.06	1.00	0.492	0.602	0.817
0.04	14.48	0.999	0.999	0.9996	1.08	1.01	0.480	0.592	0.810
0.06	9.67	0.997	0.998	0.999	1.10	1.01	0.468	0.582	0.805
0.08	7.26	0.996	0.997	0.999	1.12	1.01	0.457	0.571	0.799
0.10	5.82	0.993	0.995	0.998	1.14	1.02	0.445	0.561	0.794
0.12	4.86	0.990	9.993	0.997	1.16	1.02	0.434	0.551	0.788
0.14	4.18	0.986	0.990	0.996	1.18	1.02	0.423	0.541	0.782
0.16	3.67	0.982	0.987	0.995	1.20	1.03	0.412	0.531	0.776
0.18	3.28	0.978	0.984	0.994	1.22	1.04	0.402	0.521	0.771
0.20	2.96	0.973	0.980	0.992	1.24	1.04	0.391	0.512	0.765
0.22	2.71	0.967	0.976	0.990	1.26	1.05	0.381	0.502	0.759
0.24	2.50	0.961	0.972	0.989	1.28	1.06	0.371	0.492	0.753
0.26	2.32	0.954	0.967	0.987	1.30	1.07	0.361	0.483	0.747
0.28	2.17	0.947	0.962	0.985	1.32	1.08	0.351	0.474	0.742
0.30	2.04	0.939	0.956	0.982	1.34	1.08	0.342	0.464	0.736
0.32	1.92	0.932	0.951	0.980	1.36	1.09	0.332	0.455	0.730
0.34	1.82	0.923	0.944	0.977	1.38	1.10	0.323	0.446	0.724
0.36	1.74	0.914	0.938	0.975	1.40	1.11	0.314	0.437	0.718
0.38	1.66	0.905	0.931	0.972	1.42	1.13	0.305	0.429	0.713
0.40	1.59	0.896	0.924	0.969	1.44	1.14	0.297	0.420	0.707
0.42	1.53	0.886	0.917	0.966	1.46	1.15	0.289	0.412	0.701
0.44	1.47	0.876	0.909	0.963	1.48	1.16	0.280	0.403	0.695

0.46	1.42	0.865	0.902	0.959	1.50	1.18	0.272	0.395	0.690
0.48	1.38	0.854	0.893	0.956	1.52	1.19	0.265	0.387	0.684
0.50	1.34	0.843	0.885	0.952	1.54	1.20	0.257	0.379	0.678
0.52	1.30	0.832	0.877	0.949	1.56	1.22	0.250	0.371	0.672
0.54	1.27	0.820	0.868	0.945	1.58	1.23	0.242	0.363	0.667
0.56	1.24	0.808	0.859	0.941	1.60	1.25	0.235	0.356	0.661
0.58	1.21	0.796	0.850	0.937	1.62	1.27	0.228	0.348	0.656
0.60	1.19	0.784	0.840	0.933	1.64	1.28	0.222	0.341	0.650
0.62	1.17	0.772	0.831	0.929	1.66	1.30	0.215	0.334	0.645
0.64	1.16	0.759	0.821	0.924	1.68	1.32	0.209	0.327	0.639
0.66	1.13	0.747	0.812	0.920	1.70	1.34	0.203	0.320	0.634
0.68	1.12	0.734	0.802	0.915	1.72	1.36	0.197	0.313	0.628
0.70	1.09	0.721	0.792	0.911	1.74	1.38	0.191	0.306	0.623
0.72	1.08	0.708	0.781	0.906	1.76	1.40	0.185	0.300	0.617
0.74	1.07	0.695	0.771	0.901	1.78	1.42	0.179	0.293	0.612
0.76	1.06	0.682	0.761	0.896	1.80	1.44	0.174	0.287	0.607
0.78	1.05	0.669	0.750	0.891	1.82	1.46	0.169	0.281	0.602
0.80	1.04	0.656	0.740	0.886	1.84	1.48	0.164	0.275	0.596
0.82	1.03	0.643	0.729	0.881	1.86	1.51	0.159	0.269	0.591
0.84	1.02	0.630	0.719	0.876	1.88	1.53	0.154	0.263	0.586
0.86	1.02	0.617	0.708	0.871	1.90	1.56	0.149	0.257	0.581
0.88	1.01	0.604	0.698	0.865	1.92	1.58	0.145	0.251	0.576
0.90	1.01	0.591	0.687	0.860	1.94	1.61	0.140	0.246	0.571
0.92	1.01	0.578	0.676	0.855	1.96	1.63	0.136	0.240	0.566
0.94	1.00	0.566	0.666	0.850	1.98	1.66	0.132	0.235	0.561
0.96	1.00	0.553	0.655	0.844	2.00	1.69	0.128	0.230	0.556
0.98	1.00	0.541	0.645	0.839	2.02	1.72	0.124	0.225	0.551
1.00	1.00	0.528	0.632	0.833	2.04	1.75	0.120	0.220	0.546

(Continued)

TABLE A.6-8 One-Dimensional Isentropic Relations (*Continued*)

M	A/A^*	p/p_0	ρ/ρ_0	T/T_0	M	A/A^*	p/p_0	ρ/ρ_0	T/T_0
2.06	1.78	0.116	0.215	0.541	2.54	2.74	0.055	0.126	0.437
2.08	1.81	0.113	0.210	0.536	2.56	2.79	0.053	0.123	0.433
2.10	1.84	0.109	0.206	0.531	2.58	2.84	0.052	0.121	0.429
2.12	1.87	0.106	0.201	0.526	2.60	2.90	0.050	0.118	0.425
2.14	1.90	0.103	0.197	0.522	2.62	2.95	0.049	0.115	0.421
2.16	1.94	0.100	0.192	0.517	2.64	3.01	0.047	0.113	0.418
2.18	1.97	0.097	0.188	0.513	2.66	3.06	0.046	0.110	0.414
2.20	2.01	0.094	0.184	0.508	2.68	3.12	0.044	0.108	0.410
2.22	2.04	0.091	0.180	0.504	2.70	3.18	0.043	0.106	0.407
2.24	2.08	0.088	0.176	0.499	2.72	3.24	0.042	0.103	0.403
2.26	2.12	0.085	0.172	0.495	2.74	3.31	0.040	0.101	0.400
2.28	2.15	0.083	0.168	0.490	2.76	3.37	0.039	0.099	0.396
2.30	2.19	0.080	0.165	0.486	2.78	3.43	0.038	0.097	0.393
2.32	2.23	0.078	0.161	0.482	2.80	3.50	0.037	0.095	0.389
2.34	2.27	0.075	0.157	0.477	2.82	3.57	0.036	0.093	0.386
2.36	2.32	0.073	0.154	0.473	2.84	3.64	0.035	0.091	0.383
2.38	2.36	0.071	0.150	0.469	2.86	3.71	0.034	0.089	0.379
2.40	2.40	0.068	0.147	0.465	2.88	3.78	0.033	0.087	0.376
2.42	2.45	0.066	0.144	0.461	2.90	3.85	0.032	0.085	0.373
2.44	2.49	0.064	0.141	0.456	2.92	3.92	0.031	0.083	0.370
2.46	2.54	0.062	0.138	0.452	2.94	4.00	0.030	0.081	0.366
2.48	2.59	0.060	0.135	0.448	2.96	4.08	0.029	0.080	0.363
2.50	2.64	0.059	0.132	0.444	2.98	4.15	0.028	0.078	0.360
2.52	2.69	0.057	0.129	0.441	3.00	4.23	0.027	0.076	0.357

TABLE A.6-9 One-Dimensional Normal-Shock Relations (for a Perfect Gas with $k = 1.4$)

M_1	M_2	$\frac{p_2}{p_1}$	$\frac{T_2}{T_1}$	$\frac{p_{0,2}}{p_{0,1}}$	M_1	M_2	$\frac{p_2}{p_1}$	$\frac{T_2}{T_1}$	$\frac{p_{0,2}}{p_{0,1}}$
1.00	1.000	1.000	1.000	1.000	1.40	0.740	2.120	1.255	0.958
1.02	0.980	1.047	1.013	1.000	1.42	0.731	2.186	1.268	0.953
1.04	0.962	1.095	1.026	1.000	1.44	0.723	2.253	1.281	0.948
1.06	0.944	1.144	1.039	1.000	1.46	0.716	2.320	1.294	0.942
1.08	0.928	1.194	1.052	0.999	1.48	0.708	2.389	1.307	0.936
1.10	0.912	1.245	1.065	0.999	1.50	0.701	2.458	1.320	0.930
1.12	0.896	1.297	1.078	0.998	1.52	0.694	2.529	1.334	0.923
1.14	0.882	1.350	1.090	0.997	1.54	0.687	2.600	1.347	0.917
1.16	0.868	1.403	1.103	0.996	1.56	0.681	2.673	1.361	0.910
1.18	0.855	1.458	1.115	0.995	1.58	0.675	2.746	1.374	0.903
1.20	0.842	1.513	1.128	0.993	1.60	0.668	2.820	1.388	0.895
1.22	0.830	1.570	1.140	0.991	1.62	0.663	2.895	1.402	0.888
1.24	0.818	1.627	1.153	0.988	1.64	0.657	2.971	1.416	0.880
1.26	0.807	1.686	1.166	0.986	1.66	0.651	3.048	1.430	0.872
1.28	0.796	1.745	1.178	0.983	1.68	0.646	3.126	1.444	0.864
1.30	0.786	1.805	1.191	0.979	1.70	0.641	3.205	1.458	0.856
1.32	0.776	1.866	1.204	0.976	1.72	0.635	3.285	1.473	0.847
1.34	0.766	1.928	1.216	0.972	1.74	0.631	3.366	1.487	0.839
1.36	0.757	1.991	1.229	0.968	1.76	0.626	3.447	1.502	0.830
1.38	0.748	2.055	1.242	0.963	1.78	0.621	3.530	1.517	0.821

(Continued)

TABLE A.6-9 One-Dimensional Normal-Shock Relations (for a Perfect Gas with $k = 1.4$)

M_1	M_2	$\frac{p_2}{p_1}$	$\frac{T_2}{T_1}$	$\frac{p_{0,2}}{p_{0,1}}$	M_1	M_2	$\frac{p_2}{p_1}$	$\frac{T_2}{T_1}$	$\frac{p_{0,2}}{p_{0,1}}$
1.80	0.617	3.613	1.532	0.813	2.44	0.519	6.779	2.079	0.523
1.82	0.612	3.698	1.547	0.804	2.46	0.517	6.894	2.098	0.515
1.84	0.608	3.783	1.562	0.795	2.48	0.515	7.009	2.118	0.507
1.86	0.604	3.869	1.577	0.786	2.50	0.513	7.125	2.138	0.499
1.88	0.600	3.957	1.592	0.777	2.52	0.511	7.242	2.157	0.491
1.90	0.596	4.045	1.608	0.767	2.54	0.509	7.360	2.177	0.483
1.92	0.592	4.134	1.624	0.758	2.56	0.507	7.479	2.198	0.475
1.94	0.588	4.224	1.639	0.749	2.58	0.506	7.599	2.218	0.468
1.96	0.584	4.315	1.655	0.740	2.60	0.504	7.720	2.238	0.460
1.98	0.581	4.407	1.671	0.730	2.62	0.502	7.842	2.260	0.453
2.00	0.577	4.500	1.688	0.721	2.64	0.500	7.965	2.280	0.445
2.02	0.574	4.594	1.704	0.711	2.66	0.499	8.088	2.301	0.438
2.04	0.571	4.689	1.720	0.702	2.68	0.497	8.213	2.322	0.431
2.06	0.567	4.784	1.737	0.693	2.70	0.496	8.338	2.343	0.424
2.08	0.564	4.881	1.754	0.683	2.72	0.494	8.465	2.364	0.417
2.10	0.561	4.978	1.770	0.674	2.74	0.493	8.592	2.396	0.410

2.12	0.558	5.077	1.787	0.665	2.76	0.491	8.721	2.407	0.403
2.14	0.555	5.176	1.805	0.656	2.78	0.490	8.850	2.429	0.396
2.16	0.553	5.277	1.822	0.646	2.80	0.488	8.980	2.451	0.389
2.18	0.550	5.378	1.839	0.637	2.82	0.487	9.111	2.473	0.383
2.20	0.547	5.480	1.857	0.628	2.84	0.485	9.243	2.496	0.376
2.22	0.544	5.583	1.875	0.619	2.86	0.484	9.376	2.518	0.370
2.24	0.542	5.687	1.892	0.610	2.88	0.483	9.510	2.541	0.364
2.26	0.539	5.792	1.910	0.601	2.90	0.481	9.645	2.563	0.358
2.28	0.537	5.898	1.929	0.592	2.92	0.480	9.781	2.586	0.352
2.30	0.534	6.005	1.947	0.583	2.94	0.479	9.918	2.609	0.346
2.32	0.532	6.113	1.965	0.575	2.96	0.478	10.055	2.632	0.340
2.34	0.530	6.222	1.984	0.566	2.98	0.476	10.194	2.656	0.334
2.36	0.527	6.331	2.003	0.557	3.00	0.475	10.333	2.679	0.328
2.38	0.525	6.442	2.021	0.549	—	—	—	—	—
2.40	0.523	6.553	2.040	0.540	—	—	—	—	—
2.42	0.521	6.666	2.060	0.532	—	—	—	—	—

Source. From *Mechanics of Fluids* by I. H. Shames. Copyright © 1962 by McGraw-Hill. Used with permission of McGraw-Hill Book Company.

TABLE A.6-10*a* Property Values of Air at Standard Atmospheric Pressure (English Units)

T (F)	ρ (lb_m/ft^3)	c_p (Btu/lb_m-F)	μ (lb_m/ft-s)	k_t (Btu/h-ft-F)	Pr
−280	0.2248	0.2452	0.4653×10^{-5}	0.005342	0.770
−190	0.1478	0.2412	0.6910×10^{-5}	0.007936	0.753
−100	0.1104	0.2403	0.8930×10^{-5}	0.01045	0.739
−10	0.0882	0.2401	1.074×10^{-5}	0.01287	0.722
80	0.0735	0.2402	1.241×10^{-5}	0.01516	0.708
170	0.0623	0.2410	1.394×10^{-5}	0.01735	0.697
260	0.0551	0.2422	1.536×10^{-5}	0.01944	0.689
350	0.0489	0.2438	1.669×10^{-5}	0.02142	0.683
440	0.0440	0.2459	1.795×10^{-5}	0.02333	0.680
530	0.0401	0.2482	1.914×10^{-5}	0.02519	0.680
620	0.0367	0.2520	2.028×10^{-5}	0.02692	0.680
710	0.0339	0.2540	2.135×10^{-5}	0.02862	0.682
800	0.0314	0.2568	2.239×10^{-5}	0.03022	0.684
890	0.0294	0.2593	2.339×10^{-5}	0.03183	0.686
980	0.0275	0.2622	2.436×10^{-5}	0.03339	0.689

Source. From *Analysis of Heat and Mass Transfer* by E. R. G. Eckert and R. M. Drake, Jr. Copyright © 1972 by McGraw-Hill. Used with permission of McGraw-Hill Book Company.

TABLE A.6-10*b* Property Values of Air at Standard Atmospheric Pressure (SI Units)

T (K)	ρ (kg/m^3)	c_p (J/kg-K)	μ (kg/m-s)	k_t (W/m-K)	Pr
100	3.6010	1.0266×10^3	0.6924×10^{-5}	0.009246	0.770
150	2.3675	1.0099×10^3	1.0283×10^{-5}	0.013735	0.753
200	1.7684	1.0061×10^3	1.3289×10^{-5}	0.01809	0.739
250	1.4128	1.0053×10^3	1.488×10^{-5}	0.02227	0.722
300	1.1774	1.0057×10^3	1.983×10^{-5}	0.02624	0.708
350	0.9980	1.0090×10^3	2.075×10^{-5}	0.03003	0.697
400	0.8826	1.0140×10^3	2.286×10^{-5}	0.03365	0.689
450	0.7833	1.0207×10^3	2.484×10^{-5}	0.03707	0.683
500	0.7048	1.0295×10^3	2.671×10^{-5}	0.04038	0.680
550	0.6423	1.0392×10^3	2.848×10^{-5}	0.04360	0.680
600	0.5879	1.0551×10^3	3.018×10^{-5}	0.04659	0.680
650	0.5430	1.0635×10^3	3.177×10^{-5}	0.04953	0.682
700	0.5030	1.0752×10^3	3.332×10^{-5}	0.05230	0.684
750	0.4709	1.0856×10^3	3.481×10^{-5}	0.05509	0.686
800	0.4405	1.0978×10^3	3.625×10^{-5}	0.05779	0.689
850	0.4149	1.1095×10^3	3.765×10^{-5}	0.06028	0.692
900	0.3925	1.1212×10^3	3.899×10^{-5}	0.06279	0.696
950	0.3716	1.1321×10^3	4.023×10^{-5}	0.06525	0.699
1000	0.3524	1.1417×10^3	4.152×10^{-5}	0.06752	0.702

Source. From *Analysis of Heat and Mass Transfer* by E. R. G. Eckert and R. M. Drake, Jr. Copyright © 1972 by McGraw-Hill. Used with permission of McGraw-Hill Book Company.

The viscosity of gases can be approximated by the Sutherland empirical formula, which follows

$$\mu = \frac{b\sqrt{T}}{1 + a/T}$$

where b and a are constants. For air
English units:

$$b = 7.14 \times 10^{-7} \frac{\text{lb}_\text{m}}{\text{ft-s-(R)}^{1/2}}$$

$$a = 180 \text{ R}$$

SI units:

$$b = 1.458 \times 10^{-6} \frac{\text{kg}}{\text{m-s-(K)}^{1/2}}$$

$$a = 110.4 \text{ K}$$

References

Batchelor, G. K., *An Introduction to Fluid Dynamics*, Cambridge University Press, New York, 1967.

Callen, H. B., *Thermodynamics*, Wiley, New York, 1960.

Davis, H. F., *Introduction to Vector Analysis*, 2nd ed., Allyn and Bacon, Boston, 1967.

Eskinazi, S., *Principles of Fluid Mechanics*, 2nd ed., Allyn and Bacon, Boston, 1968.

Fox, R. W. and McDonald, A. T., *Introduction to Fluid Mechanics*, Wiley, New York, 1973.

Glauert, H., *The Elements of Aerofoil and Airscrew Theory*, 2nd ed., Cambridge University Press, New York, 1948.

Hunsacker, J. C., and Rightmire, B. G., *Engineering Applications of Fluid Mechanics*, McGraw-Hill, New York, 1947.

John, J. E. A. and Haberman, W., *Introduction to Fluid Mechanics*, Prentice-Hall, Englewood Cliffs, N.J., 1971.

John, J. E. A., *Gas Dynamics*, Allyn and Bacon, Boston, 1969.

Karlekar, B. V. and Desmond, R. M., *Engineering Heat Transfer*, West Publishing Co., New York, 1977.

Kenyon, R. A., *Principles of Fluid Mechanics*, Ronald Press, New York, 1960.

Krasnov, N. F., *Aerodynamics of Bodies of Revolution*, edited by Morris, D. N., American Elsevier, New York, 1970.

Landau, L. D. and Lifshitz, E. M., *Fluid Mechanics*, Pergamon Press, Reading, Mass., 1959.

Langhaar, H. L., *Dimensional Analysis and Theory of Models*, Wiley, New York, 1951.

Li, W. H. and Lam, S. H., *Principles of Fluid Mechanics*, Addison-Wesley, Reading, Mass., 1964.

Milne-Thomson, L. M., *Theoretical Hydrodynamics*, 4th ed., MacMillan, New York, 1960.

Olson, R. M., *Essentials of Engineering Fluid Mechanics*, 3rd ed., Intext Press, New York, 1973.

Owczarek, J. A., *Introduction to Fluid Mechanics*, International Textbook, Scranton, Pa., 1968.

Pao, R. H., *Fluid Dynamics*, Merrill Books, Columbus, Ohio, 1967.

Potter, M. C. and Foss, J. H., *Fluid Mechanics*, Ronald Press, New York, 1975.

Sabersky, R. H., Acosta, A. J., and Hauptmann, E. G., *Fluid Flow—A First Course in Fluid Mechanics*, 2nd ed., Macmillan, New York, 1971.

Schlichting, H., *Boundary-Layer Theory*, 6th ed., McGraw-Hill, New York, 1968.

Shames, I. H., *Mechanics of Fluids*, McGraw-Hill, New York, 1962.

Shapiro, A. H., *The Dynamics and Thermodynamics of Compressible Flow*, Ronald Press, New York, 1953.

Shepherd, D. G., *Elements of Fluid Mechanics*, Harcourt, Brace and World, New York, 1965.

Streeter, V. L. and Wylie, E. S., *Fluid Mechanics*, Prentice-Hall, Englewood Cliffs, N.J., 1968.

Sutton, G. P., *Rocket Propulsion Elements, An Introduction to the Engineering of Rockets*, 2nd ed., Wiley, New York, 1965.

The U.S. Standard Atmosphere (1962), Washington, D.C., U.S. Government Printing Office, 1962.

Vennard, J. K., *Elementary Fluid Mechanics*, 4th ed., Wiley, New York, 1966.

Whitaker, S., *Introduction to Fluid Mechanics*, Prentice-Hall, Englewood Cliffs, N.J., 1968.

Index